T0329197

LONDON MATHEMATICAL SOCIETY LECTURE NOTE SERIES

Managing Editor: Professor M. Reid, Mathematics Institute,
University of Warwick, Coventry CV4 7AL, United Kingdom

The titles below are available from booksellers, or from Cambridge University Press at
www.cambridge.org/mathematics

London Mathematical Society Lecture Note Series: 450

Partial Differential Equations arising from Physics and Geometry

A Volume in Memory of Abbas Bahri

Edited by

MOHAMED BEN AYED
Université de Sfax, Tunisia

MOHAMED ALI JENDOUBI
Université de Carthage, Tunisia

YOMNA RÉBAÏ
Université de Carthage, Tunisia

HASNA RIAHI
École Nationale d'Ingénieurs de Tunis, Tunisia

HATEM ZAAG
Université de Paris XIII

CAMBRIDGE
UNIVERSITY PRESS

CAMBRIDGE
UNIVERSITY PRESS

Shaftesbury Road, Cambridge CB2 8EA, United Kingdom

One Liberty Plaza, 20th Floor, New York, NY 10006, USA

477 Williamstown Road, Port Melbourne, VIC 3207, Australia

314–321, 3rd Floor, Plot 3, Splendor Forum, Jasola District Centre, New Delhi – 110025, India

103 Penang Road, #05–06/07, Visioncrest Commercial, Singapore 238467

Cambridge University Press is part of Cambridge University Press & Assessment,
a department of the University of Cambridge.

We share the University's mission to contribute to society through the pursuit of
education, learning and research at the highest international levels of excellence.

www.cambridge.org
Information on this title: www.cambridge.org/9781108431637

DOI: 10.1017/9781108367639

© Cambridge University Press & Assessment 2019

First published 2019

A catalogue record for this publication is available from the British Library

ISBN   978-1-108-43163-7   Paperback

Contents

Preface

From March 20 to 29, 2015, a conference bearing the book's name took place in Hammamet, Tunisia.[1]

It was organized by MIMS[2] and CIMPA,[3] and it gave us the opportunity to celebrate the 60th birthday of Professor Abbas Bahri, Rutgers University. Shortly after, Professor Bahri passed away, on January 10, 2016, after a long struggle against sickness. His death caused deep sadness among the academic community and beyond, particularly in Tunisia, France and the United States of America, given the great influence he had in those countries. In Tunisia he created a new school of thought in PDEs, by supervising several students who continue to develop that innovative style. Several memorial tributes took place after his death and many obituaries were published. He will be missed a lot. In this book, we include a chapter presenting a short biography of Professor Bahri, concentrating on his scientific achievements.

Following the Hammamet conference, and given the high quality of the presentations, we felt we should record those contributions by publishing the proceedings of the conference as a book. The majority of the speakers agreed to participate, and we are very grateful to them for their participation in the conference and their commitment to this book.

After the death of Professor Bahri, the book, which was undergoing the refereeing process, suddenly acquired a deeper meaning for all of us, editors and authors: it changed from the status of a simple conference proceedings to that of a tribute to Professor Bahri, dedicated to his memory.

The book's contents reflect, to some extent, the conference talks and courses, which present the state of the art in PDEs, in connection with Professor Bahri's

---

[1] http://archive.schools.cimpa.info/archivesecoles/20160922162631/
[2] http://www.mims.tn/
[3] https://www.cimpa.info/

contributions. Accordingly, the main speakers at the conference were among the best in their field, mainly from France, the USA, Italy and Tunisia.

MIMS is the Mediterranean Institute for the Mathematical Sciences, founded in Tunis in 2012 to promote mathematics education and research in Tunisia and in the Mediterranean area. It was designed to be a bridge between countries from the North and the South promoting better cooperation.

CIMPA is the International Center for Pure and Applied Mathematics based in Nice, France. It is funded by France, Spain, Switzerland and Norway, together with UNESCO. It promotes mathematical research in developing countries by enhancing North–South cooperation.

Given that many of our contributors are leaders in their field, we expect the book to attract readers from the community of researchers in PDEs interested in interactions with geometry and physics.

We also aim to attract PhD students as readers, since some papers in the book are lecture notes from the six-hour courses given during the conference. We would like to stress the fact that lecturers made the effort of making their courses accessible to PhD students with a basic background in PDEs, as required by CIMPA.

Before closing this preface, we would like to warmly thank again the authors for their valuable contributions. Our thanks go also to Cambridge University Press, for its support with this project, and for carefully considering our submission. We also thank all the production team for handling our LATEX files with a lot of care and patience. We would also like to acknowledge financial support we received from various institutions, which made the Hammamet conference possible: the Commission for Developing Countries (CDC) of the International Mathematical Union (IMU), the French Embassy in Tunis, University of Carthage, University of Paris 13, University of Tunis El-Manar, University of Sfax, the Tunisian Mathematical Society (SMT) and the Tunisian Association for Applied and Industrial Mathematics (ATMAI).

Paris, June 11, 2017
The editors

Abbas Bahri: A Dedicated Life

This volume is dedicated to the memory of Abbas Bahri. Most of the contributors to this book participated in the conference organized in Hammamet, Tunisia in March 2015, on the occasion of his 60th birthday. A short while later, Abbas passed away on January 10, 2016 after a long illness. In this note, we would like to pay tribute to him, stressing in particular his mathematical achievements and influence.

Abbas Bahri was a leading figure in Nonlinear Analysis and Conformal Geometry. As a matter of fact, he played a fundamental role in our understanding of the lack of compactness arising in some variational problems. For example, his book entitled *Critical Points at Infinity in Some Variational Problems* [3] had a tremendous influence on researchers working in the field of Nonlinear Partial Differential Equations involving critical Sobolev exponents. In particular, he performed in that book the finite-dimensional reduction for Yamabe type problems and the related shadow flow for an appropriate pseudogradient. He also gave the accurate expansion of the Euler–Lagrange functional and its gradient. All these techniques later became widely-used tools in the field.

0.1 A short biography

Abbas Bahri was born on January 1, 1955 in Tunis, Tunisia. At the age of 16 he moved to Paris, where he was admitted to the prestigious École Normale Supérieure, Rue d'Ulm at the age of 19. He later obtained his *Agrégation* in mathematics, then defended a *Thèse d'État* in 1981 at the age of 26 in Université Pierre et Marie Curie (Paris 6), under the direction of Professor Haim Brezis.

Starting his career as a Research Assistant in CNRS between 1979 and 1981, he later obtained other positions in the University of Chicago, École

Polytechnique, Palaiseau, and École Nationale d'Ingénieurs (ENIT), Tunis. In 1988 he was appointed Professor at Rutgers University. As director of the Center for Nonlinear Analysis he organized many seminars and supervised a number of PhD students. He also received many prestigious invitations all over the world. His remarkable achievements have been widely recognized. He was awarded the Langevin and Fermat prizes in 1989 for "introducing new tools in the calculus of variation" and he received in 1990 the Board of Trustees Award for Excellence, Rutgers University's highest honor for outstanding research.

Beyond his mathematical achivements, which will be discussed in the next section, we would like to pay tribute to Abbas Bahri for two other reasons.

The first reason, which is connected to research, concerns his total commitment to "transmission", in particular in his homeland, Tunisia. The decisive act began in the early 1990s, when he started supervising about ten PhD students in Tunisia, including two Mauritanians. He devoted much energy and time to this, dividing his holidays in Tunisia between his family and his students. He was in fact establishing a new "mathematical tradition" in Tunisia, a tradition which is proudly continued by his students who hold many outstanding positions in Tunisia and abroad. More recently, despite his illness, he displayed tremendous courage and went on a "math tour" in Tunisia in 2014–2015, giving lectures at many universities, including École Polytechnique de Tunisie, La Marsa, the University of Kairouan, and the University of Sfax.

The second reason concerns his commitment to progress, democracy, and social justice in the world. He particularly believed in, and fought for, the democratization of his country of origin, where free rational thinking would prevail, and he was confident in the intellectual potential of the Tunisian people.

Besides being a gifted mathematician with an exceptional sense of originality and depth, Abbas Bahri was also interested in – among other things – history, art, music, literature, philosophy, and politics. He believed in the contribution of Arab and Muslim culture to the development of human knowledge and intellect, and as a source of inspiration for progress. He also viewed this contribution as a way to transcend cultural differences. Abbas Bahri valued diversity and nurtured friendships all over the world.

0.2 Mathematical contributions

Abbas Bahri's mathematical interests were very broad, ranging from nonlinear PDEs arising from geometry and physics to systems of differential equations of Celestial Mechanics. However, his research focused mainly on fundamental problems in Contact Forms and Conformal Geometry. Bahri's contributions are various and he published many important results in collaboration with a number of authors.

Bahri was fascinated by variational problems arising in Contact Geometry at the beginning of his career, and he continued to work on this topic for the rest of his life. He was, in particular, motivated by the Weinstein conjecture about the existence of periodic orbits of the Reeb vector field ξ of a given contact form α defined in M^3, a three-dimensional closed and oriented manifold. Although this problem features a variational structure, its corresponding variational form is defined on the loop space of M, $H^1(S^1, M)$, by

$$J(x) := \int_0^1 \alpha(\dot{x}) dt, \quad x \in H^1(S^1, M).$$

In fact, the critical points of J are the periodic orbits of ξ.

J is a very bad variational problem on $H^1(S^1, M)$ because the variational flows do not seem to be Fredholm and the critical points of J have an infinite Morse index.

It is in this framework that Bahri developed the concept of *critical points at infinity*. In fact, he discovered that the ω-limit set of non-compact orbits of the gradient flow behave like a *usual critical point*, once a Morse reduction in the neighborhood of such geometric objects is performed. In particular, one can associate with such *asymptotes* a Morse index as well as stable and unstable manifolds.

To study the functional J, Bahri tried to restrict the variations of the curve. In order to do so, taking a non-vanishing vector field v in $\ker \alpha$ and denoting $\beta(.) := d\alpha(v,.)$, he defined the subspace $C_\beta := \{x \in H^1(S^1, M) : \beta(\dot{x}) = 0 \text{ and } \alpha(\dot{x}) = a\}$, where a is a positive constant (which may depend on x). Assuming that β is a contact form with the same orientation as α, he proved (in collaboration with D. Bennequin [1]) that:

J is a C^2 function on C_β whose critical points are the periodic orbits of ξ.
Moreover, those orbits have a finite Morse index.

We notice that the curve in C_β can be expressed in a simple way, that is, if $x \in C_\beta$ then $\dot{x} = a\xi + bv$, where a is a positive constant (depending on x) and therefore $J(x) = a > 0$.

It is easy to see that J does not satisfy the Palais–Smale (PS) condition since it just controls the value a of the curve but the b-component along v is free. Therefore, it can have any behavior along a PS sequence.

Crucially, Bahri used the deformation of the level sets of the associated functional J. For this purpose, in general, the used vector field is $-\nabla J$. In the case of the contact form α, taking $\overline{w}$ such that $d\alpha(v,\overline{w}) = 1$, if $z :=$ $\lambda\xi + \mu v + \eta\overline{w}$ belongs to $T_x C_\beta$ (eventually λ, μ and η have to satisfy some conditions (see (2.7) of [8])) then

$$\nabla J(x).z = -\int_0^1 b\eta dt.$$

In view of this formula, there is a "natural decreasing pseudo-gradient" that can be derived by taking $\eta = b$ in the formula above (the other variables λ and μ will be computed using (2.7) of [8]). This flow has several remarkable geometric properties. One of them is that the linking of two curves under the J-decreasing evolution through the flow of z (with $\eta = b$) never decreases. However, this flow has several "undesired" blow-ups and it is therefore difficult to define a homology related to the periodic orbits of ξ with this pseudo-gradient.

Bahri's main idea was to use a special (constructed) decreasing pseudo-gradient Z for (J, C_β). This program was done in several of his works (in particular [11]) since he required many properties to be satisfied. In particular, the new vector-field Z blows up only along the stratified set $\cup_k \Gamma_{2k}$, where $\Gamma_{2k} := \{$curves made of $k-$pieces of $\xi-$orbits, alternating with $k-$ pieces of $\pm v-$orbits$\}$. Furthermore, along its (semi)-flow-lines, the number of zeros of b (the v-component of $\dot{x}$) never increases and the L^1-norm of b is bounded. The two points are very important in the study of the PS sequences.

Bahri extended this pseudo-gradient on $\cup\Gamma_{2k}$ and he defined the functional

$$J_\infty(x) := \sum_{k=1}^{\infty} a_k, \quad x \in \cup\Gamma_{2k},$$

where a_k is the length of the kth piece of ξ. The critical points of J_∞ are what Bahri called *critical points at infinity*. This precise pseudo-gradient allowed him to understand the lack of compactness and to characterize the critical points at infinity [11, 9]. These points are characterized as follows. A curve in $\cup\Gamma_{2k}$ is a critical point at infinity if it satisfies one of the following assertions.

1. The v-jumps are between conjugate points (conjugate points are points on the same v-orbit such that the form α is transported onto itself by the transport map along v). These critical points are called *true critical points at infinity*.

2. The ξ-pieces have characteristic length and, in addition, the v-jumps send $\ker \alpha$ to itself (a ξ-piece $[x_0, x_1]$ is characteristic if v completes an exact number $n \in \mathbb{Z}$ of half revolutions from x_0 to x_1).

Furthermore, in [13], Bahri proved that the linking property is conserved, that is: for any decreasing flow-line C_s, originating at a periodic orbit and ending at another periodic orbit O' (contractible in M) of ξ with a difference of indices equal to 1, the linking number $lk(C_s, O')$ never decreases with s.

The properties required for the constructed pseudo-gradient Z allowed Bahri to define an intersection operator ∂ for the variational problem (J, C_β), and he was therefore able to define a kind of homology for the critical points (at infinity) of J [9, 12]. However, he noticed that the critical points (at infinity) do not change the topology of the level sets of J (because J is not Fredholm) and this is a serious difficulty to overcome.

In his last paper [14] Bahri used these properties, combined with the *Fadel–Rabinowitz Morse index*, to present a new beautiful proof of the *Weinstein conjecture* on S^3. This new proof combines the case of the tight contact structure on S^3 and the case of all the over-twisted ones and could therefore lead to a better understanding of the existence process for periodic orbits of ξ. It could also possibly lead to multiplicity results on all three-dimensional closed manifolds with finite fundamental group. Moreover, it can be extended to closed manifolds M^{2n+1} with $n \geq 1$ satisfying some topological assumptions (with some technical difficulties).

When Bahri developed the *theory of critical points at infinity* he applied it to many problems. In collaboration with P. Rabinowitz he studied the 3-body problem in Celestial Mechanics. This problem is modeled by the following Hamiltonian system: $m_i \ddot{q}_i + V_{q_i}(q) = 0$, $i = 1, 2, 3$, where $m_i > 0$, $q_i \in \mathbb{R}^3$, and $V(q) = \sum_{i \neq j, i, j = 1}^{3} m_i m_j / |q_i - q_j|^\alpha$, with $\alpha > 0$. This problem has a variational structure. Its T-periodic solutions correspond to critical points of the functional $I(q) := \int_0^T (\frac{1}{2} \sum m_i |\dot{q}_i|^2 - V(q)) dt$ defined on the class of T-periodic functions. In [5] Bahri proved the existence of infinitely many T-periodic solutions of this problem (with $\alpha \geq 2$). The proof of this remarkable result is based on the understanding of the lack of compactness of I. In fact, sections 7 and 8 of [5] are devoted to the analysis of the critical points at infinity of I. This new object allowed the authors to prove the result.

Moreover, Bahri applied his new theory to the Yamabe and scalar curvature problems. For this program, he collected in the monograph [3] many delicate and difficult estimates needed to understand the lack of compactness and to characterize the critical points at infinity of the associated variational functional.

It is known that, in the region where the Palais–Smale condition fails, the functions have to be decomposed as the sum of some bubbles. In collaboration with J.M. Coron, Bahri studied this lack of compactness of the scalar curvature problem on S^3 [4] and he gave a criterion (depending on the scalar function to be prescribed) for the existence of a solution for this problem. This criterion was extended by various authors in other situations and equations with lack of compactness. Furthermore, Bahri used the *theory of critical points at infinity* to give another proof for the Yamabe conjecture for a locally conformally flat manifold [7].

For a bounded domain Ω, the Yamabe problem ($\Delta u + u^{(n+2)/(n-2)}, u > 0$ on Ω; $u = 0$ in $\partial\Omega$) becomes more difficult. The associated variational functional is defined by $J(u) := 1/\int_\Omega |u|^{2n/(n-2)}$ for $u \in \Sigma = \{u \in H_0^1(\Omega) : \|u\| = 1\}$. In collaboration with J.M. Coron, Bahri proved that if Ω has a non-trivial topology then this problem has at least one solution [2]. In fact, the proof is based on combining some analysis and algebraic topology arguments. The analysis part consists of (i) characterizing the levels where the lack of compactness occurs and (ii) proving that there is no difference of topology between the level sets J^a and J^b for a and b large, where $J^a := \{u \in \Sigma : J(u) < a, u > 0\}$. Concerning the algebraic topology argument, since Ω has no trivial topology they were able to find a non-trivial class in the homology of the bottom level set. Furthermore, they proved an intrinsic argument which shows that, for $c_1 < c_2 < c_3$ three consecutive levels (where the lack of compactness occurs), starting from a non-trivial class in the homology of the pair (J^{c_2}, J^{c_1}), if there is no solution, they can find another non-trivial class in the homology of the pair (J^{c_3}, J^{c_2}). Thus, by induction, they are able to find non-trivial classes in the homology of all the pairs $(J^{c_k}, J^{c_{k-1}})$ where the c_is are the levels where the lack of compactness occurs. This gives a contradiction with item (ii) of the analysis part.

Concerning the subcritical case, in collaboration with P.L. Lions, taking a bounded and regular domain $\Omega \subset \mathbb{R}^n$, $n \geq 2$, Bahri studied the following PDE: (P): $-\Delta u = f(x, u)$ in Ω; $u = 0$ on $\partial\Omega$ with $|f(x, s)| \leq C(1 + |s|^p)$ ($1 < p < (n+2)/(n-2)$ if $n \geq 3$) and other assumptions on f. In [6], Bahri introduced another type of result. He proved that, taking a sequence of solutions (u_k) of (P), the boundness of $|u_k|_\infty$ is related to the Morse index of (u_k). In fact, the authors proved that the sequence $(|u_k|_\infty)$ is bounded if and only if the sequence of the Morse index of u_k is bounded. The proof relies on some blow-up analysis. The limit problem becomes: $-\Delta u = |u|^{p-1}u$ in $\mathbb{R}^n$ (or $-\Delta u = |u|^{p-1}u$ in Π: a half space with $u = 0$ on $\partial\Pi$). These problems were studied in the positive case, but there is no result for the changing sign solution. Using a bootstrap argument, Bahri proved that these limit problems do not have non-trivial bounded solutions with bounded Morse index. This new criterion (the notion of the Morse index) becomes very useful for classifying the solutions of other equations.

In his last book [10] Bahri studied, in the first part, the changing-sign Yamabe problem. He considered the case of $\mathbb{R}^3$ (or equivalently S^3):

$$\Delta u + u^5 = 0 \qquad \text{in } \mathbb{R}^3. \tag{0.1}$$

In this case the solutions are known to exist, in fact in infinite number. Moreover, if we impose the positivity of the solutions then we see that the only solutions are given by the family $\delta_{(a,\lambda)} := c\sqrt{\lambda}/\sqrt{(1+\lambda^2|.-a|^2)}$. As for changing-sign solutions, we only know their asymptotic behavior at infinity. In this case Bahri studied the asymptotes generated by these solutions and their combinations under the action of the conformal group. As he said in his monograph, *"The Yamabe problem, without the positivity assumption, is a simpler model of less explicit non-compactness phenomena. The equation (0.1) is "un cas d'école""*. In fact, this work provides a family of estimates and techniques by which the problem of finding infinitely many solutions to the changing-sign Yamabe-type problem on domains of $\mathbb{R}^n$, $n \geq 3$, can be studied. Moreover, using the compactness result of Uhlenberg, the ideas introduced in this work could be useful in the study of the Yang–Mills equations. As a matter of fact, this was the topic of the course Bahri gave in February 2015 in the *Faculté des Sciences de Sfax*.

An interesting idea used in [10] consists of deriving an a priori estimate for the remainder term for a PS sequence. Let Ω be a bounded domain and let v be the unique solution of $-\Delta v = f(v)$ on Ω, $v = 0$ in $\partial\Omega$. To prove that $|v(x)| \leq c\varphi(x)$ for each $x \in \Omega$, where φ is a given function, Bahri introduces the following PDE: $-\Delta\tilde{v} = f(\min(|\tilde{v}|,\varphi)\text{sign }\tilde{v})$ on Ω, $\tilde{v} = 0$ in $\partial\Omega$. By studying the new function $\tilde{v}$, he proves that $|\tilde{v}|$ is small with respect to φ and therefore it satisfies $-\Delta\tilde{v} = f(\tilde{v})$ on Ω, $\tilde{v} = 0$ in $\partial\Omega$, which implies that $\tilde{v} = v$ and therefore, v is small with respect to φ. The aim in introducing the new function $\tilde{v}$ is to overcome some difficulties arising from some non-linear terms of f. Bahri introduced this idea to get some a priori estimate on the remainder function of a PS sequence for the Yamabe sign-changing problem.

As mentioned earlier, Abbas Bahri paved the way for future generations by introducing a new "mathematical tradition" that is now being continued by his students. His contributions go beyond mathematics and his influence has reached many, in Tunisia and all over the world. He is missed not only by his family and friends, but also by many people who met him and appreciated his human qualities and research achievements.

References

[1] Bahri A., *Pseudo-orbits of Contact Forms*, Pitman Research Notes in Mathematics Series, 173. Longman Scientific & Technical, Harlow, 1988.

[2] Bahri A. and Coron J.M., *On a nonlinear elliptic equation involving the critical Sobolev exponent: the effect of the topology of the domain*, Comm. Pure Appl. Math. 41–3, 253–294, 1988.

[3] Bahri A., *Critical points at infinity in some variational problems*, Research Notes in Mathematics, 182, Longman-Pitman, London, 1989.

[4] Bahri A. and Coron J.M., *The scalar-curvature problem on the standard three-dimensional sphere*, J. Funct. Anal. 95, no. 1, 106–172, 1991.

[5] Bahri, A. and Rabinowitz, P., *Periodic orbits of hamiltonian systems of three body type*. Ann. Inst. H. Poincaré Anal. Non linéaire 8, 561–649, 1991.

[6] Bahri A and Lions P.L., *Solutions of superlinear elliptic equations and their Morse indices*, Comm. Pure Appl. Math. 45, 1205–1215, 1992.

[7] Bahri A., *Another proof of the Yamabe conjecture for locally conformally flat manifolds*, Nonlinear Anal. 20, no. 10, 1261–1278, 1993.

[8] Bahri A., *Classical and quantic periodic motions of multiply polarized spin particles*, Pitman Research Notes in Mathematics Series, 378. Longman, Harlow, 1998.

[9] Bahri A., *Flow lines and algebraic invariants in contact form geometry, progress in nonlinear differential equations and their applications*, 53. Birkhauser Boston, Inc., Boston, MA, 2003.

[10] Bahri A. and Y. Xu, *Recent Progress in Conformal Geometry*, ICP Advances Texts in Mathematics 1, London: Imperial College Press, 2007.

[11] Bahri A., *Compactness*, Adv. Nonlinear Studies, 8 (3), 465–568, 2008.

[12] Bahri A., *Homology computation*, Adv. Nonlinear Studies, 8, 1–17, 2008.

[13] Bahri A., *Linking numbers in contact form geometry, with an application to the computation of the intersection operator for the first contact form of J. Gonzalo and F. Varela*, Arab J. Math, 3, 199–210, 2014.

[14] Bahri A., *A Linking/S^1-equivariant variational argument in the space of dual legendrian curves and the proof of the weinstein conjecture on S^3 "in the large"*, Adv. Nonlinear Studies, 15, 497–526, 2015.

Mohamed Ben Ayed, University of Sfax

1

Blow-up Rate for a Semilinear Wave Equation with Exponential Nonlinearity in One Space Dimension

Asma Azaiez*, Nader Masmoudi† and Hatem Zaag‡

We consider in this paper blow-up solutions of the semilinear wave equation in one space dimension, with an exponential source term. Assuming that initial data are in $H^1_{\text{loc}} \times L^2_{\text{loc}}$ or sometimes in $W^{1,\infty} \times L^\infty$, we derive the blow-up rate near a non-characteristic point in the smaller space, and give some bounds near other points. Our results generalize those proved by Godin under high regularity assumptions on initial data.

1.1 Introduction

We consider the one dimensional semilinear wave equation:

$$\begin{cases} \partial_t^2 u = \partial_x^2 u + e^u, \\ u(0) = u_0 \text{ and } \partial_t u(0) = u_1, \end{cases} \tag{1.1}$$

where $u(t) : x \in \mathbb{R} \to u(x,t) \in \mathbb{R}, u_0 \in H^1_{\text{loc},u}$ and $u_1 \in L^2_{\text{loc},u}$. We may also add more restrictions on initial data by assuming that $(u_0, u_1) \in W^{1,\infty} \times L^\infty$. The Cauchy problem for equation (1.1) in the space $H^1_{\text{loc},u} \times L^2_{\text{loc},u}$ follows from fixed point techniques (see Section 1.2).

If the solution is not global in time, we show in this paper that it blows up (see Theorems 1.1 and 1.2). For that reason, we call it a blow-up solution. The existence of blow-up solutions is guaranteed by ODE techniques and the finite speed of propagation.

More blow-up results can be found in Kichenassamy and Littman [12], [13], where the authors introduce a systematic procedure for reducing nonlinear wave equations to characteristic problems of Fuchsian type and construct

* This author is supported by the ERC Advanced Grant no. 291214, BLOWDISOL.
† This author is partially supported by NSF grant DMS- 1211806.
‡ This author is supported by the ERC Advanced Grant no. 291214, BLOWDISOL and by ANR project ANAÉ ref. ANR-13-BS01-0010-03.

singular solutions of general semilinear equations which blow up on a non-characteristic surface, provided that the first term of an expansion of such solutions can be found.

The case of the power nonlinearity has been understood completely in a series of papers, in the real case (in one space dimension) by Merle and Zaag [16], [17], [20] and [21] and in Côte and Zaag [6] (see also the note [18]), and in the complex case by Azaiez [3]. Some of those results have been extended to higher dimensions for conformal or subconformal p:

$$1 < p \leq p_c \equiv 1 + \frac{4}{N-1}, \tag{1.2}$$

under radial symmetry outside the origin in [19]. For non-radial solutions, we would like to mention [14] and [15] where the blow-up rate was obtained. We also mention the recent contribution of [23] and [22] where the blow-up behavior is given, together with some stability results.

In [5] and [4], Caffarelli and Friedman considered semilinear wave equations with a nonlinearity of power type. If the space dimension N is at most 3, they showed in [5] the existence of solutions of Cauchy problems which blow up on a C^1 spacelike hypersurface. If $N = 1$ and under suitable assumptions, they obtained in [4] a very general result which shows that solutions of Cauchy problems either are global or blow up on a C^1 spacelike curve. In [11] and [10], Godin shows that the solutions of Cauchy problems either are global or blow up on a C^1 spacelike curve for the following mixed problem ($\gamma \neq 1$, $|\gamma| \geq 1$):

$$\begin{cases} \partial_t^2 u = \partial_x^2 u + e^u, x > 0, \\ \partial_x u + \gamma \, \partial_t u = 0 \text{ if } x = 0. \end{cases} \tag{1.3}$$

In [11], Godin gives sharp upper and lower bounds on the blow-up rate for initial data in $C^4 \times C^3$. It so happens that his proof can be extended for initial data $(u_0, u_1) \in H^1_{\text{loc},u} \times L^2_{\text{loc},u}$ (see Proposition 1.15).

Let us consider u a blow-up solution of (1.1). Our aim in this paper is to derive upper and lower estimates on the blow-up rate of $u(x,t)$. In particular, we first give general results (see Theorem 1.1), then, considering only non-characteristic points, we give better estimates in Theorem 1.2.

From Alinhac [1], we define a continuous curve Γ as the graph of a function $x \mapsto T(x)$ such that the domain of definition of u (or the maximal influence domain of u) is

$$D = \{(x,t) | 0 \leq t < T(x)\}. \tag{1.4}$$

From the finite speed of propagation, T is a 1-Lipschitz function. The graph Γ is called the blow-up graph of u.

Let us introduce the following non-degeneracy condition for Γ. If we introduce for all $x \in \mathbb{R}$, $t \leq T(x)$ and $\delta > 0$, the cone

$$\mathcal{C}_{x,t,\delta} = \{(\xi, \tau) \neq (x,t) \mid 0 \leq \tau \leq t - \delta|\xi - x|\}, \tag{1.5}$$

then our non-degeneracy condition is the following: x_0 is a non-characteristic point if

$$\exists \delta_0 = \delta_0(x_0) \in (0,1) \text{ such that } u \text{ is defined on } \mathcal{C}_{x_0,T(x_0),\delta_0}. \tag{1.6}$$

If condition (1.6) is not true, then we call x_0 a characteristic point. We denote by $\mathcal{R} \subset \mathbb{R}$ (resp. $\mathcal{S} \subset \mathbb{R}$) the set of non-characteristic (resp. characteristic) points.

We also introduce for each $a \in \mathbb{R}$ and $T \leq T(a)$ the following similarity variables:

$$w_{a,T}(y,s) = u(x,t) + 2\log(T - t), \quad y = \frac{x - a}{T - t}, \quad s = -\log(T - t). \tag{1.7}$$

If $T = T(a)$, we write w_a instead of $w_{a,T(a)}$.

From equation (1.1), we see that $w_{a,T}$ (or w for simplicity) satisfies, for all $s \geq -\log T$, and $y \in (-1, 1)$,

$$\partial_s^2 w - \partial_y((1 - y^2)\partial_y w) - e^w + 2 = -\partial_s w - 2y\partial_{y,s}^2 w. \tag{1.8}$$

In the new set of variables (y,s), deriving the behavior of u as $t \to T$ is equivalent to studying the behavior of w as $s \to +\infty$.

Our first result gives rough blow-up estimates. Introducing the following set:

$$D_R \equiv \{(x,t) \in (\mathbb{R}, \mathbb{R}_+), |x| < R - t\}, \tag{1.9}$$

where $R > 0$, we have the following result.

Theorem 1.1 (Blow-up estimates near any point) *We claim the following:*

(i) **(Upper bound)** *For all $R > 0$ and $a \in \mathbb{R}$ such that $(a, T(a)) \in D_R$, it holds that:*

$$\forall |y| < 1, \forall s \geq -\log T(a), \ w_a(y,s) \leq -2\log(1 - |y|) + C(R),$$

$$\forall t \in [0, T(a)), \ e^{u(a,t)} \leq \frac{C(R)}{d((a,t),\Gamma)^2} \leq \frac{C(R)}{(T(a) - t)^2},$$

where $d((x,t),\Gamma)$ is the (Euclidean) distance from (x,t) to Γ.

(ii) **(Lower bound)** *For all $R > 0$ and $a \in \mathbb{R}$ such that $(a, T(a)) \in D_R$, it holds that*

$$\frac{1}{T(a) - t} \int_{I(a,t)} e^{-u(x,t)} dx \leq C(R)\sqrt{d((a,t),\Gamma)} \leq C(R)\sqrt{T(a) - t}.$$

If, in addition, $(u_0, u_1) \in W^{1,\infty} \times L^{\infty}$ then

$$\forall t \in [0, T(a)), e^{u(a,t)} \geq \frac{C(R)}{d((a,t), \Gamma)} \geq \frac{C(R)}{T(a) - t}.$$

(iii) **(Lower bound on the local energy "norm")** *There exists $\epsilon_0 > 0$ such that for all $a \in \mathbb{R}$, and $t \in [0, T(a))$,*

$$\frac{1}{T(a) - t} \int_{I(a,t)} ((u_t(x,t))^2 + (u_x(x,t))^2 + e^{u(x,t)}) dx \geq \frac{\epsilon_0}{(T(a) - t)^2}, \quad (1.10)$$

where $I(a,t) = (a - (T(a) - t), a + (T(a) - t))$.

Remark The upper bound in item (i) was already proved by Godin [11], for more regular initial data. Here, we show that Godin's strategy works even for less regular data. We refer to the integral in (1.10) as the local energy "norm", since it is like the local energy as in Shatah and Struwe [24], though with the "+" sign in front of the nonlinear term. Note that the lower bound in item (iii) is given by the solution of the associated ODE $u'' = e^u$. However, the lower bound in (ii) doesn't seem to be optimal, since it does not obey the ODE behavior. Indeed, we expect the blow-up for equation (1.1) in the "ODE style", in the sense that the solution is comparable to the solution of the ODE $u'' = e^u$ at blow-up. This is in fact the case with regular data, as shown by Godin [11].

If, in addition, $a \in \mathcal{R}$, we have optimal blow-up estimates.

Theorem 1.2 (An optimal bound on the blow-up rate near a non-characteristic point in a smaller space) *Assume that $(u_0, u_1) \in W^{1,\infty} \times L^{\infty}$. Then, for all $R > 0$, for any $a \in \mathcal{R}$ such that $(a, T(a)) \in D_R$, we have the following:*

(i) **(Uniform bounds on w)** *For all $s \geq -\log T(a) + 1$,*

$$|w_a(y,s)| + \int_{-1}^{1} ((\partial_s w_a(y,s))^2 + (\partial_y w_a(y,s))^2) \, dy \leq C(R),$$

where w_a is defined in (1.7).

(ii) **(Uniform bounds on u)** *For all $t \in [0, T(a))$,*

$$|u(x,t) + 2\log(T(a) - t)| + (T(a) - t) \int_{I} (\partial_x u(x,t))^2 + (\partial_t u(x,t))^2 \, dx \leq C(R).$$

In particular, we have

$$\frac{1}{C(R)} \leq e^{u(x,t)} (T(a) - t)^2 \leq C(R).$$

Remark This result implies that the solution indeed blows up on the curve Γ.

Remark Note that when $a \in \mathcal{R}$, Theorem 1.1 already holds and directly follows from Theorem 1.2. Accordingly, Theorem 1.1 is completely meaningful when $a \in \mathcal{S}$.

Following Antonini, Merle and Zaag in [2] and [15], we would like to mention the existence of a Lyapunov functional in similarity variables. More precisely, let us define

$$E(w(s)) = \int_{-1}^{1} \left(\frac{1}{2}(\partial_s w)^2 + \frac{1}{2}(1 - y^2)(\partial_y w)^2 - e^w + 2w \right) dy. \qquad (1.11)$$

We claim that the functional E defined by (1.11) is a decreasing function of time for solutions of (1.8) on $(-1,1)$.

Proposition 1.3 (A Lyapunov functional for equation (1.1)) *For all $a \in \mathbb{R}$, $T \le T(a)$, $s_2 \ge s_1 \ge -\log T$, the following identities hold for $w = w_{a,T}$:*

$$E(w(s_2)) - E(w(s_1)) = - \int_{s_1}^{s_2} (\partial_s w(-1,s))^2 + (\partial_s w(1,s))^2 ds.$$

Remark The existence of such an energy in the context of the nonlinear heat equation has been introduced by Giga and Kohn in [7], [8] and [9].

Remark As for the semilinear wave equation with conformal power nonlinearity, the dissipation of the energy $E(w)$ degenerates to the boundary ± 1.

This paper is organized as follows:

In Section 1.2, we solve the local in time Cauchy problem.

Section 1.3 is devoted to some energy estimates.

In Section 1.4, we give and prove upper and lower bounds, following the strategy of Godin [11].

Finally, Section 1.5 is devoted to the proofs of Theorem 1.1, Theorem 1.2 and Proposition 1.3.

1.2 The Local Cauchy Problem

In this section, we solve the local Cauchy problem associated with (1.1) in the space $H^1_{\text{loc},u} \times L^2_{\text{loc},u}$. In order to do so, we will proceed in three steps.

(1) In Step 1, we solve the problem in $H^1_{\text{loc},u} \times L^2_{\text{loc},u}$, for some uniform $T > 0$ small enough.

(2) In Step 2, we consider $x_0 \in \mathbb{R}$, and use Step 1 and a truncation to find a local solution defined in some cone $\mathcal{C}_{x_0, \tilde{T}(x_0), 1}$ for some $\tilde{T}(x_0) > 0$. Then,

by a covering argument, the maximal domain of definition is given by
$D = \cup_{x_0 \in \mathbb{R}} \mathcal{C}_{x_0, \bar{T}(x_0), 1}$.

(3) In Step 3, we consider some approximation of equation (1.1), and discuss the convergence of the approximating sequence.

Step 1: The Cauchy problem in $H^1_{\text{loc},u} \times L^2_{\text{loc},u}$

In this step, we will solve the local Cauchy problem associated with (1.1) in the space $H = H^1_{\text{loc},u} \times L^2_{\text{loc},u}$. In order to do so, we will apply a fixed point technique. We first introduce the wave group in one space dimension:

$$S(t) : H \to H,$$

$$(u_0, u_1) \mapsto S(t)(u_0, u_1)(x),$$

$$S(t)(u_0, u_1)(x) = \begin{pmatrix} \frac{1}{2}(u_0(x+t) + u_0(x-t)) + \frac{1}{2}\int_{x-t}^{x+t} u_1 dt \\ \frac{1}{2}(u_0'(x+t) - u_0'(x-t)) + \frac{1}{2}(u_1(x+t) + u_1(x-t)) \end{pmatrix}.$$

Clearly, $S(t)$ is well defined in H, for all $t \in \mathbb{R}$, and more precisely, there is a universal constant C_0 such that

$$\|S(t)(u_0, u_1)\|_H \leq C_0(1+t)\|(u_0, u_1)\|_H. \tag{1.12}$$

This is the aim of the step.

Lemma 1.4 (Cauchy problem in $H^1_{\text{loc},u} \times L^2_{\text{loc},u}$) *For all $(u_0, u_1) \in H$, there exists $T > 0$ such that there exists a unique solution of the problem (1.1) in $C([0,T], H)$.*

Proof Consider $T > 0$ (to be chosen later) small enough in terms of $\|(u_0, u_1)\|_H$.

We first write the Duhamel formulation for our equation:

$$u(t) = S(t)(u_0, u_1) + \int_0^t S(t-\tau)(0, e^{u(\tau)})d\tau. \tag{1.13}$$

Introducing

$$R = 2C_0(1+T)\|(u_0, u_1)\|_H, \tag{1.14}$$

we will work in the Banach space $E = C([0,T], H)$ equipped with the norm $\|u\|_E = \sup_{0 \leq t \leq T} \|u\|_H$. Then, we introduce

$$\Phi : E \to E$$

$$V(t) = \begin{pmatrix} v(t) \\ v_1(t) \end{pmatrix} \mapsto S(t)(u_0, u_1) + \int_0^t S(t-\tau)(0, e^{v(t)})d\tau$$

and the ball $B_E(0, R)$.

We will show that for $T > 0$ small enough, Φ has a unique fixed point in $B_E(0,R)$. To do so, we have to check two points:

1. Φ maps $B_E(0,R)$ to itself;
2. Φ is k-Lipschitz with $k < 1$ for T small enough.

- Proof of 1: Let $V = \begin{pmatrix} v \\ v_1 \end{pmatrix} \in B_E(0,R)$; this means that:

$$\forall t \in [0,T], \ v(t) \in H^1_{loc,u}(\mathbb{R}) \subset L^\infty(\mathbb{R})$$

and that

$$||v(t)||_{L^\infty(\mathbb{R})} \leq C_* R.$$

Therefore

$$||(0,e^v)||_E = \sup_{0 \leq t \leq T} ||e^{v(t)}||_{L^2_{loc,u}}$$
$$\leq e^{C_* R \sqrt{2}}. \tag{1.15}$$

This means that

$$\forall \tau \in [0,T] \ (0,e^{v(\tau)}) \in H,$$

hence $S(t-\tau)(0,e^{v(\tau)})$ is well defined from (1.12) and so is its integral between 0 and t. So Φ is well defined from E to E.

Let us compute $||\Phi(v)||_E$.

Using (1.12), (1.14) and (1.15) we write for all $t \in [0,T]$,

$$||\Phi(v)(t)||_H \leq ||S(t)(u_0,u_1)||_H + \int_0^t ||S(t-\tau)(0,e^{v(\tau)})||_H d\tau$$
$$\leq \frac{R}{2} + \int_0^T C_0(1+T)\sqrt{2}e^{C_* R} d\tau$$
$$\leq \frac{R}{2} + C_0 T(1+T)\sqrt{2}e^{C_* R}. \tag{1.16}$$

Choosing T small enough so that

$$\frac{R}{2} + C_0 T(1+T)\sqrt{2}e^{C_* R} \leq R$$

or

$$T(1+T) \leq \frac{Re^{-C_* R}}{2\sqrt{2}C_0}$$

guarantees that Φ goes from $B_E(0,R)$ to $B_E(0,R)$.

- Proof of 2: Let $V, \bar{V} \in B_E(0,R)$. We have

$$\Phi(V) - \Phi(\bar{V}) = \int_0^T S(t-\tau)(0, e^{v(t)} - e^{\bar{v}(t)}) d\tau.$$

Since $||v(t)||_{L^\infty(\mathbb{R})} \le C_* R$ and the same for $||\bar{v}(t)||_{L^\infty(\mathbb{R})}$, we write

$$|e^{v(\tau)} - e^{\bar{v}(\tau)}| \le e^{C_* R} |v(\tau) - \bar{v}(\tau)|,$$

hence

$$||e^{v(\tau)} - e^{\bar{v}(\tau)}||_{L^2_{\text{loc},u}} \le e^{C_* R} ||v(\tau) - \bar{v}(\tau)||_{L^2_{\text{loc},u}}$$

$$\le e^{C_* R} ||V - \bar{V}||_E. \tag{1.17}$$

Applying $S(t-\tau)$ we write from (1.12), for all $0 \le \tau \le t \le T$,

$$||S(t-\tau)(0, e^{v(\tau)} - e^{\bar{v}(\tau)})||_H \le C_0(1+T)||(0, e^{v(\tau)} - e^{\bar{v}(\tau)})||_H$$

$$\le C_0(1+T)||e^{v(\tau)} - e^{\bar{v}(\tau)}||_{L^2_{\text{loc},u}}$$

$$\le C_0(1+T)e^{C_* R} ||V - \bar{V}||_E. \tag{1.18}$$

Integrating, we end up with

$$||\Phi(V) - \Phi(\bar{V})||_E \le C_0 T(1+T)e^{C_* R} ||V - \bar{V}||_E. \tag{1.19}$$

$k = C_0 T(1+T)e^{C_* R}$ can be made < 1 if T is small.

Conclusion From points 1 and 2, Φ has a unique fixed point $u(t)$ in $B_E(0,R)$. This fixed point is the solution of the Duhamel formulation (1.13) and of our equation (1.1). This concludes the proof of Lemma 1.4. $\qquad\square$

Step 2: The Cauchy problem in a larger region

Let $(u_0, u_1) \in H^1_{\text{loc},u} \times L^2_{\text{loc},u}$ be initial data for the problem (1.1). Using the finite speed of propagation, we will localize the problem and reduces it to the case of initial data in $H^1_{\text{loc},u} \times L^2_{\text{loc},u}$ already treated in Step 1. For $(x_0, t_0) \in \mathbb{R} \times (0, +\infty)$, we will check the existence of the solution in the cone $\mathcal{C}_{x_0, t_0, 1}$. In order to do so, we introduce χ, a C^∞ function with compact support such that $\chi(x) = 1$ if $|x - x_0| < t_0$; let also $(\bar{u}_0, \bar{u}_1) = (u_0 \chi, u_1 \chi)$ (note that $\bar{u}_0$ and $\bar{u}_1$ depend on (x_0, t_0) but we omit this dependence in the indices for simplicity). So, $(\bar{u}_0, \bar{u}_1) \in H^1_{\text{loc},u} \times L^2_{\text{loc},u}$. From Step 1, if $\bar{u}$ is the corresponding solution of equation (1.1), then, by the finite speed of propagation, $u = \bar{u}$ in the intersection of their domains of definition with the cone $\mathcal{C}_{x_0, t_0, 1}$. As $\bar{u}$ is defined for all (x,t) in $\mathbb{R} \times [0,T)$ from Step 1 for some $T = T(x_0, t_0)$, we get the existence of u locally in $\mathcal{C}_{x_0, t_0, 1} \cap \mathbb{R} \times [0,T)$. Varying (x_0, t_0) and covering $\mathbb{R} \times (0, +\infty[$ by an infinite number of cones, we prove the existence and the uniqueness of the solution in a union of backward light cones, which is either the whole half-space $\mathbb{R} \times (0,+\infty)$, or the subgraph of a 1-Lipschitz function $x \mapsto T(x)$. We have just proved the following.

Lemma 1.5 (The Cauchy problem in a larger region) *Consider* $(u_0, u_1) \in H^1_{\text{loc},u} \times L^2_{\text{loc},u}$. *Then, there exists a unique solution defined in D, a subdomain of* $\mathbb{R} \times [0,+\infty)$, *such that for any* $(x_0, t_0) \in D, (u, \partial_t u)_{(t_0)} \in H^1_{\text{loc}} \times L^2_{\text{loc}}(D_{t_0})$, *with* $D_{t_0} = \{x \in \mathbb{R} | (x, t_0) \in D\}$. *Moreover,*

- *either* $D = \mathbb{R} \times [0,+\infty)$,
- *or* $D = \{(x,t) | 0 \leq t < T(x)\}$ *for some 1-Lipschitz function* $x \mapsto T(x)$.

Step 3: Regular approximations for equation (1.1)

Consider $(u_0, u_1) \in H^1_{\text{loc},u} \times L^2_{\text{loc},u}$, u its solution constructed in Step 2, and assume that it is non-global, hence defined under the graph of a 1-Lipschitz function $x \mapsto T(x)$. Consider for any $n \in \mathbb{N}$ a regularized increasing truncation of F satisfying

$$F_n(u) = \begin{cases} e^u & \text{if } u \leq n, \\ e^n & \text{if } u \geq n+1 \end{cases} \tag{1.20}$$

and $F_n(u) \leq \min(e^u, e^{n+1})$. Consider also a sequence $(u_{0,n}, u_{1,n}) \in (C^\infty(\mathbb{R}))^2$ such that $(u_{0,n}, u_{1,n}) \to (u_0, u_1)$ in $H^1 \times L^2(-R, R)$ as $n \to \infty$, for any $R > 0$.

Then, we consider the problem

$$\begin{cases} \partial_t^2 u_n = \partial_x^2 u_n + F_n(u_n), \\ (u_n(0), \partial_t u_n(0)) = (u_{0,n}, u_{1,n}) \in H^1_{\text{loc},u} \times L^2_{\text{loc},u}. \end{cases} \tag{1.21}$$

Since Steps 1 and 2 clearly extend to locally Lipschitz nonlinearities, we get a unique solution u_n defined in the half-space $\mathbb{R} \times (0,+\infty)$, or in the subgraph of a 1-Lipschitz function. Since $F_n(u) \leq e^{n+1}$, for all $u \in \mathbb{R}$, it is easy to see that in fact u_n is defined for all $(x,t) \in \mathbb{R} \times [0,+\infty)$. From the regularity of F_n, $u_{0,n}$ and $u_{1,n}$, it is clear that u_n is a strong solution in $C^2(\mathbb{R}, [0, \infty))$. Introducing the following sets:

$$K^+(x,t) = \{(y,s) \in (\mathbb{R}, \mathbb{R}_+), |y - x| < s - t\}, \tag{1.22}$$

$$K^-(x,t) = \{(y,s) \in (\mathbb{R}, \mathbb{R}_+), |y - x| < t - s\},$$

and

$$K_R^\pm(x,t) = K^\pm(x,t) \cap \overline{D_R}.$$

We claim the following.

Lemma 1.6 (Uniform bounds on variations of u_n in cones) *Consider $R > 0$; one can find $C(R) > 0$ such that if $(x,t) \in D \cap \overline{D_R}$, then $\forall n \in \mathbb{N}$:*

$$u_n(y,s) \geq u_n(x,t) - C(R), \; \forall(y,s) \in \overline{K_R^+(x,t)},$$
$$u_n(y,s) \leq u_n(x,t) + C(R), \; \forall(y,s) \in \overline{K^-(x,t)}.$$

Remark Of course C depends also on initial data, but we omit that dependence, since we never change initial data in this setting. Note that since $(x,t) \in \overline{D_R}$, it follows that $K_R^-(x,t) = K^-(x,t)$.

Proof We will prove the first inequality, the second one can be proved in the same way. For more details see page 74 of [11].

Let $R > 0$, consider (x,t) fixed in $D \cap \overline{D_R}$, and (y,s) in $D \cap \overline{K_R^+(x,t)}$. We introduce the following change of variables:

$$\xi = (y-x) - (s-t), \eta = -(y-x) - (s-t), \bar{u}_n(\xi,\eta) = u_n(y,s). \quad (1.23)$$

From (1.21), we see that $\bar{u}_n$ satisfies:

$$\partial_{\xi\eta} \bar{u}_n(\xi,\eta) = \frac{1}{4} F_n(\bar{u}_n) \geq 0. \quad (1.24)$$

Let $(\bar{\xi}, \bar{\eta})$ be the new coordinates of (y,s) in the new set of variables. Note that $\bar{\xi} \leq 0$ and $\bar{\eta} \leq 0$. We note that there exists $\xi_0 \geq 0$ and $\eta_0 \geq 0$ such that the points $(\xi_0, \bar{\eta})$ and $(\bar{\xi}, \eta_0)$ lie on the horizontal line $\{s = 0\}$ and have as original coordinates respectively $(y^*, 0)$ and $(\tilde{y}, 0)$ for some y^* and $\tilde{y}$ in $[-R, R]$. We note also that in the new set of variables, we have:

$$u_n(y,s) - u_n(x,t) = \bar{u}_n(\bar{\xi}, \bar{\eta}) - \bar{u}_n(0,0) = \bar{u}_n(\bar{\xi}, \bar{\eta}) - \bar{u}_n(\bar{\xi}, 0) + \bar{u}_n(\bar{\xi}, 0) - \bar{u}_n(0,0)$$

$$= -\int_{\bar{\eta}}^0 \partial_\eta \bar{u}_n(\bar{\xi}, \eta) d\eta - \int_{\bar{\xi}}^0 \partial_\xi \bar{u}_n(\xi, 0) d\xi. \quad (1.25)$$

From (1.24), $\partial_\eta \bar{u}_n$ is monotonic in ξ. So, for example for $\eta = \bar{\eta}$, as $\bar{\xi} \leq 0 \leq \xi_0$, we have:

$$\partial_\eta \bar{u}_n(\bar{\xi}, \bar{\eta}) \leq \partial_\eta \bar{u}_n(0, \bar{\eta}) \leq \partial_\eta \bar{u}_n(\xi_0, \bar{\eta}).$$

Similarly, for any $\eta \in (\bar{\eta}, 0)$, we can bound from above the function $\partial_\eta \bar{u}_n(\bar{\xi}, \eta)$ by its value at the point $(\xi^*(\eta), \eta)$, which is the projection of $(\bar{\xi}, \eta)$ on the axis $\{s = 0\}$ in parallel to the axis ξ (as $\bar{\xi} \leq 0 \leq \xi^*(\eta)$).

In the same way, from (1.24), $\partial_\xi \bar{u}_n$ is monotonic in η. As $\bar{\eta} \leq 0 \leq \eta_0$, we can bound, for $\xi \in (\bar{\xi}, 0)$, $\partial_\xi \bar{u}_n(\xi, 0)$ by its value at the point $(\xi, \eta^*(\xi))$, which is the projection of $(\xi, 0)$ on the axis $\{s = 0\}$ in parallel to the axis η $(0 < \eta^*(\xi))$. So it follows that:

$$\partial_\eta \bar{u}_n(\bar{\xi}, \eta) \leq \partial_\eta \bar{u}_n(\xi^*(\eta), \eta), \ \forall \eta \in (\bar{\eta}, 0),$$
$$\partial_\xi \bar{u}_n(\xi, 0) \leq \partial_\xi \bar{u}_n(\xi, \eta^*(\xi)), \ \forall \xi \in (\bar{\xi}, 0).$$
(1.26)

By a straightforward geometrical construction, we see that the coordinates of $(\xi^*(\eta), \eta)$ and $(\xi, \eta^*(\xi))$, in the original set of variables $\{y, s\}$, are respectively $(x + t - \eta\sqrt{2}, 0)$ and $(x - t + \eta\sqrt{2}, 0)$. Both points are in $[-R, R]$.

Furthermore, we have from (1.23):

$$\partial_\eta \bar{u}_n(\xi^*(\eta), \eta) = \frac{1}{2}(-\partial_t u_n - \partial_x u_n)(x + t - \eta\sqrt{2}, 0)$$
$$= \frac{1}{2}(-u_{1,n} - \partial_x u_{0,n})(x + t - \eta\sqrt{2}),$$

$$\partial_\xi \bar{u}_n(\xi, \eta^*(\xi)) = \frac{1}{2}(-\partial_t u_n + \partial_x u_n)(x - t + \eta\sqrt{2}, 0) \qquad (1.27)$$
$$= \frac{1}{2}(-u_{1,n} + \partial_x u_{0,n})(x - t + \eta\sqrt{2}).$$

Using (1.27), the Cauchy–Schwarz inequality and the fact that $u_{1,n}$ and $\partial_x u_{0,n}$ are uniformly bounded in $L^2(-R, R)$ since they are convergent, we have:

$$\int_{\bar{\eta}}^0 \partial_\eta \bar{u}_n(\xi^*(\eta), \eta) d\eta \leq C(R),$$
$$\int_{\bar{\xi}}^0 \partial_\xi \bar{u}_n(\xi, \eta^*(\xi)) d\xi \leq C(R).$$
(1.28)

Using (1.25), (1.26) and (1.28), we reach the conclusion of Lemma 1.6. $\qquad \square$

Let us show the following.

Lemma 1.7 (Convergence of u_n as $n \to \infty$) *Consider $(x, t) \in \mathbb{R} \times [0, +\infty)$. We have the following:*

- *if $t > T(x)$, then $u_n(x, t) \to +\infty$,*
- *if $t < T(x)$, then $u_n(x, t) \to u(x, t)$.*

Proof We claim that it is enough to show the convergence for a subsequence. Indeed, this is clear from the fact that the limit is explicit and doesn't depend on the subsequence. Consider $(x, t) \in \mathbb{R} \times [0, +\infty)$; up to extracting a subsequence, there is an $l(x, t) \in \overline{\mathbb{R}}$ such that $u_n(x, t) \to l(x, t)$ as $n \to \infty$.

Let us show that $l \neq -\infty$. Since $F_n(u) \geq 0$, it follows that $u_n(x, t) \geq \underline{u}_n(x, t)$, where

$$\begin{cases} \partial_t^2 \underline{u}_n = \partial_x^2 \underline{u}_n, \\ \underline{u}_n(0) = u_{0,n} \text{ and } \partial_t \underline{u}_n(0) = u_{1,n}. \end{cases} \qquad (1.29)$$

Since $\underline{u}_n \in L^\infty_{\text{loc}}(\mathbb{R}^+, H^1(-R,R)) \subset L^\infty_{\text{loc}}(\mathbb{R}^+, L^\infty(-R,R))$, for any $R > 0$, from the fact that $(u_{0,n}, u_{1,n})$ is convergent in $H^1_{\text{loc}} \times L^2_{\text{loc}}$, it follows that $l(x,t) \geq \limsup_{n \to +\infty} \underline{u}_n(x,t) > -\infty$.

Note from the fact that $F_n(u) \leq e^u$ that we have

$$\forall x \in \mathbb{R}, t < T(x), u_n(x,t) \leq u(x,t). \tag{1.30}$$

Introducing $R = |x| + t + 1$, we see by definition (1.9) of D_R that $(x,t) \in D_R$. Let us handle two cases in the following.

Case 1: $t < T(x)$

Let us introduce v_n, the solution of

$$\begin{cases} \partial_t^2 v_n = \partial_x^2 v_n + e^{v_n}, \\ v_n(0) = u_{0,n} \text{ and } \partial_t v_n(0) = u_{1,n} \in H^1_{\text{loc},u} \times L^2_{\text{loc},u}. \end{cases}$$

From the local Cauchy theory in $H^1_{\text{loc},u} \times L^2_{\text{loc},u}$ and the Sobolev embedding, we know that

$$v_n \to u \text{ uniformly as } n \to \infty \text{ in compact sets of } D. \tag{1.31}$$

Let us consider

$$\tilde{K} = \overline{K_-(x, (t+T(x))/2)}$$

and $\tilde{M} = \max_{(y,s) \in \tilde{K}} |u(y,s)| < +\infty$, since $\tilde{K}$ is a compact set in D.

From (1.31), we may assume n large enough, so that

$$||u_{0,n} - u_0||_{L^\infty(\tilde{K} \cap \{t=0\})} \leq 1,$$

$$\sup_{(y,s) \in \tilde{K}} |v_n(y,s)| \leq \tilde{M} + 1 \tag{1.32}$$

and

$$n \geq \tilde{M} + 3. \tag{1.33}$$

In particular,

$$||u_{0,n}||_{L^\infty(\tilde{K} \cap \{t=0\})} \leq \tilde{M} + 1. \tag{1.34}$$

We claim that

$$\forall (y,s) \in \tilde{K}, |u_n(y,s)| \leq \tilde{M} + 2. \tag{1.35}$$

Indeed, arguing by contradiction, we may assume from (1.34) and continuity of u_n that

$$\forall s \in [0, \tilde{t}_n], ||u_n(s)||_{L^\infty(\tilde{K} \cap \{t=s\})} \leq \tilde{M} + 2 \tag{1.36}$$

and

$$\|u_n(\tilde{t}_n)\|_{L^\infty(\tilde{K}\cap\{t=\tilde{t}_n\})} = \tilde{M} + 2, \tag{1.37}$$

for some $\tilde{t}_n \in (0, \frac{t+T(x)}{2})$.

From (1.33), (1.36) and the definition (1.20) of F_n, we see that

$$\forall (y,s) \in \tilde{K} \text{ with } s \le \tilde{t}_n, F_n(u_n(y,s)) = e^{u_n(y,s)}.$$

Therefore, u_n and v_n satisfy the same equation with the same initial data on $\tilde{K} \cap \{s \le \tilde{t}_n\}$. From uniqueness of the solution to the Cauchy problem, we see that

$$\forall (y,s) \in \tilde{K} \text{ with } s \le \tilde{t}_n, u_n(y,s) = v_n(y,s).$$

A contradiction then follows from (1.32) and (1.37). Thus, (1.35) holds.

Again, from the choice of n in (1.33), we see that

$$\forall (y,s) \in \tilde{K}, F_n(u_n(y,s)) = e^{u_n(y,s)},$$

hence, from uniqueness,

$$\forall (y,s) \in \tilde{K}, u_n(y,s) = v_n(y,s).$$

From (1.31), and since $(x,t) \in \tilde{K}$, it follows that $u_n(x,t) \to u(x,t)$ as $n \to \infty$.

Case 2: $t > T(x)$

Assume by contradiction that $l < +\infty$. From Lemma 1.6, it follows that

$$\forall (y,s) \in \overline{K^-(x,t)}, u_n(y,s) \le u_n(x,t) + C(R).$$

For $n \ge n_0$ large enough, this gives $u_n(y,s) \le l+1+C(R)$.

If $M = E(l+1+C(R))+1$, then

$$\forall n \ge \max(M, n_0), \forall (y,s) \in \overline{K^-(x,t)}, F_n(u_n(y,s)) = e^{u_n(y,s)},$$

and u_n satisfies (1.1) in $\overline{K^-(x,t)}$ with initial data $(u_{0,n}, u_{1,n}) \to (u_0, u_1) \in H^1 \times L^2(K^-(x,t) \cap \{t=0\})$. From the finite speed of propagation and the continuity of solutions to the Cauchy problem with respect to the initial data, it follows that u_n and u are both defined in $K^-(x,t)$ for n large enough, in particular u is defined at (x,s) with $T(x) < s < t$ with $u = u_n$ in $\overline{K^-(x,t)}$. This gives a contradiction with the expression of the domain of definition (1.4) of u. $\qquad\square$

1.3 Energy Estimates

In this section, we use some localized energy techniques from Shatah and Struwe [24] to derive a non-blow-up criterion which will give the lower bound in Theorem 1.1. More precisely, we give the following.

Proposition 1.8 (Non-blow-up criterion for a semilinear wave equation)
$\forall c_0 > 0$, *there exist* $M_0(c_0) > 0$ *and* $M(c_0) > 0$ *such that, if*

$$(H) : \begin{cases} ||\partial_x u_0||^2_{L^2(-1,1)} + ||u_1||^2_{L^2(-1,1)} \le c_0^2 \\ \forall |x| < 1, \ u_0(x) \le M_0, \end{cases} \tag{1.38}$$

then equation (1.1) with initial data (u_0, u_1) *has a unique solution* $(u, \partial_t u) \in C([0,1), H^1 \times L^2(|x| < 1-t))$ *such that for all* $t \in [0,1)$ *we have:*

$$||\partial_x u(t)||^2_{L^2(|x|<1-t)} + ||\partial_t u(t)||^2_{L^2(|x|<1-t)} \le 2c_0^2 \tag{1.39}$$

and

$$\forall |x| < 1-t, \ u(x,t) \le M. \tag{1.40}$$

Note that here we work in the space $H^1_{\text{loc}} \times L^2_{\text{loc}}$ which is larger than the space $H^1_{\text{loc},u} \times L^2_{\text{loc},u}$ which is adopted elsewhere for equation (1.1). Before giving the proof of this result, let us first give the following corollary, which is a direct consequence of Proposition 1.8.

Corollary 1.9 *There exists* $\bar{\epsilon}_0 > 0$ *such that if*

$$\int_{-1}^{1} (u_1(x))^2 + (\partial_x u_0(x))^2 + e^{u_0(x)} \, dx \le \bar{\epsilon}_0, \tag{1.41}$$

then the solution u *of equation (1.1) with initial data* (u_0, u_1) *doesn't blow up in the cone* $C_{0,1,1}$.

Let us first derive Corollary 1.9 from Proposition 1.8.

Proof of Corollary 1.9 assuming that Proposition 1.8 holds
From (1.41), if $\bar{\epsilon}_0 \le 1$ we see that

$$\int_{-1}^{1} \left((u_1(x))^2 + (\partial_x u_0(x))^2 \right) dx \le \bar{\epsilon}_0 \le 1, \tag{1.42}$$

$$\int_{-1}^{1} e^{u_0(x)} dx \le \bar{\epsilon}_0.$$

Therefore, for some $x_0 \in (-1, 1)$, we have $2e^{u_0(x_0)} = \int_{-1}^{1} e^{u_0(x)} dx \leq \bar{\epsilon}_0$, hence $u_0(x_0) \leq \log \frac{\bar{\epsilon}_0}{2}$. Using (1.42), we see that for all $x \in (-1, 1)$,

$$u_0(x) = u_0(x_0) + \int_{x_0}^{x} \partial_x u_0 \leq u_0(x_0) + \sqrt{2} \left(\int_{-1}^{1} (\partial_x u_0(x))^2 dx \right)^{\frac{1}{2}}$$

$$\leq \log \frac{\bar{\epsilon}_0}{2} + \sqrt{2\bar{\epsilon}_0} \leq M_0(1),$$

defined in Proposition 1.8, provided that $\bar{\epsilon}_0$ is small enough. Therefore, the hypothesis (H) of Proposition 1.8 holds with $c_0 = 1$, and so does its conclusion. This concludes the proof of Corollary 1.9, assuming that Proposition 1.8 holds.

□

Now, we give the proof of Proposition 1.8.

Proof of Proposition 1.8 Consider $c_0 > 0$ and introduce

$$M_0 = \log \left(\frac{c_0^2}{16} \right) - c_0\sqrt{2} - \frac{c_0^2}{8} \text{ and } M(c_0) = \log \left(\frac{c_0^2}{16} \right).$$

Then, we consider (u_0, u_1) satisfying hypothesis (H). From the solution of the Cauchy problem in $H_{\text{loc}}^1 \times L_{\text{loc}}^2$, which follows exactly by the same argument as in the space $H_{\text{loc},u}^1 \times L_{\text{loc},u}^2$ presented in Section 1.2, there exists $t^* \in (0, 1]$ such that equation (1.1) has a unique solution with $(u, \partial_t u) \in C([0, t^*), H^1 \times L^2(|x| < 1 - t))$. Our aim is to show that $t^* = 1$ and that (1.39) and (1.40) hold for all $t \in [0, 1)$.

Clearly, from the solution of the Cauchy problem, it is enough to show that (1.39) and (1.40) hold for all $t \in [0, t^*)$, so we only do that in the following.

Arguing by contradiction, we assume that there exists at least some time $t \in [0, t^*)$ such that either (1.39) or (1.40) doesn't hold. If $\bar{t}$ is the lowest possible t, then we have from continuity either

$$\|\partial_x u(\bar{t})\|_{L^2(|x| < 1 - \bar{t})}^2 + \|\partial_t u(\bar{t})\|_{L^2(|x| < 1 - t)}^2 = 2c_0,$$

or

$$\exists |x_0| < 1 - \bar{t}, \text{such that } u(x_0, \bar{t}) = M.$$

Note that since (1.39) holds for all $t \in [0, \bar{t})$, it follows that

$$\forall t \in [0, \bar{t}), \forall |x| < 1 - t, u(x, t) \leq M = \log \left(\frac{c_0^2}{16} \right). \tag{1.43}$$

Following the alternative on $\bar{t}$, two cases arise in the following.

Case 1: $||\partial_x u(\bar{t})||^2_{L^2(|x|<1-\bar{t})} + ||\partial_t u(\bar{t})||^2_{L^2(|x|<1-\bar{t})} = 2c_0^2.$

Referring to Shatah and Struwe [24], we see that:

$$\int_{|x|<1-\bar{t}} (\tfrac{1}{2}(\partial_x u^2 + \partial_t u^2) - e^u)\, dx - \int_{|x|<1} (\tfrac{1}{2}(\partial_x u_0^2 + u_1^2) - e^{u_0})\, dx$$
$$= \int_{\Gamma} (e^u - \tfrac{1}{2}|\partial_x u - \frac{x}{|x|}\partial_t u|^2)\, d\sigma, \quad (1.44)$$

where

$$\Gamma = \{(x,t) \in \mathbb{R} \times \mathbb{R}_+, \text{ such that } |x| = 1 - t\} \cap [0,\bar{t}].$$

Using (1.43), it follows that

$$\int_{|x|<1-\bar{t}} e^{u(x,\bar{t})} dx \le \int_{|x|<1-\bar{t}} e^M \le \frac{c_0^2}{8}.$$

$$\int_{\Gamma} e^u d\sigma \le \int_0^{\bar{t}} (e^{u(1-t,t)} + e^{u(t-1,t)}) dt \le \frac{c_0^2}{8}.$$

Therefore, from (1.44) and (1.38), we write

$$\int_{|x|<1-\bar{t}} ((\partial_x u)^2 + (\partial_t u)^2) dx \le \int_{|x|<1} (\partial_x u_0)^2 dx + (u_1)^2 + \int_{|x|<1-\bar{t}} e^{u(x,\bar{t})} dx + \int_{\Gamma} e^u d\sigma$$
$$\le c_0^2 + \frac{3}{8}c_0^2 < 2c_0^2,$$

which is a contradiction.

Case 2: $\exists x_0 \in (-(1-\bar{t}), 1-\bar{t}), u(x_0,\bar{t}) = M.$

Recall Duhamel's formula:

$$\forall |x| < 1 - \bar{t},$$

$$u(x,\bar{t}) = \frac{1}{2}(u_0(x-\bar{t}) + u_0(x+\bar{t})) + \frac{1}{2}\int_{x-\bar{t}}^{x+\bar{t}} u_1(z)dz$$
$$+ \frac{1}{2}\int_0^{\bar{t}}\int_{x-\bar{t}+\tau}^{x+\bar{t}-\tau} e^{u(z,\tau)}\, dzd\tau. \quad (1.45)$$

From (H), we write

$$\int_{x-\bar{t}}^{x+\bar{t}} u_1\, dx \le \left(\int_{-1}^1 u_1^2\right)^{\frac{1}{2}} 2\sqrt{2} \le c_0\sqrt{2}.$$

From (1.43), we write

$$\int_0^{\bar{t}}\int_{z-\bar{t}+\tau}^{z+\bar{t}-\tau} e^{u(z,\tau)}dzd\tau \le \int_0^{\bar{t}}\int_{z-\bar{t}+\tau}^{z+\bar{t}-\tau} \frac{c_0^2}{16} \le \frac{c_0^2}{8}.$$

Since $u_0(x \pm \bar{t}) \leq M_0 = \log(\frac{c_0^2}{16}) - c_0\sqrt{2} - \frac{c_0^2}{8}$, it follows from (1.45) that

$$u(x,\bar{t}) \leq M_0 + c_0\frac{\sqrt{2}}{2} + \frac{c_0^2}{16} < \log\left(\frac{c_0^2}{16}\right) = M,$$

and a contradiction follows.

This concludes the proof of Proposition 1.8. Since we have already derived Corollary 1.9 from Proposition 1.8, this is also the conclusion of the proof of Corollary 1.9. □

1.4 ODE Type Estimates

In this section, we extend the work of Godin in [11]. In fact, we show that his estimates hold for more general initial data. As in the introduction, we consider $u(x,t)$ a non-global solution of equation (1.1) with initial data $(u_0, u_1) \in H^1_{\text{loc},u} \times L^2_{\text{loc},u}$. This section is organized as follows.

In the first subsection, we give some preliminary results and we show that the solution goes to $+\infty$ on the graph Γ.

In the second subsection, we give and prove upper and lower bounds on the blow-up rate.

1.4.1 Preliminaries

In this subsection, we first give some geometrical estimates on the blow-up curve (see Lemmas 1.10, 1.11 and 1.12). Then, we use equation (1.1) to derive a kind of maximum principle in light cones (see Lemma 1.13), then a lower bound on the blow-up rate (see Proposition 1.14).

We first give the following geometrical property concerning the distance to $\{t = T(x)\}$, the boundary of the domain of definition of $u(x,t)$.

Lemma 1.10 (Estimate for the distance to the blow-up boundary) *For all* $(x,t) \in D$, *we have*

$$\frac{1}{\sqrt{2}}(T(x) - t) \leq d((x,t),\Gamma) \leq T(x) - t, \tag{1.46}$$

where $d((x,t),\Gamma)$ *is the distance from* (x,t) *to* Γ.

Proof Note first by definition that

$$d((x,t),\Gamma) \leq d((x,t),(x,T(x))) = T(x) - t.$$

Then, from the finite speed of propagation, Γ is above $\mathcal{C}_{x,T(x),1}$, the backward light cone with vertex $(x,T(x))$. Since $(x,t) \in \mathcal{C}_{x,T(x),1}$, it follows that

$$d((x,t),\Gamma) \geq d((x,t),\mathcal{C}_{x,T(x),1}) = \frac{\sqrt{2}}{2}(T(x) - t).$$

This concludes the proof of Lemma 1.10. $\qquad\square$

Now, we give a geometrical property concerning distances, specific for non-characteristic points.

Lemma 1.11 (A geometrical property for non-characteristic points) *Let $a \in \mathcal{R}$. There exists $c := C(\delta)$, where $\delta = \delta(a)$ is given by (1.6), such that for all $(x,t) \in \mathcal{C}_{a,T(a),1}$,*

$$\frac{1}{c} \leq \frac{T(x) - t}{T(a) - t} \leq c.$$

Remark From Lemma 1.10, it follows that

$$\frac{1}{\bar{c}} \leq \frac{d((x,t),\Gamma)}{d((a,t),\Gamma)} \leq \bar{c}$$

whenever $a \in \mathcal{R}$ and $(x,t) \in \mathcal{C}_{a,T(a),1}$.

Proof Let a be a non-characteristic point. We recall from condition (1.6) that

$$\exists \delta = \delta(a) \in (0,1) \text{ such that } u \text{ is defined on } \mathcal{C}_{a,T(a),\delta}.$$

Let (x,t) be in the light cone with vertex $(a,T(a))$. Using the fact that the blow-up graph is above the cone $\mathcal{C}_{a,T(a),\delta}$ and the fact that $(x,t) \in \mathcal{C}_{a,T(a),1}$, we see that

$$T(x) - t \geq T(a) - \delta|x - a| - t \geq (T(a) - t)(1 - \delta). \tag{1.47}$$

In addition, as Γ is a 1-Lipschitz graph, we have

$$T(x) \leq T(a) + |x - a|,$$

so, for all $(x,t) \in \mathcal{C}_{a,T(a),1}$,

$$T(x) - t \leq T(a) + |x - a| - t \leq 2(T(a) - t). \tag{1.48}$$

From (1.47) and (1.48), there exists $c = c(\delta)$ such that

$$\frac{1}{c} \leq \frac{T(x) - t}{T(a) - t} \leq c.$$

This concludes the proof of Lemma 1.11. $\qquad\square$

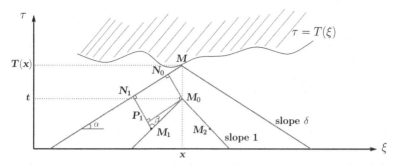

Figure 1.1. Illustration for the proof of (1.49)

Finally, we give the following coercivity estimate on the distance to the blow-up curve, still specific for non-characteristic points.

Lemma 1.12 *Let $x \in \mathcal{R}$ and $t \in [0, T(x))$. For all $\tau \in [0, t)$ and $j = 1, 2$, we have*

$$d((z_j, w_j), \Gamma) \geq \frac{1}{C}(d((x,t), \Gamma) + |(x,t) - (z_j, w_j)|), \tag{1.49}$$

where $(z_1, w_1) = (x + t - \tau, \tau)$ and $(z_2, w_2) = (x - t + \tau, \tau)$.

Remark Note that (z_j, w_j) for $j = 1, 2$ lie on the backward light cone with vertex (x, t).

Proof Consider $x \in \mathcal{R}$, $t \in [0, T(x))$. By definition, there exists $\delta \in (0, 1)$ such that $\mathcal{C}_{x,T(x),\delta} \subset D$. We will prove the estimate for $j = 1$ and $\tau \in [0, t)$, since the the estimate for $j = 2$ follows by symmetry. In order to do so, we introduce the following notations, as illustrated in Figure 1.1: $M = (x, T(x))$, $M_0 = (x, t)$ and $M_1 = (z_1, w_1) = (x + t - \tau, \tau)$, which is on the left boundary of the backward light cone $\mathcal{C}_{x,t,1}$; N_1 the orthogonal projection of M_1 on the left boundary of the cone $\mathcal{C}_{x,T(x),\delta}$; P_1 the orthogonal projection of M_0 on $[N_1, M_1]$. Note that the quadrangle $M_0 N_0 N_1 P_1$ is a rectangle. If α is such that $\tan \alpha = \delta$ and $\beta = \widehat{PM_1M_0}$, then we see from elementary considerations on angles that $\beta = \alpha + \frac{\pi}{4}$ and $\widehat{N_0M_0M} = \alpha$.

Therefore, using Lemma 1.10, and the angles on the triangle M_0N_0M, we see that:

$$d((x,t), \Gamma) \leq T(x) - t = MM_0 = \frac{M_0N_0}{\cos\alpha} = \frac{N_1P_1}{\cos\alpha}. \tag{1.50}$$

Moreover, since the blow-up graph is above the cone $\mathcal{C}_{x,T(x),\delta}$, it follows that

$$d((z_1, w_1), \Gamma) \geq M_1N_1.$$

In particular,

$$M_1 N_1 = N_1 P_1 + P_1 M_1 = N_1 P_1 + \cos(\tfrac{\pi}{4} + \alpha) M_1 M_0. \qquad (1.51)$$

Since $0 < \delta < 1$, hence $0 < \alpha < \tfrac{\pi}{4}$, it follows that $\cos(\tfrac{\pi}{4} + \alpha) > 0$. Since $M_1 M_0 = |(z_1, w_1) - (x, t)|$, the result follows from (1.50) and (1.51).

In the same way, we can prove this for the other point $M_2 = (z_2, w_2)$, which gives (1.49). $\qquad\square$

Now, we give the following corollary from the approximation procedure in Lemmas 1.6 and 1.7.

Lemma 1.13 (Uniform bounds on variations of u in cones) *For any $R > 0$, there exists a constant $C(R) > 0$ such that if $(x,t) \in D \cap \overline{D_R}$ then*

$$u(y,s) \geq u(x,t) - C(R), \ \forall (y,s) \in D \cap \overline{K_R^+(x,t)},$$

$$u(y,s) \leq u(x,t) + C(R), \ \forall (y,s) \in \overline{K^-(x,t)},$$

where the cones $K^\pm$ and $K_R^\pm$ are defined in (1.22).

Remark The constant $C(R)$ depends also on u_0 and u_1, but we omit this dependence in the sequel.

In the following, we give a lower bound on the blow-up rate and we show that $u(x,t) \to +\infty$ as $t \to T(x)$.

Proposition 1.14 (A general lower bound on the blow-up rate)

(i) If $(u_0, u_1) \in W^{1,\infty} \times L^\infty$, then for all $R > 0$, there exists $C(R) > 0$ such that for all $(x,t) \in D \cap D_R$,

$$d((x,t), \Gamma) e^{u(x,t)} \geq C.$$

In particular, for all $(x,t) \in D \cap \overline{D_R}$, $u(x,t) \to +\infty$ as $d((x,t), \Gamma) \to 0$.

(ii) If we only have $(u_0, u_1) \in H^1_{\mathrm{loc},u} \times L^2_{\mathrm{loc},u}$, then for all $R > 0$, there exists $C(R) > 0$ such that for all $(x_0, t) \in D \cap D_R$,

$$\frac{1}{T(x_0) - t} \int_{|x - x_0| < T(x_0) - t} e^{-u(x,t)} dx \leq C(R) \sqrt{d(x_0, t)}.$$

In particular, e^{-u} converges to 0 on average over slices of the light cone, as $d(x_0, t) \to 0$.

Remark Near non-characteristic points, we are able to derive the optimal lower bound on the blow-up rate. See item *(ii)* of Proposition 1.15.

Proof of Proposition 1.14

(i) Clearly, the last sentence in item (i) follows from the first, hence, we only prove the first.

Let $R > 0$ and $(x,t) \in D \cap D_R$. Using the approximation procedure defined in (1.21), we write $u_n = \underline{u}_n + \tilde{u}_n$ with:

$$\underline{u}_n(x,t) = \frac{1}{2}\left(u_{0,n}(x-t) + u_{0,n}(x+t)\right) + \frac{1}{2}\int_{x-t}^{x+t} u_{1,n}(\xi)d\xi,$$

$$\tilde{u}_n(x,t) = \frac{1}{2}\int_0^t \int_{x-t+\tau}^{x+t-\tau} F_n(u_n(z,\tau))dzd\tau.$$

(Note that $\underline{u}_n$ was already defined in (1.29).)

Since $F_n \geq 0$ from (1.20), it follows that

$$u_n(x,t) \geq \underline{u}_n(x,t) \geq -C(R) \text{ for all } (x,t) \in D \cap D_R. \tag{1.52}$$

Differentiating $\underline{u}_n$, we see that

$$\partial_t\underline{u}_n(x,t) = \frac{1}{2}\left(\partial_x u_{0,n}(x+t) - \partial_x u_{0,n}(x-t)\right) + \frac{1}{2}\left(u_{1,n}(x+t) + u_{1,n}(x-t)\right)$$

$$\leq \|\partial_x u_{0,n}\|_{L^\infty(-R,R)} + \|u_{1,n}\|_{L^\infty(-R,R)} \leq C(R). \tag{1.53}$$

Differentiating $\tilde{u}_n$, we get

$$\partial_t\tilde{u}_n(x,t) = \frac{1}{2}\int_0^t \left(F_n(u_n(x-t+\tau,\tau), +F_n(u_n(x+t-\tau,\tau))\right)d\tau$$

$$\leq \frac{1}{2}\int_0^t \left(e^{u_n(x-t+\tau,\tau)} + e^{u_n(x+t-\tau,\tau)}\right)d\tau$$

since $F_n(u) \leq e^u$. Since $u_n(x-t+\tau,\tau) \leq u_n(x,t) + C(R)$ and $u_n(x+t-\tau,\tau) \leq u_n(x,t) + C(R)$ from Lemma 1.6, it follows that

$$\partial_t\tilde{u}_n(x,t) \leq cte^{u_n(x,t)} \leq C(R)e^{u_n(x,t)}. \tag{1.54}$$

Therefore, using (1.52) we see that

$$\partial_t u_n(x,t) = \partial_t\underline{u}_n(x,t) + \partial_t\tilde{u}_n(x,t) \leq C(R) + C(R)e^{u_n(x,t)} \leq C(R)e^{u_n(x,t)},$$

hence

$$\partial_t u_n(x,t)e^{-u_n(x,t)} \leq C(R). \tag{1.55}$$

Integrating (1.55) on any interval $[t_1,t_2]$ with $0 \leq t_1 < T(x) < t_2$, we get $e^{-u_n(x,t_1)} - e^{-u_n(x,t_2)} \leq C(t_2 - t_1)$. Making $n \to \infty$ and using Lemma 1.7 we see that $e^{-u(x,t_1)} \leq C(t_2 - t_1)$.

Taking $t_1 = t$ and making $t_2 \to T(x)$, we get $e^{-u(x,t)} \leq C(T(x) - t)$. Using Lemma 1.10 concludes the proof of item (i) of Proposition 1.14.

(ii) If $(u_0, u_1) \in H^1_{loc,u} \times L^2_{loc,u}$, then a small modification in the argument of item (i) gives the result. Indeed, if $t_0 \in [0, T(x_0))$,

$$a_0 = x_0 - (T(x_0) - t_0), \quad b_0 = x_0 + (T(x_0) - t_0),$$

and $t \geq 0$, we write from (1.52) and (1.53)

$$\int_{a_0}^{b_0} \partial_t \underline{u}_n(x,t) e^{-u_n(x,t)} dx \leq C(R)\sqrt{b_0 - a_0}.$$

Furthermore, from (1.54) we write

$$\int_{a_0}^{b_0} \partial_t \tilde{u}_n(x,t) e^{-u_n(x,t)} dx \leq C(R)(b_0 - a_0).$$

Therefore, it follows that

$$-\frac{d}{dt} \int_{a_0}^{b_0} e^{-u_n(x,t)} dx = \int_{a_0}^{b_0} \partial_t u_n(x,t) e^{-u_n(x,t)} dx \leq C(R)\sqrt{b_0 - a_0}. \qquad (1.56)$$

Integrating (1.56) on an interval (t_0, t_0'), where

$$t_0' = 2T(x_0) - t_0, \qquad (1.57)$$

we get

$$\int_{a_0}^{b_0} e^{-u_n(x,t_0)} dx - \int_{a_0}^{b_0} e^{-u_n(x,t_0')} dx \leq C(R)\sqrt{b_0 - a_0}(t_0' - t_0)$$

$$= 2\sqrt{2}C(R)(T(x_0) - t_0)^{\frac{3}{2}}. \qquad (1.58)$$

Since $x \mapsto T(x)$ is 1-Lipschitz and $T(x_0)$ is the middle of $[t_0, t_0']$, we clearly see that the segment $[a_0, b_0] \times \{t_0'\}$ lies outside the domain of definition of $u(x,t)$; using Lemma 1.7, we see that

$$\int_{a_0}^{b_0} e^{-u_n(x,t_0')} dx \to 0 \text{ as } n \to +\infty,$$

on the one hand (we use the Lebesgue Lemma together with the bound (1.52)). On the other hand, similarly, we see that

$$\int_{a_0}^{b_0} e^{-u_n(x,t_0)} dx \to \int_{a_0}^{b_0} e^{-u(x,t_0)} dx \text{ as } n \to +\infty.$$

Thus, the conclusion follows from (1.58), together with Lemma 1.10, this concludes the proof of Proposition 1.14. □

1.4.2 The Blow-up Rate

This subsection is devoted to bounding the solution u. We have obtained the following result.

Proposition 1.15 *For any $R > 0$, there exists $C(R) > 0$, such that:*

(i) (Upper bound on u) for all $(x,t) \in D \cap D_R$ we have

$$e^u d((x,t),\Gamma)^2 \le C;$$

(ii) (Lower bound on u) if, in addition, $(u_0, u_1) \in W^{1,\infty} \times L^\infty$ and x is a non-characteristic point, then for all $(x,t) \in D \cap \overline{D_R}$,

$$e^u d((x,t),\Gamma)^2 \ge \frac{1}{C}.$$

Remark In [11] Godin didn't use the notion of characteristic point, but the regularity of initial data was fundamental to achieve the result. In this work, our initial data are less regular, so we have focused on the case of a non-characteristic point in order to get his result.

Proof of Proposition 1.15
(i) Consider $R > 0$. We will show the existence of some $C(R) > 0$ such that for any $(x,t_1) \in D \cap D_R$, we have

$$e^u d((x,t),\Gamma)^2 \le C(R).$$

Consider then $(x,t_1) \in D \cap D_R$. Since $x \mapsto T(x)$ is 1-Lipschitz, we clearly see that

$$(x,T(x)) \in D_{\bar{R}} \text{ with } \bar{R} = 2R + T(a) + 1. \tag{1.59}$$

Consider now $t_2 \in (t_1, T(x))$, to be fixed later. We introduce the square domain with vertices $(x,t_1), (x + \frac{t_2 - t_1}{2}, \frac{t_1 + t_2}{2}), (x,t_2), (x - \frac{t_2 - t_1}{2}, \frac{t_1 + t_2}{2})$. Let

$$T_{\sup} = \{(\xi,t) \mid \frac{t_1 - t_2}{2} < t < t_2, \ |x - \xi| < t_2 - t\},$$

$$T_{\inf} = \{(\xi,t) \mid t_1 < t < \frac{t_1 - t_2}{2}, \ |x - \xi| < t - t_1\},$$

respectively the upper and lower half of the considered square. From Duhamel's formula, we write:

$$u(x,t_2) = \frac{1}{2}u\left(x + \frac{t_2 - t_1}{2}, \frac{t_2 + t_1}{2}\right) + \frac{1}{2}u\left(x - \frac{t_2 - t_1}{2}, \frac{t_2 + t_1}{2}\right)$$
$$+ \frac{1}{2}\int_{x - \frac{t_2 - t_1}{2}}^{x + \frac{t_2 - t_1}{2}} \partial_t u\left(\xi, \frac{t_2 - t_1}{2}\right) d\xi + \frac{1}{2}\int_{T_{\sup}} e^{u(\xi,t)} d\xi \, dt$$

and

$$u(x,t_1) = \frac{1}{2}u\left(x + \frac{t_2 - t_1}{2}, \frac{t_2 + t_1}{2}\right) + \frac{1}{2}u\left(x - \frac{t_2 - t_1}{2}, \frac{t_2 + t_1}{2}\right)$$
$$- \frac{1}{2}\int_{x - \frac{t_2 - t_1}{2}}^{x + \frac{t_2 - t_1}{2}} \partial_t u\left(\xi, \frac{t_2 - t_1}{2}\right) d\xi + \frac{1}{2}\int_{T_{\inf}} e^{u(\xi,t)} d\xi\, dt.$$

So,

$$u(x,t_1) + u(x,t_2) \tag{1.60}$$
$$= u\left(x + \frac{t_2 - t_1}{2}, \frac{t_2 + t_1}{2}\right) + u\left(x - \frac{t_2 - t_1}{2}, \frac{t_2 + t_1}{2}\right) + \frac{1}{2}\int_{T_{\inf} \cup T_{\sup}} e^{u(\xi,t)} d\xi\, dt.$$

Since the square $T_{\inf} \cup T_{\sup} \subset D_{\bar{R}}$ from (1.59), applying Lemma 1.13, we have for all $(\tilde{x}, \tilde{t}) \in T_{\inf} \cup T_{\sup}$ and for some $C(R) > 0$:

$$u(\tilde{x}, \tilde{t}) \geq u(x,t_1) - C.$$

Applying this to (1.60), we get

$$u(x,t_2) \geq u(x,t_1) - 2C + \frac{(t_2 - t_1)^2}{4} e^{(u(x,t_1) - C)}.$$

Now, choosing $t_2 = t_1 + \sigma e^{-u(x,t_1)/2}$, where $\sigma = 2e^{\frac{c}{2}}\sqrt{\eta + 2C}$, we see that either $(x,t) \notin D$ or $(x,t_2) \in D$ and $u(x,t_2) \geq u(x,t_1) + 1$ by the above-given analysis.

In the second case, we may proceed similarly and define for $n \geq 3$ a sequence

$$t_n = t_{n-1} + \sigma e^{-u(x,t_{n-1})/2}, \tag{1.61}$$

as long as $(x, t_{n-1}) \in D$. Clearly, the sequence (t_n) is increasing whenever it exists. Repeating between t_n and t_{n-1}, for $n \geq 3$, the argument we first wrote for t_1 and t_2, we see that

$$u(x,t_n) \geq u(x,t_{n-1}) + 1, \tag{1.62}$$

as long as $(x, t_n) \in D$. Two cases arise then.

Case 1: The sequence (t_n) can be defined for all $n \geq 1$, which means that

$$(x, t_n) \in D, \ \forall n \in \mathbb{N}^*. \tag{1.63}$$

In particular, (1.61) and (1.62) hold for all $n \geq 2$.

If $t_\infty = \lim_{n \to \infty} t_n$, then, from (1.63), we see that $t_\infty \leq T(x)$. Since $u(x,t_n) \to +\infty$ as $n \to \infty$ from (1.62), we need to have

$$t_\infty = T(x),$$

from the Cauchy theory. Therefore, using Lemma 1.10, (1.61) and (1.62), we see that

$$d((x,t_1),\Gamma) \leq T(x) - t_1 = \sum_{n=1}^{\infty}(t_{n+1} - t_n) \leq \sigma \sum_{n=1}^{\infty} e^{-(u(x,t_1)+(n-1))/2}$$

$$\equiv C(R)e^{-u(x,t_1)/2},$$

which is the desired estimate.

Case 2: The sequence (t_n) exists only for all $n \in [1,k]$ for some $k \geq 2$. This means that $(x,t_k) \notin D$, that is $t_k \geq T(x)$.

Moreover, (1.61) holds for all $n \in [2,k]$, and (1.62) holds for all $n \in [2,k-1]$ (in particular, it is never true if $k = 2$). As for Case 1, we use Lemma 1.10, (1.61) and (1.62) to write

$$d((x,t_1),\Gamma) \leq T(x) - t_1 \leq t_k - t_1 = \sum_{n=1}^{k-1}(t_{n+1} - t_n) \leq \sigma \sum_{n=1}^{k-1} e^{-(u(x,t_1)+(n-1))/2}$$

$$\leq \sigma e^{-(u(x,t_1))/2} \sum_{n=1}^{k-1} e^{(n-1)/2} \equiv C(R)e^{-u(x,t_1)/2},$$

which is the desired estimate. This concludes the proof of item (i) of Proposition 1.15.

(ii) Consider $R > 0$ and x a non-characteristic point such that $(x,t) \in D_R \cap D$. We dissociate u into two parts $u = \bar{u} + \tilde{u}$ with:

$$\bar{u}(x,t) = \frac{1}{2}(u_0(x-t) + u_0(x+t)) + \frac{1}{2}\int_{x-t}^{x+t} u_1(\xi)d\xi,$$

$$\tilde{u}(x,t) = \frac{1}{2}\int_0^t \int_{x-t+\tau}^{x+t-\tau} e^{u(z,\tau)}dzd\tau.$$

Differentiating $\bar{u}$, we see that

$$\partial_t\bar{u}(x,t) = \frac{1}{2}(\partial_x u_0(x+t) - \partial_x u_0(x-t)) + \frac{1}{2}(u_1(x+t) + u_1(x-t))$$

$$\leq ||\partial_x u_0||_{L^\infty(-R,R)} + ||u_1||_{L^\infty(-R,R)} \leq C(R),$$

since $(x,t) \in D_R$. Consider now an arbitrary $a \in (\frac{1}{2},1)$. Since $u(x,t) \to +\infty$ as $d((x,t),\Gamma) \to 0$ (see Proposition 1.14), it follows that

$$e^{(a-1)u(x,t)}\partial_t\bar{u}(x,t) \leq C(R)d((x,t),\Gamma)^{-2a+1}. \tag{1.64}$$

Now, we will prove a similar inequality for $\tilde{u}$. Differentiating $\tilde{u}$, we see that

$$\partial_t\tilde{u} = \frac{1}{2}\int_0^t (e^{u(x-t+\tau,\tau)} + e^{u(x+t-\tau,\tau)})d\tau. \tag{1.65}$$

Using the upper bound in Proposition 1.15, part (i), which is already proved, and Lemma 1.13, we see that for all $(y,s) \in \overline{K^-(x,t)}$,

$$u(y,s) = (1-a)u(y,s) + au(y,s) \leq (1-a)(u(x,t)+C)$$
$$+ a(\log C - 2\log d((y,s), \Gamma)).$$

So,

$$e^{u(y,s)} \leq Ce^{(1-a)u(x,t)}(d((y,s), \Gamma))^{-2a}. \tag{1.66}$$

Since x is a non-characteristic point, there exists $\delta_0 \in (0,1)$ such that the cone $C_{\bar{x}, T(\bar{x}), \delta_0}$ is below the blow-up graph Γ.

Applying (1.66) and Lemma 1.12 to (1.65), and using the fact that $|(x,t) - (z_1, w_1)|^2 = 2(\tau - t)^2$, we write (recall that $\frac{1}{2} < a < 1$):

$$\partial_t \tilde{u} = \frac{1}{2} \int_0^t e^{u(z_1, w_1)} + e^{u(z_2, w_2)} d\tau$$

$$\leq \frac{1}{2} \int_0^t Ce^{(1-a)u(x,t)}(d((z_1, w_1), \Gamma)^{-2a} + d((z_2, w_2), \Gamma)^{-2a}) d\tau$$

$$\leq Ce^{(1-a)u(x,t)} \int_0^t (d((x,t), \Gamma) + |(x,t) - (z_1, w_1)|)^{-2a}$$

$$+ (d((x,t), \Gamma) + |(x,t) - (z_1, w_1)|)^{-2a} d\tau$$

$$\leq Ce^{(1-a)u(x,t)} \int_0^t \left(d((x,t), \Gamma) + \sqrt{2}(t - \tau) \right)^{-2a} d\tau$$

$$\leq Ce^{(1-a)u(x,t)} \frac{1}{\sqrt{2}(2a-1)} d((x,t), \Gamma)^{-2a+1},$$

which yields

$$e^{(a-1)u(x,t)} \partial_t \tilde{u}(x,t) \leq Cd((x,t), \Gamma)^{-2a+1}. \tag{1.67}$$

In conclusion, we have from (1.64), (1.67) and Lemma 1.10:

$$e^{(a-1)u(x,t)} \partial_t u(x,t) \leq Cd((x,t), \Gamma)^{-2a+1} \leq C(T(x) - t)^{-2a+1}. \tag{1.68}$$

Since $u(x,t) \to +\infty$ as $d((x,t), \Gamma) \to 0$ from Proposition 1.14, integrating (1.68) between t and $T(x)$, we see that

$$e^{(a-1)u(x,t)} \leq C(T(x) - t)^{2-2a}.$$

Using Lemma 1.10 again, we complete the proof of part (ii) of Proposition 1.15.   □

1.5 Blow-up Estimates for Equation (1.1)

In this section, we prove the three results of our paper: Theorem 1.1, Theorem 1.2 and Proposition 1.3. Each proof is given in a separate subsection.

1.5.1 Blow-up Estimates in the General Case

In this subsection, we use energy and ODE type estimates from previous sections and give the proof of Theorem 1.1.

Proof of Theorem 1.1 (*i*) Let $R > 0$, $a \in \mathbb{R}$ such that $(a, T(a)) \in D_R$ and $(x, t) \in C_{a,T(a),1}$. Consider (ξ, τ), the closest point of $C_{a,T(a),1}$ to (x, t). This means that

$$||(x,t) - (\xi,\tau)|| = \inf_{x' \in \mathbb{R}} \{||(x,t) - (x', |x'| = 1 - t)||\} = d((x,t), C_{a,T(a),1}).$$

By a simple geometrical construction, we see that it satisfies the following:

$$\begin{cases} \tau = T(a) - (\xi - a), \\ \tau = t + (\xi - x), \end{cases} \tag{1.69}$$

hence, $T(a) - t - 2\xi + a + x = 0$, so

$$\xi - x = \frac{1}{2}((T(a) - t) + (a - x)). \tag{1.70}$$

Using the second equation of (1.69) and (1.70) we see that:

$$||(x,t) - (\xi,\tau)|| = \sqrt{(\xi - x)^2 + (\tau - t)^2} = \sqrt{2}|\xi - x|$$
$$= \frac{\sqrt{2}}{2}|(T(a) - t) + (a - x)|.$$

Thus,

$$d((x,t), \Gamma) \geq d((x,t), C_{a,T(a),1}) = \frac{\sqrt{2}}{2}|(T(a) - t) + (a - x)|. \tag{1.71}$$

From Proposition 1.15, (1.71) and the similarity transformation (1.7) we have:

$$e^{w_a(y,s)} \leq (T(a) - t)^2 e^{u(x,t)} \leq C \frac{(T(a) - t)^2}{d((x,t), \Gamma)^2} \leq C \left(\frac{T(a) - t}{T(a) - t - |x - a|} \right)^2$$
$$\leq \frac{C}{(1 - |y|)^2},$$

which gives the first inequality of (*i*). The second one is given by Proposition 1.15 and Lemma 1.10.

(*ii*) This is a direct consequence of Proposition 1.14 and Lemma 1.10.

(*iii*) Now, we will use Section 1.3 to prove this. Arguing by contradiction, we assume that u is not global and that $\forall \epsilon_0 > 0, \exists x_0 \in \mathbb{R}, \exists t_0 \in [0, T(x_0))$, such that

$$\frac{1}{T(x_0) - t_0} \int_I ((u_t(x, t_0))^2 + (u_x(x, t_0))^2 + e^{u(x, t_0)}) dx < \frac{\epsilon_0}{(T(x_0) - t_0)^2},$$

$$\text{where } I = (x_0 - (T(x_0) - t_0), x_0 + (T(x_0) - t_0)).$$

We introduce the following change of variables:

$$v(\xi, \tau) = u(x, t) + \log(T(x_0) - t_0), \text{ with } x = x_0 + \xi(T(x_0) - t_0),$$

$$t = t_0 + \tau(T(x_0) - t_0).$$

Note that v satisfies equation (1.1). For $\epsilon_0 = \bar{\epsilon}_0$, v satisfies (1.41), so, by Corollary 1.9, v doesn't blow-up in $\{(\xi, \tau) \,|\, |\xi| < 1 - \tau, \tau \in [0, 1)\}$, thus, u doesn't blow-up in $\{(x, t) \,|\, |x - x_0| < T(x_0) - t, t \in [t_0, T(x_0))\}$, which is a contradiction. This concludes the proof of Theorem 1.1. $\qquad\square$

1.5.2 Blow-up Estimates in the Non-characteristic Case

In this subsection, we prove Theorem 1.2. We give first the following corollary of Proposition 1.15.

Corollary 1.16 *Assume that* $(u_0, u_1) \in W^{1,\infty} \times L^\infty$. *Then, for all* $R > 0$, $a \in \mathcal{R}$ *such that* $(a, T(a)) \in D_R$, *we have for all* $s \geq -\log T(a)$ *and* $|y| < 1$

$$|w_a(y, s)| \leq C(R).$$

Proof of Corollary 1.16 Assume that $(u_0, u_1) \in W^{1,\infty} \times L^\infty$ and consider $R > 0$ and $a \in \mathcal{R}$ such that $(a, T(a)) \in D_R$. On the one hand, we recall from Proposition 1.15 that

$$\forall (x, t) \in \mathcal{C}_{a, T(a), 1}, \frac{1}{C} \leq e^u d((x, t), \Gamma)^2 \leq C.$$

Using (1.7), we see that

$$\frac{1}{C} \leq \left(\frac{d((x, t), \Gamma)}{T(a) - t}\right)^2 e^{w_a(y, s)} \leq C, \text{ with } y = \frac{x - a}{T(a) - t} \text{ and } s = -\log(T(a) - t).$$

Since

$$\frac{1}{C} \leq \frac{d((x, t), \Gamma)}{T(x) - t} \leq C,$$

from Lemmas 1.10 and 1.11, this yields the conclusion of Corollary 1.16. $\quad\square$

Now, we give the proof of Theorem 1.2.

Proof of Theorem 1.2 Consider $R > 0$ and $a \in \mathcal{R}$ such that $(a, T(a)) \in D_R$. We note first that the fact that $|w_a(y,s)| \le C$ for all $|y| < 1$ and $s \ge -\log T(a)$ follows from Corollary 1.16. It remains only to show that $\int_{-1}^{1}((\partial_s w_a(y,s))^2 + (\partial_y w_a(y,s))^2)\,dy \le C$. From Proposition 1.15 and Lemmas 1.10 and 1.11, we have:

$$\forall (x,t) \in \mathcal{C}_{a,T(a),1}, \; e^{u(x,t)} \le \frac{C(R)}{(T(a)-t)^2}. \tag{1.72}$$

We define $\mathcal{E}_a$, the energy of equation (1.1), by

$$\mathcal{E}_a(t) = \frac{1}{2}\int_{|x-a|<T(a)-t}((\partial_t u(x,t))^2 + (\partial_x u(x,t))^2)\,dx - \int_{|x-a|<T(a)-t} e^{u(x,t)}\,dx. \tag{1.73}$$

From Shatah and Struwe [24], we have

$$\frac{\mathrm{d}}{\mathrm{d}t}\mathcal{E}_a(t) \le Ce^{u(a-(T(a)-t),t)} + Ce^{u(a+(T(a)-t),t)}.$$

Integrating it over $[0,t]$ and using (1.72), we see that

$$\mathcal{E}_a(t) \le \mathcal{E}_a(0) + C\int_0^t e^{u(a-(T(a)-s),s)}\,ds + C\int_0^t e^{u(a+(T(a)-s),s)}\,ds$$

$$\le C(a, \|(u_0,u_1)\|_{H^1 \times L^2(-R,R)}) + C\int_0^t \frac{ds}{(T(a)-s)^2} \le C + \frac{C}{(T(a)-t)}.$$

Thus,

$$\mathcal{E}_a(t) \le \frac{C}{(T(a)-t)}. \tag{1.74}$$

Now, using (1.72) and (1.74) to bound the two first terms in the definition of $\mathcal{E}_a(t)$, (1.73), we get

$$\frac{1}{2}\int_{|x-a|<T(a)-t}((\partial_t u(x,t))^2 + (\partial_x u(x,t))^2)\,dx \le \int_{|x-a|<T(a)-t} e^u\,dx + \frac{C}{(T(a)-t)}$$

$$\le \int_{|x-a|<T(a)-t} \frac{C}{(T(a)-t)^2}\,dx + \frac{C}{(T(a)-t)} \le \frac{C}{(T(a)-t)}. \tag{1.75}$$

Writing inequality (1.75) in similarity variables, we get for all $s \ge -\log T(a)$,

$$\int_{-1}^{1}(\partial_s w_a(y,s))^2 + \int_{-1}^{1}(\partial_y w_a(y,s))^2 \le C(R).$$

This yields the conclusion of Theorem 1.2. $\qquad\square$

Now, we give the proof of Proposition 1.3.

Proof of Proposition 1.3 Multiplying (1.8) by $\partial_s w$, and integrating over $(-1,1)$, we see that

$$\int_{-1}^{1} \partial_s w \partial_s^2 w \, dy - \int_{-1}^{1} \partial_s w \partial_y ((1-y^2)\partial_y w) \, dy - \int_{-1}^{1} \partial_s w e^w \, dy + 2 \int_{-1}^{1} \partial_s w \, dy$$

$$= -\int_{-1}^{1} (\partial_s w)^2 \, dy + 2 \int_{-1}^{1} y \partial_s w \partial_{y,s}^2 w \, dy.$$

Thus,

$$\frac{d}{ds} \left(\int_{-1}^{1} \frac{1}{2} (\partial_s w)^2 + 2w - e^w \, dy \right) + I_1 = -\int_{-1}^{1} (\partial_s w)^2 \, dy + I_2,$$

where

$$I_1 = -\int_{-1}^{1} \partial_y \left((1-y^2)\partial_y w \right) \partial_s w \, dy = \int_{-1}^{1} \left((1-y^2)\partial_y w \right) \partial_{ys}^2 w \, dy$$

$$= \frac{1}{2} \frac{d}{ds} \left(\int_{-1}^{1} (1-y^2)(\partial_s w)^2 \, dy \right)$$

and

$$I_2 = -2 \int_{-1}^{1} y \partial_s w \partial_{y,s}^2 w \, dy = -\int_{-1}^{1} y \partial_y (\partial_s w)^2 \, dy$$

$$= \int_{-1}^{1} (\partial_s w)^2 \, dy - (\partial_s w(-1,s))^2 - (\partial_s w(1,s))^2.$$

Thus,

$$\frac{d}{ds} \left(\int_{-1}^{1} \left[\frac{1}{2} (\partial_s w)^2 + \frac{1}{2} (1-y^2)(\partial_y w)^2 - e^w + 2w \right] dy \right)$$

$$= -(\partial_s w(-1,s))^2 - (\partial_s w(1,s))^2,$$

which yields the conclusion of Proposition 1.3 by integration in time. □

References

[1] S. Alinhac. *Blowup for nonlinear hyperbolic equations*. Progress in Nonlinear Differential Equations and their Applications, 17. Birkhäuser Boston Inc., Boston, MA, 1995.

[2] C. Antonini and F. Merle. Optimal bounds on positive blow-up solutions for a semilinear wave equation. *Internat. Math. Res. Notices*, (21):1141–1167, 2001.

[3] A. Azaiez. Blow-up profile for the complex-valued semilinear wave equation. *Trans. Amer. Math. Soc.*, 367(8):5891–5933, 2015.

[4] L. A. Caffarelli and A. Friedman. Differentiability of the blow-up curve for one-dimensional nonlinear wave equations. *Arch. Rational Mech. Anal.*, 91(1):83–98, 1985.

[5] L. A. Caffarelli and A. Friedman. The blow-up boundary for nonlinear wave equations. *Trans. Amer. Math. Soc.*, 297(1):223–241, 1986.

[6] R. Côte and H. Zaag. Construction of a multisoliton blowup solution to the semilinear wave equation in one space dimension. *Comm. Pure Appl. Math.*, 66(10):1541–1581, 2013.

[7] Y. Giga and R. V. Kohn. Asymptotically self-similar blow-up of semilinear heat equations. *Comm. Pure Appl. Math.*, 38(3):297–319, 1985.

[8] Y. Giga and R. V. Kohn. Characterizing blowup using similarity variables. *Indiana Univ. Math. J.*, 36(1):1–40, 1987.

[9] Y. Giga and R. V. Kohn. Nondegeneracy of blowup for semilinear heat equations. *Comm. Pure Appl. Math.*, 42(6):845–884, 1989.

[10] P. Godin. The blow-up curve of solutions of mixed problems for semilinear wave equations with exponential nonlinearities in one space dimension. II. *Ann. Inst. H. Poincaré Anal. Non Linéaire*, 17(6):779–815, 2000.

[11] P. Godin. The blow-up curve of solutions of mixed problems for semilinear wave equations with exponential nonlinearities in one space dimension. I. *Calc. Var. Partial Differential Equations*, 13(1):69–95, 2001.

[12] S. Kichenassamy and W. Littman. Blow-up surfaces for nonlinear wave equations. I. *Comm. Partial Differential Equations*, 18(3–4):431–452, 1993.

[13] S. Kichenassamy and W. Littman. Blow-up surfaces for nonlinear wave equations. II. *Comm. Partial Differential Equations*, 18(11):1869–1899, 1993.

[14] F. Merle and H. Zaag. Determination of the blow-up rate for the semilinear wave equation. *Amer. J. Math.*, 125(5):1147–1164, 2003.

[15] F. Merle and H. Zaag. Determination of the blow-up rate for a critical semilinear wave equation. *Math. Ann.*, 331(2):395–416, 2005.

[16] F. Merle and H. Zaag. Existence and universality of the blow-up profile for the semilinear wave equation in one space dimension. *J. Funct. Anal.*, 253(1):43–121, 2007.

[17] F. Merle and H. Zaag. Openness of the set of non-characteristic points and regularity of the blow-up curve for the 1 D semilinear wave equation. *Comm. Math. Phys.*, 282(1):55–86, 2008.

[18] F. Merle and H. Zaag. Points caractéristiques à l'explosion pour une équation semilinéaire des ondes. In *"Séminaire X-EDP"*. École Polytech., Palaiseau, 2010.

[19] F. Merle and H. Zaag. Blow-up behavior outside the origin for a semilinear wave equation in the radial case. *Bull. Sci. Math.*, 135(4):353–373, 2011.

[20] F. Merle and H. Zaag. Existence and classification of characteristic points at blow-up for a semilinear wave equation in one space dimension. *Amer. J. Math.*, 134(3):581–648, 2012.

[21] F. Merle and H. Zaag. Isolatedness of characteristic points at blowup for a 1-dimensional semilinear wave equation. *Duke Math. J*, 161(15):2837–2908, 2012.

[22] F. Merle and H. Zaag. On the stability of the notion of non-characteristic point and blow-up profile for semilinear wave equations. *Comm. Math. Phys.*, 333(3):1529–1562, 2015.

[23] F. Merle and H. Zaag. Dynamics near explicit stationary solutions in similarity variables for solutions of a semilinear wave equation in higher dimensions. *Trans. Amer. Math. Soc.*, 368(1):27–87, 2016.

[24] J. Shatah and M. Struwe. *Geometric wave equations*, volume 2 of *Courant Lecture Notes in Mathematics*. New York University Courant Institute of Mathematical Sciences, New York, 1998.

* Université de Cergy-Pontoise, Laboratoire Analyse Géometrie Modélisation, CNRS-UMR 8088, 2 avenue Adolphe Chauvin 95302, Cergy-Pontoise, France. asma.azaiez@u-cergy.fr
† Courant Institute, NYU, 251 Mercer Street, NY 10012, New York. masmoudi@cims.nyu.edu
‡ Université Paris 13, Institut Galilée, Laboratoire Analyse Géometrie et Applications, CNRS-UMR 7539, 99 avenue J.B. Clément 93430, Villetaneuse, France. zaag@math.univ-paris13.fr

2

On the Role of Anisotropy in the Weak Stability of the Navier–Stokes System

Hajer Bahouri*, Jean-Yves Chemin† and Isabelle Gallagher‡

In this article, we investigate the weak stability of the three-dimensional incompressible Navier–Stokes system. Because of the invariances of this system, a positive answer in general to this question would imply global regularity for any data. Thus some restrictions have to be imposed if we hope to prove such a weak openness result. The result we prove in this paper solves this issue under an anisotropy assumption. To achieve our goal, we write a new kind of profile decomposition and establish global existence results for the Navier–Stokes system associated with new classes of arbitrarily large initial data, generalizing the examples dealt with in [13, 14, 15].

Key words and phrases. Navier–Stokes equations; anisotropy; Besov spaces; profile decomposition; weak stability

2.1 Introduction and Statement of Results

2.1.1 Setting of the Problem

We are interested in the Cauchy problem for the three-dimensional, incompressible Navier–Stokes system

$$
\text{(NS)} : \begin{cases} \partial_t u + u \cdot \nabla u - \Delta u = -\nabla p & \text{in } \mathbb{R}^+ \times \mathbb{R}^3, \\ \operatorname{div} u = 0, \\ u_{|t=0} = u_0, \end{cases}
$$

where $u(t,x)$ and $p(t,x)$ are respectively the velocity and the pressure of the fluid at time $t \geq 0$ and position $x \in \mathbb{R}^3$.

An important point in the study of (NS) is its scale invariance, which reads as follows: defining the scaling operators, for any positive real number λ and

33

any point x_0 of $\mathbb{R}^3$,

$$\Lambda_{\lambda,x_0}\phi(t,x) \stackrel{\text{def}}{=} \frac{1}{\lambda}\phi\left(\frac{t}{\lambda^2},\frac{x-x_0}{\lambda}\right) \quad \text{and} \quad \Lambda_\lambda\phi(t,x) \stackrel{\text{def}}{=} \frac{1}{\lambda}\phi\left(\frac{t}{\lambda^2},\frac{x}{\lambda}\right), \quad (2.1)$$

if u solves (NS) with data u_0, then $\Lambda_{\lambda,x_0}u$ solves (NS) with data $\Lambda_{\lambda,x_0}u_0$. Note in particular that in two space dimensions, $L^\infty(\mathbb{R}^+;L^2(\mathbb{R}^2))$ is scale invariant, while in three space dimensions that is the case for $L^\infty(\mathbb{R}^+;L^3(\mathbb{R}^3))$ or the family of spaces $L^\infty(\mathbb{R}^+;B_{p,q}^{-1+\frac{3}{p}}(\mathbb{R}^3))$,[1] with $1 \leq p < \infty$ and $0 < q \leq \infty$.

Let us also emphasize that the (NS) system formally conserves the energy, in the sense that smooth enough solutions satisfy the following equality for all times $t \geq 0$:

$$\frac{1}{2}\|u(t)\|^2_{L^2(\mathbb{R}^3)} + \int_0^t \|\nabla u(t')\|^2_{L^2(\mathbb{R}^3)}\,dt' = \frac{1}{2}\|u_0\|^2_{L^2(\mathbb{R}^3)}. \quad (2.2)$$

The energy equality (2.2) can easily be derived by observing that, thanks to the divergence-free condition, the nonlinear term is skew-symmetric in L^2: one has indeed if u and p are smooth enough and decaying at infinity that

$$\left(u(t)\cdot\nabla u(t) + \nabla p(t)|u(t)\right)_{L^2} = 0.$$

The mathematical study of the Navier–Stokes system has a long history beginning with the founding paper [35] of J. Leray in 1933. In this article, J. Leray proved that any finite energy initial data (meaning square-integrable data) generates a (possibly non-unique) global in time weak solution; and in any dimension $d \geq 2$. He moreover proved in [36] the uniqueness of the solution in two space dimensions, but in dimension three and more, the question of the uniqueness of Leray's solutions is still an open problem. Actually the difference between dimension two and higher dimensions is linked to the fact that $\|u(t)\|_{L^2(\mathbb{R}^2)}$ is both scale invariant and bounded globally in time thanks to the energy estimate, while it is not the case in dimension $d \geq 3$ (since $L^2(\mathbb{R}^3)$ is not scale invariant).

Recall that $u \in L^2_{\text{loc}}([0,T] \times \mathbb{R}^3)$ is said to be a *weak solution* of (NS) associated with the data u_0 if for any compactly supported, divergence-free vector field ϕ in $\mathcal{C}^\infty([0,T] \times \mathbb{R}^3)$ the following holds for all $t \leq T$:

$$\int_{\mathbb{R}^3} u\cdot\phi(t,x)dx = \int_{\mathbb{R}^3} u_0(x)\cdot\phi(0,x)dx + \int_0^t\int_{\mathbb{R}^3}(u\cdot\Delta\phi+u\otimes u:\nabla\phi+u\cdot\partial_t\phi)dxdt',$$

---

[1] Here $B_{p,q}^{-1+\frac{3}{p}}(\mathbb{R}^3)$ denotes the usual homogeneous Besov space (see [2], [9] or [46] for a precise definition).

with

$$u \otimes u : \nabla \phi \overset{\text{def}}{=} \sum_{1 \leq j,k \leq 3} u^j u^k \partial_k \phi^j .$$

Weak solutions satisfying the energy inequality

$$\frac{1}{2} \|u(t)\|_{L^2(\mathbb{R}^3)}^2 + \int_0^t \|\nabla u(t')\|_{L^2(\mathbb{R}^3)}^2 \, dt' \leq \frac{1}{2} \|u_0\|_{L^2(\mathbb{R}^3)}^2 \qquad (2.3)$$

are said to be *turbulent solutions*, following the terminology of J. Leray [35].

In what follows, we say that a family $(X_T)_{T>0}$ of spaces of distributions over $[0, T] \times \mathbb{R}^3$ is *scaling invariant* if for all $T > 0$ one has, under Notation (2.1):

$$\forall \lambda > 0, \forall x_0 \in \mathbb{R}^3 , u \in X_T \Longleftrightarrow \Lambda_{\lambda,x_0} u \in X_{\lambda^{-2}T} \quad \text{with} \quad \|u\|_{X_T} = \|\Lambda_{\lambda,x_0} u\|_{X_{\lambda^{-2}T}} .$$

Similarly, a space X_0 of distributions defined on $\mathbb{R}^3$ will be said to be scaling invariant if

$$\forall \lambda > 0, \forall x_0 \in \mathbb{R}^3 , u_0 \in X_0 \Longleftrightarrow \Lambda_{\lambda,x_0} u_0 \in X_0 \quad \text{with} \quad \|u_0\|_{X_0} = \|\Lambda_{\lambda,x_0} u_0\|_{X_0} .$$

This leads to the definition of a scaled solution, which will be the notion of solution we consider throughout this article.

Definition 2.1 *A vector field u is said to be a scaled solution to* (NS) *associated with the data u_0 if it is a weak solution, belonging to a family of scaling invariant spaces.*

After Leray's results, the question of the global wellposedness of the Navier–Stokes system in dimension $d \geq 3$ was raised, and has been open ever since, although several partial answers to the construction of a global unique solution were established since (we refer for instance to [2] or [34] and the references therein for recent surveys on the subject). Let us simply recall the best result known to this day on the uniqueness of solutions to (NS), which is due to H. Koch and D. Tataru in [33]: if

$$\|u_0\|_{\text{BMO}^{-1}(\mathbb{R}^3)} \overset{\text{def}}{=} \|u_0\|_{B_{\infty,\infty}^{-1}(\mathbb{R}^3)} + \sup_{\substack{x \in \mathbb{R}^3 \\ R>0}} \frac{1}{R^{\frac{3}{2}}} \left(\int_{[0,R^2] \times B(x,R)} |(e^{t\Delta} u_0)(t,y)|^2 \, dydt \right)^{\frac{1}{2}}$$

is small enough, then there is a global, unique solution to (NS), lying in $\text{BMO}^{-1} \cap X$ for all times, with X a scale invariant space to be specified – we shall not be using that space in the sequel. Note that the space BMO^{-1} is invariant by the scaling operator Λ_{λ,x_0} and that the norm in $B_{\infty,\infty}^{-1}(\mathbb{R}^3)$ denotes a Besov norm. Actually, the Besov space $B_{\infty,\infty}^{-1}(\mathbb{R}^3)$ is the largest space in which any scale and translation invariant Banach space of tempered distributions embeds (see [39]). However, it was proved in [10] and [22] that (NS) is ill-posed for initial data in $B^{-1}_{\infty,\infty}(\mathbb{R}^3)$.

Our goal in this paper is to investigate the *stability* of global solutions. Let us recall that strong stability results have been achieved. Specifically, it was proved in [1] (see [19] for the Besov setting) that the set of initial data generating a global solution is open in BMO^{-1}. More precisely, denoting by VMO^{-1} the closure of smooth functions in BMO^{-1}, it was established in [1] that if u_0 belongs to VMO^{-1} and generates a global, smooth solution to (NS), then any sequence $(u_{0,n})_{n\in\mathbb{N}}$ converging to u_0 in the BMO^{-1} norm also generates a global smooth solution as soon as n is large enough.

In this paper we would like to address the question of *weak stability*.

If $(u_{0,n})_{n\in\mathbb{N}}$, bounded in some scale invariant space X_0, converges to u_0 in the sense of distributions, with u_0 giving rise to a global smooth solution, is it the case for $u_{0,n}$ when n is large enough ?

Because of the invariances of the (NS) system, a positive answer in general to this question would imply global regularity for any data and so would solve the question of the possible blow-up in finite time of solutions to (NS), which is actually one of the Millennium Prize Problems in Mathematics. Indeed, consider for instance the sequence

$$u_{0,n} = \lambda_n \Phi_0(\lambda_n \cdot) = \Lambda_{\lambda_n} \Phi_0 \quad \text{with} \quad \lim_{n\to\infty}\left(\lambda_n + \frac{1}{\lambda_n}\right) = \infty, \qquad (2.4)$$

with Φ_0 any smooth divergence-free vector field. If the weak stability result were true, then since the weak limit of $(u_{0,n})_{n\in\mathbb{N}}$ is zero (which gives rise to the unique, global solution which is identically zero) then for n large enough $u_{0,n}$ would give rise to a unique, global solution. By scale invariance then so would Φ_0, for any Φ_0, so that would solve the global regularity problem for (NS). Another natural example is the sequence

$$u_{0,n} = \Phi_0(\cdot - x_n) = \Lambda_{1,x_n}\Phi_0, \qquad (2.5)$$

with $(x_n)_{n\in\mathbb{N}}$ a sequence of $\mathbb{R}^3$ whose norm goes to infinity. Thus sequences built by rescaling fixed divergence-free vector fields according to the invariances of the equations have to be excluded from our analysis, since solving the (NS) system for any smooth initial data seems out of reach.

Thus clearly some restrictions have to be imposed if we hope to prove such a weak openness result. Let us note that a first step in that direction was achieved in [4], under two additional assumptions on the weak convergence. The first one is an assumption on the asymptotic separation of the horizontal and vertical spectral supports of the sequence $(u_{0,n})_{n\in\mathbb{N}}$, while the second one requires that some of the profiles involved in the profile decomposition of $(u_{0,n})_{n\in\mathbb{N}}$ vanish at zero. In this paper, we remove the second assumption and give a positive answer to the question of weak stability, provided that the convergence of the sequence $(u_{0,n})_{n\in\mathbb{N}}$ towards u_0 holds "anisotropically" in frequency space (see Definition 2.4). The main ingredient which enables us to eliminate the second assumption required in [4] is a novel form of anisotropic profile decomposition. This new profile decomposition enables us to decompose the sequence of initial data $u_{0,n}$, up to a small remainder term, into a finite sum of orthogonal sequences of divergence-free vector fields; these sequences are obtained from the classical anisotropic profile decompositions by grouping together all the profiles having the same horizontal scale. The price to pay is that the profiles are no longer fixed functions as in the classical case, but bounded sequences. To carry out the strategy of proof developed in [4] in this framework, we are led to establishing global existence results for (NS) associated with new classes of arbitrarily large initial data, generalizing the examples dealt with in [13, 14, 15], and where regularity is sharply estimated.

2.1.2 Statement of the Main Result

We prove in this article a weak stability result for the (NS) system under an anisotropy assumption. This leads us naturally to introducing anisotropic Besov spaces. These spaces generalize the more usual isotropic Besov spaces, which are studied for instance in [2, 9, 46].

Definition 2.2 *Let $\widehat{\chi}$ (the Fourier transform of χ) be a radial function in $\mathcal{D}(\mathbb{R})$ such that $\widehat{\chi}(t) = 1$ for $|t| \leq 1$ and $\widehat{\chi}(t) = 0$ for $|t| > 2$. For $(j,k) \in \mathbb{Z}^2$, the horizontal truncations are defined by*

$$\widehat{S_k^h f}(\xi) \stackrel{def}{=} \widehat{\chi}\big(2^{-k}|(\xi_1,\xi_2)|\big)\widehat{f}(\xi) \quad and \quad \Delta_k^h \stackrel{def}{=} S_{k+1}^h - S_k^h,$$

and the vertical truncations by

$$\widehat{S_j^v f} \stackrel{def}{=} \widehat{\chi}(2^{-j}|\xi_3|)\widehat{f}(\xi) \quad and \quad \Delta_j^v \stackrel{def}{=} S_{j+1}^v - S_j^v.$$

For all p in $[1,\infty]$ and q in $]0,\infty]$, and all (s,s') in $\mathbb{R}^2$, with $s < 2/p, s' < 1/p$ (or $s \leq 2/p$ and $s' \leq 1/p$ if $q = 1$), the anisotropic homogeneous Besov space $B_{p,q}^{s,s'}$ is defined as the space of tempered distributions f such that

$$\|f\|_{B_{p,q}^{s,s'}} \stackrel{def}{=} \left\| 2^{ks+js'} \|\Delta_k^h \Delta_j^v f\|_{L^p} \right\|_{\ell^q} < \infty.$$

In all other cases of indices s and s', the Besov space is defined similarly, up to taking the quotient with polynomials.

Remark 2.3 The Besov spaces $B_{p,q}^{s,s'}$ (for $s \neq s'$) are anisotropic in essence, which, as pointed out above, will be an important feature of our analysis. These spaces have properties which look very much like the ones of classical Besov spaces. We refer for instance to [2], [17], [23] and [41] for all necessary details. By construction, these spaces are defined using an anisotropic Littlewood–Paley decomposition. It is useful to point out that the horizontal and vertical truncations S_k^h, Δ_k^h, S_j^v and Δ_j^v, introduced in Definition 2.2, map L^p into L^p with norms independent of k, j and p. For our purpose, it is crucial to recall the following inequalities, known as Bernstein inequalities: if $1 \leq p_1 \leq p_2 \leq \infty$, then for any $\alpha \in \mathbb{N}^2$ and $m \in \mathbb{N}$

$$\|\partial_{(x_1,x_2)}^\alpha \Delta_k^h f\|_{L^{p_2}(\mathbb{R}^2;L^r(\mathbb{R}))} \lesssim 2^{k(|\alpha|+2(1/p_1-1/p_2))} \|\Delta_k^h f\|_{L^{p_1}(\mathbb{R}^2;L^r(\mathbb{R}))} \quad (2.6)$$

$$\text{and} \quad \|\partial_{x_3}^m \Delta_j^v f\|_{L^r(\mathbb{R}^2;L^{p_2}(\mathbb{R}))} \lesssim 2^{j(m+1/p_1-1/p_2)} \|\Delta_j^v f\|_{L^r(\mathbb{R}^2;L^{p_1}(\mathbb{R}))}, \quad (2.7)$$

as well as the action of the heat flow on frequency localized distributions in an anisotropic context, namely for any p in $[1,\infty]$

$$\|e^{t\Delta} \Delta_k^h \Delta_j^v f\|_{L^p} \lesssim e^{-ct(2^{2k}+2^{2j})} \|\Delta_k^h \Delta_j^v f\|_{L^p}. \quad (2.8)$$

Notation For clarity, in what follows we denote by $\mathcal{B}^{s,s'}$ the space $B_{2,1}^{s,s'}$, by $\mathcal{B}^s$ the space $\mathcal{B}^{s,\frac{1}{2}}$ and by $\mathcal{B}_{p,q}$ the space $B_{p,q}^{-1+\frac{2}{p},\frac{1}{p}}$. In particular $\mathcal{B}_{2,1} = \mathcal{B}^0$.

Let us point out that the scaling operators (2.1) enjoy the following invariances:

$$\|\Lambda_{\lambda,x_0}\varphi\|_{\mathcal{B}_{p,q}} = \|\varphi\|_{\mathcal{B}_{p,q}}$$

$$\text{and} \quad \forall r \in [1,\infty], \quad \|\Lambda_{\lambda,x_0}\Phi\|_{L^r(\mathbb{R}^+;B_{p,q}^{-1+2/p+2/r,1/p})} = \|\Phi\|_{L^r(\mathbb{R}^+;B_{p,q}^{-1+2/p+2/r,1/p})},$$

and also the following scaling property:

$$\forall r \in [1,\infty], \forall \sigma \in \mathbb{R},$$

$$\|\Lambda_{\lambda,x_0}\Phi\|_{L^r(\mathbb{R}^+;B_{p,q}^{-1+2/p+2/r-\sigma,1/p})} \sim \lambda^\sigma \|\Phi\|_{L^r(\mathbb{R}^+;B_{p,q}^{-1+2/p+2/r-\sigma,1/p})}. \quad (2.9)$$

The Navier–Stokes system in anisotropic spaces has been studied in a number of frameworks. We refer, for instance, to [4], [17], [23], [25] and [41]. In particular, in [4] it is proved that if u_0 belongs to $\mathcal{B}^0$, then there is a unique solution (global in time if the data are small enough) in $L^2([0,T];\mathcal{B}^1)$. That norm controls the equation, in the sense that as soon as the solution belongs to $L^2([0,T];\mathcal{B}^1)$, then it lies in fact in $L^r([0,T];\mathcal{B}^{\frac{2}{r}})$ for all $1 \leq r \leq \infty$. The space $\mathcal{B}^1$ is included in L^∞ and since the seminal work [35] of J. Leray, it is known that the $L^2([0,T];L^\infty(\mathbb{R}^3))$ norm controls the propagation of regularity and also ensures weak uniqueness among turbulent solutions. Thus the space $\mathcal{B}^0$ is natural in this context.

As mentioned above, the result we establish in this paper involves an anisotropy assumption on the sequence $(u_{0,n})_{n\in\mathbb{N}}$ of initial data. Let us introduce this assumption that we call the notion of *anisotropically oscillating sequences*, and which is a natural adaptation to our setting of the vocabulary of P. Gérard in [21].

Definition 2.4 *Let $0 < q \le \infty$ be given. We say that a sequence $(f_n)_{n\in\mathbb{N}}$, bounded in $\mathcal{B}_{1,q}$, is anisotropically oscillating if the following property holds. There exists $p \ge 2$ such that for all sequences (k_n, j_n) in $\mathbb{Z}^{\mathbb{N}} \times \mathbb{Z}^{\mathbb{N}}$,*

$$\liminf_{n\to\infty} 2^{k_n(-1+\frac{2}{p})+\frac{j_n}{p}} \|\Delta_{k_n}^h \Delta_{j_n}^v f_n\|_{L^p(\mathbb{R}^3)} = C > 0 \implies \lim_{n\to\infty} |j_n - k_n| = \infty.$$
(2.10)

Remark 2.5 In view of the Bernstein inequalities (2.6) and (2.7), it is easy to see that any function f in $\mathcal{B}_{1,q}$ belongs also to $\mathcal{B}_{p,\infty}$ for any $p \ge 1$, hence

$$f \in \mathcal{B}_{1,q} \implies \sup_{(k,j)\in\mathbb{Z}^2} 2^{k(-1+\frac{2}{p})+\frac{j}{p}} \|\Delta_k^h \Delta_j^v f\|_{L^p} < \infty.$$

The left-hand side of (2.10) indicates which ranges of frequencies are predominant in the sequence (f_n): if $\liminf_{n\to\infty} 2^{k_n(-1+\frac{2}{p})+\frac{j_n}{p}} \|\Delta_{k_n}^h \Delta_{j_n}^v f_n\|_{L^p}$ is zero for a couple of frequencies $(2^{k_n}, 2^{j_n})$, then the sequence $(f_n)_{n\in\mathbb{N}}$ is "unrelated" to those frequencies, with the vocabulary of P. Gérard in [21]. The right-hand side of (2.10) is then an *anisotropy* property. Indeed one sees easily that a sequence such as $(u_{0,n})_{n\in\mathbb{N}}$ defined in (2.4) is precisely not anisotropically oscillating: for the left-hand side of (2.10) to hold for this example one would need $j_n \sim k_n \sim n$, which is precisely not the condition required on the right-hand side of (2.10). A typical sequence satisfying Assumption (2.10) is rather (for $a \in \mathbb{R}^3$)

$$f_n(x) := 2^{\alpha n} f\big(2^{\alpha n}(x_1 - a_1), 2^{\alpha n}(x_2 - a_2), 2^{\beta n}(x_3 - a_3)\big), \quad (\alpha, \beta) \in \mathbb{R}^2, \quad \alpha \ne \beta,$$

with f smooth.

Our main result is stated as follows.

Theorem 2.6 *Let q be given in $]0, 1[$ and let u_0 in $\mathcal{B}_{1,q}$ generate a unique global solution to (NS) in $L^2(\mathbb{R}^+; \mathcal{B}^1)$. Let $(u_{0,n})_{n\in\mathbb{N}}$ be a sequence of divergence free vector fields converging towards u_0 in the sense of distributions, and such that $(u_{0,n} - u_0)_{n\in\mathbb{N}}$ is anisotropically oscillating. Then for n large enough, $u_{0,n}$ generates a unique, global solution to the (NS) system in the space $L^2(\mathbb{R}^+; \mathcal{B}^1)$.*

Remark 2.7 One can see from the proof of Theorem 2.6 that the solution $u_n(t)$ associated with $u_{0,n}$ converges for all times, in the sense of distributions to the solution associated with u_0. In this sense the Navier–Stokes system is stable by weak convergence.

The proof of Theorem 2.6 enables us to infer easily the following result, which generalizes the statement of Theorem 2.6 to the case when the solution to the (NS) system generated by u_0 is assumed to blow up in finite time (for a strategy of proof, one can consult [4]).

Corollary 2.8 *Let $(u_{0,n})_{n\in\mathbb{N}}$ be a sequence of divergence-free vector fields bounded in the space $\mathcal{B}_{1,q}$ for some $0 < q < 1$, converging towards some u_0 in $\mathcal{B}_{1,q}$ in the sense of distributions, with $u_0 - (u_{0,n})_{n\in\mathbb{N}}$ anisotropically oscillating. Let u be the solution to the Navier–Stokes system associated with u_0 and assume that the life span of u is $T^* < \infty$. Then for all positive times $T < T^*$, there is a subsequence such that the life span of the solution associated with $u_{0,n}$ is at least T.*

Remark 2.9 As explained above, the natural space in our context would be $\mathcal{B}^0$. For technical reasons, we assume in our result more smoothness on the sequence of initial data, since obviously by Bernstein inequalities (2.6) and (2.7), we have $\mathcal{B}_{1,q} \hookrightarrow \mathcal{B}^0$.

2.1.3 Layout

The proof of Theorem 2.6 is addressed in Section 2.2. In Subsection 2.2.2, we provide a new kind of "anisotropic profile decomposition" of the sequence of initial data, whose proof can be found in Section 2.3. This enables us to replace the sequence of Cauchy data, up to an arbitrarily small remainder term, by a finite (but large) sum of orthogonal sequences of divergence-free vector fields. In Subsection 2.2.3, we state that each individual element involved in the decomposition derived in Subsection 2.2.2 gives rise to a unique global solution to the (NS) system (the proof is postponed to Section 2.4). Subsection 2.2.4 is devoted to the proof of the fact that the sum of each individual profile does provide an approximate solution to the Navier–Stokes system, thanks to an orthogonality argument, which completes the proof of Theorem 2.6.

For all points $x = (x_1, x_2, x_3)$ in $\mathbb{R}^3$ and all vector fields $u = (u^1, u^2, u^3)$, we denote by

$$x_\text{h} \stackrel{\text{def}}{=} (x_1, x_2) \quad \text{and} \quad u^\text{h} \stackrel{\text{def}}{=} (u^1, u^2)$$

their horizontal parts. We also define horizontal differentiation operators $\nabla^\text{h} \stackrel{\text{def}}{=} (\partial_1, \partial_2)$ and $\text{div}_\text{h} \stackrel{\text{def}}{=} \nabla^\text{h}\cdot$, as well as $\Delta_\text{h} \stackrel{\text{def}}{=} \partial_1^2 + \partial_2^2$.

We also use the following shorthand notation: $X_h Y_v := X(\mathbb{R}^2; Y(\mathbb{R}))$, where X is a function space defined on $\mathbb{R}^2$ and Y is defined on $\mathbb{R}$.

As we shall be considering functions which have different types of variations in the x_3 variable and the x_h variable, the following notation will be used:

$$[f]_\beta (x) \stackrel{\text{def}}{=} f(x_h, \beta x_3). \tag{2.11}$$

Clearly, for any function f, we have the following identity which will be of constant use throughout this paper:

$$\left\| [f]_\beta \right\|_{B^{s_1,s_2}_{p,1}} \sim \beta^{s_2 - \frac{1}{p}} \|f\|_{B^{s_1,s_2}_{p,1}}. \tag{2.12}$$

Finally, we denote by C a constant which does not depend on the various parameters appearing in this paper, and which may change from line to line. We also denote sometimes $x \leq Cy$ by $x \lesssim y$.

2.2 Proof of the Main Theorem

2.2.1 General Scheme of the Proof

The main arguments leading to Theorem 2.6 are the following: by a profile decomposition argument, the sequence of initial data is decomposed into the weak limit u_0 and the sum of sequences of divergence-free vector fields, up to a small remainder term. Then to prove that each individual element of the decomposition generates a unique global solution to (NS), it is necessary to estimate sharply the regularity in scaling invariant (anisotropic) norms. The mutual orthogonality of each term in the decomposition of the initial data implies finally that the sum of the solutions associated with each element is itself an approximate solution to (NS), globally in time, which concludes the proof of the result.

2.2.2 Anisotropic Profile Decomposition

The study of the lack of compactness in critical Sobolev embeddings has attracted a lot of attention in the past decades, both for its interesting geometric features and for its applications to nonlinear partial differential equations. This study originates in the works of P.-L. Lions (see [37] and [38]) by means of defect measures, and earlier decompositions of bounded sequences into a sum of "profiles" can be found in the studies by H. Brézis and J.-M. Coron in [11] and M. Struwe in [45]. Our source of inspiration here is the work [21] of P. Gérard in which the defect of compactness of the critical Sobolev embeddings (for L^2-based Sobolev spaces) in Lebesgue spaces is described by means of an asymptotic, orthogonal decomposition in terms of rescaled and translated profiles. This was generalized to L^p-based Sobolev spaces by S. Jaffard in [26], to Besov spaces by G. Koch [32], and finally to general critical embeddings by H. Bahouri, A. Cohen and G. Koch in [3] (see also [6, 7, 8] for the limiting case of Sobolev embeddings in Orlicz spaces and [44] for an abstract, functional analytic presentation of the concept in various settings).

In the pioneering works [5] (for the critical 3D wave equation) and [40] (for the critical 2D Schrödinger equation), it was highlighted that this type of decomposition provides applications to the study of nonlinear partial differential equations. The ideas of [5] were revisited in [31] and [18] in the context of the Schrödinger equations and Navier–Stokes system, respectively, with an aim of describing the structure of bounded sequences of solutions to those equations. These profile decomposition techniques have since been used successfully to study the possible blow-up of solutions to nonlinear partial differential equations, in various contexts; we refer for instance to [20], [24], [27], [28], [29], [30], [42], [43].

The first step in the proof of Theorem 2.6 consists of writing down an anisotropic profile decomposition of the sequence of initial data $(u_{0,n})_{n\in\mathbb{N}}$ (see Theorem 2.12). To state our result in a clear way, let us start by introducing some definitions and notations.

Definition 2.10 *We say that two sequences of positive real numbers $(\lambda_n^1)_{n\in\mathbb{N}}$ and $(\lambda_n^2)_{n\in\mathbb{N}}$ are orthogonal if*

$$\frac{\lambda_n^1}{\lambda_n^2} + \frac{\lambda_n^2}{\lambda_n^1} \to \infty, \quad n \to \infty.$$

A family of sequences $\left((\lambda_n^j)_{n\in\mathbb{N}}\right)_j$ is said to be a family of scales if $\lambda_n^0 \equiv 1$ and if $(\lambda_n^j)_{n\in\mathbb{N}}$ and $(\lambda_n^k)_{n\in\mathbb{N}}$ are orthogonal when $j \neq k$.

Definition 2.11 *Let μ be a positive real number less than $1/2$, fixed from now on.*

We define $D_\mu \overset{def}{=} [-2+\mu, 1-\mu] \times [1/2, 7/2]$ and $\widetilde{D}_\mu \overset{def}{=} [-1+\mu, 1-\mu] \times [1/2, 3/2]$. We denote by S_μ the space of functions a belonging to $\bigcap_{(s,s')\in D_\mu} B^{s,s'}$ such that

$$\|a\|_{S_\mu} \overset{def}{=} \sup_{(s,s')\in D_\mu} \|a\|_{B^{s,s'}} < \infty.$$

Notation In all that follows, θ is a given function in $\mathcal{D}(B_{\mathbb{R}^3}(0,1))$ which has value 1 near $B_{\mathbb{R}^3}(0,1/2)$. For any positive real number η, we denote

$$\theta_\eta(x) \overset{\text{def}}{=} \theta(\eta x) \quad \text{and} \quad \theta_{h,\eta}(x_h) \overset{\text{def}}{=} \theta_\eta(x_h, 0). \tag{2.13}$$

In order to make notations as clear as possible, the letter v (possibly with indices) will always denote a two-component divergence free vector field, which may depend on the vertical variable x_3.

The following result, the proof of which is postponed to Section 2.3, is in the spirit of the *profile decomposition* theorem of P. Gérard in [21] concerning the critical Sobolev embedding in Lebesgue spaces.

Theorem 2.12 *Under the assumptions of Theorem 2.6 and up to the extraction of a subsequence, the following holds. There is a family of scales $\left((\lambda_n^j)_{n\in\mathbb{N}}\right)_{j\in\mathbb{N}}$ and for all $L \geq 1$ there is a family of sequences $\left((h_n^j)_{n\in\mathbb{N}}\right)_{j\in\mathbb{N}}$ going to zero such that for any real number α in $]0,1[$, there are families of sequences of divergence-free vector fields (for j ranging from 1 to L), $(v_{n,\alpha,L}^j)_{n\in\mathbb{N}}$, $(w_{n,\alpha,L}^j)_{n\in\mathbb{N}}$, $(v_{0,n,\alpha,L}^{0,\infty})_{n\in\mathbb{N}}$, $(w_{0,n,\alpha,L}^{0,\infty})_{n\in\mathbb{N}}$, $(v_{0,n,\alpha,L}^{0,\text{loc}})_{n\in\mathbb{N}}$ and $(w_{0,n,\alpha,L}^{0,\text{loc}})_{n\in\mathbb{N}}$, all belonging to S_μ, and a smooth, compactly supported function $u_{0,\alpha}$ such that the sequence $(u_{0,n})_{n\in\mathbb{N}}$ can be written in the form*

$$u_{0,n} \equiv u_{0,\alpha} + \left[\left(v_{0,n,\alpha,L}^{0,\text{loc}} + h_n^0 w_{0,n,\alpha,L}^{0,\text{loc,h}}, w_{0,n,\alpha,L}^{0,\text{loc},3}\right)\right]_{h_n^0} + \left[\left(v_{0,n,\alpha,L}^{0,\infty} + h_n^0 w_{0,n,\alpha,L}^{0,\infty,h}, w_{0,n,\alpha,L}^{0,\infty,3}\right)\right]_{h_n^0}$$

$$+ \sum_{j=1}^{L} \Lambda_{\lambda_n^j} \left[\left(v_{n,\alpha,L}^j + h_n^j w_{n,\alpha,L}^{j,\text{h}}, w_{n,\alpha,L}^{j,3}\right)\right]_{h_n^j} + \rho_{n,\alpha,L},$$

where $u_{0,\alpha}$ approximates u_0 in the sense that

$$\lim_{\alpha\to 0} \|u_{0,\alpha} - u_0\|_{\mathcal{B}_{1,q}} = 0, \tag{2.14}$$

where the remainder term satisfies

$$\lim_{L\to\infty} \lim_{\alpha\to 0} \limsup_{n\to\infty} \|e^{t\Delta}\rho_{n,\alpha,L}\|_{L^2(\mathbb{R}^+;\mathcal{B}^1)} = 0, \tag{2.15}$$

while the following uniform bounds hold:

$$\mathcal{M} \overset{\text{def}}{=} \sup_{L\geq 1} \sup_{\alpha\in]0,1[} \sup_{n\in\mathbb{N}} \left(\left\|\left(v_{0,n,\alpha,L}^{0,\infty}, w_{0,n,\alpha,L}^{0,\infty,3}\right)\right\|_{\mathcal{B}^0} + \left\|\left(v_{0,n,\alpha,L}^{0,\text{loc}}, w_{0,n,\alpha,L}^{0,\text{loc},3}\right)\right\|_{\mathcal{B}^0}\right.$$

$$\left. + \|u_{0,\alpha}\|_{\mathcal{B}^0} + \sum_{j=1}^{L}\left\|\left(v_{n,\alpha,L}^j, w_{n,\alpha,L}^{j,3}\right)\right\|_{\mathcal{B}^0}\right) < \infty \tag{2.16}$$

and for all α in $]0,1[$,

$$\mathcal{M}_\alpha \overset{def}{=} \sup_{L\geq 1} \sup_{\substack{1\leq j\leq L \\ n\in\mathbb{N}}} \left(\left\| (v^{0,\infty}_{0,n,\alpha,L}, w^{0,\infty,3}_{0,n,\alpha,L}) \right\|_{S_\mu} + \left\| (v^{0,loc}_{0,n,\alpha,L}, w^{0,loc,3}_{0,n,\alpha,L}) \right\|_{S_\mu} \right.$$

$$\left. + \|u_{0,\alpha}\|_{S_\mu} + \left\| (v^j_{n,\alpha,L}, w^{j,3}_{n,\alpha,L}) \right\|_{S_\mu} \right) \tag{2.17}$$

is finite. Finally, we have

$$\lim_{L\to\infty} \lim_{\alpha\to 0} \limsup_{n\to\infty} \left\| (v^{0,loc}_{0,n,\alpha,L}, w^{0,loc,3}_{0,n,\alpha,L})(\cdot,0) \right\|_{B^0_{2,1}(\mathbb{R}^2)} = 0,$$

$$\tag{2.18}$$

$$\forall (\alpha,L), \exists \eta(\alpha,L) / \forall \eta \leq \eta(\alpha,L), \forall n \in \mathbb{N}, \; (1-\theta_{h,\eta})(v^{0,loc}_{0,n,\alpha,L}, w^{0,loc,3}_{0,n,\alpha,L}) = 0 \; and$$

$$\tag{2.19}$$

$$\forall (\alpha,L,\eta), \; \exists n(\alpha,L,\eta) / \forall n \geq n(\alpha,L,\eta), \; \theta_{h,\eta}(v^{0,\infty}_{0,n,\alpha,L}, w^{0,\infty,3}_{0,n,\alpha,L}) = 0. \tag{2.20}$$

Theorem 2.12 states that the sequence $u_{0,n}$ is equal, up to a small remainder term, to a finite sum of orthogonal sequences of divergence-free vector fields. These sequences are obtained from the profile decomposition derived in [4] (see Proposition 2.4 in [4]) by grouping together all the profiles having the same horizontal scale λ_n, and the form they take depends on whether the scale λ_n is identically equal to one or not.

Note that in contrast with classical profile decompositions (see for instance [21]), cores of concentration do not appear in the profile decomposition given in Theorem 2.12 since all the profiles with the same horizontal scale are grouped together, and thus the decomposition is written in terms of scales only. The price to pay is that the profiles are no longer fixed functions, but bounded sequences. To carry out the strategy of proof developed in [4] in this framework, we have to establish that each element involved in the decomposition of Theorem 2.12 generates a global solution to the (NS) system as soon as n is large enough. Since we deal with bounded sequences, it is necessary to sharply estimate the regularity.

Let us emphasize that in the case when λ_n goes to 0 or infinity, these sequences are of the type

$$\Lambda_{\lambda_n} \left[(v^h_{0,n} + h_n w^h_{0,n}, w^3_{0,n}) \right]_{h_n}, \tag{2.21}$$

where we used Notation (2.11), and with h_n a sequence going to zero. It is essential (to establish our result) that the profiles that must be considered in that case are only profiles of type (2.21) with h_n tending to zero. Actually the divergence-free assumption on $u_{0,n}$ allows us to include the terms of type (2.21) with h_n tending to infinity into the remainder term; and the anisotropically oscillating assumption for $(u_0 - u_{0,n})_{n \in \mathbb{N}}$ allows us to exclude in the profile decomposition of $u_{0,n}$ sequences of type (2.21) with $h_n \equiv 1$.

In the case when λ_n is identically equal to one, we deal with three types of orthogonal sequence: the first one consists in $u_{0,\alpha}$ an approximation of the weak limit u_0, the second one is of type (2.21) with $\lambda_n \equiv 1$ and h_n tending to zero, and is uniformly localized in the horizontal variable and vanishes at $x_3 = 0$, while the third one is also of type (2.21) with $\lambda_n \equiv 1$ and h_n converging to zero, and its support in the horizontal variable goes to infinity. Note that, contrary to the case when the horizontal scale λ_n tends to 0 or infinity, all the profiles involved in the anisotropic decomposition of the sequence $(u_0 - u_{0,n})_{n \in \mathbb{N}}$ having the same horizontal scale $\lambda_n \equiv 1$ are not grouped together: the sum of these profiles is divided into two parts depending on whether the horizontal cores of concentration escape to infinity or not. This splitting plays a key role in establishing our result under the only assumption of anisotropic oscillation, by removing the second assumption required in [4].

2.2.3 Propagation of Profiles

The second step of the proof of Theorem 2.6 consists of proving that each individual profile involved in the decomposition of Theorem 2.12 generates a global solution to (NS) as soon as n is large enough. This is mainly based on the following results concerning respectively profiles of the type

$$\Lambda_{\lambda_n^j} \big[(v_{n,\alpha,L}^j + h_n^j w_{n,\alpha,L}^{j,\mathrm{h}}, w_{n,\alpha,L}^{j,3}) \big]_{h_n^j}$$

with λ_n^j going to 0 or infinity and h_n^j converging to zero, and the profiles of horizontal scale one, see respectively Theorems 2.14 and 2.15.

In order to state these theorems, let us begin by defining the function spaces we shall be working with.

Definition 2.13 *We define the space* $\mathcal{A}^{s,s'} = L^\infty(\mathbb{R}^+; \mathcal{B}^{s,s'}) \cap L^2(\mathbb{R}^+; \mathcal{B}^{s+1,s'})$ *equipped with the norm*

$$\|a\|_{\mathcal{A}^{s,s'}} \overset{def}{=} \|a\|_{L^\infty(\mathbb{R}^+; \mathcal{B}^{s,s'})} + \|a\|_{L^2(\mathbb{R}^+; \mathcal{B}^{s+1,s'})},$$

and we denote $\mathcal{A}^s = \mathcal{A}^{s,\frac{1}{2}}$.

We denote by $\mathcal{F}^{s,s'}$ *any function space such that*

$$\|L_0 f\|_{L^2(\mathbb{R}^+; \mathcal{B}^{s+1,s'})} \lesssim \|f\|_{\mathcal{F}^{s,s'}},$$

where, for any non-negative real number τ, $L_\tau f$ denotes the solution of the heat equation

$$\begin{cases} \partial_t L_\tau f - \Delta L_\tau f = f, \\ \quad L_\tau f_{|t=\tau} = 0. \end{cases}$$

We denote $\mathcal{F}^s = \mathcal{F}^{s,\frac{1}{2}}$.

Examples Using the smoothing effect of the heat flow, it is easy to prove that the spaces $\widetilde{L}^2(\mathbb{R}^+; \mathcal{B}^{s-1,s'})$, $\widetilde{L}^2(\mathbb{R}^+; \mathcal{B}^{s,s'-1})$ are $\mathcal{F}^{s,s'}$ spaces, as well as the spaces $L^1(\mathbb{R}^+; \mathcal{B}^{s,s'})$ and $L^1(\mathbb{R}^+; \mathcal{B}^{s+1,s'-1})$. Actually, recalling that $L_0 f = \int_0^t e^{(t-t')\Delta} f(t') \, dt'$ and taking advantage of (2.8), we get for any function in $\widetilde{L}^2(\mathbb{R}^+; \mathcal{B}^{s-1,s'})$

$$\|\Delta_k^h \Delta_j^v L_0 f\|_{L^2} \lesssim \int_0^t e^{-ct'(2^{2k}+2^{2j})} \|\Delta_k^h \Delta_j^v f(t')\|_{L^2} \, dt',$$

where we make use of notations of Definition 2.2. We deduce that there is a sequence $d_{j,k}(t')$ in the sphere of $\ell^1(\mathbb{Z} \times \mathbb{Z}; L^2(\mathbb{R}^+))$ such that

$$\|\Delta_k^h \Delta_j^v L_0 f\|_{L^2} \lesssim \|f\|_{\widetilde{L}^2(\mathbb{R}^+; \mathcal{B}^{s-1,s'})} 2^{-k(s-1)} 2^{-js'} \int_0^t e^{-ct'(2^{2k}+2^{2j})} d_{j,k}(t') \, dt'.$$

Young's inequality in time therefore gives

$$\|\Delta_k^h \Delta_j^v L_0 f\|_{L^2(\mathbb{R}^+; L^2)} \lesssim \|f\|_{\widetilde{L}^2(\mathbb{R}^+; \mathcal{B}^{s-1,s'})} 2^{-k(s-1)-js'} d_{j,k},$$

where $d_{j,k}$ is a generic sequence in the sphere of $\ell^1(\mathbb{Z} \times \mathbb{Z})$, which ends the proof of the result in the case when f belongs to $\widetilde{L}^2(\mathbb{R}^+; \mathcal{B}^{s-1,s'})$. The argument is similar in the other cases.

Notation In the following we designate by $\mathcal{T}_0(A,B)$ a generic constant depending only on the quantities A and B. We denote by $\mathcal{T}_1$ a generic non-decreasing function from $\mathbb{R}^+$ into $\mathbb{R}^+$ such that

$$\limsup_{r \to 0} \frac{\mathcal{T}_1(r)}{r} < \infty, \tag{2.22}$$

and by $\mathcal{T}_2$ a generic locally bounded function from $\mathbb{R}^+$ into $\mathbb{R}^+$. All those functions may vary from line to line. Let us notice that for any positive sequence $(a_n)_{n\in\mathbb{N}}$ belonging to ℓ^1, we have

$$\sum_n \mathcal{T}_1(a_n) \le \mathcal{T}_2\Big(\sum_n a_n\Big). \tag{2.23}$$

As in the isotropic case, the following space-time (quasi)-norms, first introduced by J.-Y. Chemin and N. Lerner in [16]:

$$\|f\|_{\widetilde{L}^r([0,T];B_{p,q}^{s,s'})} \stackrel{\text{def}}{=} \left\| 2^{ks+js'} \|\Delta_k^{\text{h}}\Delta_j^{\text{v}}f\|_{L^r([0,T];L^p)} \right\|_{\ell^q}, \tag{2.24}$$

are very useful in the context of the Navier–Stokes system, and will be of constant use all along this paper. Notice that of course $\widetilde{L}^r([0,T];B_{p,r}^{s,s'}) = L^r([0,T];B_{p,r}^{s,s'})$, and by Minkowski's inequality, we have the embedding $\widetilde{L}^r([0,T];B_{p,q}^{s,s'}) \subset L^r([0,T];B_{p,q}^{s,s'})$ if $r \geq q$.

Our first theorem of global existence for the Navier–Stokes system, which concerns profiles with horizontal scales going to 0 or infinity, generalizes the example considered in [13].

Theorem 2.14 *A locally bounded function ε_1 from $\mathbb{R}^+$ into $\mathbb{R}^+$ exists, which satisfies the following. For any (v_0, w_0^3) in S_μ (see Definition 2.11), for any positive real number β such that $\beta \leq \varepsilon_1(\|(v_0, w_0^3)\|_{S_\mu})$, the divergence-free vector field*

$$\Phi_0 \stackrel{\text{def}}{=} \left[(v_0 - \beta\nabla^{\text{h}}\Delta_{\text{h}}^{-1}\partial_3 w_0^3, w_0^3) \right]_\beta$$

generates a global solution Φ_β to (NS), which satisfies

$$\|\Phi_\beta\|_{\mathcal{A}^0} \leq \mathcal{T}_1(\|(v_0, w_0^3)\|_{\mathcal{B}^0}) + \beta\,\mathcal{T}_2(\|(v_0, w_0^3)\|_{S_\mu}). \tag{2.25}$$

Moreover, for any (s, s') in $[-1 + \mu, 1 - \mu] \times [1/2, 7/2]$, we have, for any r in $[1, \infty]$,

$$\|\Phi_\beta\|_{L^r(\mathbb{R}^+;\mathcal{B}^{s+\frac{2}{r}})} + \frac{1}{\beta^{s'-\frac{1}{2}}}\|\Phi_\beta\|_{L^r(\mathbb{R}^+;\mathcal{B}^{\frac{2}{r},s'})} \leq \mathcal{T}_2(\|(v_0, w_0^3)\|_{S_\mu}). \tag{2.26}$$

The proof of Theorem 2.14 is provided in Subsection 2.4.1.

The existence of a global regular solution for the set of profiles associated with the horizontal scale 1 is ensured by the following theorem, which can be viewed as a generalization of Theorem 3 of [14] and of Theorem 2 of [15].

Theorem 2.15 *With the notation of Theorem 2.12, let us consider the initial data:*

$$\Phi_{0,n,\alpha,L}^0 \stackrel{\text{def}}{=} u_{0,\alpha} + \left[\left(v_{0,n,\alpha,L}^{0,\infty} + h_n^0 w_{0,n,\alpha,L}^{0,\infty,\text{h}}, w_{0,n,\alpha,L}^{0,\infty,3} \right) \right]_{h_n^0}$$
$$+ \left[\left(v_{0,n,\alpha,L}^{0,\text{loc}} + h_n^0 w_{0,n,\alpha,L}^{0,\text{loc},\text{h}}, w_{0,n,\alpha,L}^{0,\text{loc},3} \right) \right]_{h_n^0}.$$

There is a constant ε_0, depending only on u_0 and on $\mathcal{M}_\alpha$, such that if $h_n^0 \leq \varepsilon_0$, then the initial data $\Phi_{0,n,\alpha,L}^0$ generates a global smooth solution $\Phi_{n,\alpha,L}^0$ which *satisfies for all s in $[-1+\mu, 1-\mu]$ and all r in $[1,\infty]$,*

$$\|\Phi^0_{n,\alpha,L}\|_{L^r(\mathbb{R}^+;\mathcal{B}^{s+\frac{2}{r}})} \leq \mathcal{T}_0(u_0, \mathcal{M}_\alpha). \tag{2.27}$$

The proof of Theorem 2.15 is provided in Subsection 2.4.2.

2.2.4 End of the Proof of the Main Theorem

To end the proof of Theorem 2.6, we need to check that the sum of the propagation of the remainder term through the transport-diffusion equation and the solutions to (NS) associated with each individual profile (provided by Theorems 2.14 and 2.15) is an approximate solution to the Navier–Stokes system. This can be achieved by proving that the nonlinear interactions of all the solutions are negligible, thanks to the orthogonality between the scales. For that purpose, let us look at the profile decomposition given by Theorem 2.12. For a given positive and small ε, Assertion (2.15) allows us to choose α, L and N_0 (depending of course on ε) such that

$$\forall n \geq N_0, \ \|e^{t\Delta}\rho_{n,\alpha,L}\|_{L^2(\mathbb{R}^+;\mathcal{B}^1)} \leq \varepsilon. \tag{2.28}$$

The parameters α and L are fixed so that (2.28) holds. Let us consider the two functions ε_1, $\mathcal{T}_1$ and $\mathcal{T}_2$ (resp. ε_0 and $\mathcal{T}_0$) which appear in the statement of Theorem 2.14 (resp. Theorem 2.15). Since each sequence $(h^j_n)_{n\in\mathbb{N}}$, for $0 \leq j \leq L$, goes to zero as n goes to infinity, one can choose an integer N_1 greater than or equal to N_0 such that

$$\forall n \geq N_1, \ \forall j \in \{0,\ldots,L\}, \ h^j_n \leq \min\left\{\varepsilon_1(\mathcal{M}_\alpha), \varepsilon_0, \frac{\varepsilon}{L\mathcal{T}_2(\mathcal{M}_\alpha)}\right\}. \tag{2.29}$$

Now for $1 \leq j \leq L$ (resp. $j = 0$), let us denote by $\Phi^j_{n,\varepsilon}$ (resp. $\Phi^0_{n,\varepsilon}$) the global solution of (NS) associated with the initial data:

$$\left[(v^j_{n,\alpha,L} + h^j_n w^{j,h}_{n,\alpha,L}, w^{j,3}_{n,\alpha,L})\right]_{h^j_n}$$

$$\left(\text{resp.} \quad u_{0,\alpha} + \left[(v^{0,\infty}_{0,n,\alpha,L} + h^0_n w^{0,\infty,h}_{0,n,\alpha,L}, w^{0,\infty,3}_{0,n,\alpha,L})\right]_{h^0_n}\right.$$

$$\left. + \left[(v^{0,\mathrm{loc}}_{0,n,\alpha,L} + h^0_n w^{0,\mathrm{loc},h}_{0,n,\alpha,L}, w^{0,\mathrm{loc},3}_{0,n,\alpha,L})\right]_{h^0_n}\right)$$

given by Theorem 2.14 (resp. Theorem 2.15). We look for the global solution associated with $u_{0,n}$ in the form

$$u_n = u^{\mathrm{app}}_{n,\varepsilon} + R_{n,\varepsilon} \quad \text{with} \quad u^{\mathrm{app}}_{n,\varepsilon} \overset{\mathrm{def}}{=} \sum_{j=0}^{L} \Lambda_{\lambda^j_n} \Phi^j_{n,\varepsilon} + e^{t\Delta}\rho_{n,\alpha,L}. \tag{2.30}$$

In view of the scaling invariance of the Navier–Stokes system, $\Lambda_{\lambda_n^j} \Phi_{n,\varepsilon}^j$ solves (NS) with the initial data $\Lambda_{\lambda_n^j} \big[(v_{n,\alpha,L}^j + h_n^j w_{n,\alpha,L}^{j,h}, w_{n,\alpha,L}^{j,3}) \big]_{h_n^j}$. This gives the following equation on $R_{n,\varepsilon}$:

$$\partial_t R_{n,\varepsilon} - \Delta R_{n,\varepsilon} + \operatorname{div} \big(R_{n,\varepsilon} \otimes R_{n,\varepsilon} + R_{n,\varepsilon} \otimes u_{n,\varepsilon}^{\mathrm{app}} + u_{n,\varepsilon}^{\mathrm{app}} \otimes R_{n,\varepsilon} \big) + \nabla p_{n,\varepsilon}$$

$$= F_{n,\varepsilon} \overset{\mathrm{def}}{=} F_{n,\varepsilon}^1 + F_{n,\varepsilon}^2 + F_{n,\varepsilon}^3$$

with
$$F_{n,\varepsilon}^1 \overset{\mathrm{def}}{=} - \operatorname{div} \big(e^{t\Delta} \rho_{n,\alpha,L} \otimes e^{t\Delta} \rho_{n,\alpha,L} \big),$$

$$F_{n,\varepsilon}^2 \overset{\mathrm{def}}{=} - \sum_{j=0}^{L} \operatorname{div} \big(\Lambda_{\lambda_n^j} \Phi_{n,\varepsilon}^j \otimes e^{t\Delta} \rho_{n,\alpha,L} + e^{t\Delta} \rho_{n,\alpha,L} \otimes \Lambda_{\lambda_n^j} \Phi_{n,\varepsilon}^j \big)$$

and
$$F_{n,\varepsilon}^3 \overset{\mathrm{def}}{=} - \sum_{\substack{0 \leq j,k \leq L \\ j \neq k}} \operatorname{div} \big(\Lambda_{\lambda_n^j} \Phi_{n,\varepsilon}^j \otimes \Lambda_{\lambda_n^k} \Phi_{n,\varepsilon}^k \big), \tag{2.31}$$

and where $\big(\operatorname{div}(u \otimes v) \big)^j = \sum_{k=1}^{3} \partial_k (u^j v^k)$.

In order to establish that the function u_n defined by (2.30) provides a global solution to the (NS) system, it suffices to prove that there exist some space $\mathcal{F}^0$ as in Definition 2.13 and an integer $N \geq N_1$ such that

$$\forall n \geq N, \quad \| F_{n,\varepsilon} \|_{\mathcal{F}^0} \leq C\varepsilon, \tag{2.32}$$

where C depends only on L and $\mathcal{M}_\alpha$. In the next estimates we omit the dependence of all constants on α and L, which are fixed. Indeed, if (2.32) holds, then $R_{n,\varepsilon}$ exists globally thanks to strong stability in $\mathcal{B}^0$ (see [4] for the setting of $\mathcal{B}_{1,1}$).

Let us start with the estimate of $F_{n,\varepsilon}^1$. Using the fact that $\mathcal{B}^1$ is an algebra, we have

$$\big\| e^{t\Delta} \rho_{n,\alpha,L}^h \otimes e^{t\Delta} \rho_{n,\alpha,L} \big\|_{L^1(\mathbb{R}^+;\mathcal{B}^1)} \lesssim \big\| e^{t\Delta} \rho_{n,\alpha,L} \big\|_{L^2(\mathbb{R}^+;\mathcal{B}^1)}^2,$$

so
$$\big\| \operatorname{div}_h \big(e^{t\Delta} \rho_{n,\alpha,L}^h \otimes e^{t\Delta} \rho_{n,\alpha,L} \big) \big\|_{L^1(\mathbb{R}^+;\mathcal{B}^0)} \lesssim \big\| e^{t\Delta} \rho_{n,\alpha,L} \big\|_{L^2(\mathbb{R}^+;\mathcal{B}^1)}^2$$

and
$$\big\| \partial_3 \big(e^{t\Delta} \rho_{n,\alpha,L}^3 e^{t\Delta} \rho_{n,\alpha,L} \big) \big\|_{L^1(\mathbb{R}^+;\mathcal{B}^{1,-\frac{1}{2}})} \lesssim \big\| e^{t\Delta} \rho_{n,\alpha,L} \big\|_{L^2(\mathbb{R}^+;\mathcal{B}^1)}^2.$$

According to Inequality (2.28), this gives rise to

$$\forall n \geq N_1, \quad \| F_{n,\varepsilon}^1 \|_{\mathcal{F}^0} \lesssim \varepsilon^2. \tag{2.33}$$

Now let us consider $F_{n,\varepsilon}^2$. By the scaling invariance of the operators $\Lambda_{\lambda_n^j}$ in $L^2(\mathbb{R}^+;\mathcal{B}^1)$ and again the fact that $\mathcal{B}^1$ is an algebra, we get

$$\left\| \Lambda_{\lambda_n^j} \Phi_{n,\varepsilon}^j \otimes e^{t\Delta}\rho_{n,\alpha,L} + e^{t\Delta}\rho_{n,\alpha,L} \otimes \Lambda_{\lambda_n^j} \Phi_{n,\varepsilon}^j \right\|_{L^1(\mathbb{R}^+;\mathcal{B}^1)} \tag{2.34}$$
$$\lesssim \left\| \Phi_{n,\varepsilon}^j \right\|_{L^2(\mathbb{R}^+;\mathcal{B}^1)} \left\| e^{t\Delta}\rho_{n,\alpha,L} \right\|_{L^2(\mathbb{R}^+;\mathcal{B}^1)}.$$

Making use of Estimates (2.25) and (2.27), we infer that

$$\sum_{j=0}^{L} \left\| \Phi_{n,\varepsilon}^j \right\|_{L^2(\mathbb{R}^+;\mathcal{B}^1)} \leq \mathcal{T}_0(u_0,\mathcal{M}_\alpha) + \mathcal{T}_2(\mathcal{M}) + \sum_{j=1}^{L} h_n^j \mathcal{T}_2(\mathcal{M}_\alpha),$$

which in view of Condition (2.29) on the sequences $(h_n^j)_{n\in\mathbb{N}}$ implies that

$$\left\| \sum_{j=0}^{L} \Phi_{n,\varepsilon}^j \right\|_{L^2(\mathbb{R}^+;\mathcal{B}^1)} \leq \mathcal{T}_0(u_0,\mathcal{M}_\alpha) + \mathcal{T}_2(\mathcal{M}) + \varepsilon.$$

It follows (of course up to a change of $\mathcal{T}_2$) that for small enough ε

$$\left\| \sum_{j=0}^{L} \Phi_{n,\varepsilon}^j \right\|_{L^2(\mathbb{R}^+;\mathcal{B}^1)} \leq \mathcal{T}_0(u_0,\mathcal{M}_\alpha) + \mathcal{T}_2(\mathcal{M}). \tag{2.35}$$

Thanks to (2.28) and (2.34), this gives rise to

$$\forall n \geq N_1, \quad \left\| F_{n,\varepsilon}^2 \right\|_{\mathcal{F}^0} \leq \varepsilon \left(\mathcal{T}_0(u_0,\mathcal{M}_\alpha) + \mathcal{T}_2(\mathcal{M}) \right). \tag{2.36}$$

Finally let us consider $F_{n,\varepsilon}^3$. Using the fact that $\mathcal{B}^1$ is an algebra along with the Hölder inequality, we infer that for a small enough γ in $]0,1[$,

$$\left\| \Lambda_{\lambda_n^j} \Phi_{n,\varepsilon}^j \otimes \Lambda_{\lambda_n^k} \Phi_{n,\varepsilon}^k \right\|_{L^1(\mathbb{R}^+;\mathcal{B}^1)} \leq \left\| \Lambda_{\lambda_n^j} \Phi_{n,\varepsilon}^j \right\|_{L^{\frac{2}{1+\gamma}}(\mathbb{R}^+;\mathcal{B}^1)} \left\| \Lambda_{\lambda_n^k} \Phi_{n,\varepsilon}^k \right\|_{L^{\frac{2}{1-\gamma}}(\mathbb{R}^+;\mathcal{B}^1)}.$$

The scaling invariance (2.9) gives

$$\left\| \Lambda_{\lambda_n^j} \Phi_{n,\varepsilon}^j \right\|_{L^{\frac{2}{1+\gamma}}(\mathbb{R}^+;\mathcal{B}^1)} \sim (\lambda_n^j)^\gamma \left\| \Phi_{n,\varepsilon}^j \right\|_{L^{\frac{2}{1+\gamma}}(\mathbb{R}^+;\mathcal{B}^1)} \quad \text{and}$$

$$\left\| \Lambda_{\lambda_n^k} \Phi_{n,\varepsilon}^k \right\|_{L^{\frac{2}{1-\gamma}}(\mathbb{R}^+;\mathcal{B}^1)} \sim \frac{1}{(\lambda_n^k)^\gamma} \left\| \Phi_{n,\varepsilon}^k \right\|_{L^{\frac{2}{1-\gamma}}(\mathbb{R}^+;\mathcal{B}^1)}.$$

For small enough γ, Theorems 2.14 and 2.15 imply that

$$\left\| \Lambda_{\lambda_n^j} \Phi_{n,\varepsilon}^j \otimes \Lambda_{\lambda_n^k} \Phi_{n,\varepsilon}^k \right\|_{L^1(\mathbb{R}^+;\mathcal{B}^1)} \lesssim \left(\frac{\lambda_n^j}{\lambda_n^k} \right)^\gamma.$$

We deduce that

$$\left\| F_{n,\varepsilon}^3 \right\|_{\mathcal{F}^0} \lesssim \sum_{\substack{0 \leq j,k \leq L \\ j \neq k}} \min \left\{ \frac{\lambda_n^j}{\lambda_n^k}, \frac{\lambda_n^k}{\lambda_n^j} \right\}^\gamma.$$

As the sequences $(\lambda_n^j)_{n\in\mathbb{N}}$ and $(\lambda_n^k)_{n\in\mathbb{N}}$ are orthogonal (see Definition 2.10), we have for any j and k such that $j \neq k$ that

$$\lim_{n\to\infty} \min\left\{ \frac{\lambda_n^j}{\lambda_n^k}, \frac{\lambda_n^k}{\lambda_n^j} \right\} = 0.$$

Thus an integer N_2 greater than or equal to N_1 exists such that

$$\forall n \geq N_2, \quad \|F_{n,\varepsilon}^3\|_{\mathcal{F}^0} \lesssim \varepsilon.$$

Together with (2.33) and (2.36), this implies that

$$n \geq N_2 \Longrightarrow \|F_{n,\varepsilon}\|_{\mathcal{F}^0} \lesssim \varepsilon,$$

which proves (2.32) and thus concludes the proof of Theorem 2.6. $\qquad\square$

2.3 Profile Decomposition of the Sequence of Initial Data: Proof of Theorem 2.12

The proof of Theorem 2.12 is structured as follows. First, in Section 2.3.1 we write down the profile decomposition of any bounded sequence of *anisotropically oscillating* divergence-free vector fields, following the results of [4]. Next we reorganize the profile decomposition by grouping together all profiles having the same horizontal scale and we check that all the conclusions of Theorem 2.12 hold: this is performed in Section 2.3.2.

2.3.1 Profile Decomposition of Anisotropically Oscillating, Divergence-free Vector Fields

In this section we start by recalling the result of [4], where an anisotropic profile decomposition of sequences of $\mathcal{B}_{1,q}$ anisotropically oscillating is achieved. Let us first define anisotropic scaling operators, similar to the operators defined in (2.1): for any two sequences of positive real numbers $(\varepsilon_n)_{n\in\mathbb{N}}$ and $(\gamma_n)_{n\in\mathbb{N}}$, and for any sequence $(x_n)_{n\in\mathbb{N}}$ of points in $\mathbb{R}^3$, we denote

$$\Lambda_{\varepsilon_n,\gamma_n,x_n}\phi(x) \stackrel{\text{def}}{=} \frac{1}{\varepsilon_n}\phi\left(\frac{x_{\mathrm{h}} - x_{n,\mathrm{h}}}{\varepsilon_n}, \frac{x_3 - x_{n,3}}{\gamma_n} \right).$$

Let us also introduce the definition of orthogonal triplets of sequences, analogous to Definition 2.10.

Definition 2.16 *We say that two triplets of sequences $(\varepsilon_n^\ell, \gamma_n^\ell, x_n^\ell)_{n\in\mathbb{N}}$ with ℓ belonging to $\{1,2\}$, where $(\varepsilon_n^\ell, \gamma_n^\ell)_{n\in\mathbb{N}}$ are two sequences of positive real* *numbers and x_n^ℓ are sequences in $\mathbb{R}^3$, are orthogonal if, when n tends to infinity,*

$$either \quad \frac{\varepsilon_n^1}{\varepsilon_n^2} + \frac{\varepsilon_n^2}{\varepsilon_n^1} + \frac{\gamma_n^1}{\gamma_n^2} + \frac{\gamma_n^2}{\gamma_n^1} \to \infty$$

$$or \quad (\varepsilon_n^1, \gamma_n^1) \equiv (\varepsilon_n^2, \gamma_n^2) \quad and \quad |(x_n^1)^{\varepsilon_n^1, \gamma_n^1} - (x_n^2)^{\varepsilon_n^1, \gamma_n^1}| \to \infty,$$

where we have denoted $(x_n^\ell)^{\varepsilon_n^k, \gamma_n^k} \stackrel{def}{=} \left(\dfrac{x_{n,h}^\ell}{\varepsilon_n^k}, \dfrac{x_{n,3}^\ell}{\gamma_n^k} \right)$. A family of sequences $((\varepsilon_n^j, \gamma_n^j, x_n^j)_{n\in\mathbb{N}})_{j\geq 0}$ is said to be a family of scales and cores if $\varepsilon_n^0 \equiv \gamma_n^0 \equiv 1$, $x_n^0 \equiv 0$, and if $(\varepsilon_n^\ell, \gamma_n^\ell, x_n^\ell)_{n\in\mathbb{N}}$ and $(\varepsilon_n^k, \gamma_n^k, x_n^k)_{n\in\mathbb{N}}$ are orthogonal when $\ell \neq k$.

Now, let us recall without proof the following result.

Proposition 2.17 ([4]) *Under the assumptions of Theorem 2.6, the following holds. For all integers $\ell \geq 0$ there is a triplet of scales and cores in the sense of Definition 2.16, denoted by $(\varepsilon_n^\ell, \gamma_n^\ell, x_n^\ell)_{n\in\mathbb{N}}$, and for all α in $]0,1[$ there are arbitrarily smooth divergence-free vector fields $(\widetilde{\phi}_\alpha^{h,\ell}, 0)$ and $(-\nabla^h \Delta_h^{-1}\partial_3 \phi_\alpha^\ell, \phi_\alpha^\ell)$ with $\widetilde{\phi}_\alpha^{h,\ell}$ and ϕ_α^ℓ compactly supported, and such that, up to extracting a subsequence, one can write the sequence $(u_{0,n})_{n\in\mathbb{N}}$ in the following form, for each $L \geq 1$:*

$$u_{0,n} = u_0 + \sum_{\ell=1}^{L} \Lambda_{\varepsilon_n^\ell, \gamma_n^\ell, x_n^\ell} \left(\widetilde{\phi}_\alpha^{h,\ell} + \widetilde{r}_\alpha^{h,\ell} - \frac{\varepsilon_n^\ell}{\gamma_n^\ell} \nabla^h \Delta_h^{-1} \partial_3 (\phi_\alpha^\ell + r_\alpha^\ell), \phi_\alpha^\ell + r_\alpha^\ell \right)$$
$$+ \left(\widetilde{\psi}_n^{h,L} - \nabla^h \Delta_h^{-1} \partial_3 \psi_n^L, \psi_n^L \right),$$
(2.37)

where $\widetilde{\psi}_n^{h,L}$ and ψ_n^L are independent of α and satisfy

$$\limsup_{n\to\infty} \left(\|\widetilde{\psi}_n^{h,L}\|_{\mathcal{B}^0} + \|\psi_n^L\|_{\mathcal{B}^0} \right) \to 0, \quad L \to \infty,$$
(2.38)

while $\widetilde{r}_\alpha^{h,\ell}$ and r_α^ℓ are independent of n and L and satisfy for each $\ell \in \mathbb{N}$

$$\|\widetilde{r}_\alpha^{h,\ell}\|_{\mathcal{B}_{1,q}} + \|r_\alpha^\ell\|_{\mathcal{B}_{1,q}} \leq \alpha.$$
(2.39)

Moreover, the following properties hold:

$$\forall \ell \geq 1, \quad \lim_{n\to\infty} (\gamma_n^\ell)^{-1}\varepsilon_n^\ell \in \{0, \infty\},$$
$$and \quad \lim_{n\to\infty} (\gamma_n^\ell)^{-1}\varepsilon_n^\ell = \infty \implies \phi_\alpha^\ell \equiv r_\alpha^\ell \equiv 0,$$
(2.40)

as well as the following stability result, which is uniform in α:

$$\sum_{\ell \geq 1} \left(\|\widetilde{\phi}_\alpha^{h,\ell}\|_{\mathcal{B}_{1,q}} + \|\widetilde{r}_\alpha^{h,\ell}\|_{\mathcal{B}_{1,q}} + \|\phi_\alpha^\ell\|_{\mathcal{B}_{1,q}} + \|r_\alpha^\ell\|_{\mathcal{B}_{1,q}} \right) \lesssim \sup_n \|u_{0,n}\|_{\mathcal{B}_{1,q}} + \|u_0\|_{\mathcal{B}_{1,q}}.$$
(2.41)

Remark 2.18 *As pointed out in [4, Section 2], if two scales appearing in the above decomposition are not orthogonal, then they can be chosen to be equal. We shall therefore assume from now on that this is the case: two sequences of scales are either orthogonal or equal.*

2.3.2 Regrouping of Profiles According to Horizontal Scales

In order to proceed with the re-organization of the profile decomposition provided in Proposition 2.17, we introduce some more definitions, keeping the notation of Proposition 2.17. For a given $L \geq 1$ we define recursively an increasing (finite) sequence of indices $\ell_k \in \{1, \ldots, L\}$ by

$$\ell_0 \overset{\text{def}}{=} 0, \quad \ell_{k+1} \overset{\text{def}}{=} \min \left\{ \ell \in \{\ell_k + 1, \ldots, L\} \, / \, \frac{\varepsilon_n^\ell}{\gamma_n^\ell} \to 0 \text{ and } \ell \notin \bigcup_{k'=0}^{k} \Gamma^L(\varepsilon_n^{\ell_{k'}}) \right\},$$
(2.42)

where for $0 \leq \ell \leq L$, we define (recalling that by Remark 2.18 if two scales are not orthogonal, then they are equal),

$$\Gamma^L(\varepsilon_n^\ell) \overset{\text{def}}{=} \left\{ \ell' \in \{1, \ldots, L\} \, / \, \varepsilon_n^{\ell'} \equiv \varepsilon_n^\ell \text{ and } \varepsilon_n^{\ell'} (\gamma_n^{\ell'})^{-1} \to 0, n \to \infty \right\}. \quad (2.43)$$

We call $\mathcal{L}(L)$ the largest index of the sequence (ℓ_k) and we may then introduce the following partition:

$$\left\{ \ell \in \{1, \ldots, L\} \, / \, \varepsilon_n^\ell (\gamma_n^\ell)^{-1} \to 0 \right\} = \bigcup_{k=0}^{\mathcal{L}(L)} \Gamma^L(\varepsilon_n^{\ell_k}). \quad (2.44)$$

We shall now regroup profiles in the decomposition (2.37) of $u_{0,n}$ according to the value of their horizontal scale. We fix from now on an integer $L \geq 1$.

Construction of the Profiles for $\ell = 0$

Before going into the technical details of the construction, let us discuss an example explaining the computations of this paragraph. Consider the particular case when $u_{0,n}$ is given by

$$u_{0,n}(x) = u_0(x) + \left(v_0^0(x_{\mathrm{h}}, 2^{-n}x_3) + w_0^{0,\mathrm{h}}(x_{\mathrm{h}}, 2^{-2n}x_3), 0 \right) + \left(v_0^0(x_1 + 2^n, x_2, 2^{-n}x_3), 0 \right),$$

with v_0^0 and $w_0^{0,\mathrm{h}}$ smooth (say in $B_{1,q}^{s,s'}$ for all s, s' in $\mathbb{R}$) and compactly supported. Let us assume that $u_{0,n}$ converges towards u_0 in the sense of distributions, and that $(u_{0,n} - u_0)_{n \in \mathbb{N}}$ is anisotropically oscillating. Then we can write

$$u_{0,n}(x) = u_0(x) + \left(v_{0,n}^{0,\mathrm{loc}}(x_{\mathrm{h}}, 2^{-n}x_3), 0 \right) + \left(v_{0,n}^{0,\infty}(x_{\mathrm{h}}, 2^{-n}x_3), 0 \right),$$

with $v_{0,n}^{0,\mathrm{loc}}(y) := v_0^0(y) + w_0^{0,\mathrm{h}}(y_h, 2^{-n}y_3)$ and $v_{0,n}^{0,\infty}(y) = v_0^0(y_1 + 2^n, y_2, y_3)$. Now since $u_{0,n} \rightharpoonup u_0$ as n goes to infinity, we have that $v_0^0(x_h, 0) + w_0^{\mathrm{h}}(x_h, 0) \equiv 0$, hence $v_{0,n}^{0,\mathrm{loc}}(x_h, 0) = 0$. The initial data $u_{0,n}$ has therefore been re-written as

$$u_{0,n}(x) = u_0(x) + \left(v_{0,n}^{0,\mathrm{loc}}(x_h, 2^{-n}x_3), 0\right) + \left(v_{0,n}^{0,\infty}(x_h, 2^{-n}x_3), 0\right)$$
$$\text{with} \quad v_{0,n}^{0,\mathrm{loc}}(x_h, 0) = 0$$

and where the support in x_h of $v_{0,n}^{0,\mathrm{loc}}(x_h, 2^{-n}x_3)$ is in a fixed compact set whereas the support in x_h of $v_{0,n}^{0,\infty}(x_h, 2^{-n}x_3)$ escapes to infinity. This is of the same form as in the statement of Theorem 2.12.

When considering all the profiles having the same horizontal scale (1 here), the point is therefore to choose the smallest vertical scale (2^n here) and to write the decomposition in terms of that scale only. Of course this implies that, contrary to usual profile decompositions, the profiles are no longer fixed functions in $\mathcal{B}_{1,q}$, but sequences of functions, bounded in $\mathcal{B}_{1,q}$.

In view of the above example, let ℓ_0^- be an integer such that $\gamma_n^{\ell_0^-}$ is the smallest vertical scale going to infinity, associated with profiles for $1 \le \ell \le L$, having 1 for horizontal scale. More precisely, we ask that

$$\gamma_n^{\ell_0^-} = \min_{\ell \in \Gamma^L(1)} \gamma_n^\ell,$$

where according to (2.43),

$$\Gamma^L(1) = \left\{\ell' \in \{1,\ldots,L\} / \varepsilon_n^{\ell'} \equiv 1 \quad \text{and} \quad \gamma_n^{\ell'} \to \infty, n \to \infty\right\}.$$

Notice that the minimum of the sequences γ_n^ℓ is well defined in our context, thanks to the fact that due to Remark 2.18, either two sequences are orthogonal in the sense of Definition 2.16, or they are equal. Observe also that ℓ_0^- is by no means unique, as several profiles may have the same horizontal scale as well as the same vertical scale (in which case the concentration cores must be orthogonal).

Now we denote

$$h_n^0 \stackrel{\text{def}}{=} (\gamma_n^{\ell_0^-})^{-1}, \tag{2.45}$$

and we notice that h_n^0 goes to zero as n goes to infinity for each L. Note also that h_n^0 depends on L through the choice of ℓ_0^-, since if L increases then ℓ_0^- may also increase; this dependence is omitted in the notation for simplicity. Let us define (up to a subsequence extraction)

$$a^\ell \stackrel{\text{def}}{=} \lim_{n\to\infty} \left(x_{n,\mathrm{h}}^\ell, \frac{x_{n,3}^\ell}{\gamma_n^\ell}\right). \tag{2.46}$$

We then define the divergence-free vector fields

$$v_{0,n,\alpha,L}^{0,\mathrm{loc}}(y) \stackrel{\mathrm{def}}{=} \sum_{\substack{\ell \in \Gamma^L(1) \\ a_\mathrm{h}^\ell \in \mathbb{R}^2}} \widetilde{\phi}_\alpha^{h,\ell}\left(y_\mathrm{h} - x_{n,\mathrm{h}}^\ell, \frac{y_3}{h_n^0 \gamma_n^\ell} - \frac{x_{n,3}^\ell}{\gamma_n^\ell}\right) \tag{2.47}$$

and

$$w_{0,n,\alpha,L}^{0,\mathrm{loc}}(y) \stackrel{\mathrm{def}}{=} \sum_{\substack{\ell \in \Gamma^L(1) \\ a_\mathrm{h}^\ell \in \mathbb{R}^2}} \left(-\frac{1}{h_n^0 \gamma_n^\ell} \nabla^h \Delta_\mathrm{h}^{-1} \partial_3 \phi_\alpha^\ell, \phi_\alpha^\ell\right)\left(y_\mathrm{h} - x_{n,\mathrm{h}}^\ell, \frac{y_3}{h_n^0 \gamma_n^\ell} - \frac{x_{n,3}^\ell}{\gamma_n^\ell}\right).$$

$$\tag{2.48}$$

By construction we have

$$w_{0,n,\alpha,L}^{0,\mathrm{loc},h} = -\nabla^h \Delta_\mathrm{h}^{-1} \partial_3 w_{0,n,\alpha,L}^{0,\mathrm{loc},3}.$$

Similarly, we define

$$v_{0,n,\alpha,L}^{0,\infty}(y) \stackrel{\mathrm{def}}{=} \sum_{\substack{\ell \in \Gamma^L(1) \\ |a_\mathrm{h}^\ell| = \infty}} \widetilde{\phi}_\alpha^{h,\ell}\left(y_\mathrm{h} - x_{n,\mathrm{h}}^\ell, \frac{y_3}{h_n^0 \gamma_n^\ell} - \frac{x_{n,3}^\ell}{\gamma_n^\ell}\right) \tag{2.49}$$

and

$$w_{0,n,\alpha,L}^{0,\infty}(y) \stackrel{\mathrm{def}}{=} \sum_{\substack{\ell \in \Gamma^L(1) \\ |a_\mathrm{h}^\ell| = \infty}} \left(-\frac{1}{h_n^0 \gamma_n^\ell} \nabla^h \Delta_\mathrm{h}^{-1} \partial_3 \phi_\alpha^\ell, \phi_\alpha^\ell\right)\left(y_\mathrm{h} - x_{n,\mathrm{h}}^\ell, \frac{y_3}{h_n^0 \gamma_n^\ell} - \frac{x_{n,3}^\ell}{\gamma_n^\ell}\right).$$

$$\tag{2.50}$$

By construction we have again

$$w_{0,n,\alpha,L}^{0,\infty,h} = -\nabla^h \Delta_\mathrm{h}^{-1} \partial_3 w_{0,n,\alpha,L}^{0,\infty,3}.$$

Moreover, recalling the notation

$$[f]_{h_n^0}(x) \stackrel{\mathrm{def}}{=} f(x_\mathrm{h}, h_n^0 x_3)$$

and

$$\Lambda_{\varepsilon_n, \gamma_n, x_n} \phi(x) \stackrel{\mathrm{def}}{=} \frac{1}{\varepsilon_n} \phi\left(\frac{x_\mathrm{h} - x_{n,\mathrm{h}}}{\varepsilon_n}, \frac{x_3 - x_{n,3}}{\gamma_n}\right),$$

one can compute that

$$\sum_{\substack{\ell \in \Gamma^L(1) \\ a_\mathrm{h}^\ell \in \mathbb{R}^2}} \Lambda_{1, \gamma_n^\ell, x_n^\ell} \left(\widetilde{\phi}_\alpha^{h,\ell} - \frac{1}{\gamma_n^\ell} \nabla^h \Delta_\mathrm{h}^{-1} \partial_3 \phi_\alpha^\ell, \phi_\alpha^\ell\right) = \left[(v_{0,n,\alpha,L}^{0,\mathrm{loc}} + h_n^0 w_{0,n,\alpha,L}^{0,\mathrm{loc},h}, w_{0,n,\alpha,L}^{0,\mathrm{loc},3})\right]_{h_n^0}$$

$$\tag{2.51}$$

and

$$\sum_{\substack{\ell \in \Gamma^L(1) \\ |a_h^\ell|=\infty}} \Lambda_{1,\gamma_n^\ell,x_n^\ell} \left(\widetilde{\phi}_\alpha^{h,\ell} - \frac{1}{\gamma_n^\ell} \nabla^h \Delta_h^{-1} \partial_3 \phi_\alpha^\ell, \phi_\alpha^\ell \right) = \left[(v_{0,n,\alpha,L}^{0,\infty} + h_n^0 w_{0,n,\alpha,L}^{0,\infty,h}, w_{0,n,\alpha,L}^{0,\infty,3}) \right]_{h_n^0}.$$

(2.52)

Let us now check that $v_{0,n,\alpha,L}^{0,\text{loc}}$, $w_{0,n,\alpha,L}^{0,\text{loc}}$, $v_{0,n,\alpha,L}^{0,\infty}$ and $w_{0,n,\alpha,L}^{0,\infty}$ satisfy the bounds given in the statement of Theorem 2.12. We shall study only $v_{0,n,\alpha,L}^{0,\text{loc}}$ and $w_{0,n,\alpha,L}^{0,\text{loc}}$ as the other study is very similar. By translation and scale invariance of $\mathcal{B}^0$ and using definitions (2.47) and (2.48), we get

$$\|v_{0,n,\alpha,L}^{0,\text{loc},h}\|_{\mathcal{B}^0} \leq \sum_{\ell \geq 1} \|\widetilde{\phi}_\alpha^{h,\ell}\|_{\mathcal{B}^0} \quad \text{and} \quad \|w_{0,n,\alpha,L}^{0,\text{loc},3}\|_{\mathcal{B}^0} \leq \sum_{\ell \geq 1} \|\phi_\alpha^\ell\|_{\mathcal{B}^0}. \quad (2.53)$$

According to (2.41) and the Sobolev embedding $\mathcal{B}_{1,q} \hookrightarrow \mathcal{B}^0$, this gives rise to

$$\|v_{0,n,\alpha,L}^0\|_{\mathcal{B}^0} + \|w_{0,n,\alpha,L}^{0,\text{loc},3}\|_{\mathcal{B}^0} \leq C \quad \text{uniformly in } \alpha, L, n. \quad (2.54)$$

Moreover, for each given α, the profiles are as smooth as needed, and since in the above sums by construction $\gamma_{n,L}^{\ell_0^-} \leq \gamma_n^\ell$, one gets also, after an easy computation

$$\forall s \in \mathbb{R}, \forall s' \geq 1/2, \quad \|v_{0,n,\alpha,L}^{0,\text{loc}}\|_{\mathcal{B}^{s,s'}} + \|w_{0,n,\alpha,L}^{0,\text{loc},3}\|_{\mathcal{B}^{s,s'}} \leq C(\alpha) \quad \text{uniformly in } n, L.$$

(2.55)

Estimates (2.54) and (2.55) easily give (2.16) and (2.17).

Finally, let us estimate $v_{0,n,\alpha,L}^{0,\text{loc},h}(\cdot,0)$ and $w_{0,n,\alpha,L}^{0,\text{loc},3}(\cdot,0)$ in $B_{2,1}^0(\mathbb{R}^2)$ and prove (2.18). On the one hand, by assumption we know that $u_{0,n} \rightharpoonup u_0$ in the sense of distributions. On the other hand, we can take weak limits in the decomposition of $u_{0,n}$ provided by Proposition 2.17. We recall that by (2.40), if $\varepsilon_n^\ell/\gamma_n^\ell \to \infty$ then $\phi_\alpha^\ell \equiv r_\alpha^\ell \equiv 0$. Then we notice that clearly

$$\varepsilon_n^\ell \to 0 \text{ or } \varepsilon_n^\ell \to \infty \implies \Lambda_{\varepsilon_n^\ell,\gamma_n^\ell,x_n^\ell} f \rightharpoonup 0$$

for any value of the sequences γ_n^ℓ, x_n^ℓ and any function f. Moreover,

$$\gamma_n^\ell \to 0 \implies \Lambda_{1,\gamma_n^\ell,x_n^\ell} f \rightharpoonup 0$$

for any sequence of cores x_n^ℓ and any function f, so we are left with the study of profiles such that $\varepsilon_n^\ell \equiv 1$ and $\gamma_n^\ell \to \infty$. Then we also notice that if $\gamma_n^\ell \to \infty$, then with Notation (2.46),

$$|a_h^\ell| = \infty \implies \Lambda_{1,\gamma_n^\ell,x_n^\ell} f \rightharpoonup 0. \quad (2.56)$$

In that case, in view of (2.38) and (2.41),

$$\sum_{\ell=1}^{L} \Lambda_{\varepsilon_n^\ell, \gamma_n^\ell, x_n^\ell} \frac{\varepsilon_n^\ell}{\gamma_n^\ell} \nabla^h \Delta_h^{-1} \partial_3 (\phi_\alpha^\ell + r_\alpha^\ell) + \nabla^h \Delta_h^{-1} \partial_3 \psi_n^L \rightharpoonup 0.$$

Consequently, for each $L \geq 1$ and each α in $]0,1[$, we have in view of (2.37), as n goes to infinity

$$-\psi_n^L - \sum_{\ell \in \Gamma^L(1)} r_\alpha^\ell (\cdot - x_{n,h}^\ell, \frac{\cdot - x_{n,3}^\ell}{\gamma_n^\ell}) \rightharpoonup \sum_{\substack{\ell \in \Gamma^L(1) \\ s.t. a_h^\ell \in \mathbb{R}^2}} \phi_\alpha^\ell (\cdot - a_h^\ell, 0)$$

$$\tag{2.57}$$

$$-\tilde{\psi}_n^{h,L} - \sum_{\ell \in \Gamma^L(1)} \tilde{r}_\alpha^{h,\ell} (\cdot - x_{n,h}^\ell, \frac{\cdot - x_{n,3}^\ell}{\gamma_n^\ell}) \rightharpoonup \sum_{\substack{\ell \in \Gamma^L(1) \\ s.t. a_h^\ell \in \mathbb{R}^2}} \tilde{\phi}_\alpha^{h,\ell} (\cdot - a_h^\ell, 0).$$

Now let $\eta > 0$ be given. Then, thanks to (2.38) and (2.39), there is an $L_0 \geq 1$ such that for all $L \geq L_0$ there is an $\alpha_0 \leq 1$ (depending on L) such that for all $L \geq L_0$ and $\alpha \leq \alpha_0$, uniformly in $n \geq n(L_0, \eta)$,

$$\left\| (\tilde{\psi}_n^{h,L}, \psi_n^L) \right\|_{\mathcal{B}^0} + \left\| \sum_{\ell \in \Gamma^L(1)} (\tilde{r}_\alpha^{h,\ell}, r_\alpha^\ell)(\cdot - x_{n,h}^\ell, \frac{\cdot - x_{n,3}^\ell}{\gamma_n^\ell}) \right\|_{\mathcal{B}^0} \leq \eta.$$

Using the fact that $\mathcal{B}^0$ is embedded in $L^\infty(\mathbb{R}; B_{2,1}^0(\mathbb{R}^2))$, we infer from (2.57) that for $L \geq L_0$ and $\alpha \leq \alpha_0$

$$\left\| \sum_{\substack{\ell \in \Gamma^L(1) \\ s.t. a_h^\ell \in \mathbb{R}^2}} \tilde{\phi}_\alpha^{h,\ell} (\cdot - a_h^\ell, 0) \right\|_{B_{2,1}^0(\mathbb{R}^2)} \leq \eta \tag{2.58}$$

and

$$\left\| \sum_{\substack{\ell \in \Gamma^L(1) \\ s.t. a_h^\ell \in \mathbb{R}^2}} \phi_\alpha^\ell (\cdot - a_h^\ell, 0) \right\|_{B_{2,1}^0(\mathbb{R}^2)} \leq \eta. \tag{2.59}$$

But by (2.47), we have

$$v_{0,n,\alpha,L}^{0,\text{loc},h}(\cdot, 0) = \sum_{\substack{\ell \in \Gamma^L(1) \\ a_h^\ell \in \mathbb{R}^2}} \tilde{\phi}_\alpha^{h,\ell} \left(\cdot - x_{n,h}^\ell, -\frac{x_{n,3}^\ell}{\gamma_n^\ell} \right)$$

and by (2.48) we have also

$$w_{0,n,\alpha,L}^{0,\text{loc},3}(\cdot, 0) = \sum_{\substack{\ell \in \Gamma^L(1) \\ a_h^\ell \in \mathbb{R}^2}} \phi_\alpha^\ell \left(\cdot - x_{n,h}^\ell, -\frac{x_{n,3}^\ell}{\gamma_n^\ell} \right).$$

It follows that we can write for all $L \geq L_0$ and $\alpha \leq \alpha_0$,

$$\limsup_{n \to \infty} \|v_{0,n,\alpha,L}^{0,\mathrm{loc},h}(\cdot,0)\|_{B_{2,1}^0(\mathbb{R}^2)} \leq \Big\| \sum_{\substack{\ell \in \Gamma^L(1) \\ a_h^\ell \in \mathbb{R}^2}} \tilde{\phi}_\alpha^{h,\ell}(\cdot - a_h^\ell, 0) \Big\|_{B_{2,1}^0(\mathbb{R}^2)}$$

$$\leq \eta$$

thanks to (2.58). A similar estimate for $w_{0,n,\alpha,L}^{0,\mathrm{loc},3}(\cdot,0)$ using (2.59) gives finally

$$\lim_{L \to \infty} \lim_{\alpha \to 0} \limsup_{n \to \infty} \Big(\|v_{0,n,\alpha,L}^{0,\mathrm{loc},h}(\cdot,0)\|_{B_{2,1}^0(\mathbb{R}^2)} + \|w_{0,n,\alpha,L}^{0,\mathrm{loc},3}(\cdot,0)\|_{B_{2,1}^0(\mathbb{R}^2)} \Big) = 0. \quad (2.60)$$

The results (2.19) and (2.20) involving the cut-off function θ are simply due to the fact that the profiles are compactly supported.

Construction of the Profiles for $\ell \geq 1$

The construction is very similar to the previous one. We start by considering a fixed integer $j \in \{1, \dots, \mathcal{L}(L)\}$.

Then we define an integer ℓ_j^- so that, up to a sequence extraction,

$$\gamma_n^{\ell_j^-} = \min_{\ell \in \Gamma^L(\varepsilon_n^{\ell_j})} \gamma_n^\ell,$$

whereas in (2.43)

$$\Gamma^L(\varepsilon_n^\ell) \overset{\mathrm{def}}{=} \Big\{ \ell' \in \{1, \dots, L\} / \varepsilon_n^{\ell'} \equiv \varepsilon_n^\ell \quad \text{and} \quad \varepsilon_n^{\ell'}(\gamma_n^{\ell'})^{-1} \to 0, n \to \infty \Big\}.$$

Notice that necessarily $\varepsilon^{\ell_j} \neq 1$. Finally, we define

$$h_n^j \overset{\mathrm{def}}{=} \varepsilon_n^{\ell_j}(\gamma_n^{\ell_j^-})^{-1}.$$

By construction we have that $h_n^j \to 0$ as $n \to \infty$ (recall that $\varepsilon_n^{\ell_j} \equiv \varepsilon_n^{\ell_j^-}$). Then we define for $j \leq \mathcal{L}(L)$

$$v_{n,\alpha,L}^{j,h}(y) \overset{\mathrm{def}}{=} \sum_{\ell \in \Gamma^L(\varepsilon_n^{\ell_j})} \tilde{\phi}_\alpha^{h,\ell}\Big(y_h - \frac{x_{n,h}^\ell}{\varepsilon_n^{\ell_j}}, \frac{\varepsilon_n^{\ell_j}}{h_n^j \gamma_n^\ell} y_3 - \frac{x_{n,3}^\ell}{\gamma_n^\ell} \Big) \quad (2.61)$$

and

$$w_{n,\alpha,L}^j(y) \overset{\mathrm{def}}{=} \sum_{\ell \in \Gamma^L(\varepsilon_n^{\ell_j})} \Big(-\frac{\varepsilon_n^{\ell_j}}{h_n^j \gamma_n^\ell} \nabla^h \Delta_h^{-1} \partial_3 \phi_\alpha^\ell, \phi_\alpha^\ell \Big) \Big(y_h - \frac{x_{n,h}^\ell}{\varepsilon_n^{\ell_j}}, \frac{\varepsilon_n^{\ell_j}}{h_n^j \gamma_n^\ell} y_3 - \frac{x_{n,3}^\ell}{\gamma_n^\ell} \Big)$$

and we choose

$$\mathcal{L}(L) < j \leq L \quad \Rightarrow \quad v_{n,\alpha,L}^{j,h} \equiv 0 \quad \text{and} \quad w_{n,\alpha,L}^j \equiv 0. \quad (2.62)$$

We notice that

$$w_{n,\alpha,L}^{j,h} = -\nabla^h \Delta_h^{-1} \partial_3 w_{n,\alpha,L}^{j,3}.$$

Defining

$$\lambda_n^j \stackrel{\text{def}}{=} \varepsilon_n^{\ell_j},$$

a computation, similar to that giving (2.51), implies directly that

$$\sum_{\ell \in \Gamma^L(\varepsilon_n^{\ell_j})} \Lambda_{\varepsilon_n^{\ell_j}, \gamma_n^\ell, x_n^\ell} \left(\tilde{\phi}_\alpha^{h,\ell} - \frac{\lambda_n^j}{\gamma_n^\ell} \nabla^h \Delta_h^{-1} \partial_3 \phi_\alpha^\ell, \phi_\alpha^\ell \right) \tag{2.63}$$

$$= \Lambda_{\lambda_n^j} \left[(v_{n,\alpha,L}^{j,h} + h_n^j w_{n,\alpha,L}^{j,h}, w_{n,\alpha,L}^{j,3}) \right]_{h_n^j}.$$

Notice that since $\varepsilon_n^{\ell_j} \not\equiv 1$ as recalled above, we have that $\lambda_n^j \to 0$ or ∞ as $n \to \infty$.

The a priori bounds for the profiles $(v_{n,\alpha,L}^{j,h}, w_{n,\alpha,L}^{j,3})_{1 \le j \le L}$ are obtained exactly as in the previous paragraph: let us prove that

$$\sum_{j \ge 1} \left(\|v_{n,\alpha,L}^{j,h}\|_{\mathcal{B}^0} + \|w_{n,\alpha,L}^{j,3}\|_{\mathcal{B}^0} \right) \le C, \quad \text{and}$$

$$\forall s \in \mathbb{R}, \quad \forall s' \ge 1/2, \quad \sum_{j \ge 1} \left(\|v_{n,\alpha,L}^{j,h}\|_{\mathcal{B}^{s,s'}} + \|w_{n,\alpha,L}^{j,3}\|_{\mathcal{B}^{s,s'}} \right) \le C(\alpha). \tag{2.64}$$

We shall detail the argument for the first inequality only, and in the case of $v_{n,\alpha,L}^{j,h}$, as the study of $w_{n,\alpha,L}^{j,3}$ is similar. We write, using the definition of $v_{n,\alpha,L}^{j,h}$ in (2.61),

$$\sum_{j=1}^L \|v_{n,\alpha,L}^{j,h}\|_{\mathcal{B}^0} = \sum_{j=1}^{\mathcal{L}(L)} \left\| \sum_{\ell \in \Gamma^L(\varepsilon_n^{\ell_j})} \tilde{\phi}_\alpha^{h,\ell} \left(y_h - \frac{x_{n,h}^\ell}{\varepsilon_n^\ell}, \frac{\varepsilon_n^{\ell_j}}{h_n^j \gamma_n^\ell} y_3 - \frac{x_{n,3}^\ell}{\gamma_n^\ell} \right) \right\|_{\mathcal{B}^0},$$

so by definition of the partition (2.44) and by scale and translation invariance of $\mathcal{B}^0$ we find, thanks to (2.41), that there is a constant C independent of L such that

$$\sum_{j=1}^L \|v_{n,\alpha,L}^{j,h}\|_{\mathcal{B}^0} \le \sum_{\ell=1}^L \|\tilde{\phi}_\alpha^{h,\ell}\|_{\mathcal{B}^0} \le C.$$

The result is proved.

Construction of the Remainder Term

With the notation of Proposition 2.17, let us first define the remainder terms

$$\tilde{\rho}_{n,\alpha,L}^{(1),h} \stackrel{\text{def}}{=} -\sum_{\ell=1}^L \frac{\varepsilon_n^\ell}{\gamma_n^\ell} \Lambda_{\varepsilon_n^\ell, \gamma_n^\ell, x_n^\ell} \nabla^h \Delta_h^{-1} \partial_3 r_\alpha^\ell - \nabla^h \Delta_h^{-1} \partial_3 \psi_n^L \tag{2.65}$$

and

$$\rho_{n,\alpha,L}^{(2)} \stackrel{\text{def}}{=} \sum_{\ell=1}^{L} \Lambda_{\varepsilon_n^\ell,\gamma_n^\ell,x_n^\ell}(\tilde{r}_\alpha^{h,\ell},0) + \sum_{\ell=1}^{L} \Lambda_{\varepsilon_n^\ell,\gamma_n^\ell,x_n^\ell}(0,r_\alpha^\ell) + (\tilde{\psi}_n^{h,L},\psi_n^L). \qquad (2.66)$$

Observe that by construction, thanks to (2.38) and (2.39) and to the fact that if $r_\alpha^\ell \neq 0$, then $\varepsilon_n^\ell/\gamma_n^\ell$ goes to zero as n goes to infinity, we have

$$\lim_{L\to\infty}\lim_{\alpha\to 0}\limsup_{n\to\infty} \|\tilde{\rho}_{\alpha,n,L}^{(1),h}\|_{\mathcal{B}^{1,-\frac{1}{2}}} = 0,$$

$$\text{and} \quad \lim_{L\to\infty}\lim_{\alpha\to 0}\limsup_{n\to\infty} \|\rho_{\alpha,n,L}^{(2)}\|_{\mathcal{B}^0} = 0. \qquad (2.67)$$

Then we notice that for each $\ell \in \mathbb{N}$ and each $\alpha \in]0,1[$, we have by a direct computation

$$\left\| \Lambda_{\varepsilon_n^\ell,\gamma_n^\ell,x_n^\ell}(\tilde{\phi}_\alpha^{h,\ell},0) \right\|_{\mathcal{B}^{1,-\frac{1}{2}}} \sim \frac{\gamma_n^\ell}{\varepsilon_n^\ell} \|\tilde{\phi}_\alpha^{h,\ell}\|_{\mathcal{B}^{1,-\frac{1}{2}}}.$$

We deduce that if $\varepsilon_n^\ell/\gamma_n^\ell \to \infty$, then $\Lambda_{\varepsilon_n^\ell,\gamma_n^\ell,x_n^\ell}(\tilde{\phi}_\alpha^{h,\ell},0)$ goes to zero in $\mathcal{B}^{1,-\frac{1}{2}}$ as n goes to infinity, hence so does the sum over $\ell \in \{1,\dots,L\}$. It follows that for each given α in $]0,1[$ and $L \geq 1$ we may define

$$\rho_{n,\alpha,L}^{(1)} \stackrel{\text{def}}{=} \tilde{\rho}_{n,\alpha,L}^{(1),h} + \sum_{\substack{\ell=1 \\ \varepsilon_n^\ell/\gamma_n^\ell \to \infty}}^{L} \Lambda_{\varepsilon_n^\ell,\gamma_n^\ell,x_n^\ell}(\tilde{\phi}_\alpha^{h,\ell},0)$$

and we have

$$\lim_{L\to\infty}\lim_{\alpha\to 0}\limsup_{n\to\infty} \|\rho_{n,\alpha,L}^{(1)}\|_{\mathcal{B}^{1,-\frac{1}{2}}} = 0. \qquad (2.68)$$

Finally, as $\mathcal{D}(\mathbb{R}^3)$ is dense in $\mathcal{B}_{1,q}$, let us choose a family $(u_{0,\alpha})_\alpha$ of functions in $\mathcal{D}(\mathbb{R}^3)$ such that $\|u_0 - u_{0,\alpha}\|_{\mathcal{B}_{1,q}} \leq \alpha$ and let us define

$$\rho_{n,\alpha,L} \stackrel{\text{def}}{=} \rho_{\alpha,n,L}^{(1)} + \rho_{n,\alpha,L}^{(2)} + u_0 - u_{0,\alpha}. \qquad (2.69)$$

Inequalities (2.67) and (2.68) give

$$\lim_{L\to\infty}\lim_{\alpha\to 0}\limsup_{n\to\infty} \|e^{t\Delta}\rho_{n,\alpha,L}\|_{L^2(\mathbb{R}^+;\mathcal{B}^1)} = 0. \qquad (2.70)$$

End of the Proof of Theorem 2.12

Let us return to the decomposition given in Proposition 2.17, and use definitions (2.65), (2.66) and (2.69), which imply that

$$u_{0,n} = u_{0,\alpha} + \sum_{\substack{\ell=1 \\ \varepsilon_n^\ell/\gamma_n^\ell \to 0}}^{L} \Lambda_{\varepsilon_n^\ell,\gamma_n^\ell,x_n^\ell}\left(\tilde{\phi}_\alpha^{h,\ell} - \frac{\varepsilon_n^\ell}{\gamma_n^\ell}\nabla^h\Delta_h^{-1}\partial_3\phi_\alpha^\ell, \phi_\alpha^\ell\right) + \rho_{n,\alpha,L}.$$

We recall that for all ℓ in $\mathbb{N}$, we have $\lim_{n\to\infty} (\gamma_n^\ell)^{-1}\varepsilon_n^\ell \in \{0,\infty\}$ and in the case where the ratio $\varepsilon_n^\ell/\gamma_n^\ell$ goes to infinity then $\phi_\alpha^\ell \equiv 0$. Next we separate the case when the horizontal scale is one from the others: with the notation (2.43) we write

$$u_{0,n} = u_{0,\alpha} + \sum_{\ell \in \Gamma^L(1)} \Lambda_{1,\gamma_n^\ell,x_n^\ell}\left(\tilde{\phi}_\alpha^{h,\ell} - \frac{1}{\gamma_n^\ell}\nabla^h \Delta_h^{-1}\partial_3\phi_\alpha^\ell, \phi_\alpha^\ell\right)$$

$$+ \sum_{\substack{\ell=1 \\ \varepsilon_n^\ell \neq 1 \\ \varepsilon_n^\ell/\gamma_n^\ell \to 0}}^{L} \Lambda_{\varepsilon_n^\ell,\gamma_n^\ell,x_n^\ell}\left(\tilde{\phi}_\alpha^{h,\ell} - \frac{\varepsilon_n^\ell}{\gamma_n^\ell}\nabla^h \Delta_h^{-1}\partial_3\phi_\alpha^\ell, \phi_\alpha^\ell\right) + \rho_{n,\alpha,L}.$$

With (2.51) this can be written

$$u_{0,n} = u_{0,\alpha} + \left[(v_{0,n,\alpha,L}^{0,\mathrm{loc},h} + h_n^0 w_{0,n,\alpha,L}^{0,\mathrm{loc},h}, w_{0,n,\alpha,L}^{0,\mathrm{loc},3})\right]_{h_n^0} + \left[(v_{0,n,\alpha,L}^{0,\infty,h} + h_n^0 w_{0,n,\alpha,L}^{0,\infty,h}, w_{0,n,\alpha,L}^{0,\infty,3})\right]_{h_n^0}$$

$$+ \sum_{\substack{\ell=1 \\ \varepsilon_n^\ell \neq 1 \\ \varepsilon_n^\ell/\gamma_n^\ell \to 0}}^{L} \Lambda_{\varepsilon_n^\ell,\gamma_n^\ell,x_n^\ell}\left(\tilde{\phi}_\alpha^{h,\ell} - \frac{\varepsilon_n^\ell}{\gamma_n^\ell}\nabla^h \Delta_h^{-1}\partial_3\phi_\alpha^\ell, \phi_\alpha^\ell\right) + \rho_{n,\alpha,L}.$$

Next we use the partition (2.44), so that with notation (2.42) and (2.43),

$$u_{0,n} = u_{0,\alpha} + \left[(v_{0,n,\alpha,L}^{0,\mathrm{loc},h} + h_n^0 w_{0,n,\alpha,L}^{0,\mathrm{loc},h}, w_{0,n,\alpha,L}^{0,\mathrm{loc},3})\right]_{h_n^0} + \left[(v_{0,n,\alpha,L}^{0,\infty,h} + h_n^0 w_{0,n,\alpha,L}^{0,\infty,h}, w_{0,n,\alpha,L}^{0,\infty,3})\right]_{h_n^0}$$

$$+ \sum_{j=1}^{\mathcal{L}(L)} \sum_{\substack{\ell \in \Gamma^L(\varepsilon_n^{\ell_j}) \\ \varepsilon_n^{\ell_j} \neq 1}} \Lambda_{\varepsilon_n^{\ell_j},\gamma_n^\ell,x_n^\ell}\left(\tilde{\phi}_\alpha^{h,\ell} - \frac{\varepsilon_n^{\ell_j}}{\gamma_n^\ell}\nabla^h \Delta_h^{-1}\partial_3\phi_\alpha^\ell, \phi_\alpha^\ell\right) + \rho_{n,\alpha,L}.$$

Then we finally use the identity (2.63) which gives

$$u_{0,n} = u_{0,\alpha} + \left[(v_{0,n,\alpha,L}^{0,\mathrm{loc},h} + h_n^0 w_{0,n,\alpha,L}^{0,\mathrm{loc},h}, w_{0,n,\alpha,L}^{0,\mathrm{loc},3})\right]_{h_n^0} + \left[(v_{0,n,\alpha,L}^{0,\infty,h} + h_n^0 w_{0,n,\alpha,L}^{0,\infty,h}, w_{0,n,\alpha,L}^{0,\infty,3})\right]_{h_n^0}$$

$$+ \sum_{j=1}^{L} \Lambda_{\lambda_n^j}\left[(v_{n,\alpha,L}^{j,h} + h_n^j w_{n,\alpha,L}^{j,h}, w_{n,\alpha,L}^{j,3})\right]_{h_n^j} + \rho_{n,\alpha,L}.$$

The end of the proof follows from the estimates (2.54), (2.55), (2.60), (2.64), along with (2.70). Theorem 2.12 is proved.

2.4 Proof of Theorems 2.14 and 2.15

2.4.1 Proof of Theorem 2.14

In order to prove that the initial data defined by

$$\Phi_0 \overset{\text{def}}{=} \left[(v_0 - \beta \nabla^h \Delta_h^{-1} \partial_3 w_0^3, w_0^3) \right]_\beta,$$

with (v_0, w_0^3) satisfying the assumptions of Theorem 2.14, gives rise to a global smooth solution for small enough β, we look for the solution in the form

$$\Phi_\beta = \Phi^{\text{app}} + \psi \quad \text{with} \quad \Phi^{\text{app}} \overset{\text{def}}{=} \left[(v + \beta w^h, w^3) \right]_\beta, \qquad (2.71)$$

where v solves the two-dimensional Navier–Stokes equations

$$(\text{NS2D})_{x_3}: \quad \begin{cases} \partial_t v + v \cdot \nabla^h v - \Delta_h v = -\nabla^h p & \text{in } \mathbb{R}^+ \times \mathbb{R}^2, \\ \text{div}_h v = 0, \\ v_{|t=0} = v_0(\cdot, x_3), \end{cases}$$

while w^3 solves the transport-diffusion equation

$$(T_\beta): \quad \begin{cases} \partial_t w^3 + v \cdot \nabla^h w^3 - \Delta_h w^3 - \beta^2 \partial_3^2 w^3 = 0 & \text{in } \mathbb{R}^+ \times \mathbb{R}^3, \\ w^3_{|t=0} = w_0^3 \end{cases}$$

and w^h is determined by the divergence-free condition on w.

In Subsection 2.4.1 (resp. 2.4.1), we establish a priori estimates on v (resp. w), and in Subsection 2.4.1, we achieve the proof of Theorem 2.14 by studying the perturbed Navier–Stokes equation satisfied by ψ.

Two-Dimensional Flows with Parameter

The goal of this section is to prove the following proposition on v, the solution of $(\text{NS2D})_{x_3}$. It is a general result on the regularity of the solution of (NS2D) when the initial data depends on a real parameter x_3, measured in terms of Besov spaces with respect to the variable x_3.

Proposition 2.19 *Let v_0 be a two-component divergence-free vector field depending on the vertical variable x_3, and belonging to S_μ. Then the unique, global solution v to $(\text{NS2D})_{x_3}$ belongs to $\mathcal{A}^0$ and satisfies the following estimate:*

$$\|v\|_{\mathcal{A}^0} \le T_1(\|v_0\|_{\mathcal{B}^0}). \qquad (2.72)$$

Moreover, for all (s, s') in D_μ, we have

$$\forall r \in [1, \infty], \ \|v\|_{\widetilde{L}^r(\mathbb{R}^+; \mathcal{B}^{s+\frac{2}{r}, s'})} \le T_2(\|v_0\|_{S_\mu}). \qquad (2.73)$$

Proof The proof of Proposition 2.19 is done in three steps. First, we deduce from the classical energy estimate for the two-dimensional Navier–Stokes system a stability result in the spaces $L^r(\mathbb{R}^+; H^{s+\frac{2}{r}}(\mathbb{R}^2))^2$ with r in $[2, \infty]$ and s in $]-1, 1[$. This is the purpose of Lemma 2.20, the proof of which uses essentially energy estimates together with paraproduct laws. Then we have to translate the stability result of Lemma 2.20 in terms of Besov spaces with respect to the third variable, seen before simply as a parameter. This is the object of Lemma 2.21, the proof of which relies on the equivalence of two definitions of Besov spaces with regularity index in $]0, 1[$: the first one involving the dyadic decomposition of the frequency space, and the other one consisting of estimating integrals in physical space. Finally, invoking the Gronwall lemma and product laws we conclude the proof of the proposition. □

Step 1: 2D-stability result Let us start by proving the following lemma.

Lemma 2.20 *For any compact set I included in $]-1, 1[$, a constant C exists such that, for any r in $[2, \infty]$ and any s in I, we have for any two solutions v_1 and v_2 of the two-dimensional Navier–Stokes equations*

$$\|v_1 - v_2\|_{L^r(\mathbb{R}^+; H^{s+\frac{2}{r}}(\mathbb{R}^2))} \lesssim \|v_1(0) - v_2(0)\|_{H^s(\mathbb{R}^2)} E_{12}(0), \qquad (2.74)$$

where we define

$$E_{12}(0) \overset{def}{=} \exp C\big(\|v_1(0)\|_{L^2}^2 + \|v_2(0)\|_{L^2}^2\big).$$

Proof Defining $v_{12}(t) \overset{def}{=} v_1(t) - v_2(t)$, we find that

$$\partial_t v_{12} + v_2 \cdot \nabla^h v_{12} - \Delta_h v_{12} = -v_{12} \cdot \nabla^h v_1 - \nabla^h p. \qquad (2.75)$$

Thus, taking the H^s scalar product with v_{12}, we get, thanks to the divergence-free condition,

$$\frac{1}{2}\frac{d}{dt}\|v_{12}(t)\|_{H^s}^2 + \|\nabla^h v_{12}(t)\|_{H^s}^2 = -\big(v_2(t) \cdot \nabla^h v_{12}(t) | v_{12}(t)\big)_{H^s}$$
$$- \big(v_{12}(t) \cdot \nabla^h v_1(t) | v_{12}(t)\big)_{H^s},$$

---

[2] Here $H^{s+\frac{2}{r}}(\mathbb{R}^2)$ denotes the usual homogeneous Sobolev space.

whence, by time integration we get

$$\|v_{12}(t)\|_{H^s}^2 + 2\int_0^t \|\nabla^h v_{12}(t')\|_{H^s}^2 dt'$$

$$= \|v_{12}(0)\|_{H^s}^2 - 2\int_0^t \left(v_2(t')\cdot\nabla^h v_{12}(t')|v_{12}(t')\right)_{H^s} dt'$$

$$- 2\int_0^t \left(v_{12}(t')\cdot\nabla^h v_1(t')|v_{12}(t')\right)_{H^s} dt'.$$

Now making use of the following estimate proved in [12, Lemma 1.1]:

$$\left(v\cdot\nabla^h a|a\right)_{H^s} \lesssim \|\nabla^h v\|_{L^2}\|a\|_{H^s}\|\nabla^h a\|_{H^s}, \qquad (2.76)$$

available uniformly with respect to s in any compact set of $]-2,1[$, we deduce that there is a positive constant C such that for any s in I, we have

$$2\left|\int_0^t \left(v_2(t')\cdot\nabla^h v_{12}(t')|v_{12}(t')\right)_{H^s} dt'\right|$$

$$\leq \frac{1}{2}\int_0^t \|\nabla^h v_{12}(t')\|_{H^s}^2 dt' + \frac{C^2}{2}\int_0^t \|v_{12}(t')\|_{H^s}^2 \|\nabla^h v_2(t')\|_{L^2}^2 dt'. \qquad (2.77)$$

Noticing that

$$\int_0^t \left(v_{12}(t')\cdot\nabla^h v_1(t')|v_{12}(t')\right)_{H^s} dt' \leq \int_0^t \|\nabla^h v_{12}(t')\|_{H^s}\|v_{12}(t')\cdot\nabla^h v_1(t')\|_{H^{s-1}} dt',$$

we deduce by the Cauchy–Schwarz inequality and product laws in Sobolev spaces on $\mathbb{R}^2$ that for s in I,

$$2\left|\int_0^t \left(v_{12}(t')\cdot\nabla^h v_1(t')|v_{12}(t')\right)_{H^s} dt'\right| \qquad (2.78)$$

$$\leq \frac{1}{2}\int_0^t \|\nabla^h v_{12}(t')\|_{H^s}^2 dt' + \frac{C^2}{2}\int_0^t \|v_{12}(t')\|_{H^s}^2 \|\nabla^h v_1(t')\|_{L^2}^2 dt'.$$

Combining (2.77) and (2.78), we get for s in I,

$$\|v_{12}(t)\|_{H^s}^2 + \int_0^t \|\nabla^h v_{12}(t')\|_{H^s}^2 dt'$$

$$\lesssim \|v_{12}(0)\|_{H^s}^2 + \int_0^t \|v_{12}(t')\|_{H^s}^2\left(\|\nabla^h v_1(t')\|_{L^2}^2 + \|\nabla^h v_2(t')\|_{L^2}^2\right) dt'.$$

Gronwall's lemma implies that there exists a positive constant C such that

$$\|v_{12}(t)\|_{H^s}^2 + \int_0^t \|\nabla^h v_{12}(t')\|_{H^s}^2 \, dt'$$

$$\lesssim \|v_{12}(0)\|_{H^s}^2 \exp C \int_0^t \left(\|\nabla^h v_1(t')\|_{L^2}^2 + \|\nabla^h v_2(t')\|_{L^2}^2 \right) dt'.$$

But for any i in $\{1,2\}$, we have by the L^2 energy estimate (2.2)

$$\int_0^t \|\nabla^h v_i(t')\|_{L^2}^2 \, dt' \leq \frac{1}{2} \|v_i(0)\|_{L^2}^2. \tag{2.79}$$

Consequently, for s in I,

$$\|v_{12}(t)\|_{H^s}^2 + \int_0^t \|\nabla^h v_{12}(t')\|_{H^s}^2 \, dt' \lesssim \|v_{12}(0)\|_{H^s}^2 E_{12}(0),$$

which leads to the result by interpolation. $\qquad\square$

Step 2: propagation of vertical regularity Thanks to Lemma 2.20, we can propagate vertical regularity as stated in the following result.

Lemma 2.21 *For any compact set I included in $]-1,1[$, a constant C exists such that, for any r in $[2,\infty]$ and any s in I, we have for any solution v to (NS2D)$_{x_3}$,*

$$\|v\|_{L^r(\mathbb{R}^+;L_v^\infty(H_h^{s+\frac{2}{r}}))} \lesssim \|v_0\|_{B^s} E(0) \quad \text{with} \quad E(0) \stackrel{def}{=} \exp\left(C\|v(0)\|_{L_v^\infty L_h^2}^2 \right).$$

Proof As mentioned above, the proof of Lemma 2.21 uses crucially the characterization of Besov spaces via differences in physical space, namely that for any Banach space X of distributions one has (see for instance Theorem 2.36 of [2])

$$\left\| \left(2^{\frac{j}{2}} \|\Delta_j^v u\|_{L^2(X)} \right)_j \right\|_{\ell^1(\mathbb{Z})} \sim \int_\mathbb{R} \frac{\|u - (\tau_{-z} u)\|_{L_v^2(X)}}{|z|^{\frac{1}{2}}} \frac{dz}{|z|}, \tag{2.80}$$

where the translation operator τ_{-z} is defined by

$$(\tau_{-z} u)(t, x_h, x_3) \stackrel{def}{=} u(t, x_h, x_3 + z).$$

Lemma 2.20 asserts that, for any r in $[2,\infty]$, any s in I and any couple (x_3, z) in $\mathbb{R}^2$, the solution v to (NS2D)$_{x_3}$ satisfies

$$\|v - \tau_{-z} v\|_{Y_r^s} \lesssim \|v_0 - \tau_{-z} v_0\|_{H_h^s} E(0) \quad \text{with} \quad Y_r^s \stackrel{def}{=} L^r(\mathbb{R}^+; H_h^{s+\frac{2}{r}}).$$

Taking the L^2 norm of the above inequality with respect to the x_3 variable and then the L^1 norm with respect to the measure $|z|^{-\frac{3}{2}} dz$ gives

$$\int_{\mathbb{R}} \frac{\|v - \tau_{-z}v\|_{L^2_v(Y^s_r)}}{|z|^{\frac{1}{2}}} \frac{dz}{|z|} \lesssim \int_{\mathbb{R}} \frac{\|v_0 - \tau_{-z}v_0\|_{L^2_v(H^s_h)}}{|z|^{\frac{1}{2}}} \frac{dz}{|z|} E(0). \qquad (2.81)$$

Now, making use of the characterization (2.80) with $X = Y^s_r$, we find that

$$\int_{\mathbb{R}} \frac{\|v - \tau_{-z}v\|_{L^2_v(Y^s_r)}}{|z|^{\frac{1}{2}}} \frac{dz}{|z|} \sim \sum_{j \in \mathbb{Z}} 2^{\frac{j}{2}} \left\| \left\| \left(2^{k(s+\frac{2}{r})} \Delta^v_j \Delta^h_k v(t, \cdot, z) \right)_k \right\|_{L^r(\mathbb{R}^+; \ell^2(\mathbb{Z}; L^2_h))} \right\|_{L^2_v}.$$

Similarly, we have

$$\int_{\mathbb{R}} \frac{\|v_0 - \tau_{-z}v_0\|_{L^2_v(H^s_h)}}{|z|^{\frac{1}{2}}} \frac{dz}{|z|} \sim \sum_{j \in \mathbb{Z}} 2^{\frac{j}{2}} \left\| \left(2^{ks} \| \Delta^v_j \Delta^h_k v_0 \|_{L^2_h} \right)_k \right\|_{\ell^2(\mathbb{Z}; L^2_v)}.$$

Thus, by the embedding from $\ell^1(\mathbb{Z})$ to $\ell^2(\mathbb{Z})$, we get

$$\int_{\mathbb{R}} \frac{\|v_0 - \tau_{-z}v_0\|_{L^2_v(H^s_h)}}{|z|^{\frac{1}{2}}} \frac{dz}{|z|} \lesssim \sum_{(j,k) \in \mathbb{Z}^2} 2^{\frac{j}{2}} 2^{ks} \| \Delta^v_j \Delta^h_k v_0 \|_{L^2(\mathbb{R}^3)}.$$

This implies that Estimate (2.81) can also be written as

$$\sum_{j \in \mathbb{Z}} 2^{\frac{j}{2}} \left\| \left\| \left(2^{k(s+\frac{2}{r})} \Delta^v_j \Delta^h_k v(t, \cdot, z) \right)_k \right\|_{L^r(\mathbb{R}^+; \ell^2(\mathbb{Z}; L^2_h))} \right\|_{L^2_v} \lesssim \|v_0\|_{\mathcal{B}^s} E(0).$$

As $r \geq 2$, Minkowski's inequality implies that

$$\sum_{j \in \mathbb{Z}} 2^{\frac{j}{2}} \left\| \left\| \left(2^{k(s+\frac{2}{r})} \Delta^v_j \Delta^h_k v(t, \cdot) \right)_k \right\|_{\ell^2(\mathbb{Z}; L^2(\mathbb{R}^3))} \right\|_{L^r(\mathbb{R}^+)} \lesssim \|v_0\|_{\mathcal{B}^s} E(0).$$

Bernstein inequalities (2.6) and (2.7) ensure that

$$\| \Delta^v_j \Delta^h_k v(t, \cdot) \|_{L^\infty_v(L^2_h)} \lesssim 2^{\frac{j}{2}} \| \Delta^h_k v(t, \cdot) \|_{L^2(\mathbb{R}^3)},$$

which gives rise to

$$\left\| \left\| \left(2^{k(s+\frac{2}{r})} \| \Delta^h_k v \|_{L^\infty_v(L^2_h)} \right)_k \right\|_{\ell^2(\mathbb{Z})} \right\|_{L^r(\mathbb{R}^+)} \lesssim \|v_0\|_{\mathcal{B}^s} E(0).$$

Permuting the ℓ^2 norm and the L^∞_v norm, thanks to Minkowski's inequality again, achieves the proof of the lemma. $\qquad \square$

Step 3: end of the proof of Proposition 2.19 Our aim is to establish (2.73) for all (s, s') in D_μ. Let us start by proving the following inequality: for any v

solving $(NS2D)_{x_3}$, for any r in $[4,\infty]$, any s in $]-\frac{1}{2},\frac{1}{2}[$ and any positive s',

$$\|v\|_{\widetilde{L}^r(\mathbb{R}^+;\mathcal{B}^{s+\frac{2}{r},s'})} \lesssim \|v_0\|_{\mathcal{B}^{s,s'}} \exp\left(\int_0^\infty C\big(\|v(t)\|^4_{L^\infty_v(L^4_h)} + \|v(t)\|^2_{L^\infty_v(H^1_h)}\big)dt\right).$$
(2.82)

For that purpose, let us introduce, for any non-negative λ, the following notation: for any function F we define

$$F_\lambda(t) \stackrel{\text{def}}{=} F(t)\exp\left(-\lambda\int_0^t \phi(t')dt'\right) \quad \text{with} \quad \phi(t) \stackrel{\text{def}}{=} \|v(t)\|^4_{L^\infty_v(L^4_h)} + \|v(t)\|^2_{L^\infty_v(H^1_h)}.$$

Combining Lemma 2.21 with the Sobolev embedding of $H^{\frac{1}{2}}(\mathbb{R}^2)$ into $L^4(\mathbb{R}^2)$, we find that

$$\int_0^t \phi(t')\,dt' \lesssim E(0)(\|v_0\|^2_{\mathcal{B}^0} + \|v_0\|^4_{\mathcal{B}^0}).$$
(2.83)

Now making use of the Duhamel formula and the action of the heat flow (see for instance Proposition B.2 in [4]), we infer that

$$\|\Delta^v_j\Delta^h_k v_\lambda(t)\|_{L^2} \leq Ce^{-c2^{2k}t}\|\Delta^v_j\Delta^h_k v_0\|_{L^2}$$
$$+ C2^k\int_0^t \exp\left(-c(t-t')2^{2k} - \lambda\int_{t'}^t \phi(t'')dt''\right)\|\Delta^v_j\Delta^h_k(v\otimes v)_\lambda(t')\|_{L^2}dt'.$$
(2.84)

Recall that $(v\otimes v)_\lambda = v\otimes v_\lambda$. Now to estimate the term $\|\Delta^v_j\Delta^h_k(v\otimes v)_\lambda(t')\|_{L^2}$, we make use of the anisotropic version of Bony's paraproduct decomposition (one can consult [2] and [41] for an introduction to anisotropic Littlewood–Paley theory), writing

$$ab = \sum_{\ell=1}^4 T^\ell(a,b) \quad \text{with}$$
$$T^1(a,b) = \sum_{j,k} S^v_j S^h_k a \Delta^v_j \Delta^h_k b,$$
(2.85)
$$T^2(a,b) = \sum_{j,k} S^v_j \Delta^h_k a \Delta^v_j S^h_{k+1} b,$$
$$T^3(a,b) = \sum_{j,k} \Delta^v_j S^h_k a S^v_{j+1} \Delta^h_k b,$$
$$T^4(a,b) = \sum_{j,k} \Delta^v_j \Delta^h_k a S^v_{j+1} S^h_{k+1} b.$$

In light of the Bernstein inequality (2.6), we have

$$\|\Delta^v_j\Delta^h_k T^1(v(t),v_\lambda(t))\|_{L^2} \lesssim 2^{\frac{k}{2}}\|\Delta^v_j\Delta^h_k T^1(v(t),v_\lambda(t))\|_{L^2_v(L^{4/3}_h)},$$

which, in view of (2.85), Hölder's inequalities and the action of the horizontal and vertical truncations on Lebesgue spaces, ensures the existence of some fixed nonzero integer N_0 such that

$$\|\Delta_j^v \Delta_k^h T^1(v(t), v_\lambda(t))\|_{L^2} \lesssim 2^{\frac{k}{2}} \sum_{\substack{j' \geq j - N_0 \\ k' \geq k - N_0}} \|S_{j'}^v S_{k'}^h v(t)\|_{L^\infty_v(L^4_h)} \|\Delta_{j'}^v \Delta_{k'}^h v_\lambda(t)\|_{L^2}$$

$$\lesssim 2^{\frac{k}{2}} \|v(t)\|_{L^\infty_v(L^4_h)} \sum_{\substack{j' \geq j - N_0 \\ k' \geq k - N_0}} \|\Delta_{j'}^v \Delta_{k'}^h v_\lambda(t)\|_{L^2}.$$

According to the definition of $\widetilde{L}^4(\mathbb{R}^+; \mathcal{B}^{s+\frac{1}{2}, s'})$, we get

$$2^{js'} 2^{ks} \|\Delta_j^v \Delta_k^h T^1(v(t), v_\lambda(t))\|_{L^2}$$
$$\lesssim \|v_\lambda\|_{\widetilde{L}^4(\mathbb{R}^+; \mathcal{B}^{s+\frac{1}{2}, s'})} \|v(t)\|_{L^\infty_v(L^4_h)} \sum_{\substack{j' \geq j - N_0 \\ k' \geq k - N_0}} 2^{-(j'-j)s'} 2^{-(k'-k)(s+\frac{1}{2})} \widetilde{f}_{j',k'}(t),$$

where $\widetilde{f}_{j',k'}(t)$, defined by

$$\widetilde{f}_{j',k'}(t) \stackrel{\text{def}}{=} \|v_\lambda\|_{\widetilde{L}^4(\mathbb{R}^+; \mathcal{B}^{s+\frac{1}{2}, s'})}^{-1} 2^{k'(s+\frac{1}{2})} 2^{j's'} \|\Delta_{j'}^v \Delta_{k'}^h v_\lambda(t)\|_{L^2},$$

is on the sphere of $\ell^1(\mathbb{Z}^2; L^4(\mathbb{R}^+))$.

Since $s > -1/2$ and $s' > 0$, it follows by Young's inequality on series that

$$2^{js'} 2^{ks} \|\Delta_j^v \Delta_k^h T^1(v(t), v_\lambda(t))\|_{L^2} \lesssim \|v_\lambda\|_{\widetilde{L}^4(\mathbb{R}^+; \mathcal{B}^{s+\frac{1}{2}, s'})} \|v(t)\|_{L^\infty_v(L^4_h)} f_{j,k}(t),$$

where $f_{j,k}(t)$ is on the sphere of $\ell^1(\mathbb{Z}^2; L^4(\mathbb{R}^+))$.

As by definition $\phi(t)$ is greater than $\|v(t)\|_{L^\infty_v(L^4_h)}^4$, we infer that

$$\mathcal{T}_{j,k,\lambda}^1(t) \stackrel{\text{def}}{=} 2^k 2^{js'} 2^{ks} \int_0^t \exp\left(-c(t-t') 2^{2k} - \lambda \int_{t'}^t \phi(t'') dt''\right)$$
$$\times \|\Delta_j^v \Delta_k^h T^1(v(t), v_\lambda(t'))\|_{L^2} dt'$$
$$\lesssim \|v_\lambda\|_{\widetilde{L}^4(\mathbb{R}^+; \mathcal{B}^{s+1/2, s'})}$$
$$\times 2^k \int_0^t \exp\left(-c(t-t') 2^{2k} - \lambda \int_{t'}^t \phi(t'') dt''\right) \phi^{\frac{1}{4}}(t') f_{j,k}(t') dt'. \tag{2.86}$$

By Hölder's inequality, this leads to

$$
T^1_{j,k,\lambda}(t) \lesssim \|v_\lambda\|_{\widetilde{L}^4(\mathbb{R}^+;\mathcal{B}^{s+\frac{1}{2},s'})} \left(\int_0^t e^{-c(t-t')2^{2k}} f^4_{j,k}(t')\mathrm{d}t' \right)^{\frac{1}{4}}
$$
$$
\times\, 2^k \left(\int_0^t \exp\left(-c(t-t')2^{2k} - \frac{4}{3}\lambda \int_{t'}^t \phi(t'')\mathrm{d}t'' \right) \phi(t')^{\frac{1}{3}}\mathrm{d}t' \right)^{\frac{3}{4}}.
$$

Finally, applying Hölder's inequality in the last term of the above inequality, we get

$$
T^1_{j,k,\lambda}(t) \lesssim \frac{1}{\lambda^{\frac{1}{4}}} \left(\int_0^t e^{-c(t-t')2^{2k}} f^4_{j,k}(t')\mathrm{d}t' \right)^{\frac{1}{4}} \|v_\lambda\|_{\widetilde{L}^4(\mathbb{R}^+;\mathcal{B}^{s+1/2,s'})}. \tag{2.87}
$$

Now let us study the term with T^2. Using again that the support of the Fourier transform of the product of two functions is included in the sum of the two supports, let us write that

$$
\|\Delta^v_j \Delta^h_k T^2(v(t),v_\lambda(t))\|_{L^2} \lesssim \sum_{\substack{j'\geq j-N_0 \\ k'\geq k-N_0}} \|S^v_{j'}\Delta^h_{k'}v(t)\|_{L^\infty_v(L^2_h)} \|\Delta^v_{j'}S^h_{k'+1}v_\lambda(t)\|_{L^2_v(L^\infty_h)}.
$$

Combining the Bernstein inequality (2.6) with the definition of the function ϕ, we get

$$
\|S^v_{j'}\Delta^h_{k'}v(t)\|_{L^\infty_v(L^2_h)} \lesssim 2^{-k'}\|v(t)\|_{L^\infty_v(H^1_h)} \lesssim 2^{-k'}\phi^{\frac{1}{2}}(t). \tag{2.88}
$$

Now let us observe that, using the Bernstein inequality again, we have

$$
\|\Delta^v_{j'}S^h_{k'+1}v_\lambda(t)\|_{L^2_v(L^\infty_h)} \lesssim \sum_{k''\leq k'} \|\Delta^v_{j'}\Delta^h_{k''}v_\lambda(t)\|_{L^2_v(L^\infty_h)}
$$
$$
\lesssim \sum_{k''\leq k'} 2^{k''} \|\Delta^v_{j'}\Delta^h_{k''}v_\lambda(t)\|_{L^2}.
$$

By definition of the $\widetilde{L}^4(\mathbb{R}^+;\mathcal{B}^{s+\frac{1}{2},s'})$ norm, we have

$$
2^{j's'}2^{k'(s-\frac{1}{2})} \|\Delta^v_{j'}S^h_{k'+1}v_\lambda(t)\|_{L^2_v(L^\infty_h)} \lesssim \|v_\lambda\|_{\widetilde{L}^4(\mathbb{R}^+;\mathcal{B}^{s+\frac{1}{2},s'})} \sum_{k''\leq k'} 2^{(k'-k'')(s-\frac{1}{2})} f_{j',k''}(t),
$$

where $f_{j',k''}(t)$, on the sphere of $\ell^1(\mathbb{Z}^2;L^4(\mathbb{R}^+))$, is defined by

$$
f_{j',k''}(t) \stackrel{\text{def}}{=} \|v_\lambda\|^{-1}_{\widetilde{L}^4(\mathbb{R}^+;\mathcal{B}^{s+1/2,s'})} 2^{j's'}2^{k''(s+1/2)} \|\Delta^v_{j'}\Delta^h_{k''}v_\lambda(t)\|_{L^2}.
$$

Since $s < \frac{1}{2}$, this ensures by Young's inequality that

$$
\|\Delta^v_{j'}S^h_{k'+1}v_\lambda(t)\|_{L^2_v(L^\infty_h)} \lesssim 2^{-j's'}2^{-k'(s-\frac{1}{2})} \|v_\lambda\|_{\widetilde{L}^4(\mathbb{R}^+;\mathcal{B}^{s+1/2,s'})} \widetilde{f}_{j',k'}(t),
$$

where $\widetilde{f}_{j',k'}(t)$ is on the sphere of $\ell^1(\mathbb{Z}^2; L^4(\mathbb{R}^+))$. Together with Inequality (2.88), this gives

$$2^{js'}2^{k(s+\frac{1}{2})}\|\Delta_j^{\mathrm{v}}\Delta_k^{\mathrm{h}}T^2(v(t),v_\lambda(t))\|_{L^2} \lesssim \phi(t)^{\frac{1}{2}}\|v_\lambda\|_{\widetilde{L}^4(\mathbb{R}^+;\mathcal{B}^{s+1/2,s'})}f_{j,k}(t),$$

where $f_{j,k}(t)$ is on the sphere of $\ell^1(\mathbb{Z}^2; L^4(\mathbb{R}^+))$. We deduce that

$$\begin{aligned}
T_{j,k,\lambda}^2(t) &\stackrel{\text{def}}{=} 2^k 2^{js'} 2^{ks} \int_0^t \exp\left(-c(t-t')2^{2k} - \lambda \int_{t'}^t \phi(t'')\mathrm{d}t''\right) \\
&\quad \times \|\Delta_j^{\mathrm{v}}\Delta_k^{\mathrm{h}}T^2(v(t'),v_\lambda(t'))\|_{L^2}\,\mathrm{d}t' \\
&\lesssim \|v_\lambda\|_{\widetilde{L}^4(\mathbb{R}^+;\mathcal{B}^{s+1/2,s'})} \\
&\quad \times 2^{\frac{k}{2}} \int_0^t \exp\left(-c(t-t')2^{2k} - \lambda \int_{t'}^t \phi(t'')\mathrm{d}t''\right)\phi(t')^{\frac{1}{2}}f_{j,k}(t')\mathrm{d}t'.
\end{aligned}$$
(2.89)

Using Hölder's inequality twice, we get

$$\begin{aligned}
T_{j,k,\lambda}^2(t) &\lesssim \|v_\lambda\|_{\widetilde{L}^4(\mathbb{R}^+;\mathcal{B}^{s+1/2,s'})}\left(\int_0^t e^{-c(t-t')2^{2k}}f_{j,k}^4(t')\mathrm{d}t'\right)^{\frac{1}{4}} \\
&\quad \times 2^{\frac{k}{2}}\left(\int_0^t \exp\left(-c(t-t')2^{2k} - \lambda \int_{t'}^t \phi(t'')\mathrm{d}t''\right)\phi(t')^{\frac{2}{3}}\mathrm{d}t'\right)^{\frac{3}{4}} \\
&\lesssim \frac{1}{\lambda^{\frac{1}{2}}}\|v_\lambda\|_{\widetilde{L}^4(\mathbb{R}^+;\mathcal{B}^{s+1/2,s'})}\left(\int_0^t e^{-c(t-t')2^{2k}}f_{j,k}^4(t')\mathrm{d}t'\right)^{\frac{1}{4}}.
\end{aligned}$$
(2.90)

As T^3 is estimated like T^1 and T^4 is estimated like T^2, this implies finally that

$$\begin{aligned}
2^{js'}2^{ks}\|\Delta_j^{\mathrm{v}}\Delta_k^{\mathrm{h}}v_\lambda(t)\|_{L^2} &\lesssim 2^{js'}2^{ks}e^{-c2^{2k}t}\|\Delta_j^{\mathrm{v}}\Delta_k^{\mathrm{h}}v_0\|_{L^2} \\
&\quad + \left(\int_0^t e^{-c(t-t')2^{2k}}f_{j,k}^4(t')\mathrm{d}t'\right)^{\frac{1}{4}}\left(\frac{1}{\lambda^{\frac{1}{4}}} + \frac{1}{\lambda^{\frac{1}{2}}}\right)\|v_\lambda\|_{\widetilde{L}^4(\mathbb{R}^+;\mathcal{B}^{s+1/2,s'})}.
\end{aligned}$$

As we have

$$\left(\int_0^\infty \left(\int_0^t e^{-c(t-t')2^{2k}}f_{j,k}^4(t')\mathrm{d}t'\right)^{\frac{1}{4}\times 4}\mathrm{d}t\right)^{\frac{1}{4}} = c^{-1}d_{j,k}2^{-\frac{k}{2}}$$

$$\text{and } \sup_{t\in\mathbb{R}^+}\left(\int_0^t e^{-c(t-t')2^{2k}}f_{j,k}^4(t')\mathrm{d}t'\right)^{\frac{1}{4}} = d_{j,k}, \quad \text{with } d_{j,k}\in\ell^1(\mathbb{Z}^2),$$

we infer that

$$\begin{aligned}
2^{js'}2^{ks}\left(\|\Delta_j^{\mathrm{v}}\Delta_k^{\mathrm{h}}v_\lambda\|_{L^\infty(\mathbb{R}^+;L^2)} + 2^{\frac{k}{2}}\|\Delta_j^{\mathrm{v}}\Delta_k^{\mathrm{h}}v_\lambda\|_{L^4(\mathbb{R}^+;L^2)}\right) \\
\lesssim 2^{js'}2^{ks}\|\Delta_j^{\mathrm{v}}\Delta_k^{\mathrm{h}}v_0\|_{L^2} + d_{j,k}\left(\frac{1}{\lambda^{\frac{1}{4}}} + \frac{1}{\lambda^{\frac{1}{2}}}\right)\|v_\lambda\|_{\widetilde{L}^4(\mathbb{R}^+;\mathcal{B}^{s+1/2,s'})}.
\end{aligned}$$

This ends the proof of (2.82) by taking the sum over j and k and choosing λ large enough.

Now to show that Estimate (2.82) remains available for $r = 2$, we start from Formula (2.84) with $\lambda = 0$. Applying again anisotropic paraproduct decomposition, we find by arguments similar to those conducted above

$$2^{js'}2^{k(s+1)}\|\Delta_j^{\mathrm{v}}\Delta_k^{\mathrm{h}}v(t)\|_{L^2} \lesssim 2^{js'}2^{k(s+1)}e^{-c2^{2k}t}\|\Delta_j^{\mathrm{v}}\Delta_k^{\mathrm{h}}v_0\|_{L^2}$$

$$+2^{2k}\|v\|_{\widetilde{L}^4(\mathbb{R}^+;\mathcal{B}^{s+\frac{1}{2},s'})}\int_0^t e^{-c(t-t')2^{2k}}\left((g_{j,k}(t')+2^{-\frac{k}{2}}h_{j,k}(t'))\right)dt',$$

where $g_{j,k}$ (resp. $h_{j,k}$) are in $\ell^1(\mathbb{Z}^2;L^2(\mathbb{R}^+))$ (resp. $\ell^1(\mathbb{Z}^2;L^{\frac{4}{3}}(\mathbb{R}^+))$), with

$$\sum_{(j,k)\in\mathbb{Z}^2}\|g_{j,k}\|_{L^2(\mathbb{R}^+)} \lesssim \|\phi\|_{L^1}^{\frac{1}{4}} \quad \text{and} \quad \sum_{(j,k)\in\mathbb{Z}^2}\|h_{j,k}\|_{L^{\frac{4}{3}}(\mathbb{R}^+)} \lesssim \|\phi\|_{L^1}^{\frac{1}{2}}.$$

Laws of convolution in the time variable, summation over j and k and (2.82) imply that

$$\|v\|_{\widetilde{L}^2(\mathbb{R}^+;\mathcal{B}^{s+1,s'})} \lesssim \|v_0\|_{\mathcal{B}^{s,s'}}\exp\left(C\int_0^\infty\phi(t)dt\right).$$

This implies by interpolation in view of (2.82) that for all r in $[2,\infty]$, all s in $]-\frac{1}{2},\frac{1}{2}[$ and all positive s'

$$\|v\|_{\widetilde{L}^r(\mathbb{R}^+;\mathcal{B}^{s+2/r,s'})} \lesssim \|v_0\|_{\mathcal{B}^{s,s'}}\exp\left(C\int_0^\infty\phi(t)dt\right), \qquad (2.91)$$

which in view of (2.83) ensures Inequality (2.72) and achieves the proof of Estimate (2.73) in the case when s belongs to $]-\frac{1}{2},\frac{1}{2}[$.

To conclude the proof of the proposition, it remains to complete the range of indices. Let us first double the interval on the index s, by proving that for any s in $]-1,1[$, any $s' \geq 1/2$ and any r in $[2,\infty]$ we have

$$\|v\|_{\widetilde{L}^r(\mathbb{R}^+;\mathcal{B}^{s+2/r,s'})} \lesssim \|v_0\|_{\mathcal{B}^{s,s'}} + \|v_0\|_{\mathcal{B}^{s/2,s'}}\|v_0\|_{\mathcal{B}^{s/2}}\exp(C\|v_0\|_{\mathcal{B}^0}E_0). \quad (2.92)$$

Anisotropic product laws (see for instance Appendix B in [4]) ensure that for any s in $]-1,1[$ and any $s' \geq 1/2$, we have

$$\|v(t)\otimes v(t)\|_{\mathcal{B}^{s,s'}} \lesssim \|v(t)\|_{\mathcal{B}^{\frac{s+1}{2},s'}}\|v(t)\|_{\mathcal{B}^{\frac{s+1}{2},s'}}.$$

According to Formula (2.84) and the smoothing effect of the horizontal heat flow, we find that, for any s belonging to $]-1,1[$, any $s' \geq 1/2$ and any r in $[2,\infty]$,

$$\|v\|_{\widetilde{L}^r(\mathbb{R}^+;\mathcal{B}^{s+2/r,s'})} \lesssim \|v_0\|_{\mathcal{B}^{s,s'}} + \|v\otimes v\|_{\widetilde{L}^2(\mathbb{R}^+;\mathcal{B}^{s,s'})}$$

$$\lesssim \|v_0\|_{\mathcal{B}^{s,s'}} + \|v\|_{\widetilde{L}^4(\mathbb{R}^+;\mathcal{B}^{\frac{s+1}{2},s'})}\|v\|_{\widetilde{L}^4(\mathbb{R}^+;\mathcal{B}^{\frac{s+1}{2},s'})}.$$

Finally, Inequality (2.82) ensures that for any s in $]-1,1[$, any $s' \geq 1/2$ and any r in $[2,\infty]$,

$$\|v\|_{\widetilde{L}^r(\mathbb{R}^+;\mathcal{B}^{s+2/r,s'})} \lesssim \|v_0\|_{\mathcal{B}^{s,s'}} + \|v_0\|_{\mathcal{B}^{s/2}}\|v_0\|_{\mathcal{B}^{s/2,s'}} \exp(C\|v_0\|_{\mathcal{B}^0}E(0)), \quad (2.93)$$

which ends the proof of Inequality (2.92), and thus for (2.73) when r is in $[2,\infty]$ and s is in $]-1,1[$.

Let us now treat the case when s belongs to $]-2,0]$ and $s' \geq 1/2$. Again by anisotropic product laws, we have

$$\|v(t) \otimes v(t)\|_{\mathcal{B}^{s+1,s'}} \lesssim \|v(t)\|_{\mathcal{B}^{s/2+1}}\|v(t)\|_{\mathcal{B}^{s/2+1,s'}},$$

which implies that

$$\|v \otimes v\|_{L^1(\mathbb{R}^+;\mathcal{B}^{s+1,s'})} \lesssim \|v\|_{L^2(\mathbb{R}^+;\mathcal{B}^{s/2+1})}\|v\|_{L^2(\mathbb{R}^+;\mathcal{B}^{s/2+1,s'})}.$$

The smoothing effect of the heat flow gives then, for any r in $[1,\infty]$, any s in $]-2,0]$ and any $s' \geq 1/2$,

$$\|v\|_{\widetilde{L}^r(\mathbb{R}^+;\mathcal{B}^{s+2/r,s'})} \lesssim \|v_0\|_{\mathcal{B}^{s,s'}} + \|v\|_{L^2(\mathbb{R}^+;\mathcal{B}^{s/2+1})}\|v\|_{L^2(\mathbb{R}^+;\mathcal{B}^{s/2+1,s'})}.$$

Inequality (2.93) implies that, for any r in $[1,\infty]$, any s in $]-2,0]$ and any $s' \geq 1/2$,

$$\|v\|_{\widetilde{L}^r(\mathbb{R}^+;\mathcal{B}^{s+2/r,s'})} \lesssim \|v_0\|_{\mathcal{B}^{s,s'}} + \|v_0\|_{\mathcal{B}^{s/2}}\big(\|v_0\|_{\mathcal{B}^{s/2,s'}} + \|v_0\|_{\mathcal{B}^{s/4}}\|v_0\|_{\mathcal{B}^{s/4,s'}}$$
$$\exp(C\|v_0\|_{\mathcal{B}^0}E_0)\big)$$
$$+ \|v_0\|^2_{\mathcal{B}^{s/4}}\big(\|v_0\|_{\mathcal{B}^{s/2,s'}} + \|v_0\|_{\mathcal{B}^{s/4}}\|v_0\|_{\mathcal{B}^{s/4,s'}}$$
$$\exp(C\|v_0\|_{\mathcal{B}^0}E_0)\big).$$

This concludes the proof of Estimate (2.73), and thus achieves the proof of Proposition 2.19. $\qquad\square$

Propagation of Regularity by a Transport-Diffusion Equation

Now let us estimate the norm of the function w^3 defined as the solution of (T_β) defined in Subsection 2.4.1. This is described in the following proposition.

Proposition 2.22 *Let v_0 and v be as in Proposition 2.19. For any non-negative real number β, let us consider w^3, the solution of*

$$(T_\beta) \quad \begin{cases} \partial_t w^3 + v \cdot \nabla^h w^3 - \Delta_h w^3 - \beta^2 \partial_3^2 w^3 = 0 & in \ \mathbb{R}^+ \times \mathbb{R}^3, \\ w^3_{|t=0} = w^3_0. \end{cases}$$

Then w^3 satisfies the following estimates, where all the constants are independent of β:

$$\|w^3\|_{\mathcal{A}^0} \lesssim \|w^3_0\|_{\mathcal{B}^0} \exp\big(\mathcal{T}_1(\|v_0\|_{\mathcal{B}^0})\big), \quad (2.94)$$

and for any s in $[-2+\mu, 0]$ and any $s' \geq 1/2$, we have

$$\|w^3\|_{\mathcal{A}^{s,s'}} \lesssim \left(\|w_0^3\|_{\mathcal{B}^{s,s'}} + \|w_0^3\|_{\mathcal{B}^0} T_2(\|v_0\|_{S_\mu})\right) \exp\left(T_1(\|v_0\|_{\mathcal{B}^0})\right). \qquad (2.95)$$

Proof Proposition 2.22 follows easily from the following lemma which is a general result about the propagation of anisotropic regularity by a transport-diffusion equation.

Lemma 2.23 *Let us consider (s, s'), a couple of real numbers, and $\mathcal{Q}$, a bilinear operator which continuously maps $\mathcal{B}^1 \times \mathcal{B}^{s+1,s'}$ into $\mathcal{B}^{s,s'}$. A constant C exists such that for any two-component vector field v in $L^2(\mathbb{R}^+; \mathcal{B}^1)$, any f in $L^1(\mathbb{R}^+; \mathcal{B}^{s,s'})$, any a_0 in $\mathcal{B}^{s,s'}$ and for any non-negative β, if $\Delta_\beta \overset{def}{=} \Delta_h + \beta^2 \partial_z^2$ and a is the solution of*

$$\partial_t a - \Delta_\beta a + \mathcal{Q}(v, a) = f \quad and \quad a_{|t=0} = a_0,$$

then a satisfies

$$\forall r \in [1, \infty], \ \|a\|_{\widetilde{L}^r(\mathbb{R}^+; \mathcal{B}^{s+2/r,s'})} \leq C\left(\|a_0\|_{\mathcal{B}^{s,s'}} + \|f\|_{L^1(\mathbb{R}^+; \mathcal{B}^{s,s'})}\right)$$
$$\exp\left(C \int_0^\infty \|v(t)\|_{\mathcal{B}^1}^2 \mathrm{d}t\right).$$

The fact that the third index of the Besov spaces is one induces some technical difficulties which lead us to work first on subintervals I of $\mathbb{R}^+$ on which $\|v\|_{L^2(I; \mathcal{B}^1)}$ is small.

Let us then start by considering any subinterval $I = [\tau_0, \tau_1]$ of $\mathbb{R}^+$. The Duhamel formula and the smoothing effect of the heat flow imply that

$$\|\Delta_k^h \Delta_j^v a(t)\|_{L^2} \leq e^{-c2^{2k}(t-\tau_0)} \|\Delta_k^h \Delta_j^v a(\tau_0)\|_{L^2}$$
$$+ C \int_{\tau_0}^t e^{-c2^{2k}(t-t')} \|\Delta_k^h \Delta_j^v \left(\mathcal{Q}(v(t'), a(t')) + f(t')\right)\|_{L^2} \mathrm{d}t'.$$

After multiplication by $2^{ks+js'}$ and using Young's inequality in the time integral, we deduce that

$$2^{ks+js'} \left(\|\Delta_k^h \Delta_j^v a\|_{L^\infty(I; L^2)} + 2^{2k} \|\Delta_k^h \Delta_j^v a\|_{L^1(I; L^2)}\right) \leq C 2^{ks+js'} \|\Delta_k^h \Delta_j^v a(\tau_0)\|_{L^2}$$
$$+ C \int_I d_{k,j}(t') \left(\|v(t')\|_{\mathcal{B}^1} \|a(t')\|_{\mathcal{B}^{s+1,s'}} + \|f(t')\|_{\mathcal{B}^{s,s'}}\right) \mathrm{d}t',$$

where for any t, $d_{k,j}(t)$ is an element of the sphere of $\ell^1(\mathbb{Z}^2)$. By summation over (k, j) and using the Cauchy–Schwarz inequality, we infer that

$$\|a\|_{\widetilde{L}^\infty(I; \mathcal{B}^{s,s'})} + \|a\|_{L^1(I; \mathcal{B}^{s+2,s'})} \leq C\|a(\tau_0)\|_{\mathcal{B}^{s,s'}} + C\|f\|_{L^1(I; \mathcal{B}^{s,s'})}$$
$$+ C\|v\|_{L^2(I; \mathcal{B}^1)} \|a\|_{L^2(I; \mathcal{B}^{s+1,s'})}. \qquad (2.96)$$

Let us define an increasing sequence $(T_m)_{0\leq m\leq M+1}$ by induction such that $T_0 = 0$, $T_{M+1} = \infty$ and

$$\forall m < M, \quad \int_{T_m}^{T_{m+1}} \|v(t)\|_{\mathcal{B}^1}^2 \, dt = c_0 \quad \text{and} \quad \int_{T_M}^{\infty} \|v(t)\|_{\mathcal{B}^1}^2 \, dt \leq c_0,$$

for some given c_0 which will be chosen later on. Obviously, we have

$$\int_0^{\infty} \|v(t)\|_{\mathcal{B}^1}^2 \, dt \geq \int_0^{T_M} \|v(t)\|_{\mathcal{B}^1}^2 \, dt = Mc_0. \tag{2.97}$$

Thus the number M of T_m's such that T_m is finite is less than $c_0^{-1} \|v\|_{L^2(\mathbb{R}^+;\mathcal{B}^1)}^2$. Applying Estimate (2.96) to the interval $[T_m, T_{m+1}]$, we get

$$\|a\|_{L^{\infty}([T_m,T_{m+1}];\mathcal{B}^{s,s'})} + \|a\|_{L^1([T_m,T_{m+1}];\mathcal{B}^{s+2,s'})}$$
$$\leq \|a\|_{L^2([T_m,T_{m+1}];\mathcal{B}^{s+1,s'})} + C\big(\|a(T_m)\|_{\mathcal{B}^{s,s'}} + C\|f\|_{L^1([T_m,T_{m+1}];\mathcal{B}^{s,s'})}\big),$$

if c_0 is chosen such that $C\sqrt{c_0} \leq 1$.

Since

$$\|a\|_{L^2([T_m,T_{m+1}];\mathcal{B}^{s+1,s'})} \leq \|a\|_{L^{\infty}([T_m,T_{m+1}];\mathcal{B}^{s,s'})}^{\frac{1}{2}} \|a\|_{L^1([T_m,T_{m+1}];\mathcal{B}^{s+2,s'})}^{\frac{1}{2}},$$

we infer that

$$\|a\|_{L^{\infty}([T_m,T_{m+1}];\mathcal{B}^{s,s'})} + \|a\|_{L^1([T_m,T_{m+1}];\mathcal{B}^{s+2,s'})}$$
$$\leq 2C\big(\|a(T_m)\|_{\mathcal{B}^{s,s'}} + \|f\|_{L^1([T_m,T_{m+1}];\mathcal{B}^{s,s'})}\big). \tag{2.98}$$

Now let us us prove by induction that

$$\|a\|_{L^{\infty}([0,T_m];\mathcal{B}^{s,s'})} \leq (2C)^m\big(\|a_0\|_{\mathcal{B}^{s,s'}} + \|f\|_{L^1([0,T_m],\mathcal{B}^{s,s'})}\big).$$

Using (2.98) and the induction hypothesis we get

$$\|a\|_{L^{\infty}([T_m,T_{m+1}];\mathcal{B}^{s,s'})} \leq 2C\big(\|a\|_{L^{\infty}([0,T_m];\mathcal{B}^{s,s'})} + \|f\|_{L^1([T_m,T_{m+1}];\mathcal{B}^{s,s'})}\big)$$
$$\leq (2C)^{m+1}\big(\|a_0\|_{\mathcal{B}^{s,s'}} + \|f\|_{L^1([0,T_{m+1}],\mathcal{B}^{s,s'})}\big),$$

provided that $2C \geq 1$, which ensures in view of (2.97) that

$$\|a\|_{L^{\infty}(\mathbb{R}^+;\mathcal{B}^{s,s'})} \leq C\big(\|a_0\|_{\mathcal{B}^{s,s'}} + \|f\|_{L^1(\mathbb{R}^+;\mathcal{B}^{s,s'})}\big)\exp\Big(C\int_0^{\infty} \|v(t)\|_{\mathcal{B}^1}^2 \, dt\Big).$$

We deduce from (2.98) that

$$\|a\|_{L^1([T_m,T_{m+1}];\mathcal{B}^{s+2,s'})} \leq C\big(\|a_0\|_{\mathcal{B}^{s,s'}} + \|f\|_{L^1(\mathbb{R}^+;\mathcal{B}^{s,s'})}\big)\exp\Big(C\int_0^{\infty} \|v(t)\|_{\mathcal{B}^1}^2 \, dt\Big)$$
$$+ C\|f\|_{L^1([T_m,T_{m+1}];\mathcal{B}^{s,s'})}.$$

Once we have noticed that $xe^{Cx^2} \leq e^{C'x^2}$, the result comes by summation over m and the fact that the total number of ms is less than or equal to $c_0^{-1}\|v\|_{L^2(\mathbb{R}^+;\mathcal{B}^1)}^2$. This ends the proof of the lemma and thus of Proposition 2.22. $\quad\square$

As w^h is defined by $w^h = -\nabla_h \Delta_h^{-1} \partial_3 w^3$, we deduce from Proposition 2.22 the following corollary.

Corollary 2.24 *For any s in $[-2+\mu, 0]$ and any $s' \geq 1/2$,*

$$\|w^h\|_{\mathcal{A}^{s+1,s'-1}} \lesssim \left(\|w_0^3\|_{\mathcal{B}^{s,s'}} + \|w_0^3\|_{\mathcal{B}^0} \mathcal{T}_2(\|v_0\|_{S_\mu})\right) \exp\left(\mathcal{T}_1(\|v_0\|_{\mathcal{B}^0})\right).$$

Conclusion of the Proof of Theorem 2.14

Using the definition of the approximate solution Φ^{app} given in (2.71), we infer from Propositions 2.19 and 2.22 and Corollary 2.24 that

$$\|\Phi^{\text{app}}\|_{L^2(\mathbb{R}^+;\mathcal{B}^1)} \leq \mathcal{T}_1(\|(v_0, w_0^3)\|_{\mathcal{B}^0}) + \beta\mathcal{T}_2(\|(v_0, w_0^3)\|_{S_\mu}). \tag{2.99}$$

Moreover, the error term ψ satisfies the following modified Navier–Stokes system, with null Cauchy data:

$$\partial_t \psi + \text{div}\left(\psi \otimes \psi + \Phi^{\text{app}} \otimes \psi + \psi \otimes \Phi^{\text{app}}\right) - \Delta\psi = -\nabla q_\beta + \sum_{\ell=1}^4 E_\beta^\ell \quad \text{with}$$

$$E_\beta^1 \stackrel{\text{def}}{=} \partial_3^2[(v,0)]_\beta + \beta(0, [\partial_3 p]_\beta),$$

$$E_\beta^2 \stackrel{\text{def}}{=} \beta\left[\left(w^3 \partial_3(v, w^3) + (\nabla^h \Delta_h^{-1} \text{div}_h \partial_3(vw^3), 0)\right)\right]_\beta,$$

$$E_\beta^3 \stackrel{\text{def}}{=} \beta\left[\left(w^h \cdot \nabla_h(v, w^3) + v \cdot \nabla^h(w^h, 0)\right)\right]_\beta \quad \text{and}$$

$$E_\beta^4 \stackrel{\text{def}}{=} \beta^2\left[\left(w^h \cdot \nabla^h(w^h, 0) + w^3 \partial_3(w^h, 0)\right)\right]_\beta. \tag{2.100}$$

If we prove that

$$\left\|\sum_{\ell=1}^4 E_\beta^\ell\right\|_{\mathcal{F}^0} \leq \beta\mathcal{T}_2(\|(v_0, w_0^3)\|_{S_\mu}), \tag{2.101}$$

then according to the fact that $\psi_{|t=0} = 0$, ψ exists globally and satisfies

$$\|\psi\|_{L^2(\mathbb{R}^+;\mathcal{B}^1)} \lesssim \beta\mathcal{T}_2(\|(v_0, w_0^3)\|_{S_\mu}). \tag{2.102}$$

This in turn implies that Φ_0 generates a global regular solution Φ_β in $L^2(\mathbb{R}^+;\mathcal{B}^1)$ which satisfies

$$\|\Phi_\beta\|_{L^2(\mathbb{R}^+;\mathcal{B}^1)} \leq \mathcal{T}_1(\|(v_0, w_0^3)\|_{\mathcal{B}^0}) + \beta\mathcal{T}_2(\|(v_0, w_0^3)\|_{S_\mu}). \tag{2.103}$$

Once this bound in $L^2(\mathbb{R}^+; \mathcal{B}^1)$ is obtained, the bound in $\mathcal{A}^0$ follows by heat flow estimates, and in $\mathcal{A}^{s,s'}$ by propagation of regularity for the Navier–Stokes system.

So all we need to do is to prove Inequality (2.101). Let us first estimate the term $\partial_3^2[(v,0)]_\beta$. This requires the use of some $\widetilde{L}^2(\mathbb{R}^+; \mathcal{B}^{s,s'})$ norms. Clearly, we have

$$\left\| \partial_3^2[v]_\beta \right\|_{\widetilde{L}^2(\mathbb{R}^+; \mathcal{B}^{0,-\frac{1}{2}})} \lesssim \left\| [v]_\beta \right\|_{\widetilde{L}^2(\mathbb{R}^+; \mathcal{B}^{0,\frac{3}{2}})},$$

which implies in view of the vertical scaling property (2.12) of the space $\mathcal{B}^{0,\frac{3}{2}}$

$$\left\| \partial_3^2[v]_\beta \right\|_{\widetilde{L}^2(\mathbb{R}^+; \mathcal{B}^{0,-\frac{1}{2}})} \lesssim \beta \left\| v \right\|_{\widetilde{L}^2(\mathbb{R}^+; \mathcal{B}^{0,\frac{3}{2}})}.$$

Therefore Proposition 2.19 ensures that

$$\left\| \partial_3^2[v]_\beta \right\|_{\widetilde{L}^2(\mathbb{R}^+; \mathcal{B}^{0,-\frac{1}{2}})} \leq \beta \, \mathcal{T}_2(\|v_0\|_{S_\mu}). \tag{2.104}$$

Now let us study the pressure term. By applying the horizontal divergence to the equation satisfied by v we get, thanks to the fact that $\mathrm{div}_\mathrm{h} v = 0$,

$$\partial_3 p = -\partial_3 \Delta_\mathrm{h}^{-1} \sum_{\ell,m=1}^{2} \partial_\ell \partial_m (v^\ell v^m).$$

Since ℓ and m belong to $\{1,2\}$, the operator $\Delta_\mathrm{h}^{-1} \partial_\ell \partial_m$ is a zero-order horizontal Fourier multiplier, which implies that

$$\left\| [\partial_3 p]_\beta \right\|_{L^1(\mathbb{R}^+; \mathcal{B}^0)} = \left\| \partial_3 p \right\|_{L^1(\mathbb{R}^+; \mathcal{B}^0)}$$
$$\lesssim \left\| v \partial_3 v \right\|_{L^1(\mathbb{R}^+; \mathcal{B}^0)}.$$

According to laws of product in anisotropic Besov spaces, we obtain

$$\left\| v(t) \partial_3 v(t) \right\|_{\mathcal{B}^0} \lesssim \left\| v(t) \right\|_{\mathcal{B}^1} \left\| \partial_3 v(t) \right\|_{\mathcal{B}^0},$$

which gives rise to

$$\left\| [\partial_3 p]_\beta \right\|_{L^1(\mathbb{R}^+; \mathcal{B}^0)} \lesssim \left\| v \right\|_{L^2(\mathbb{R}^+; \mathcal{B}^1)} \left\| \partial_3 v \right\|_{L^2(\mathbb{R}^+; \mathcal{B}^0)}$$
$$\lesssim \left\| v \right\|_{L^2(\mathbb{R}^+; \mathcal{B}^1)} \left\| v \right\|_{L^2(\mathbb{R}^+; \mathcal{B}^{0,3/2})}. \tag{2.105}$$

Combining (2.104) and (2.105), we get by virtue of Proposition 2.19

$$\left\| E_\beta^1 \right\|_{\mathcal{F}^0} \leq \beta \, \mathcal{T}_2(\|v_0\|_{S_\mu}). \tag{2.106}$$

In the same way, we treat the terms E_β^2, E_β^3 and E_β^4, achieving the proof of Estimate (2.101). This ends the proof of the fact that the solution Φ_β of (NS) with initial data

$$\Phi_0 = \big[(v_0 - \beta \nabla^h \Delta_h^{-1} \partial_3 w_0^3, w_0^3) \big]_\beta$$

is global and belongs to $L^2(\mathbb{R}^+; \mathcal{B}^1)$.

The proof of the whole of Theorem 2.14 is then achieved. $\qquad\square$

2.4.2 Proof of Theorem 2.15

The proof of Theorem 2.15 is done in three steps. First we define an approximate solution, using results proved in the previous section, and then we prove useful localization results on the different parts entering in the definition of the approximate solution. In the last step, we conclude the proof of the theorem, using those localization results.

The Approximate Solution

With the notation of Theorem 2.12, let us consider the divergence-free vector field:

$$\Phi_{0,n,\alpha,L}^0 \overset{\text{def}}{=} u_{0,\alpha} + \big[(v_{0,n,\alpha,L}^{0,\infty} + h_n^0 w_{0,n,\alpha,L}^{0,\infty,h}, w_{0,n,\alpha,L}^{0,\infty,3}) \big]_{h_n^0}$$
$$+ \big[(v_{0,n,\alpha,L}^{0,\text{loc}} + h_n^0 w_{0,n,\alpha,L}^{0,\text{loc},h}, w_{0,n,\alpha,L}^{0,\text{loc},3}) \big]_{h_n^0}.$$

Our purpose is to establish that for h_n^0 small enough, depending only on the weak limit u_0 and on $\big\| (v_{0,n,\alpha,L}^{0,\infty}, w_{0,n,\alpha,L}^{0,\infty,3}) \big\|_{S_\mu}$ as well as $\big\| (v_{0,n,\alpha,L}^{0,\text{loc}}, w_{0,n,\alpha,L}^{0,\text{loc},3}) \big\|_{S_\mu}$, there is a unique, global smooth solution to (NS) with data $\Phi_{0,n,\alpha,L}^0$.

Let us start by solving (NS) globally with the data $u_{0,\alpha}$. By using the global strong stability of (NS) in $\mathcal{B}_{1,1}$ (see [4], Corollary 3) and the convergence result (2.14), we deduce that, for α small enough, $u_{0,\alpha}$ generates a unique, global solution u_α to the (NS) system belonging to $L^2(\mathbb{R}^+; B_{1,1}^{2,\frac{1}{2}})$. Actually, in view of the Sobolev embedding of $B_{1,1}^{2,\frac{1}{2}}$ into $\mathcal{B}^1$, $u_\alpha \in L^2(\mathbb{R}^+; \mathcal{B}^1)$.

Next let us define

$$\Phi_{0,n,\alpha,L}^{0,\infty} \overset{\text{def}}{=} \big[(v_{0,n,\alpha,L}^{0,\infty} + h_n^0 w_{0,n,\alpha,L}^{0,\infty,h}, w_{0,n,\alpha,L}^{0,\infty,3}) \big]_{h_n^0}.$$

Thanks to Theorem 2.14, we know that for h_n^0 smaller than $\varepsilon_1 \big(\big\| (v_{0,n,\alpha,L}^{0,\infty}, w_{0,n,\alpha,L}^{0,\infty,3}) \big\|_{S_\mu} \big)$ there is a unique global smooth solution $\Phi_{n,\alpha,L}^{0,\infty}$ associated with $\Phi_{0,n,\alpha,L}^{0,\infty}$, which belongs to $\mathcal{A}_0$, and using the notation and results of Subsection 2.2.3, in particular (2.71) and (2.102), we can write

$$\Phi_{n,\alpha,L}^{0,\infty} \stackrel{\text{def}}{=} \Phi_{n,\alpha,L}^{0,\infty,\text{app}} + \psi_{n,\alpha,L}^{0,\infty} \quad \text{with}$$

$$\Phi_{n,\alpha,L}^{0,\infty,\text{app}} \stackrel{\text{def}}{=} \left[v_{n,\alpha,L}^{0,\infty} + h_n^0 w_{n,\alpha,L}^{0,\infty,\text{h}}, w_{n,\alpha,L}^{0,\infty,3} \right]_{h_n^0} \text{ and} \qquad (2.107)$$

$$\left\| \psi_{n,\alpha,L}^{0,\infty} \right\|_{\widetilde{L}^r(\mathbb{R}^+;\mathcal{B}^{\frac{2}{r}})} \lesssim h_n^0 \mathcal{T}_2\left(\left\| (v_{0,n,\alpha,L}^{0,\infty}, w_{0,n,\alpha,L}^{0,\infty,3}) \right\|_{S_\mu} \right),$$

for all r in $[2,\infty]$,

$$\lim_{L\to\infty} \lim_{\alpha\to 0} \limsup_{n\to\infty} \left\| \Phi_{n,\alpha,L}^{0,\text{loc}}(\cdot,0) \right\|_{\widetilde{L}^r(\mathbb{R}^+;B_{2,1}^{\frac{2}{r}}(\mathbb{R}^2))} = 0, \qquad (2.108)$$

where $v_{n,\alpha,L}^{0,\infty}$ solves (NS2D)$_{x_3}$ with data $v_{0,n,\alpha,L}^{0,\infty}$, $w_{n,\alpha,L}^{0,\infty,3}$ solves the transport-diffusion equation $(T_{h_n^0})$ defined in Subsection 2.4.1 with data $w_{0,n,\alpha,L}^{0,\infty,3}$ and where

$$w_{n,\alpha,L}^{0,\infty,\text{h}} = -\nabla^{\text{h}} \Delta_{\text{h}}^{-1} \partial_3 w_{n,\alpha,L}^{0,\infty,3}.$$

Similarly, defining

$$\Phi_{0,n,\alpha,L}^{0,\text{loc}} \stackrel{\text{def}}{=} \left[(v_{0,n,\alpha,L}^{0,\text{loc}} + h_n^0 w_{0,n,\alpha,L}^{0,\text{loc},\text{h}}, w_{0,n,\alpha,L}^{0,\text{loc},3}) \right]_{h_n^0},$$

then for h_n^0 smaller than $\varepsilon_1\left(\left\| (v_{0,n,\alpha,L}^{0,\text{loc}}, w_{0,n,\alpha,L}^{0,\text{loc},3}) \right\|_{S_\mu} \right)$ there is a unique global smooth solution $\Phi_{n,\alpha,L}^{0,\text{loc}}$ associated with $\Phi_{0,n,\alpha,L}^{0,\text{loc}}$, which belongs to $\mathcal{A}_0$, and

$$\Phi_{n,\alpha,L}^{0,\text{loc}} \stackrel{\text{def}}{=} \Phi_{n,\alpha,L}^{0,\text{loc},\text{app}} + \psi_{n,\alpha,L}^{0,\text{loc}} \quad \text{with}$$

$$\Phi_{n,\alpha,L}^{0,\text{loc},\text{app}} \stackrel{\text{def}}{=} \left[v_{n,\alpha,L}^{0,\text{loc}} + h_n^0 w_{n,\alpha,L}^{0,\text{loc},\text{h}}, w_{n,\alpha,L}^{0,\text{loc},3} \right]_{h_n^0} \text{ and for all r in } [2,\infty]$$

$$\left\| \psi_{n,\alpha,L}^{0,\text{loc}} \right\|_{\widetilde{L}^r(\mathbb{R}^+;\mathcal{B}^{\frac{2}{r}})} \lesssim h_n^0 \mathcal{T}_2\left(\left\| (v_{0,n,\alpha,L}^{0,\text{loc}}, w_{0,n,\alpha,L}^{0,\text{loc},3}) \right\|_{S_\mu} \right),$$

$$(2.109)$$

where $v_{n,\alpha,L}^{0,\text{loc}}$ solves (NS2D)$_{x_3}$ with data $v_{0,n,\alpha,L}^{0,\text{loc}}$ and $w_{n,\alpha,L}^{0,\text{loc},3}$ solves $(T_{h_n^0})$ with data $w_{0,n,\alpha,L}^{0,\text{loc},3}$. Finally we recall that $w_{n,\alpha,L}^{0,\text{loc},\text{h}} = -\nabla^{\text{h}} \Delta_{\text{h}}^{-1} \partial_3 w_{n,\alpha,L}^{0,\text{loc},3}$.

In the next step, we establish localization properties on $\Phi_{n,\alpha,L}^{0,\infty}$ and $\Phi_{n,\alpha,L}^{0,\text{loc}}$. Those localization properties will enable us to prove that the function $u_\alpha + \Phi_{n,\alpha,L}^{0,\infty} + \Phi_{n,\alpha,L}^{0,\text{loc}}$ approximates the solution to the (NS) system associated with the Cauchy data $\Phi_{0,n,\alpha,L}^0$.

Localization Properties of the Approximate Solution

In this step, we prove localization properties on $\Phi_{n,\alpha,L}^{0,\infty}$ and $\Phi_{n,\alpha,L}^{0,\text{loc}}$, namely the fact that $\Phi_{n,\alpha,L}^{0,\infty,\text{app}}$ escapes to infinity in the space variable, while $\Phi_{n,\alpha,L}^{0,\text{loc},\text{app}}$ remains localized (approximately), and we also prove that $\Phi_{n,\alpha,L}^{0,\text{loc},\text{app}}$ remains small near $x_3 = 0$. Let us recall that, as claimed by (2.18), (2.19) and (2.20), those properties are true for their respective initial data. A first part of these localization properties derives from the following result.

Proposition 2.25 *Under the assumptions of Proposition 2.19, the control of the value of v at the point $x_3 = 0$ is given by*

$$\forall r \in [1,\infty], \quad \|v(\cdot,0)\|_{\widetilde{L}^r(\mathbb{R}^+;B_{2,1}^{\frac{2}{r}}(\mathbb{R}^2))} \lesssim \|v_0(\cdot,0)\|_{B_{2,1}^0(\mathbb{R}^2)} + \|v_0(\cdot,0)\|_{L^2(\mathbb{R}^2)}^2.$$

(2.110)

Moreover, we have for all η in $]0,1[$ and γ in $\{0,1\}$,

$$\|(\gamma - \theta_{h,\eta})v\|_{\mathcal{A}^0} \le \|(\gamma - \theta_{h,\eta})v_0\|_{\mathcal{B}^0} \exp \mathcal{T}_1(\|v_0\|_{\mathcal{B}^0}) + \eta \mathcal{T}_2(\|v_0\|_{S_\mu}),$$

(2.111)

where $\theta_{h,\eta}$ is the truncation function defined by (2.13).

Proof In order to establish Proposition 2.25, let us start by pointing out that the proof of Lemma 1.1 of [12] claims that for all x_3 in $\mathbb{R}$,

$$\begin{aligned}
\left(\Delta_k^h(v(t,\cdot,x_3) \cdot \nabla^h v(t,\cdot,x_3)) \big| \Delta_k^h v(t,\cdot,x_3)\right)_{L^2} \\
\lesssim d_k(t,x_3)\|\nabla^h v(t,\cdot,x_3)\|_{L^2}^2 \|\Delta_k^h v(t,\cdot,x_3)\|_{L^2},
\end{aligned}$$

(2.112)

where $(d_k(t,x_3))_{k\in\mathbb{Z}}$ is a generic element of the sphere of $\ell^1(\mathbb{Z})$.

Taking $x_3 = 0$, we deduce by an L^2 energy estimate in $\mathbb{R}^2$

$$\begin{aligned}
\frac{1}{2}\frac{d}{dt}\|\Delta_k^h v(t,\cdot,0)\|_{L^2}^2 + c2^{2k}\|\Delta_k^h v(t,\cdot,0)\|_{L^2}^2 \\
\lesssim d_k(t)\|\nabla^h v(t,\cdot,0)\|_{L^2}^2\|\Delta_k^h v(t,\cdot,0)\|_{L^2},
\end{aligned}$$

where $(d_k(t))_{k\in\mathbb{Z}}$ belongs to the sphere of $\ell^1(\mathbb{Z})$, which after division by $\|\Delta_k^h v(t,\cdot,0)\|_{L^2}$ and time integration leads to

$$\begin{aligned}
\|\Delta_k^h v(\cdot,0)\|_{L^\infty(\mathbb{R}^+;L^2)} + c2^{2k}\|\Delta_k^h v(\cdot,0)\|_{L^1(\mathbb{R}^+;L^2)} \\
\le \|\Delta_k v_0(\cdot,0)\|_{L^2} + C\int_0^\infty d_k(t)\|\nabla^h v(t,\cdot,0)\|_{L^2}^2 dt.
\end{aligned}$$

(2.113)

By summation over k and in view of (2.79), we obtain Inequality (2.110) of Proposition 2.25.

Now to go to the proof of Inequality (2.111), let us define $v_{\gamma,\eta} \overset{\text{def}}{=} (\gamma - \theta_{h,\eta})v$ and write that

$$\partial_t v_{\gamma,\eta} - \Delta_h v_{\gamma,\eta} + \text{div}_h\left(v \otimes v_{\gamma,\eta}\right) = E_\eta(v) = \sum_{i=1}^3 E_\eta^i(v) \quad \text{with}$$

$$E_\eta^1(v) \overset{\text{def}}{=} -2\eta(\nabla^h\theta)_{h,\eta}\nabla^h v - \eta^2(\Delta_h\theta)_{h,\eta}v,$$

(2.114)

$$E_\eta^2(v) \overset{\text{def}}{=} \eta v \cdot (\nabla^h\theta)_{h,\eta}v \quad \text{and}$$

$$E_\eta^3(v) \overset{\text{def}}{=} -(\gamma - \theta_{h,\eta})\nabla^h\Delta_h^{-1}\sum_{1\le\ell,m\le2}\partial_\ell\partial_m\left(v^\ell v^m\right).$$

Lemma 2.23 applied with $s = 0$, $s' = 1/2$, $a = v_{\gamma,\eta}$, $\mathcal{Q}(v,a) = \text{div}_h(v \otimes a)$, $f = E_\eta(v)$ and $\beta = 0$ reduces the problem to the proof of the following estimate:

$$\|E_\eta(v)\|_{L^1(\mathbb{R}^+;\mathcal{B}^0)} \lesssim \eta \mathcal{T}_2(\|v_0\|_{\mathcal{S}_\mu}). \qquad (2.115)$$

Actually, in view of Inequality (2.73) applied with $r = 1$ and $s = -1$ (resp. with $r = 2$ and $s = -1/2$) this will follow from

$$\|E_\eta(v)\|_{L^1(\mathbb{R}^+;\mathcal{B}^0)} \lesssim \eta\left(\|v\|_{L^1(\mathbb{R}^+;\mathcal{B}^1)} + \|v\|_{L^2(\mathbb{R}^+;\mathcal{B}^{1/2})}^2\right). \qquad (2.116)$$

Product laws in anisotropic Besov spaces and the scaling properties of homogeneous Besov spaces give

$$\|(\nabla^h\theta)_{h,\eta}\nabla^h v(t)\|_{\mathcal{B}^0} \lesssim \|(\nabla^h\theta)_{h,\eta}\|_{B_{2,1}^1(\mathbb{R}^2)}\|\nabla^h v(t)\|_{\mathcal{B}^0}$$
$$\lesssim \|\nabla^h\theta\|_{B_{2,1}^1(\mathbb{R}^2)}\|v(t)\|_{\mathcal{B}^1}.$$

Along the same lines, we get

$$\|(\Delta_h\theta)_{h,\eta}v(t)\|_{\mathcal{B}^0} \lesssim \|(\Delta_h\theta)_{h,\eta}\|_{B_{2,1}^0(\mathbb{R}^2)}\|v(t)\|_{\mathcal{B}^1}$$
$$\lesssim \frac{1}{\eta}\|\Delta_h\theta\|_{B_{2,1}^0(\mathbb{R}^2)}\|v(t)\|_{\mathcal{B}^1}.$$

Consequently

$$\|E_\eta^1(v)\|_{L^1(\mathbb{R}^+;\mathcal{B}^0)} \lesssim \eta\|v\|_{L^1(\mathbb{R}^+;\mathcal{B}^1)}. \qquad (2.117)$$

The same arguments enable us to deal with the term $E_\eta^2(v)$ and to prove that

$$\|E_\eta^2(v)\|_{L^1(\mathbb{R}^+;\mathcal{B}^0)} \lesssim \eta\|v\|_{L^2(\mathbb{R}^+;\mathcal{B}^{1/2})}^2. \qquad (2.118)$$

Let us finally study the term $E_\eta^3(v)$ which is most challenging. To this end, we make use of the horizontal paraproduct decomposition:

$$av = T_v^h a + T_a^h v + R^h(a,b) \quad \text{with} \quad T_a^h b \overset{\text{def}}{=} \sum_k S_{k-1}^h a \Delta_k^h b$$

$$\text{and} \quad R^h(a,b) \overset{\text{def}}{=} \sum_k \tilde{\Delta}_k^h a \Delta_k^h b,$$

where $\tilde{\Delta}_k^h \overset{\text{def}}{=} \tilde{\varphi}(2^{-k}\xi_h)$ with $\tilde{\varphi}$ a smooth compactly supported (in $\mathbb{R}^2 \setminus \{0\}$) function which has value 1 near $B(0,2^{-N_0})+\mathcal{C}$, where $\mathcal{C}$ is an adequate annulus.

This allows us to write

$$E_\eta^3(v) = \sum_{\ell=1}^{3} E_\eta^{3,\ell}(v) \quad \text{with}$$

$$E_\eta^{3,1}(v) \overset{\text{def}}{=} \widetilde{T}_{\nabla^h p}^h \theta_{h,\eta} \quad \text{with} \quad \nabla^h p = \nabla^h \Delta_h^{-1} \sum_{1\le \ell,m \le 2} \partial_\ell \partial_m (v^\ell v^m),$$

$$E_\eta^{3,2}(v) \overset{\text{def}}{=} - \sum_{1\le \ell,m \le 2} \big[T_{\gamma-\theta_{h,\eta}}^h, \nabla^h \Delta_h^{-1} \partial_\ell \partial_m\big] v^\ell v^m \quad \text{and}$$

$$E_\eta^{3,3}(v) \overset{\text{def}}{=} \sum_{1\le \ell,m \le 2} \nabla^h \Delta_h^{-1} \partial_\ell \partial_m \widetilde{T}_{v^\ell v^m}^h \theta_{h,\eta},$$

(2.119)

where $\widetilde{T}_a^h b = T_a^h b + R^h(a,b)$. Combining the laws of product with scaling properties of Besov spaces, we obtain

$$\|\widetilde{T}_{\nabla^h p(t)}^h \theta_{h,\eta}\|_{\mathcal{B}^0} \lesssim \|\nabla^h p(t)\|_{\mathcal{B}^{-1}} \|\theta_{h,\eta}\|_{B_{2,1}^2(\mathbb{R}^2)}$$

$$\lesssim \eta \sup_{1\le \ell,m \le 2} \|v^\ell(t) v^m(t)\|_{\mathcal{B}^0} \|\theta\|_{B_{2,1}^2(\mathbb{R}^2)}$$

$$\lesssim \eta \|v(t)\|_{\mathcal{B}^{1/2}}^2 \|\theta\|_{B_{2,1}^2(\mathbb{R}^2)}.$$

Along the same lines we get

$$\|\nabla^h \Delta_h^{-1} \partial_\ell \partial_m \widetilde{T}_{v^\ell(t) v^m(t)}^h \theta_{h,\eta}\|_{\mathcal{B}^0} \lesssim \|\widetilde{T}_{v^\ell(t) v^m(t)}^h \theta_{h,\eta}\|_{\mathcal{B}^1}$$

$$\lesssim \|v^\ell(t) v^m(t)\|_{\mathcal{B}^0} \|\theta_{h,\eta}\|_{B_{2,1}^2(\mathbb{R}^2)}$$

$$\lesssim \eta \|v(t)\|_{\mathcal{B}^{1/2}}^2 \|\theta\|_{B_{2,1}^2(\mathbb{R}^2)}.$$

We deduce that

$$\|E_\eta^{3,1}(v) + E_\eta^{3,3}(v)\|_{L^1(\mathbb{R}^+;\mathcal{B}^0)} \lesssim \eta \|v\|_{L^2(\mathbb{R}^+;\mathcal{B}^{1/2})}^2.$$

(2.120)

Now let us estimate $E_\eta^{3,2}(v)$. By definition, we have

$$\big[T_{\gamma-\theta_{h,\eta}}^h, \nabla^h \Delta_h^{-1} \partial_\ell \partial_m\big] v^\ell v^m = \sum_k \mathcal{E}_{k,\eta}(v) \quad \text{with}$$

$$\mathcal{E}_{k,\eta}(v) \overset{\text{def}}{=} \big[S_{k-N_0}^h(\gamma-\theta_{h,\eta}), \widetilde{\Delta}_k^h \nabla^h \Delta_h^{-1} \partial_\ell \partial_m\big] \Delta_k^h(v^\ell v^m).$$

Then by commutator estimates (see for instance Lemma 2.97 in [2])

$$\|\Delta_j^v \mathcal{E}_{k,\eta}(v(t))\|_{L^2} \lesssim \|\nabla \theta_{h,\eta}\|_{L^\infty} \|\Delta_k^h \Delta_j^v(v^\ell(t) v^m(t))\|_{L^2}.$$

Noticing that $\|\nabla \theta_{h,\eta}\|_{L^\infty} = \eta \|\nabla \theta\|_{L^\infty}$, we get by virtue of the laws of product

$$\|E_\eta^{3,2}(v)\|_{L^1(\mathbb{R}^+;\mathcal{B}^0)} \lesssim \eta \|v\|_{L^2(\mathbb{R}^+;\mathcal{B}^{1/2})}^2,$$

which ends the proof of Estimate (2.115) and thus of of Proposition 2.25. $\quad\square$

A similar result holds for the solution w^3 of

$$(T_\beta) \quad \partial_t w^3 + v \cdot \nabla^h w^3 - \Delta_h w^3 - \beta^2 \partial_3^2 w^3 = 0 \quad \text{and} \quad w^3_{|t=0} = w^3_0,$$

where β is any non-negative real number. In the following statement, all the constants are independent of β.

Proposition 2.26 *Let v and w_3 be as in Proposition 2.22. The control of the value of w^3 at the point $x_3 = 0$ states as follows. For any r in $[2, \infty]$,*

$$\|w^3(\cdot, 0)\|_{\widetilde{L^r}(\mathbb{R}^+; B^{\frac{2}{r}}_{2,1}(\mathbb{R}^2))} \leq T_2(\|(v_0, w_0^3)\|_{S_\mu}) \left(\|w_0^3(\cdot, 0)\|_{B^0_{2,1}(\mathbb{R}^2)}^{\frac{1-2\mu}{4(1-\mu)}} + \beta \right). \quad (2.121)$$

Moreover, with the notations of Theorem 2.14, we have for all η in $]0, 1[$ and γ in $\{0, 1\}$,

$$\|(\gamma - \theta_{h,\eta})w^3\|_{\mathcal{A}^0} \leq \|(\gamma - \theta_{h,\eta})w_0^3\|_{B^0} \exp T_1(\|v_0\|_{B^0}) + \eta T_2(\|(v_0, w_0^3)\|_{S_\mu}). \quad (2.122)$$

The proof of Proposition 2.26 is very similar to that of Proposition 2.25 and is left to the reader.

Propositions 2.25 and 2.26 easily imply the following result, using the special form of $\Phi^{0,\infty}_{n,\alpha,L}$ and $\Phi^{0,\text{loc}}_{n,\alpha,L}$ recalled in (2.107) and (2.109), and thanks to (2.18), (2.19) and (2.20).

Corollary 2.27 *The vector fields $\Phi^{0,\text{loc}}_{n,\alpha,L}$ and $\Phi^{0,\infty}_{n,\alpha,L}$ satisfy the following: $\Phi^{0,\text{loc}}_{n,\alpha,L}$ vanishes at $x_3 = 0$, in the sense that for all r in $[2, \infty]$,*

$$\lim_{L \to \infty} \lim_{\alpha \to 0} \limsup_{n \to \infty} \|\Phi^{0,\text{loc}}_{n,\alpha,L}(\cdot, 0)\|_{\widetilde{L^r}(\mathbb{R}^+; B^{\frac{2}{r}}_{2,1}(\mathbb{R}^2))} = 0, \quad (2.123)$$

and there is a constant $C(\alpha, L)$ such that for all η in $]0, 1[$,

$$\limsup_{n \to \infty} \left(\|(1 - \theta_{h,\eta})\Phi^{0,\text{loc}}_{n,\alpha,L}\|_{\mathcal{A}^0} + \|\theta_{h,\eta}\Phi^{0,\infty}_{n,\alpha,L}\|_{\mathcal{A}^0} \right) \leq C(\alpha, L)\eta.$$

Proof In view of (2.109) and under Notation (2.11)

$$\Phi^{0,\text{loc}}_{n,\alpha,L} = \Phi^{0,\text{loc,app}}_{n,\alpha,L} + \psi^{0,\text{loc}}_{n,\alpha,L} \quad \text{with}$$

$$\Phi^{0,\text{loc,app}}_{n,\alpha,L} = \left[v^{0,\text{loc}}_{n,\alpha,L} + h^0_n w^{0,\text{loc,h}}_{n,\alpha,L}, w^{0,\text{loc,3}}_{n,\alpha,L} \right]_{h^0_n},$$

where $v^{0,\text{loc}}_{n,\alpha,L}$ solves $(\text{NS2D})_{x_3}$ with data $v^{0,\text{loc}}_{0,n,\alpha,L}$, $w^{0,\text{loc,3}}_{n,\alpha,L}$ solves the transport-diffusion equation $(T_{h^0_n})$ defined in Subsection 62 with data $w^{0,\text{loc,3}}_{0,n,\alpha,L}$ and $w^{0,\text{loc,h}}_{n,\alpha,L} = -\nabla^h \Delta_h^{-1} \partial_3 w^{0,\text{loc,3}}_{n,\alpha,L}$. Combining Property (2.18) together with Propositions 2.25 and 2.26, we infer that

$$\lim_{L \to \infty} \lim_{\alpha \to 0} \limsup_{n \to \infty} \|\Phi^{0,\text{loc,app}}_{n,\alpha,L}(\cdot, 0)\|_{\widetilde{L^r}(\mathbb{R}^+; B^{\frac{2}{r}}_{2,1}(\mathbb{R}^2))} = 0,$$

which ends the proof of (2.123) invoking (2.17) and (2.109). The argument is similar for the other estimates. □

Conclusion of the Proof of Theorem 2.15

Now, with the above notations, we look for the solution to the (NS) system associated with the Cauchy data $\Phi^0_{0,n,\alpha,L}$ in the form:

$$\Phi^0_{n,\alpha,L} \overset{\text{def}}{=} u_\alpha + \Phi^{0,\infty}_{n,\alpha,L} + \Phi^{0,\text{loc}}_{n,\alpha,L} + \psi_{n,\alpha,L}.$$

In particular the two vector fields $\Phi^{0,\text{loc}}_{n,\alpha,L}$ and $\Phi^{0,\infty}_{n,\alpha,L}$ satisfy Corollary 2.27, and furthermore, thanks to the Lebesgue theorem,

$$\lim_{\eta\to 0} \|(1-\theta_\eta)u_\alpha\|_{L^2(\mathbb{R}^+;\mathcal{B}^1)} = 0. \tag{2.124}$$

Given a small number $\varepsilon > 0$, to be selected later on, we choose L, α and $\eta = \eta(\alpha, L, u_0)$ so that thanks to Corollary 2.27 and (2.124), for all r in $[2, \infty]$, and for n large enough,

$$\|\Phi^{0,\text{loc}}_{n,\alpha,L}(\cdot,0)\|_{L^r(\mathbb{R}^+;B^{2/r}_{2,1}(\mathbb{R}^2))} + \|(1-\theta_{\text{h},\eta})\Phi^{0,\text{loc}}_{n,\alpha,L}\|_{\mathcal{A}^0} + \|(1-\theta_\eta)u_\alpha\|_{L^2(\mathbb{R}^+;\mathcal{B}^1)}$$
$$+ \|\theta_{\text{h},\eta}\Phi^{0,\infty}_{n,\alpha,L}\|_{\mathcal{A}^0} \le \varepsilon. \tag{2.125}$$

For the sake of simplicity, denote in the sequel

$$(\Phi^{0,\infty}_\varepsilon, \Phi^{0,\text{loc}}_\varepsilon, \psi_\varepsilon) \overset{\text{def}}{=} (\Phi^{0,\infty}_{n,\alpha,L}, \Phi^{0,\text{loc}}_{n,\alpha,L}, \psi_{n,\alpha,L}) \text{ and } \Phi^{\text{app}}_\varepsilon \overset{\text{def}}{=} u_\alpha + \Phi^{0,\infty}_\varepsilon + \Phi^{0,\text{loc}}_\varepsilon.$$

By straightforward computations, one can verify that the vector field ψ_ε satisfies the following equation, with null Cauchy data:

$$\partial_t \psi_\varepsilon - \Delta \psi_\varepsilon + \text{div}\big(\psi_\varepsilon \otimes \psi_\varepsilon + \Phi^{\text{app}}_\varepsilon \otimes \psi_\varepsilon + \psi_\varepsilon \otimes \Phi^{\text{app}}_\varepsilon\big) = -\nabla q_\varepsilon + E_\varepsilon,$$
$$\text{with} \quad E_\varepsilon = E^1_\varepsilon + E^2_\varepsilon \quad \text{and}$$
$$E^1_\varepsilon \overset{\text{def}}{=} \text{div}\Big(\Phi^{0,\infty}_\varepsilon \otimes (\Phi^{0,\text{loc}}_\varepsilon + u_\alpha) + (\Phi^{0,\text{loc}}_\varepsilon + u_\alpha) \otimes \Phi^{0,\infty}_\varepsilon$$
$$+ \Phi^{0,\text{loc}}_\varepsilon \otimes (1-\theta_\eta)u_\alpha + (1-\theta_\eta)u_\alpha \otimes \Phi^{0,\text{loc}}_\varepsilon\Big),$$
$$E^2_\varepsilon \overset{\text{def}}{=} \text{div}\big(\Phi^{0,\text{loc}}_\varepsilon \otimes \theta_\eta u_\alpha + \theta_\eta u_\alpha \otimes \Phi^{0,\text{loc}}_\varepsilon\big). \tag{2.126}$$

The heart of the matter consists of proving that

$$\lim_{\varepsilon\to 0} \|E_\varepsilon\|_{\mathcal{F}^0} = 0. \tag{2.127}$$

Indeed, exactly as in the proof of Theorem 2.14, this ensures that ψ_ε belongs to $L^2(\mathbb{R}^+; \mathcal{B}^1)$, with

$$\lim_{\varepsilon \to 0} \|\psi_\varepsilon\|_{L^2(\mathbb{R}^+; \mathcal{B}^1)} = 0,$$

and allows us to conclude the proof of Theorem 2.15.

So let us focus on (2.127). The term E_ε^1 is the easiest, thanks to the separation of the spatial supports. Let us first write $E_\varepsilon^1 = E_{\varepsilon,\mathrm{h}}^1 + E_{\varepsilon,3}^1$ with

$$E_{\varepsilon,\mathrm{h}}^1 \overset{\mathrm{def}}{=} \mathrm{div}_\mathrm{h}\Big((\Phi_\varepsilon^{0,\mathrm{loc}} + u_\alpha) \otimes \Phi_\varepsilon^{0,\infty,\mathrm{h}} + \Phi_\varepsilon^{0,\infty} \otimes (\Phi_\varepsilon^{0,\mathrm{loc},\mathrm{h}} + u_\alpha^\mathrm{h})$$

$$+ (1 - \theta_\eta)u_\alpha \otimes \Phi^{0,\mathrm{loc},\mathrm{h}} + \Phi^{0,\mathrm{loc}} \otimes (1 - \theta_\eta)u_\alpha^\mathrm{h} \Big) \quad \text{and}$$

$$E_{\varepsilon,3}^1 \overset{\mathrm{def}}{=} \partial_3\Big((\Phi_\varepsilon^{0,\mathrm{loc}} + u_\alpha)\Phi_\varepsilon^{0,\infty,3} + \Phi_\varepsilon^{0,\infty}(\Phi_\varepsilon^{0,\mathrm{loc},3} + u_\alpha^3)$$

$$+ (1 - \theta_\eta)u_\alpha \Phi^{0,\mathrm{loc},3} + \Phi^{0,\mathrm{loc}}(1 - \theta_\eta)u_\alpha^3 \Big).$$

Now using as usual the action of derivatives and the fact that $\mathcal{B}^1$ is an algebra, we infer that

$$\|E_{\varepsilon,\mathrm{h}}^1\|_{L^1(\mathbb{R}^+; \mathcal{B}^0)} + \|E_{\varepsilon,3}^1\|_{L^1(\mathbb{R}^+; \mathcal{B}_{2,1}^{1,-1/2})} \leq \|\theta_{\mathrm{h},\eta}\Phi_\varepsilon^{0,\infty}\|_{L^2(\mathbb{R}^+; \mathcal{B}^1)} \|\Phi_\varepsilon^{0,\mathrm{loc}}$$

$$+ u_\alpha\|_{L^2(\mathbb{R}^+; \mathcal{B}^1)} + \|(1 - \theta_{\mathrm{h},\eta})(\Phi_\varepsilon^{0,\mathrm{loc}} + u_\alpha)\|_{L^2(\mathbb{R}^+; \mathcal{B}^1)} \|\Phi_\varepsilon^{0,\infty}\|_{L^2(\mathbb{R}^+; \mathcal{B}^1)}$$

$$+ \|\Phi_\varepsilon^{0,\mathrm{loc}}\|_{L^2(\mathbb{R}^+; \mathcal{B}^1)} \|u_\varepsilon^\infty\|_{L^2(\mathbb{R}^+; \mathcal{B}^1)},$$

where we denote by u_ε^∞ the function $(1 - \theta_\eta)u_\alpha$. Thanks to (2.125) and to the a priori bounds on $\Phi_\varepsilon^{0,\infty}$, $\Phi_\varepsilon^{0,\mathrm{loc}}$ and u_α, we easily get

$$\lim_{\varepsilon \to 0} \|E_\varepsilon^1\|_{\mathcal{F}^0} = 0.$$

Next let us turn to E_ε^2. To this end, we use the following estimate (see, for instance, Lemma 3.3 of [15]):

$$\|ab\|_{\mathcal{B}^1} \lesssim \|a\|_{\mathcal{B}^1} \|b(\cdot, 0)\|_{B_{2,1}^1(\mathbb{R}^2)} + \|x_3 a\|_{\mathcal{B}^1} \|\partial_3 b\|_{\mathcal{B}^1}. \tag{2.128}$$

Defining $u_\varepsilon^{\mathrm{loc}} \overset{\mathrm{def}}{=} \theta_\eta u_\alpha$, we get, applying Estimate (2.128),

$$\|E_\varepsilon^2\|_{\mathcal{F}^0} \lesssim \|u_\varepsilon^{\mathrm{loc}}\|_{L^2(\mathbb{R}^+; \mathcal{B}^1)} \|\Phi_\varepsilon^{0,\mathrm{loc}}(\cdot, 0)\|_{L^2(\mathbb{R}^+; B_{2,1}^1(\mathbb{R}^2))}$$

$$+ \|x_3 u_\varepsilon^{\mathrm{loc}}\|_{L^2(\mathbb{R}^+; \mathcal{B}^1)} \|\partial_3 \Phi_\varepsilon^{0,\mathrm{loc}}\|_{L^2(\mathbb{R}^+; \mathcal{B}^1)}.$$

Thanks to (2.125) as well as Inequality (2.26) of Theorem 2.14, we obtain

$$\lim_{\varepsilon \to 0} \|E_\varepsilon^2\|_{\mathcal{F}^0} = 0.$$

This proves (2.127), hence Theorem 2.15. $\qquad\square$

References

[1] P. Auscher, S. Dubois and P. Tchamitchian, On the stability of global solutions to Navier–Stokes equations in the space, *Journal de Mathématiques Pures et Appliquées*, **83**, 2004, 673–697.

[2] H. Bahouri, J.-Y. Chemin and R. Danchin, *Fourier Analysis and Nonlinear Partial Differential Equations*, Grundlehren der mathematischen Wissenschaften, Springer, **343**, 2011.

[3] H. Bahouri, A. Cohen and G. Koch, A general wavelet-based profile decomposition in the critical embedding of function spaces, *Confluentes Mathematici*, **3**, 2011, 1–25.

[4] H. Bahouri and I. Gallagher, On the stability in weak topology of the set of global solutions to the Navier–Stokes equations, *Archive for Rational Mechanics and Analysis*, **209**, 2013, 569–629.

[5] H. Bahouri and P. Gérard, High frequency approximation of solutions to critical nonlinear wave equations, *American Journal of Math*, **121**, 1999, 131–175.

[6] H. Bahouri, M. Majdoub and N. Masmoudi, On the lack of compactness in the 2D critical Sobolev embedding, *Journal of Functional Analysis*, **260**, 2011, 208–252.

[7] H. Bahouri, M. Majdoub and N. Masmoudi, Lack of compactness in the 2D critical Sobolev embedding, the general case, *Journal de Mathématiques Pures et Appliquées*, **101**, 2014, 415–457.

[8] H. Bahouri and G. Perelman, A Fourier approach to the profile decomposition in Orlicz spaces, *Mathematical Research Letters*, **21**, 2014, 33–54.

[9] G. Bourdaud, La propriété de Fatou dans les espaces de Besov homogènes, *Note aux Comptes Rendus Mathematique de l'Académie des Sciences*, **349**, 2011, 837–840.

[10] J. Bourgain and N. Pavlović, Ill-posedness of the Navier–Stokes equations in a critical space in 3D, *Journal of Functional Analysis*, **255**, 2008, 2233–2247.

[11] H. Brézis and J.-M. Coron, Convergence of solutions of H-Systems or how to blow bubbles, *Archive for Rational Mechanics and Analysis*, **89**, 1985, 21–86.

[12] J.-Y. Chemin, Remarques sur l'existence globale pour le système de Navier–Stokes incompressible, *SIAM Journal on Mathematical Analysis*, **23**, 1992, 20–28.

[13] J.-Y. Chemin and I. Gallagher, Large, global solutions to the Navier–Stokes equations, slowly varying in one direction, *Transactions of the American Mathematical Society* **362**, 2010, 2859–2873.

[14] J.-Y. Chemin, I. Gallagher and C. Mullaert, The role of spectral anisotropy in the resolution of the three-dimensional Navier–Stokes equations, "Studies in Phase Space Analysis with Applications to PDEs", *Progress in Nonlinear Differential Equations and Their Applications* **84**, Birkhauser, 53–79, 2013.

[15] J.-Y. Chemin, I. Gallagher and P. Zhang, Sums of large global solutions to the incompressible Navier–Stokes equations, *Journal für die reine und angewandte Mathematik*, **681**, 2013, 65–82.

[16] J.-Y. Chemin and N. Lerner, Flot de champs de vecteurs non lipschitziens et équations de Navier–Stokes, *Journal of Differential Equations*, **121**, 1995, 314–328.

[17] J.-Y. Chemin and P. Zhang, On the global wellposedness to the 3-D incompressible anisotropic Navier–Stokes equations, *Communications in Mathematical Physics*, **272**, 2007, 529–566.

[18] I. Gallagher, Profile decomposition for solutions of the Navier–Stokes equations, *Bulletin de la Société Mathématique de France*, **129**, 2001, 285–316.

[19] I. Gallagher, D. Iftimie and F. Planchon, Asymptotics and stability for global solutions to the Navier–Stokes equations, *Annales de l'Institut Fourier*, **53**, 2003, 1387–1424.

[20] I. Gallagher, G. Koch and F. Planchon, A profile decomposition approach to the $L_t^\infty(L_x^3)$ Navier–Stokes regularity criterion, *Mathematische Annalen*, **355**, 2013, 1527–1559.

[21] P. Gérard, Description du défaut de compacité de l'injection de Sobolev, *ESAIM Control, Optimisation and Calculus of Variations*, **3**, 1998, 213–233.

[22] P. Germain, The second iterate for the Navier–Stokes equation, *Journal of Functional Analysis*, **255**, 2008, 2248–2264.

[23] G. Gui and P. Zhang, Stability to the global large solutions of 3-D Navier–Stokes equations, *Advances in Mathematics* **225**, 2010, 1248–1284.

[24] T. Hmidi and S. Keraani, Blowup theory for the critical nonlinear Schrödinger equations revisited, *International Mathematics Research Notices*, **46**, 2005, 2815–2828.

[25] D. Iftimie, Resolution of the Navier–Stokes equations in anisotropic spaces, *Revista Matematica Ibero–americana*, **15**, 1999, 1–36.

[26] S. Jaffard, Analysis of the lack of compactness in the critical Sobolev embeddings, *Journal of Functional Analysis*, **161**, 1999, 384–396.

[27] H. Jia and V. Şverák, Minimal L^3-initial data for potential Navier–Stokes singularities, *SIAM J. Math. Anal.* **45**, 2013, 1448–1459.

[28] H. Jia and V. Şverák, Minimal L^3-initial data for potential Navier–Stokes singularities, *arXiv:1201.1592*.

[29] C.E. Kenig and G. Koch, An alternative approach to the Navier–Stokes equations in critical spaces, *Annales de l'Institut Henri Poincaré (C) Non Linear Analysis*, **28**, 2011, 159–187.

[30] C. E. Kenig and F. Merle, Global well-posedness, scattering and blow-up for the energy critical focusing non-linear wave equation, *Acta Mathematica*, **201**, 2008, 147–212.

[31] S. Keraani, On the defect of compactness for the Strichartz estimates of the Schrödinger equation, *Journal of Differential equations*, **175**, 2001, 353–392.

[32] G. Koch, Profile decompositions for critical Lebesgue and Besov space embeddings, *Indiana University Mathematical Journal*, **59**, 2010, 1801–1830.

[33] H. Koch and D. Tataru, Well-posedness for the Navier–Stokes equations, *Advances in Mathematics*, **157**, 2001, 22–35.

[34] P.-G. Lemarié-Rieusset, Recent developments in the Navier–Stokes problem, *Chapman and Hall/CRC Research Notes in Mathematics*, **43**, 2002.

[35] J. Leray, Essai sur le mouvement d'un liquide visqueux emplissant l'espace, *Acta Matematica*, **63**, 1933, 193–248.

[36] J. Leray, Étude de diverses équations intégrales non linéaires et de quelques problèmes que pose l'hydrodynamique. *Journal de Mathématiques Pures et Appliquées*, **12**, 1933, 1–82.

[37] P.-L. Lions, The concentration-compactness principle in the calculus of variations. The limit case I, *Revista. Matematica Iberoamericana* **1** (1), 1985, 145–201.

[38] P.-L. Lions, The concentration-compactness principle in the calculus of variations. The limit case II, *Revista Matematica Iberoamericana* **1** (2), 1985, 45–121.

[39] Y. Meyer, *Wavelets, paraproducts and Navier–Stokes equations*, Current developments in mathematics, International Press, Boston, MA, 1997.

[40] F. Merle and L. Vega, Compactness at blow-up time for L^2 solutions of the critical nonlinear Schrödinger equation in 2D, *International Mathematical Research Notices*, **1998**, 399–425.

[41] M. Paicu, Équation anisotrope de Navier–Stokes dans des espaces critiques, *Revista Matematica Iberoamericana* **21** (1), 2005, 179–235.

[42] E. Poulon, About the possibility of minimal blow up for Navier–Stokes solutions with data in $H^s(\mathbb{R}^3)$, *arXiv:1505.06197*.

[43] W. Rusin and V. Şverák, Minimal initial data for potential Navier–Stokes singularities, *Journal of Functional Analysis* **260** (3), 2011, 879–891.

[44] I. Schindler and K. Tintarev, An abstract version of the concentration compactness principle, *Revista Matematica Computense*, **15** (2002), 417–436.

[45] M. Struwe, A global compactness result for boundary value problems involving limiting nonlinearities, *Mathematische Zeitschrift*, **187**, 1984, 511–517.

[46] H. Triebel, *Interpolation theory, function spaces, differential operators*, Second edition. Johann Ambrosius Barth, Heidelberg, 1995.

* Laboratoire d'Analyse et de Mathématiques Appliquées, Paris
† Laboratoire Jacques Louis Lions, Paris
‡ Université Paris Diderot, Paris

3

The Motion Law of Fronts for Scalar Reaction-diffusion Equations with Multiple Wells: the Degenerate Case

Fabrice Bethuel* and Didier Smets[†]

Dedicated to the memory of our friend Abbas Bahri, with our deepest admiration.

We derive a precise motion law for fronts of solutions to *scalar* one-dimensional reaction-diffusion equations with equal-depth multiple wells, in the case when the second derivative of the potential vanishes at its minimizers. We show that, *renormalizing time* in an *algebraic* way, the motion of fronts is governed by a simple system of ordinary differential equations of nearest neighbor interaction type. These interactions may be either attractive or repulsive. Our results are not constrained by the possible occurrence of collisions nor splittings. They present substantial differences with the results obtained in the case when the second derivative does not vanish at the wells, a case which has been extensively studied in the literature, and where fronts have been shown to move at exponentially small speed, with motion laws which are *not renormalizable*.

3.1 Introduction

This paper is a continuation of our previous works [4, 5] where we analyzed the behavior of solutions v of the reaction-diffusion equation of gradient type

$$(\text{PGL})_\varepsilon \qquad \frac{\partial v_\varepsilon}{\partial t} - \frac{\partial^2 v_\varepsilon}{\partial x^2} = -\frac{1}{\varepsilon^2} \nabla V(v_\varepsilon),$$

where $0 < \varepsilon < 1$ is a small parameter. In [5], we considered the case where the potential V is a smooth map from $\mathbb{R}^k$ to $\mathbb{R}$ with multiple wells of equal depth whose second derivative vanishes at the wells. The main result there, stated in Theorem 3.2 here, provides an upper bound for the speed of fronts. In the present paper we *restrict ourselves to the scalar case*, $k = 1$, and provide a precise motion law for the fronts, showing in particular that the *upper bound provided in* [5] *is sharp*. We assume throughout this paper that the potential V

is a smooth function from $\mathbb{R}$ to $\mathbb{R}$ which satisfies the following assumptions:

(A$_1$) $\inf V = 0$ and the set of minimizers $\Sigma \equiv \{y \in \mathbb{R}, V(y) = 0\}$ is finite, with at least two distinct elements, that is

$$\Sigma = \{\sigma_1, \ldots, \sigma_q\}, \ q \geq 2, \ \sigma_1 < \cdots < \sigma_q.$$

(A$_2$) There exists an integer $\theta > 1$ such that for all i in $\{1, \ldots, q\}$, we have

$$V(u) = \lambda_i (u - \sigma_i)^{2\theta} + \underset{u \to \sigma_i}{o}((u - \sigma_i)^{2\theta}), \text{where } \lambda_i > 0.$$

(A$_3$) There exist constants $\alpha_\infty > 0$ and $R_\infty > 0$ such that

$$u \cdot \nabla V(u) \geq \alpha_\infty |u|^2, \text{ if } |u| > R_\infty.$$

Whereas assumption (A$_1$) expresses the fact that the potential possesses at least two minimizers, also termed wells, and (A$_3$) describes the behavior at infinity, and is of a more technical nature, assumption (A$_2$), which is *central in the present paper*, describes the local behavior near the minimizing wells. The number θ is of course related to the order of vanishing of the derivatives near zero. Since $\theta > 1$, then $V''(\sigma_i) = 0$, and (A$_2$) holds if and only if

$$\frac{d^j}{du^j} V(\sigma_i) = 0 \ \text{ for } j = 1, \ldots, 2\theta - 1 \ \text{ and } \ \frac{d^{2\theta}}{du^{2\theta}} V(\sigma_i) > 0,$$

with

$$\lambda_i = \frac{1}{(2\theta)!} \frac{d^{2\theta}}{du^{2\theta}} V(\sigma_i).$$

A typical example of such potentials is given by $V(u) = (1 - u^2)^{2\theta} = (1 - u)^{2\theta}(1 + u)^{2\theta}$ which has two minimizers, $+1$ and -1, so that $\Sigma = \{+1, -1\}$, minimizers vanishing at order 2θ. In this paper, the order of degeneracy is an integer assumed to be the same at all wells: fractional or site-dependent orders (including non-degenerate) may presumably be handled with the same tools, however at the cost of more complicated statements.

   We recall that equation (PGL)$_\varepsilon$ corresponds to the L^2 gradient-flow of the energy functional $\mathcal{E}_\varepsilon$ which is defined for a function $u : \mathbb{R} \mapsto \mathbb{R}$ by the formula

$$\mathcal{E}_\varepsilon(u) = \int_\mathbb{R} e_\varepsilon(u) = \int_\mathbb{R} \frac{\varepsilon |\dot{u}|^2}{2} + \frac{V(u)}{\varepsilon}.$$

As in [4, 5], we consider only *finite energy solutions*. More precisely, we fix an arbitrary constant $M_0 > 0$ and we consider the condition

(H$_0$) $\mathcal{E}_\varepsilon(u) \leq M_0 < +\infty.$

Besides the assumptions on the potential, the main assumption is on the initial data $v_\varepsilon^0(\cdot) = v_\varepsilon(\cdot, 0)$, assumed to satisfy (H$_0$) independently of ε. In particular, in view of the classical energy identity

$$\mathcal{E}_\varepsilon(v_\varepsilon(\cdot, T_2)) + \varepsilon \int_{T_1}^{T_2} \int_{\mathbb{R}} \left| \frac{\partial v_\varepsilon}{\partial t} \right|^2 (x, t) \mathrm{d}x \, \mathrm{d}t = \mathcal{E}_\varepsilon(v_\varepsilon(\cdot, T_1)) \quad \forall 0 \leq T_1 \leq T_2,$$

(3.1)

we have

$$\mathcal{E}_\varepsilon(v_\varepsilon(\cdot, t)) \leq M_0, \ \forall t \geq 0.$$

This implies in particular that for every given $t \geq 0$, we have $V(v_\varepsilon(x, t)) \to 0$ as $|x| \to \infty$. It is then quite straightforward to deduce from assumption (H$_0$), (A$_1$), (A$_2$) as well as the energy identity (3.1), that $v_\varepsilon(x, t) \to \sigma_\pm$ as $x \to \pm\infty$, where $\sigma_\pm \in \Sigma$ do not depend on t. In other words, for any time, our solutions connect two given minimizers of the potential.

3.1.1 Main Results: Fronts and Their Speed

The notion of fronts is central in the dynamics. For a map $u : \mathbb{R} \mapsto \mathbb{R}$, the set

$$\mathcal{D}(u) \equiv \{x \in \mathbb{R}, \ \mathrm{dist}(u(x), \Sigma) \geq \mu_0\},$$

is termed throughout *the front set* of u. The constant μ_0 which appears in its definition is fixed once for all, sufficiently small so that

$$\frac{\lambda_i}{2}(u - \sigma_i)^{2\theta} \leq V(u) \leq \frac{1}{\theta} V'(u)(u - \sigma_i) \leq 4V(u) \leq 8\lambda_i(u - \sigma_i)^{2\theta}, \quad (3.2)$$

for each $i \in \{1, \ldots, q\}$ and whenever $|u - \sigma_i| \leq \mu_0$. The front set corresponds to the set of points where u is "far" from the minimizers σ_i, and hence where transitions from one minimizer to the other may occur. A straightforward analysis yields the following.

Lemma 3.1 (see e.g. [4]) *Assume that u verifies* (H$_0$)*. Then there exist ℓ points $x_1, \ldots, x_\ell$ in $\mathcal{D}(u)$ such that*

$$\mathcal{D}(u) \subset \bigcup_{k=1}^{\ell} [x_k - \varepsilon, x_k + \varepsilon],$$

with a bound $\ell \leq \frac{M_0}{\eta_0}$ on the number of points, η_0 being some constant depending only on V.

In view of Lemma 3.1, the measure of the front sets is of order ε, and corresponds to a small neighborhood of order ε of the points x_i. Notice that if $(u_\varepsilon)_{\varepsilon > 0}$ is a family of functions satisfying (H$_0$) then it is well known that the family is locally bounded in $BV(\mathbb{R}, \mathbb{R})$ and hence locally compact in $L^1(\mathbb{R}, \mathbb{R})$. Passing to a subsequence if necessary, we may assert that

$$u_\varepsilon \to u^\star \text{ in } L^1_{\text{loc}}(\mathbb{R}),$$

where $u^\star$ takes values in Σ and is a step function. More precisely, there exist an integer $\ell \le \frac{M_0}{\eta_0}$, ℓ points $a_1 < \cdots < a_\ell$ and a function $\hat{\imath} : \{\frac{1}{2}, \ldots, \frac{1}{2} + \ell\} \to \{1, \ldots, q\}$ such that

$$u^\star = \sigma_{\hat{\imath}(k+\frac{1}{2})} \text{ on } (a_k, a_{k+1}),$$

for $k = 0, \ldots, \ell$, and where we use the convention $a_0 := -\infty$ and $a_{\ell+1} := +\infty$. The points a_k, for $k = 1 \ldots, \ell$, are the limits as ε shrinks to 0 of the points x_i provided by Lemma 3.1 (the number and the positions of which are of course ε dependent), so that the front set $\mathcal{D}(u_\varepsilon)$ shrinks as ε tends to 0 to a finite set. In the sequel, we shall refer to step functions with values into Σ as steep front chains and we will write

$$u^\star = u^\star(\ell, \hat{\imath}, \{a_k\})$$

to determine them unambiguously.

We go back to equation (PGL)$_\varepsilon$ and consider a family of functions $(v_\varepsilon)_{\varepsilon > 0}$ defined on $\mathbb{R} \times \mathbb{R}^+$ which are solutions to the equation (PGL)$_\varepsilon$ and satisfy the energy bound (H$_0$). We set

$$\mathcal{D}_\varepsilon(t) = \mathcal{D}(v_\varepsilon(\cdot, t)).$$

The evolution of the front set $\mathcal{D}_\varepsilon(t)$ when ε tends to 0 is the main focus of our paper. The following result[1] has been proved in [5].

Theorem 3.2 ([5]) *There exist constants* $\rho_0 > 0$ *and* $\alpha_0 > 0$, *depending only on the potential V and on* M_0, *such that if* $r \ge \alpha_0 \varepsilon$, *then*

$$\mathcal{D}_\varepsilon(t + \Delta t) \subset \mathcal{D}_\varepsilon(t) + [-r, r], \qquad \text{for every } t \ge 0, \tag{3.3}$$

provided $0 \le \Delta t \le \rho_0 r^2 \left(\frac{r}{\varepsilon}\right)^{\frac{\theta+1}{\theta-1}}$.

As a matter of fact, it follows from this result that the average speed of the front set at that length-scale should not exceed

$$c_{\text{ave}} \simeq \frac{r}{(\Delta t)_{\max}} \le \rho_0^{-1} r^{-(\omega+1)} \varepsilon^\omega, \tag{3.4}$$

where

$$\omega = \frac{\theta+1}{\theta-1}. \tag{3.5}$$

---

[1] which holds also more generally for systems.

Notice that $1 < \omega < +\infty$ and that the upper bound provided by (3.4) decreases with θ, that is, the more degenerate the minimizers of V are, the higher the possible speed allowed by the bound (3.4). In contrast, the speed is at most exponentially small in the case of non-degenerate potentials (see e.g. [9], [4] and the references therein). One aim of the present paper is to show that the *upper bound* provided by the estimate (3.4) is in fact optimal[2] and actually to derive a precise motion law for the fronts. An important fact, on which our results are built, is the following observation[3]:

Equation (PGL)$_\varepsilon$ *is renormalizable.*

This assertion means that, rescaling time in an appropriate way, the evolution of fronts in the asymptotic limit $\varepsilon \to 0$ is governed by an ordinary differential equation which *does not involve the parameter* ε. More precisely, we accelerate time by the factor $\varepsilon^{-\omega}$ and consider the new time $s = \varepsilon^\omega t$. In the accelerated time, we consider the map

$$\mathfrak{v}_\varepsilon(x,s) = v_\varepsilon(x, s\varepsilon^{-\omega}), \text{ and set } \mathfrak{D}_\varepsilon(s) = \mathcal{D}(\mathfrak{v}_\varepsilon(\cdot, s)). \tag{3.6}$$

It follows from Theorem 3.2 that for given $r \geq \alpha_0 \varepsilon$,

$$\mathfrak{D}_\varepsilon(s + \Delta s) \subset \mathfrak{D}_\varepsilon(s) + [-r, r], \qquad \text{for every } s \geq 0, \tag{3.7}$$

provided that $0 \leq \Delta s \leq \rho_0 r^{\omega+2}$.

Concerning the initial data, we will assume that there exists a steep front chain $v^\star(\ell_0, \hat{\imath}_0, \{a_k^0\})$ such that

$$(\mathrm{H}_1) \quad \begin{cases} v_\varepsilon^0 \longrightarrow v^\star(\ell_0, \hat{\imath}_0, \{a_k^0\}) \text{ in } L^1_{\mathrm{loc}}(\mathbb{R}), \\ \mathfrak{D}_\varepsilon(0) \longrightarrow \{a_k^0\}_{1 \leq k \leq \ell_0}, \text{locally in the sense of the Hausdorff distance,} \end{cases}$$

as $\varepsilon \to 0$. Let us emphasize that *assumption* (H_1) *is not restrictive*, since it follows assuming only the energy bound (H_0) and passing possibly to a subsequence (see above). In our first result, we will impose the additional condition

$$(\mathrm{H}_{\min}) \quad |\hat{\imath}_0(k + \tfrac{1}{2}) - \hat{\imath}_0(k - \tfrac{1}{2})| = 1 \text{ for } 1 \leq k \leq \ell_0.$$

This assumption could be rephrased as a "multiplicity one" condition: it means that the jumps consist of exactly one transition between consecutive minimizers σ_i and σ_{i+1}. To each transition point a_k^0 we may assign a sign,

---

[2] at least in the scalar case considered here.
[3] which to our knowledge has not been observed before, even using formal arguments.

denoted by $\dagger_k \in \{+, -\}$, in the following way:

$$\dagger_k = +\ \text{if}\ \sigma_{\hat{\imath}_0(k+\frac{1}{2})} = \sigma_{\hat{\imath}_0(k-\frac{1}{2})} + 1\ \text{ and }\ \dagger_k = -\ \text{if}\ \sigma_{\hat{\imath}_0(k+\frac{1}{2})} = \sigma_{\hat{\imath}_0(k-\frac{1}{2})} - 1.$$

We consider next the system of ordinary differential equations

$$(\mathcal{S}) \qquad \mathfrak{S}_k \frac{d}{ds} a_k = \frac{\Gamma_k^+}{\left(a_{k+1} - a_k\right)^{\omega+1}} - \frac{\Gamma_k^-}{\left(a_k - a_{k-1}\right)^{\omega+1}},$$

for $1 \leq k \leq \ell_0$, where we implicitly set $a_0 = -\infty$ and $a_{\ell_0+1} = +\infty$, $\mathfrak{S}_k$ stands for the energy of the corresponding stationary front, namely

$$\mathfrak{S}_k = \int_{\sigma_{\hat{\imath}_0(k-\frac{1}{2})}}^{\sigma_{\hat{\imath}_0(k+\frac{1}{2})}} \sqrt{2V(u)}\,du, \tag{3.8}$$

and where we have set,[4] for $k = 1, \ldots, \ell_0$,

$$\Gamma_k^{\pm} = \begin{cases} -2^{\omega}\left(\lambda_{\hat{\imath}_0(k\pm\frac{1}{2})}\right)^{-\frac{1}{\theta-1}} \mathcal{A}_{\theta} & \text{if } \dagger_k = -\dagger_{k\pm 1}, \\[2ex] -2^{\omega}\left(\lambda_{\hat{\imath}_0(k\pm\frac{1}{2})}\right)^{-\frac{1}{\theta-1}} \mathcal{B}_{\theta} & \text{if } \dagger_k = \dagger_{k\pm 1}. \end{cases} \tag{3.9}$$

In (3.9), $\lambda_{\hat{\imath}_0(k+\frac{1}{2})}$ is defined in (A_2) and the constants $\mathcal{A}_{\theta} < 0$ and $\mathcal{B}_{\theta} > 0$, depending only on θ, are defined in (A.9) of Appendix A, they are related to the unique solutions of the two singular boundary value problems

$$\begin{cases} -\dfrac{d^2\mathcal{U}}{dx^2} + \mathcal{U}^{2\theta-1} = 0 & \text{on } (-1,1), \\[2ex] \mathcal{U}(-1) = \pm\infty, \qquad \mathcal{U}(1) = +\infty. \end{cases}$$

Note in particular that $(\mathcal{S})$ is fully determined by the pair $(\ell_0, \hat{\imath}_0)$, and we shall therefore sometimes refer to it as $\mathcal{S}_{\ell_0, \hat{\imath}_0}$. Our first result is Theorem 3.3.

Theorem 3.3 *Assume that the initial data* $(v_{\varepsilon}(0))_{0<\varepsilon<1}$ *satisfy conditions* (H_0), (H_1), *and* $(H_{\min})$, *and let* $0 < S_{\max} \leq +\infty$ *denote the maximal time of existence for the system* $\mathcal{S}_{\ell_0, \hat{\imath}_0}$ *with initial data* $a_k(0) = a_k^0$. *Then, for* $0 < s < S_{\max}$,

$$\mathfrak{v}_{\varepsilon}(s) \longrightarrow v^{\star}(\ell_0, \hat{\imath}_0, \{a_k(s)\}) \tag{3.10}$$

in $L_{\text{loc}}^{\infty}(\mathbb{R} \setminus \cup_{k=1}^{\ell_0}\{a_k(s)\})$, *as* $\varepsilon \to 0$. *In particular,*

$$\mathfrak{D}_{\varepsilon}(s) \longrightarrow \cup_{k=1}^{\ell_0}\{a_k(s)\} \tag{3.11}$$

locally in the sense of the Hausdorff distance, as $\varepsilon \to 0$.

---

[4] In view of our definition of a_0 and a_{ℓ_0+1} the quantities Γ_0^- and $\Gamma_{\ell_0}^+$ need not be well defined.

We consider now the more general situation where (H_{min}) is not verified, and for $1 \leq k \leq \ell_0$ we denote by m_k^0 the algebraic multiplicity of a_k^0, that is, we set

$$m_k^0 = \hat{\imath}(k + \frac{1}{2}) - \hat{\imath}(k - \frac{1}{2}). \qquad (3.12)$$

The case $m_k^0 = 0$ corresponds to *ghost fronts*, whereas $|m_k^0| \geq 2$ corresponds to *multiple fronts*. The total number of fronts that will eventually emerge from such initial data is given by

$$\ell_1 = \sum_{k=1}^{\ell_0} |m_k^0|,$$

and their ordering is obtained by splitting multiple fronts according to the order in Σ. More precisely, we define the function $\hat{\imath}_1$ by

$$\begin{cases} \hat{\imath}_1(\frac{1}{2}) = \hat{\imath}_0(\frac{1}{2}), \\ \hat{\imath}_1(M_k^0 + p + \frac{1}{2}) = \hat{\imath}_0(k + \frac{1}{2}) + p, \text{ for } p = 0, \ldots, |m_k^0| - 1 \text{ if } m_k^0 > 0, \quad (3.13) \\ \hat{\imath}_1(M_k^0 + p + \frac{1}{2}) = \hat{\imath}_0(k + \frac{1}{2}) - p, \text{ for } p = 0, \ldots, |m_k^0| - 1 \text{ if } m_k^0 < 0, \end{cases}$$

where $k = 1, \ldots, \ell_0$ and $M_k^0 := \sum_{p=1}^{k-1} |m_p^0|$. We say that $(\ell_1, \hat{\imath}_1)$ is the splitting of $(\ell_0, \hat{\imath}_0)$.

Definition 3.4 *A splitting solution of (S) with initial data $(\ell_0, \hat{\imath}_0, \{a_k^0\})$ on the interval $[0, S]$ is a solution $a \equiv (a_1, \ldots, a_{\ell_1}) : (0, S) \to \mathbb{R}^{\ell_1}$ of $S_{\ell_1, \hat{\imath}_1}$ such that*

$$\lim_{s \to 0^+} a_k(s) = a_j^0 \qquad for \qquad k = M_j^0, \ldots, M_j^0 + |m_j^0| - 1,$$

for any $j = 1, \ldots, \ell_0$, where $(\ell_1, \hat{\imath}_1)$ is the splitting of $(\ell_0, \hat{\imath}_0)$.

We are now in a position to complete Theorem 3.3 by relaxing assumption (H_{min}).

Theorem 3.5 *Assume that the initial data $(v_\varepsilon^0)_{0 < \varepsilon < 1}$ satisfy conditions (H_0) and (H_1). Then there exist a subsequence $\varepsilon_n \to 0$, and a splitting solution of (S) with initial data $(\ell_0, \hat{\imath}_0, \{a_k^0\})$, defined on its maximal time of existence $[0, S_{max})$, and such that for any $0 < s < S_{max}$*

$$\mathfrak{v}_{\varepsilon_n}(s) \longrightarrow v^\star(\ell_1, \hat{\imath}_1, \{a_k(s)\}) \qquad (3.14)$$

in $L_{loc}^\infty(\mathbb{R} \setminus \cup_{k=1}^{\ell_1} \{a_k(s)\})$, as $n \to +\infty$. In particular,

$$\mathfrak{D}_{\varepsilon_n}(s) \longrightarrow \cup_{j=1}^{\ell_1} \{a_k(s)\} \qquad (3.15)$$

locally in the sense of the Hausdorff distance, as $n \to +\infty$.

Remark 3.6 Local existence of splitting solutions can be established in different ways (including in particular using Theorem 3.5 !). To our knowledge, *uniqueness is not known,* unless of course $|m_k^0| \leq 1$ for all k, this is the main reason why convergence is only obtained for a subsequence in Theorem 3.5 whereas it was for the full sequence in Theorem 3.3.

So far, our results are constrained by the maximal time of existence $S_{\max}$ of the differential equation (S), which is related to the occurrence of collisions. To pursue the analysis past collisions, we first briefly discuss some properties of the system of equations (S), we refer to Appendix B for more details. The system (S) describes nearest neighbor interactions with an interaction law of the form $\pm d^{-(\omega+1)}$, d standing for the distance between fronts. The sign of the interactions is crucial, since the system may contain both repulsive forces leading to spreading and attractive forces leading to collisions, yielding the maximal time of existence $S_{\max}$. In order to take signs into account, we set

$$\epsilon_{k+\frac{1}{2}} = \text{sign}\,(\Gamma_{k+\frac{1}{2}}) = -\,\dagger_k\,\dagger_{k+1}, \text{ for } k = 0,\ldots,\ell_0 - 1. \qquad (3.16)$$

The case $\epsilon_{k+\frac{1}{2}} = -1$ corresponds to *repulsive forces* between a_k and a_{k+1}, whereas the case $\epsilon_{k+\frac{1}{2}} = +1$ corresponds to *attractive forces* between a_k and a_{k+1}, leading to collisions. As a matter of fact, in this last case a_{k+1} corresponds to the *anti-front* of a_k. In order to describe the magnitude of the forces, we introduce the subsets $J^{\pm}$ of $\{1,\ldots,\ell_0\}$ defined by $J^{\pm} = \{k \in \{1,\ldots,\ell_0 - 1\}, \text{ such that } \epsilon_{k+\frac{1}{2}} = \mp 1\}$ and the quantities

$$\begin{cases} \delta_a(s) = \inf\{|a_k(s) - a_{k+1}(s)|, \text{ for } k \in 1,\ldots,\ell_0 - 1\}, \\ \delta_a^{\pm}(s) = \inf\{|a_k(s) - a_{k+1}(s)|, \text{ for } k \in J^{\pm}\}. \end{cases} \qquad (3.17)$$

Proposition 3.7 *There are positive constants S_1, S_2, S_3 and S_4 depending only on the coefficients of the equation (S), such that for any time $s \in [0, S_{\max})$ we have*

$$\begin{cases} \delta_a^+(s) \geq \left(S_1 s + S_2 \delta_a^+(0)^{\omega+2}\right)^{\frac{1}{\omega+2}}, \\ \delta_a^-(s) \leq \left(S_3 \delta_a^-(0)^{\omega+2} - S_4 t\right)^{\frac{1}{\omega+2}}. \end{cases} \qquad (3.18)$$

If for every $k = 1,\ldots,\ell_0$ we have $\epsilon_{k+\frac{1}{2}} = -1$, then $S_{\max} = +\infty$. Otherwise, we have the estimate

$$S_{\max} \leq \frac{S_3}{S_4}\left(\delta_a^-(0)\right)^{\omega+2} \equiv \mathcal{K}_0 \left(\delta_a^-(0)\right)^{\omega+2}. \qquad (3.19)$$

This result shows that the maximal time of existence for solutions to (S) is related to the value of $\delta_a^-(0)$, the minimal distance between fronts and anti-fronts at time 0. By the semi-group property, the same can be said about $\eth_a^-(s)$, that is

$$S_{max} - s \lesssim \eth_a^-(s)^{\omega+2}.$$

On the other hand, in view of $(\mathcal{S})$, $\eth_a^-(s)$ provides an upper bound for the speeds $\dot{a}_k(s)$ in case of collision, that is

$$|\frac{d}{ds}a_k(s)| \lesssim \eth_a^-(s)^{-(\omega+1)}.$$

It follows that

$$\int_0^{S_{max}} |\frac{d}{ds}a_k(s)|\, ds \lesssim \int_0^{S_{max}} (S_{max} - s)^{-\frac{\omega+1}{\omega+2}}\, ds < +\infty$$

and therefore that the trajectories are absolutely continuous up to the collision time. Also, since $\eth_a^+$ remains bounded from below by a positive constant, each front can only enter into collision with its anti-front (but there could be multiple copies of both). From a heuristic point of view, it is therefore rather simple to extend solutions past the collision time: it suffices to remove the colliding pairs from the collection of points, so that the total number of points has been decreased by an even number. More precisely, we have the following.

Corollary 3.8 *Let $\ell_1, \hat{\imath}_1$, $a \equiv (a_1, \ldots, a_{\ell_1})$ and S_{max} be as in Theorem 3.5. Then, there exists $\ell_2 \in \mathbb{N}$ such that $\ell_1 - \ell_2 \in 2\mathbb{N}_*$, and there exist ℓ_2 points $b_1 < \cdots < b_{\ell_2}$ such that for all $k = 1, \ldots, \ell_1$*

$$\lim_{s \to S_{max}^-} a_k(s) = b_{j(k)} \qquad \text{for some } j(k) \in \{1, \ldots, \ell_2\}.$$

Moreover, if we set $\hat{\imath}_2(\frac{1}{2}) = \hat{\imath}_1(\frac{1}{2})$ and

$$\hat{\imath}_2(q + \tfrac{1}{2}) = \hat{\imath}_1(k(q) + \tfrac{1}{2}), \quad \text{where} \quad k(q) = max\{k \in \{1, \ldots, \ell_1\} \text{ s.t. } j(k) = q\},$$

for $q = 1, \ldots, \ell_2$, then

$$\hat{\imath}_2(q + \tfrac{1}{2}) - \hat{\imath}_2(q - \tfrac{1}{2}) \in \{+1, -1, 0\}$$

for all $q = 1, \ldots, \ell_2$.

We stress that Corollary 3.8 is obtained from Theorem 3.5 using only properties of the system of ODEs $(\mathcal{S})$, in particular Proposition 3.7.

We are now in position to state our last result.

Theorem 3.9 *Under the assumptions of Theorem 3.5, we have as $n \to +\infty$,*

$$\mathfrak{v}_{\varepsilon_n}(S_{max}) \longrightarrow v^*(\ell_2, \hat{\imath}_2, \{b_k\}) \quad \text{in} \quad L_{loc}^\infty(\mathbb{R} \setminus \cup_{k=1}^{\ell_2} \{b_k\}), \qquad (3.20)$$

where ℓ_2, $\hat{\imath}_2$ and $b_1 < \cdots < b_{\ell_2}$ are given by Corollary 3.8. In particular the sequence $(\mathfrak{v}_{\varepsilon_n}(S_{\max}))_{n \in \mathbb{N}}$, considered as initial data, satisfies the assumptions (H_0) and (H_1) with $\ell_0 := \ell_2$ and $\{a_k^0\} := \{b_k^0\}$.

We may therefore apply Theorem 3.5 to the sequence of initial data $(\mathfrak{v}_{\varepsilon_n}(S_{\max}))_{n \in \mathbb{N}}$, and therefore, using the semi-group property of (3.1), extend the analysis past $S_{\max}$. Notice that since the multiplicities given by $\hat{\imath}_2$ are equal to either ± 1 or 0, no further subsequences are needed to pass through the collision times. Finally, since the total number of fronts is decreased at least by 2 at each collision time, the latter are finitely many.

Some comments on the results Motion of fronts for one-dimensional *scalar* reaction-diffusion equations has already quite a long history. Until recently, most efforts have been devoted to the case where the potential possesses *only two wells* with non-vanishing second derivative: such potentials are often referred to as *Allen–Cahn potentials*. Under suitable preparedness assumptions on the initial datum, the precise motion law for the fronts has been derived by Carr and Pego in their seminal work [9] (see also Fusco and Hale [10]). They showed that the front points are moved, up to the first collision time, according to a first-order differential equation of nearest neighbor interaction type, with interaction terms proportional to $\exp(-\varepsilon^{-1}(a_{k+1}^\varepsilon(t) - a_k^\varepsilon(t)))$. These results present substantial differences from the results in the present paper, in particular we wish to emphasize the following points:

- only attractive forces leading eventually to the annihilation of *fronts* with *anti-fronts* forces are present;
- the equation is *not* renormalizable. Indeed, the various forces $\exp(-\varepsilon^{-1}(a_{k+1}^\varepsilon(t) - a_k^\varepsilon(t)))$ for different values of k may be of very different orders of magnitude, and hence not commensurable.

Besides this, the essence of their method is quite different: it relies on a careful study of the linearized problem around the stationary front, in particular from the spectral point of view. This type of approach is also sometimes termed the *geometric approach* (see e.g. [8]). At least two other methods have been applied successfully to the Allen–Cahn equation. First, the method of subsolutions and supersolutions turns out to be extremely powerful and allowed us to handle larger classes of initial data and also to extend the analysis past collisions: this is for instance achieved by Chen in [8]. Another direction is given by the global energy approach initiated by Bronsard and Kohn [7]. We refer to [4] for more references on these methods.

Several ideas and concepts presented here are influenced by our earlier work on the motion of vortices in the two-dimensional parabolic Ginzburg–Landau equation [2, 3]. As a matter of fact, this equation yields another remarkable example of *renormalizable slow motion*, as proved by Lin or Jerrard and Soner ([13, 11]). Our interest in the questions studied in this paper was certainly driven by the possibility of finding an analogous situation in one space dimension.

This paper belongs to a series of papers we have written on the slow motion phenomenon for reaction-diffusion equation of gradient type with multiple wells (see [4, 5, 6]). Common to all of these papers is a general approach based on the following ingredients.

• A *localized version of the energy identity* (see Subsection 3.1.3). Fronts are then handled as concentration points of the energy, so that the evolution of local energies yields also the motion of fronts. Besides dissipation, this localized energy identity contains a flux term, involving the *discrepancy* function, which has a simple interpretation for stationary solutions. Using test functions which are *affine near the fronts*, the flux term does not see the core of the front, only its tail.
• Parabolic estimates *away* from the fronts.
• Handling the time derivative as a *perturbation* of the one-dimensional elliptic equations, hence allowing elementary tools as Gronwall's identities.

Parallel to this paper, we have also *revisited the scalar non-degenerate case* in [6], considering in particular the case were there are more than two wells, leading as mentioned to repulsive forces which are not present in the Allen–Cahn case. Several tools are shared by the two papers, for instance we rely on related definitions and properties of regularized fronts, and the properties of the ordinary differential equations are quite similar. From a technical point of view, differences appear at the level of the magnitudes of energies as well as of the parameter δ involved in the definition of regular fronts, and more crucially on the nature of the parabolic estimates of the front sets. Whereas in [6] we rely essentially on *linear* estimates, in the degenerate case considered here our estimates are *truly non-linear*, obtained mainly through extensive use of the comparison principle.

Finally, it is presumably worth mentioning that the situation in higher dimension is very different: the dynamics is dominated by *mean-curvature* effects. The phenomena considered in the present paper are therefore of lower order, and do not appear in the limiting equations.

Among the problems left open in our paper, we would like to emphasize again the question of *uniqueness* of splitting solutions for (S), as well as the possibility of interpreting our convergence results in terms of a *Gamma-limit* involving a renormalized energy (see e.g [15] for related results on the Ginzburg–Landau equation).

3.1.2 Regularized Fronts

The notion of regularized fronts is central in our description of the dynamics of equation $(PGL)_\varepsilon$. It is intended to describe in a quantitative way chains of stationary solutions which are well-separated and suitably glued together. It also allows us to pass from *front sets* to *front points*, a notion which is more accurate and therefore requires improved estimates. Recall first that for $i \in \{1, \ldots, q - 1\}$, there exists a unique (up to translations) solution ζ_i^+ to the stationary equation with $\varepsilon = 1$,

$$- v_{xx} + V'(v) = 0 \text{ on } \mathbb{R}, \tag{3.21}$$

with, as conditions at infinity, $v(-\infty) = \sigma_i$ and $v(+\infty) = \sigma_{i+1}$. Set, for $i \in \{1, \ldots, q - 1\}$, $\zeta_i^-(\cdot) \equiv \zeta_i(-\cdot)$, so that ζ_i^- is the unique (up to translations) solution to (3.21) such that $v(+\infty) = \sigma_i$ and $v(-\infty) = \sigma_{i+1}$. A remarkable yet elementary fact, related to the scalar nature of the equation, is that there are no other non-trivial finite energy solutions to equation (3.21) than the solutions $\zeta_i^\pm$ and their translates: in particular there are no solutions connecting minimizers which are not nearest neighbors. For $i = 1, \ldots, q - 1$, we fix a point z_i in the interval (σ_i, σ_{i+1}) where the potential V restricted to $[\sigma_i, \sigma_{i+1}]$ achieves its maximum and we set $\mathcal{Z} = \{z_1, \ldots, z_{q-1}\}$. Again, since we consider only the one-dimensional scalar case, any solution ζ_i takes once and only once the value z_i.

We next describe a *local* notion of well-preparedness.[5] For an arbitrary $r > 0$, we denote by I_r the interval $[-r, r]$.

Definition 3.10 *Let $L > 0$ and $\delta > 0$. We say that a map u verifying (H_0) satisfies the preparedness assumption $\mathcal{WP}_\varepsilon^L(\delta)$ if the following two conditions are fulfilled.*

- $\left(\text{WPI}_\varepsilon^L(\delta)\right)$ *We have*

$$\mathcal{D}(u) \cap I_{2L} \subset I_L \tag{3.22}$$

  and there exists a collection of points $\{a_k\}_{k \in J}$ in I_L, with $J = \{1, \ldots, \ell\}$, such that

$$\mathcal{D}(u) \cap I_{2L} \subset \bigcup_{k \in J} I_k, \text{ where } I_k = [a_k - \delta, a_k + \delta]. \tag{3.23}$$

---

[5] By local, we mean with respect to the interval $[-L, L]$. In contrast the related notion introduced in [6] is global on the whole of $\mathbb{R}$.

For $k \in J$, there exists a number $i(k) \in \{1, \ldots, q-1\}$ such that $u(a_k) = z_{i(k)}$ and a symbol $\dagger_k \in \{+, -\}$ such that

$$\left\| u(\cdot) - \zeta_{i(k)}^{\dagger_k} \left(\frac{\cdot - a_k}{\varepsilon} \right) \right\|_{C_\varepsilon^1(I_k)} \leq \exp\left(-\frac{\delta}{\varepsilon} \right), \qquad (3.24)$$

where $\|u\|_{C_\varepsilon^1(I_k)} = \|u\|_{L^\infty(I_k)} + \varepsilon \|u'\|_{L^\infty(I_k)}$.

- $\left(\text{WPO}_\varepsilon^L(\delta)\right)$ *Set $\Omega_L = I_{2L} \setminus \overset{\ell}{\underset{k=1}{\cup}} I_k$. We have the energy estimate*

$$\int_{\Omega_L} e_\varepsilon(u(x)) \, dx \leq C_w M_0 \left(\frac{\varepsilon}{\delta} \right)^\omega. \qquad (3.25)$$

In the above definition $C_w > 0$ denotes a constant whose exact value is fixed once for all by Proposition 3.22, and which depends only on V. Condition $WPI_\varepsilon^L(\delta)$ corresponds to an *inner matching* of the map with stationary fronts, it is only really meaningful if $\delta \gg \varepsilon$. In the sequel we always assume that

$$\frac{L}{2} \geq \delta \geq \alpha_1 \varepsilon, \qquad (3.26)$$

where α_1 is larger than the α_0 of Theorem 3.2 and also sufficiently large so that if $WPI_\varepsilon^L(\delta)$ holds then the points a_k and the indices $i(k)$ and $\dagger_k$ are *uniquely* and therefore *unambigously* determined and the intervals I_k are disjoints. In particular, the quantity $d_{\min}^{\varepsilon,L}(s)$, defined by

$$d_{\min}^{\varepsilon,L}(s) := \min\left\{ a_{k+1}^\varepsilon(s) - a_k^\varepsilon(s), \quad k = 1, \ldots, \ell(s) - 1 \right\}$$

if $\ell(s) \geq 2$, and $d_{\min}^{\varepsilon,L}(s) = 2L$ otherwise, satisfies $d_{\min}^{\varepsilon,L}(s) \geq 2\delta$. Condition $WPO_\varepsilon^L(\delta)$ is in some weak sense an *outer matching*: it is crucial for some of our energy estimates and its form is motivated by energy decay estimates for stationary solutions. Note that condition $WPI_\varepsilon^L(\delta)$ makes sense on its own, whereas condition $WPO_\varepsilon^L(\delta)$ only makes sense if condition $WPI_\varepsilon^L(\delta)$ is fulfilled. Note also that the larger δ is, the stronger condition $WPI_\varepsilon^L(\delta)$ is. The same is not obviously true for condition $WPO_\varepsilon^L(\delta)$, since the set of integration Ω_L increases with δ. As a matter of fact, the constant C_w in (3.25) is chosen sufficiently big[6] so that $WPO_\varepsilon^L(\delta)$ also becomes stronger when δ is larger. We next specify Definition 3.10 for the maps $x \mapsto \mathfrak{v}_\varepsilon(x, s)$.

Definition 3.11 *For $s \geq 0$, we say that the assumption $\mathcal{WP}_\varepsilon^L(\delta, s)$ (resp. $\mathrm{WPI}_\varepsilon^L(\delta, s)$) holds if the map $x \mapsto \mathfrak{v}_\varepsilon(x, s)$ satisfies $\mathcal{WP}_\varepsilon^L(\delta)$ (resp. $\mathrm{WPI}_\varepsilon^L(\delta)$).*

---

[6] In view of $\mathrm{WPI}_\varepsilon^L(\delta)$, how big it needs to be is indeed related to energy decay estimates for the fronts ζ_i.

When assumption $\text{WPI}_\varepsilon^L(\delta, s)$ holds, then all symbols will be indexed according to s. In particular, we write[7] $\ell(s) = \ell$, $J(s) = J$, and $a_k^\varepsilon(s) = a_k$. The points $a_k^\varepsilon(s)$ for $k \in J(s)$, are now termed the *front points*. Whereas in [6] we are able, due to parabolic regularization, to establish under suitable conditions that $\mathcal{WP}_\varepsilon^L(\delta, s)$ is fulfilled for a length of the same order as the minimal distance between the front points, this *is not the case* in the present situation. More precisely, two orders of magnitude for δ will be considered, namely

$$\delta_{\log}^\varepsilon = \frac{1}{\rho_w} \varepsilon \left| \log\left(4M_0^2 \frac{\varepsilon}{L}\right)\right| \quad \text{and} \quad \delta_{\log\log}^\varepsilon = \frac{\omega}{\rho_w} \varepsilon \log\left(\frac{1}{\rho_w} \left|\log\left(4M_0^2 \frac{\varepsilon}{L}\right)\right|\right). \tag{3.27}$$

In (3.27), the constant ρ_w (given by Lemma 3.25) depends only on V. The main property for our purposes is that $\delta_{\log\log}^\varepsilon / \varepsilon$ and $\delta_{\log}^\varepsilon / \delta_{\log\log}^\varepsilon$ both tend to $+\infty$ as ε/L tends to 0.

In many places, it is useful to rely on a slightly stronger version of the confinement condition (3.22), which we assume to hold on some interval of time. More precisely, for positive L, S we consider the condition

$$(\mathcal{C}_{L,S}) \qquad \mathfrak{D}_\varepsilon(s) \cap I_{4L} \subset I_L, \qquad \forall\, 0 \le s \le S.$$

For given $L_0 > 0$ and $S > 0$, it follows easily from assumption (H_1) and Theorem 3.2 that there exists $L \ge L_0$ for which the first condition in $(\mathcal{C}_{L,S})$ is satisfied. Under condition $(\mathcal{C}_{L,S})$, the estimate

$$\mathcal{E}_\varepsilon(\mathfrak{v}_\varepsilon(s), I_{3L} \setminus I_{\frac{3}{2}L}) \le C_e \left(\frac{\varepsilon}{L}\right)^\omega, \qquad \forall\, s \in [\varepsilon^\omega L^2, S], \tag{3.28}$$

where $C_e > 0$ depends only on V, follows from the following regularizing effect, which was obtained in [5].

Proposition 3.12 ([5]) *Let $\mathfrak{v}_\varepsilon$ be a solution to $(\text{PGL})_\varepsilon$, let $x_0 \in \mathbb{R}$, $r > 0$ and $0 \le s_0 < S$ be such that*

$$\mathfrak{v}_\varepsilon(y, s) \in B(\sigma_i, \mu_0) \quad \text{for all } (y, s) \in [x_0 - r, x_0 + r] \times [s_0, S], \tag{3.29}$$

for some $i \in \{1, \dots, \mathfrak{q}\}$. Then we have for $s_0 < s \le S$

$$\varepsilon^{-\omega} \int_{x_0 - 3r/4}^{x_0 + 3r/4} e_\varepsilon(\mathfrak{v}_\varepsilon(x, s))\, dx \le \frac{1}{10} C \left(1 + \left(\frac{\varepsilon^\omega r^2}{s - s_0}\right)^{\frac{\theta}{\theta - 1}}\right) \left(\frac{1}{r}\right)^\omega \tag{3.30}$$

---

[7] In principle and at this stage, all those symbols depend also upon ε. Since eventually ℓ and J will be ε-independent, at least for ε sufficiently small, we do not explicitly index them with ε.

as well as

$$|v_\varepsilon(y,s) - \sigma_i| \le \frac{1}{10} C \varepsilon^{\frac{1}{\theta-1}} \left(\left(\frac{1}{r}\right)^{\frac{1}{\theta-1}} + \left(\frac{\varepsilon^\omega}{s-s_0}\right)^{\frac{1}{2(\theta-1)}} \right), \tag{3.31}$$

for $y \in [x_0 - 3r/4, x_0 + 3r/4]$, where the constant $C > 0$ depends only on V.

Our first ingredient is the following.

Proposition 3.13 *There exists $\alpha_1 > 0$, depending only on M_0 and V, such that if $L \ge \alpha_1 \varepsilon$ and if $(\mathcal{C}_{L,S})$ holds, then each subinterval of $[0, S]$ of length $\varepsilon^{\omega+2}(L/\varepsilon)$ contains at least one time s for which $\mathcal{WP}_\varepsilon^L(\delta_{\log}^\varepsilon, s)$ holds.*

The idea behind Proposition 3.13 is that, $(PGL)_\varepsilon$ being a gradient flow, on a sufficiently large interval of time one may find some time where the dissipation of energy is small. Using elliptic tools, and viewing the time derivative as a forcing term, one may then establish property $\mathcal{WP}_\varepsilon^L(\delta_{\log}^\varepsilon, s)$ (see Sections 3.2 and 3.3).

The next result expresses the fact that the equation preserves to some extent the well-preparedness assumption.

Proposition 3.14 *Assume that $(\mathcal{C}_{L,S})$ holds, that $\varepsilon^\omega L^2 \le s_0 \le S$ is such that $\mathcal{WP}_\varepsilon^L(\delta_{\log}^\varepsilon, s_0)$ holds, and assume moreover that*

$$d_{min}^{\varepsilon,L}(s_0) \ge 16 \left(\frac{L}{\rho_0\varepsilon}\right)^{\frac{1}{\omega+2}} \varepsilon. \tag{3.32}$$

Then $\mathcal{WP}_\varepsilon^L(\delta_{\log\log}^\varepsilon, s)$ holds for all times $s_0 + \varepsilon^{2+\omega} \le s \le T_0^\varepsilon(s_0)$, where

$$T_0^\varepsilon(s_0) = \max \left\{ s \in [s_0 + \varepsilon^{2+\omega}, S] \quad s.t. \right.$$

$$\left. d_{min}^{\varepsilon,L}(s') \ge 8 \left(\frac{L}{\rho_0\varepsilon}\right)^{\frac{1}{\omega+2}} \varepsilon \quad \forall s' \in [s_0 + \varepsilon^{\omega+2}, s] \right\}.$$

For such an s we have $J(s) = J(s_0)$ and for any $k \in J(s_0)$ we have $\sigma_{i(k\pm\frac{1}{2})}(s) = \sigma_{i(k\pm\frac{1}{2})}(s_0)$ and $\dagger_k(s) = \dagger_k(s_0)$.

Given a family of solution $(v_\varepsilon)_{0<\varepsilon<1}$, we introduce the additional condition

$$d_{min}^*(s_0) \equiv \liminf_{\varepsilon \to 0} d_{min}^{\varepsilon,L}(s_0) > 0, \tag{3.33}$$

which makes sense if $\mathcal{WP}_\varepsilon^L(\alpha_1\varepsilon, s_0)$ holds and expresses the fact that the fronts stay uniformly well-separated. The first step in our proof, which is stated in Proposition 3.19, is to establish the conclusion of Theorem 3.3 under these stronger assumptions on the initial datum. From the inclusion (3.7) and Proposition 3.14 we will obtain:

Corollary 3.15 *Assume also that* $(C_{L,S})$ *holds, let* $s_0 \in [0, S]$ *and assume that* $\mathcal{WP}^L_\varepsilon(\alpha_1 \varepsilon, s_0)$ *holds for all* ε *sufficiently small and that (3.33) is satisfied. Then, for* ε *sufficiently small,*

$$\mathcal{WP}^L_\varepsilon(\delta^\varepsilon_{loglog}, s) \qquad and \qquad d^{\varepsilon,L}_{min}(s) \geq \frac{1}{2} d^*_{min}(s_0) \qquad (3.34)$$

are satisfied for any

$$s \in I^\varepsilon(s_0) \equiv \left[s_0 + 2L^2 \varepsilon^\omega, s_0 + \rho_0 \left(\frac{d^*_{min}(s_0)}{8} \right)^{\omega+2} \right] \cap [0, S],$$

as well as the identities $J(s) = J(s_0)$, $\sigma_{i(k\pm\frac{1}{2})}(s) = \sigma_{i(k\pm\frac{1}{2})}(s_0)$ *and* $\dagger_k(s) = \dagger_k(s_0)$, *for any* $k \in J(s_0)$.

Hence, the collection of front points $\{a^\varepsilon_k(s)\}_{k \in J}$ is well defined, and the approximating regularized fronts $\zeta^{\dagger_k}_{i(k)}$ do not depend on s (otherwise than through their position), on the full time interval $I^\varepsilon(s_0)$.

3.1.3 Paving the Way to the Motion Law

As in [4], we use extensively the localized version of (3.1), a tool which turns out to be perfectly adapted to tracking the evolution of fronts. Let χ be an arbitrary smooth test function with compact support. Set, for $s \geq 0$,

$$\mathcal{I}_\varepsilon(s, \chi) = \int_\mathbb{R} e_\varepsilon(\mathfrak{v}_\varepsilon(x, s)) \chi(x) dx. \qquad (3.35)$$

In integrated form the localized version of the energy identity is written as

$$\mathcal{I}_\varepsilon(s_2, \chi) - \mathcal{I}_\varepsilon(s_1, \chi) + \int_{s_1}^{s_2} \int_\mathbb{R} \varepsilon^{1+\omega} \chi(x) |\partial_s \mathfrak{v}_\varepsilon(x, s)|^2 dx ds$$

$$= \varepsilon^{-\omega} \int_{s_1}^{s_2} \mathcal{F}_S(s, \chi, v_\varepsilon) ds, \qquad (3.36)$$

where the term $\mathcal{F}_S$ is given by

$$\mathcal{F}_S(s, \chi, \mathfrak{v}_\varepsilon) = \int_{\mathbb{R}\times\{s\}} \left(\left[\varepsilon \frac{\dot{\mathfrak{v}}_\varepsilon^2}{2} - \frac{V(\mathfrak{v}_\varepsilon)}{\varepsilon} \right] \ddot{\chi}(x) \right) dx \equiv \int_{\mathbb{R}\times\{s\}} \xi_\varepsilon(\mathfrak{v}_\varepsilon(\cdot, s)) \ddot{\chi} dx. \qquad (3.37)$$

The last integral on the left-hand side of Identity (3.36) stands for local dissipation, whereas the right-hand side is a flux. The quantity ξ_ε is defined for a scalar function u by

$$\xi_\varepsilon(u) \equiv \varepsilon \frac{\dot{u}^2}{2} - \frac{V(u)}{\varepsilon}, \qquad (3.38)$$

and is referred to as **the discrepancy term**. It is constant for solutions to the stationary equation $-u_{xx} + \varepsilon^{-2} V'(u) = 0$ on some given interval I and vanishes for finite energy solutions on $I = \mathbb{R}$. Notice that $|\xi_\varepsilon(u)| \leq e_\varepsilon(u)$. We set for two given times $s_2 \geq s_1 \geq 0$ and $L \geq 0$

$$\text{dissip}_\varepsilon^L[s_1, s_2] = \varepsilon \int_{I_{\frac{5}{3}L} \times [s_1 \varepsilon^{-\omega}, s_2 \varepsilon^{-\omega}]} |\frac{\partial v_\varepsilon}{\partial t}|^2 dxdt = \varepsilon^{1+\omega} \int_{I_{\frac{5}{3}L} \times [s_1, s_2]} |\frac{\partial v_\varepsilon}{\partial s}|^2 dxds.$$

$$(3.39)$$

Identity (3.36) then yields the estimate, if we assume that $\text{supp}\chi \subset I_{\frac{5}{3}L}$,

$$\left| \mathcal{I}_\varepsilon(s_2, \chi) - \mathcal{I}_\varepsilon(s_1, \chi) - \varepsilon^{-\omega} \int_{s_1}^{s_2} \mathcal{F}_S(s, \chi, v_\varepsilon) ds \right| \leq \text{dissip}_\varepsilon^L[s_1, s_2] \|\chi\|_{L^\infty(\mathbb{R})}.$$

$$(3.40)$$

We will show that under suitable assumptions, the right-hand side of (3.40) is small (see Step 3 in the proof of Proposition 3.19), so that the term $\varepsilon^{-\omega} \int_{s_1}^{s_2} \mathcal{F}_S(s, \chi, v_\varepsilon) ds$ provides a good approximation of $\mathcal{I}_\varepsilon(s_2, \chi) - \mathcal{I}_\varepsilon(s_1, \chi)$. On the other hand, it follows from the properties of regularized maps proved in Section 3.2.2 (see Proposition 3.22) that if $\mathcal{WP}_\varepsilon^L(\delta_{\text{loglog}}^\varepsilon, s)$ holds then

$$\left| \mathcal{I}_\varepsilon(s, \chi) - \sum_{k \in J} \chi(a_k^\varepsilon(s)) \mathfrak{S}_{i(k)} \right| \leq CM_0 \left((\frac{\varepsilon}{\delta_{\text{loglog}}^\varepsilon})^\omega \|\chi\|_\infty + \varepsilon \|\chi'\|_\infty \right), \quad (3.41)$$

where $\mathfrak{S}_{i(k)}$ stands for the energy of the corresponding stationary front. Set

$$\mathfrak{F}_\varepsilon(s_1, s_2, \chi) \equiv \varepsilon^{-\omega} \int_{s_1}^{s_2} \mathcal{F}_S(s, \chi, v_\varepsilon) ds \equiv \int_{s_1}^{s_2} \varepsilon^{-\omega} \xi_\varepsilon(v_\varepsilon(\cdot, s)) \ddot{\chi}(\cdot) ds.$$

Combining (3.40) and (3.41) shows that, if $\mathcal{WP}_\varepsilon^L(\delta_{\text{loglog}}^\varepsilon, s)$ holds for any $s \in (s_1, s_2)$, then we have

$$\left| \sum_{k \in J} [\chi(a_k^\varepsilon(s_2)) - \chi(a_k^\varepsilon(s_1))] \mathfrak{S}_{i(k)} - \mathfrak{F}_\varepsilon(s_1, s_2, \chi) \right|$$

$$\leq CM_0 \left((\log |\log \frac{\varepsilon}{L}|)^{-\omega} \|\chi\|_\infty + \varepsilon \|\chi'\|_\infty \right) + \text{dissip}_\varepsilon^L[s_1, s_2] \|\chi\|_\infty.$$

$$(3.42)$$

If the test function χ is chosen to be affine near a given front point a_{k_0} and zero near the other front points in the collection, then the first term on the left-hand side yields a measure of the motion of a_{k_0} between times s_1 and s_2, whereas the second, namely $\mathfrak{F}_\varepsilon(s_1, s_2, \chi)$, is hence a good approximation of the measure of this motion, *provided we are able to estimate the dissipation* dissip$_\varepsilon^L[s_1, s_2]$. Our previous discussion suggests that

$$a_{k_0}^\varepsilon(s_2) - a_{k_0}^\varepsilon(s_1) \simeq \frac{1}{\chi'(a_{k_0}^\varepsilon)\mathfrak{S}_{i(k_0)}}\mathfrak{F}_\varepsilon(s_1, s_2, \chi).$$

It turns out that the computation of $\mathfrak{F}_\varepsilon(s_1, s_2, \chi)$ can be performed with satisfactory accuracy if the test function χ is affine (and hence has vanishing second derivatives) close to the front set, this is the object of the next subsections.

3.1.4 A First Compactness Result

A first step in deriving the motion law for the fronts is to obtain rough bounds from above for both dissip$_\varepsilon^L[s_1, s_2]$ and $\mathfrak{F}_\varepsilon(s_1, s_2, \chi)$. To obtain these, and under the assumptions of Corollary 3.15, notice that if $\ddot{\chi}$ vanishes on the set $\{a_k^\varepsilon(s_0)\}_{k\in J} + [-d_{\min}^*(s_0)/4, d_{\min}^*(s_0)/4]$, then from the inequality $|\xi_\varepsilon(u)| \leq e_\varepsilon(u)$, from Corollary 3.15 and from (3.30) of Proposition 3.12, we derive that for $s_1 \leq s_2$ in $I^\varepsilon(s_0)$,

$$|\mathfrak{F}_\varepsilon(s_1, s_2, \chi)| \leq Cd_{\min}^*(s_0)^{-\omega}\|\ddot{\chi}\|_{L^\infty(\mathbb{R})}(s_2 - s_1). \tag{3.43}$$

Going back to (3.36), and choosing the test function χ so that $\chi \equiv 1$ on $I_{\frac{5}{3}L}$ with compact support on I_{2L}, Estimate (3.43) combined with (3.41) yields in turn a first rough upper bound on the dissipation dissip$_\varepsilon^L[s_1, s_2]$. Combining these estimates we obtain the following.

Proposition 3.16 *Under the assumptions of Corollary 3.15, for $s_1 \leq s_2 \in I^\varepsilon(s_0)$ we have*

$$|a_k^\varepsilon(s_1) - a_k^\varepsilon(s_2)| \leq C\left(d_{\min}^*(s_0)^{-(\omega+1)}(s_2 - s_1)\right.$$
$$\left. + M_0\left((\log|\log\frac{\varepsilon}{L}|)^{-\omega}d_{\min}^*(s_0) + \varepsilon\right)\right). \tag{3.44}$$

As an easy consequence, we deduce the following compactness property, setting

$$I^*(s_0) = \left(s_0, s_0 + \rho_0\left(\frac{d_{\min}^*(s_0)}{8}\right)^{\omega+2}\right) \cap (0, S).$$

Corollary 3.17 *Under the assumptions of Corollary 3.15, there exists a subsequence $(\varepsilon_n)_{n\in\mathbb{N}}$ converging to 0 such that for any $k \in J$ the function $a_k^{\varepsilon_n}(\cdot)$ converges uniformly on any compact interval of $I^*(s_0)$ to a Lipschitz continuous function $a_k(\cdot)$.*

3.1.5 Refined Estimates Off the Front Set and the Motion Law

In order to derive the *precise motion law*, we have to provide an accurate asymptotic value for the discrepancy term *off the front set*. In other words, for a given index $k \in J$ we need to provide a uniform limit of the function $\varepsilon^{-\omega}\xi_\varepsilon$ near the points

$$a^\varepsilon_{k+\frac{1}{2}}(s) \equiv \frac{a^\varepsilon_k(s) + a^\varepsilon_{k+1}(s)}{2} \quad \text{and } a^\varepsilon_{k-\frac{1}{2}}(s) \equiv \frac{a^\varepsilon_{k-1}(s) + a^\varepsilon_k(s)}{2}.$$

We notice first that $\mathfrak{v}_\varepsilon$ takes values close to $\sigma_{i(k+\frac{1}{2})}$ near $a^\varepsilon_{k+\frac{1}{2}}(s)$. In view of Estimate (3.30), we introduce the functions

$$\mathfrak{w}_\varepsilon(\cdot, s) = \mathfrak{w}^k_\varepsilon(\cdot, s) = \mathfrak{v}_\varepsilon - \sigma_{i(k+\frac{1}{2})} \text{ and } \mathfrak{W}_\varepsilon = \mathfrak{W}^k_\varepsilon \equiv \varepsilon^{-\frac{1}{\theta-1}}\mathfrak{w}^k_\varepsilon$$

$$= \varepsilon^{-\frac{1}{\theta-1}}\left(\mathfrak{v}_\varepsilon - \sigma_{i(k+\frac{1}{2})}\right). \tag{3.45}$$

As a consequence of inequality (3.31) and Corollary 3.15 we have a uniform bound.

Lemma 3.18 *Under the assumptions of Corollary 3.15, we have*

$$|\mathfrak{W}_\varepsilon(x,s)| \le C\big(d(x,s)\big)^{-\frac{1}{\theta-1}} \tag{3.46}$$

for any $x \in (a_k(s) + \delta^\varepsilon_{\text{loglog}}, a_{k+1}(s) - \delta^\varepsilon_{\text{loglog}})$ and any $s \in I^\varepsilon(s_0)$, where we have set $d(x,s) := \text{dist}(x, \{a^\varepsilon_k(s), a^\varepsilon_{k+1}(s)\})$ and where $C > 0$ depends only on V and M_0. Moreover, we also have

$$\begin{cases} -\text{sign}(\dagger_k)\mathfrak{W}_\varepsilon\left(a^\varepsilon_k(s) + \delta^\varepsilon_{\text{loglog}}\right) \ge \dfrac{1}{C}\left(\delta^\varepsilon_{\text{loglog}}\right)^{-\frac{1}{\theta-1}}, \\[2mm] \text{sign}(\dagger_{k+1})\mathfrak{W}_\varepsilon\left(a^\varepsilon_{k+1}(s) - \delta^\varepsilon_{\text{loglog}}\right) \ge \dfrac{1}{C}\left(\delta^\varepsilon_{\text{loglog}}\right)^{-\frac{1}{\theta-1}}. \end{cases} \tag{3.47}$$

We describe next on a formal level how to obtain the desired asymptotics for $\varepsilon^{-\omega}\xi_\varepsilon$, as $\varepsilon \to 0$, near the point $a_{k+\frac{1}{2}}(s)$. Going back to the limiting points $\{a_k(s)\}_{k\in J}$ defined in Proposition 3.16, we consider the subset of $\mathbb{R} \times \mathbb{R}^+$

$$\mathcal{V}_k(s_0) = \bigcup_{s \in I^*(s_0)} (a_k(s), a_{k+1}(s)) \times \{s\}. \tag{3.48}$$

It follows from the uniform bounds established in Lemma 3.18, that, passing possibly to a further subsequence, we may assume that

$$\mathfrak{W}_{\varepsilon_n} \rightharpoonup \mathfrak{W}_* \text{ in } L^p_{\text{loc}}(\mathcal{V}_k(s_0)), \text{ for any } 1 \le p < \infty.$$

On the other hand, thanks to estimate (3.46), for a given point $(x,s) \in \mathcal{V}_k(s_0)$ we expand $(\text{PGL})_\varepsilon$ near (x,s) as

$$\varepsilon^\omega \frac{\partial \mathfrak{W}_\varepsilon}{\partial s} - \frac{\partial^2 \mathfrak{W}_\varepsilon}{\partial x^2} + 2\theta \lambda_{i(k+\frac{1}{2})} \mathfrak{W}_\varepsilon^{2\theta-1} = O(\varepsilon^{\frac{1}{\theta-1}}). \tag{3.49}$$

Passing to the limit $\varepsilon_n \to 0$, we expect that for every $s \in I^*(s_0)$, $\mathfrak{W}_*$ solves

$$\begin{cases} -\dfrac{\partial^2 \mathfrak{W}_*}{\partial x^2}(s,\cdot) + 2\theta \lambda_{i(k+\frac{1}{2})} \mathfrak{W}_*^{2\theta-1}(s,\cdot) = 0 \text{ on } (a_k(s), a_{k+1}(s)), \\ \mathfrak{W}_*(a_k(s)) = -\text{sign}(\dagger_k)\infty \text{ and } \mathfrak{W}_*(a_{k+1}(s)) = \text{sign}(\dagger_k)\infty, \end{cases} \tag{3.50}$$

the boundary conditions being a consequence of the asymptotics (3.47). It turns out, in view of Lemma 3.59 of Appendix A, that the *boundary value problem* (3.50) has a *unique* solution. By scaling, and setting $r_k(s) = \frac{1}{2}(a_{k+1}(s) - a_k(s))$, we obtain

$$\begin{cases} \mathfrak{W}_*(x,s) = \pm r_k(s)^{-\frac{1}{\theta-1}} \left(\lambda_{i(k+\frac{1}{2})}\right)^{-\frac{1}{2(\theta-1)}} \overset{\vee+}{u}\left(\dfrac{x - a_{k+\frac{1}{2}}}{r_k(s)}\right), & \text{if } \dagger_k = -\dagger_{k+1}, \\[3mm] \mathfrak{W}_*(x,s) = \pm r_k(s)^{-\frac{1}{\theta-1}} \left(\lambda_{i(k+\frac{1}{2})}\right)^{-\frac{1}{2(\theta-1)}} \overset{\triangleright}{u}\left(\dfrac{x - a_{k+\frac{1}{2}}}{r_k(s)}\right), & \text{if } \dagger_k = \dagger_{k+1}, \end{cases}$$

where $\overset{\vee+}{u}$ (resp. $\overset{\triangleright}{u}$) are the unique solutions to the problems

$$\begin{cases} -\mathcal{U}_{xx} + 2\theta \mathcal{U}^{2\theta-1} = 0 & \text{on } (-1, +1), \\ \mathcal{U}(-1) = +\infty \text{ (resp. } \mathcal{U}(-1) = -\infty) \text{ and } \mathcal{U}(+1) = +\infty. \end{cases} \tag{3.51}$$

Still on a formal level, we deduce therefore the corresponding values of the discrepancy

$$\begin{cases} \varepsilon^{-\omega}\xi_\varepsilon(\mathfrak{v}_\varepsilon) \simeq \xi(\mathfrak{W}_*) = -\lambda_{i(k+\frac{1}{2})}^{-\frac{1}{\theta-1}} r_k(s)^{-(\omega+1)} \mathcal{A}_\theta & \text{if } \dagger_k = -\dagger_{k+1}, \\[3mm] \varepsilon^{-\omega}\xi_\varepsilon(\mathfrak{v}_\varepsilon) \simeq \xi(\mathfrak{W}_*) = \lambda_{i(k+\frac{1}{2})}^{-\frac{1}{\theta-1}} r_k(s)^{-(\omega+1)} \mathcal{B}_\theta & \text{if } \dagger_k = \dagger_{k+1}, \end{cases} \tag{3.52}$$

where the numbers $\mathcal{A}_\theta$ and $\mathcal{B}_\theta$ are positive, depend only on θ, and correspond to the absolute value of the discrepancy of $\overset{\vee+}{u}$ and $\overset{\triangleright}{u}$ respectively. Notice that the signs in (3.52) are different, the first case yields attractive forces whereas the second yields repulsive ones. Inserting this relation in (3.42) and arguing as for (3.44), we will derive the motion law.

The previous formal discussion can be put on a sound mathematical ground, relying on comparison principles and the construction of appropriate upper and lower solutions (see Section 3.5). This leads to the central result of this paper.

Proposition 3.19 *Assume that conditions* (H$_0$) *and* (H$_1$) *are fulfilled. Let* $0 < S < S_{max}$ *be given and set*

$$L_0 := 3\max\left\{|a_k^0|, 1 \leq k \leq \ell_0, \left(\frac{S}{\rho_0}\right)^{\frac{1}{\omega+2}}\right\}.$$

Assume that WPI$_\varepsilon^{L_0}(\alpha_1\varepsilon, 0)$ *holds as well as (3.33) at time* $s = 0$. *Then* $J(s) = \{1,\ldots,\ell_0\}$ *and the functions* $a_k^\varepsilon(\cdot)$ *are well defined and converge uniformly on any compact interval of* $(0,S)$ *to the solution* $a_k(\cdot)$ *of (S) supplemented with the initial condition* $a_k(0) = a_k^0$.

Notice that the combination of assumptions WPI$_\varepsilon^L(\alpha_1\varepsilon, 0)$, (H$_1$) and (3.33) at $s = 0$ implies the multiplicity-one condition (H$_{min}$). Whereas the conclusion of Proposition 3.19 is similar to the one of Theorem 3.3, the assumptions of Proposition 3.19 are more restrictive. Indeed, on one hand we assume the well-preparedness condition WPI$_\varepsilon^L$, and on the other hand we impose (3.33) which is far more constraining than (H$_{min}$): it excludes in particular the possibility of having small pairs of fronts and anti-fronts. Our next efforts are hence devoted to handling this type of situation: Proposition 3.19, through rescaling arguments, will nevertheless be the main building block for that task.

In order to prove Theorems 3.3, 3.5 and 3.9 we need to relax the assumptions on the initial data, in particular we need to analyze the behavior of data with small pairs of *fronts and anti-fronts*, and show that they are going to annihilate on a short interval of time. For that purpose we will consider the following situation, corresponding to confinement of the front set at initial time. Assume that for a collection of points $\{b_q^\varepsilon\}_{q\in J_0}$ in $\mathbb{R}$ we have

$$\mathfrak{D}_\varepsilon(0) \cap I_{5L} \subset \bigcup_{q\in J_0}[b_q^\varepsilon - r, b_q^\varepsilon + r] \subset I_{\kappa_0 L} \quad \text{and} \quad b_p^\varepsilon - b_q^\varepsilon \geq 3R \qquad \text{for } p \neq q \in J_0,$$

$$(3.53)$$

for some $\kappa_0 \leq \frac{1}{2}$ *and* $\alpha_1\varepsilon \leq r \leq R/2 \leq L/4$. It follows from (3.7) that if $0 \leq s \leq \rho_0(R-r)^{\omega+2}$ then

$$\mathfrak{D}_\varepsilon(s) \cap I_{4L} \subset \bigcup_{k\in J_0}(b_k^\varepsilon - R, b_k^\varepsilon + R) \subset I_{2\kappa_0 L}, \quad \text{where the union is disjoint.}$$

Consider next $0 \leq s \leq \rho_0(R - r)^{\omega+2}$ such that $\mathcal{WP}_\varepsilon^L(\alpha_1\varepsilon, s)$ holds, so that the front points $\{a_k^\varepsilon(s)\}_{k\in J(s)}$ are well defined. For $q \in J_0$, consider $J_q(s) = \{k \in J(s), a_k^\varepsilon(s) \in (b_q^\varepsilon - R, b_q^\varepsilon + R)\}$, set $\ell_q = \sharp J_q$, and write $J_q(s) = \{k_q, k_{q+1}, \ldots, k_{q+\ell_q-1}\}$, where $k_1 = 1$, and $k_q = \ell_1 + \cdots + \ell_{q-1} + 1$, for $q \geq 2$. Our next result shows that, after a small time, only the repulsive forces survive at the scale given by r, provided the different lengths are sufficiently distinct.

Proposition 3.20 *There exist positive constants α_* and ρ_*, depending only on V and M_0, such that if (3.53) holds and*

$$\kappa_0^{-1} \geq \alpha_*, \qquad r \geq \alpha_* \varepsilon \left(\frac{L}{\varepsilon}\right)^{\frac{2}{\omega+2}}, \qquad R \geq \alpha_* r, \tag{3.54}$$

then at time

$$s_r = \rho_* r^{\omega+2}$$

condition $\mathcal{WP}_\varepsilon^L(\alpha_1 \varepsilon, s_r)$ holds and, for any $q \in J_0$ and any $k, k' \in J_q(s_r)$ we have $\dagger_k(s_r) = \dagger'_k(s_r)$, or equivalently for any $k \in J_q(s_r) \setminus \{k_q(s_r) + \ell_q(s_r) - 1\}$, we have

$$\epsilon_{k+\frac{1}{2}}(s_r) = \dagger_k(s_r)\,\dagger_{k+1}(s_r) = +1. \tag{3.55}$$

Moreover, we have

$$d_{\min}^{\varepsilon,L}(s_r) \geq r, \tag{3.56}$$

and if $\sharp J_q(s_r) \leq 1$ for every $q \in J_0$, then we actually have $d_{\min}^{\varepsilon,L}(s_r) \geq R$.

The proofs of Theorems 3.3, 3.5 and 3.9 are then deduced from Propositions 3.19 and 3.20.

The paper is organized as follows. We describe in Section 3.2 some properties of stationary fronts, as well as for solutions to some perturbations of the stationary equations. In Section 3.3 we describe several properties related to the well-preparedness assumption $\mathcal{WP}_\varepsilon^L$, in particular the quantization of the energy, how it relates to dissipation, and its numerous implications for the dynamics. We provide in particular the proofs to Proposition 3.13, Proposition 3.14 and Corollary 3.15. In Section 3.4, we prove the compactness results stated in Proposition 3.16 and Corollary 3.17. Section 3.5, provides an expansion of the discrepancy term off the front set, from a technical point of view it is the place where the analysis differs most from the non-degenerate case. Based on this analysis, we show in Section 3.6 how the motion law follows from prepared data establishing the proof of Proposition 3.19. In Section 3.7 we analyze the clearing-out of small pairs of front–anti-front and, more generally, we present the proof of Proposition 3.20. Finally, in Section 3.8 we present the proofs of the main theorems, namely Theorems 3.3, 3.5 and 3.9. Several results concerning the first- or second-order differential equations involved in the analysis of this paper are given in separate appendices, in particular the proof of Proposition 3.7.

3.2 Remarks on Stationary Solutions

3.2.1 Stationary Solutions on $\mathbb{R}$ with Vanishing Discrepancy

Stationary solutions are described using the method of separation of variables. For u, the solution to (3.21), we multiply (3.21) by $\dot{u}$ and verify that ξ is constant. We restrict ourselves to solutions with vanishing discrepancy

$$\xi = \frac{1}{2}\dot{u}^2 - V(u) = 0, \tag{3.57}$$

and solve Equation (3.57) by separation of variables. Let γ_i be defined on (σ_i, σ_{i+1}) by

$$\gamma_i(u) = \int_{z_i}^{u} \frac{ds}{\sqrt{2V(s)}}, \text{ for } u \in (\sigma_i, \sigma_{i+1}), \tag{3.58}$$

where we recall that z_i is a fixed maximum point of V in the interval (σ_i, σ_{i+1}). The map γ_i is one-to-one from (σ_i, σ_{i+1}) to $\mathbb{R}$, so we may define its inverse map $\zeta_i^+ : \mathbb{R} \to (\sigma_i, \sigma_{i+1})$ by

$$\zeta_i^+(x) = \gamma_i^{-1}(x) \text{ as well as } \zeta_i^-(x) = \gamma_i^{-1}(-x) \text{ for } x \in \mathbb{R}. \tag{3.59}$$

In view of the definition (3.59), we have $\zeta_i^{\pm}(0) = z_i$, $\zeta_i^{+\prime}(0) = \sqrt{2V(z_i)} > 0$, whereas a change of variable shows that ζ_i has finite energy given by the formula (3.8). We verify that $\zeta_i^+ \left(\frac{\cdot}{\varepsilon}\right)$ and $\zeta_i^- \left(\frac{\cdot}{\varepsilon}\right)$ solve (3.57) and hence (3.21). The next elementary result then directly follows from uniqueness in ODEs.

Lemma 3.21 *Let u be a solution to (3.21) such that (3.57) holds, and such that $u(x_0) \in (\sigma_i, \sigma_{i+1})$, for some $x_0 \in \mathbb{R}$, and some $i \in 1, \ldots q - 1$. Then, there exists $a \in \mathbb{R}$ such that $u(x) = \zeta_i^+(x - a)$ or $u(x) = \zeta_i^-(x - a), \forall x \in \mathbb{R}$.*

We provide a few simple properties of the functions $\zeta_i^{\pm}$ which enter directly into our arguments. We expand V near σ_i for $u \geq \sigma_i$ as

$$\sqrt{V(u)} = \sqrt{\lambda_i}(u - \sigma_i)^{\theta}(1 + O(u - \sigma_i)), \quad \text{as } u \to \sigma_i.$$

Integrating, we are led to the expansion

$$\gamma_i(u) = -\frac{\theta - 1}{\sqrt{2\lambda_i}}(u - \sigma_i)^{-\theta+1}(1 + O(u - \sigma_i)), \quad \text{as } u \to \sigma_i,$$

and therefore also to the expansions

$$\zeta_i^{\pm}(x) = \sigma_i + \left(\frac{\sqrt{2\lambda_i}|x|}{\theta - 1}\right)^{-\frac{1}{\theta-1}}(1 + o(1)), \quad \text{as } x \to \mp\infty.$$

Similarly,

$$\zeta_i^\pm(x) = \sigma_{i+1} - \left(\frac{\sqrt{2\lambda_{i+1}}|x|}{\theta-1}\right)^{-\frac{1}{\theta-1}}(1+o(1)), \quad \text{as } x \to \pm\infty,$$

and corresponding asymptotics for the derivatives can be derived as well (e.g. using the fact that the discrepancy is zero).

For $0 < \varepsilon < 1$ given, and $i = 1,\ldots,q-1$, consider the scaled function $\zeta_{i,\varepsilon}^\pm = \zeta_i^\pm\left(\frac{\cdot}{\varepsilon}\right)$ which is a solution to

$$-u_{xx} + \varepsilon^{-2}V'(u) = 0,$$

hence a stationary solution to $(\text{PGL})_\varepsilon$. Straightforward computations based on the previous expansions show that

$$\begin{cases} e_\varepsilon\left(\zeta_{i,\varepsilon}^\pm\right)(x) = (2\lambda_i)^{-\frac{1}{\theta-1}}(\theta-1)^{\frac{2\theta}{\theta-1}}\frac{1}{\varepsilon}\left|\frac{x}{\varepsilon}\right|^{-(\omega+1)} + \underset{\frac{x}{\varepsilon}\to\mp\infty}{o}\left(\frac{1}{\varepsilon}\left|\frac{x}{\varepsilon}\right|^{-(\omega+1)}\right), \\[2mm] e_\varepsilon\left(\zeta_{i,\varepsilon}^\pm\right)(x) = (2\lambda_{i+1})^{-\frac{1}{\theta-1}}(\theta-1)^{\frac{2\theta}{\theta-1}}\frac{1}{\varepsilon}\left|\frac{x}{\varepsilon}\right|^{-(\omega+1)} + \underset{\frac{x}{\varepsilon}\to\pm\infty}{o}\left(\frac{1}{\varepsilon}\left|\frac{x}{\varepsilon}\right|^{-(\omega+1)}\right), \end{cases}$$

$$(3.60)$$

with ω defined in (3.5). Hence there is some constant $C > 0$ independent of r and ε such that

$$\mathfrak{S}_i \geq \int_{-r}^r e_\varepsilon\left(\zeta_{i,\varepsilon}^\pm\right)\mathrm{d}x \geq \mathfrak{S}_i - C\left(\frac{\varepsilon}{r}\right)^\omega. \tag{3.61}$$

3.2.2 On the Energy of Chains of Stationary Solutions

If u satisfies condition $\text{WPI}_\varepsilon^L(\delta)$ and (H_0), we set

$$\mathfrak{E}_\varepsilon^L(u) = \sum_{k\in J}\mathfrak{S}_{i(k)} \quad \text{and} \quad \mathcal{E}_\varepsilon^L(u) = \int_{I_{2L}} e_\varepsilon(u(x))\mathrm{d}x. \tag{3.62}$$

Proposition 3.22 *We have*

$$\begin{cases} \mathcal{E}_\varepsilon^L(u) \geq \mathfrak{E}_\varepsilon^L(u) - C_{\mathrm{f}}M_0\left(\frac{\varepsilon}{\delta}\right)^\omega & \text{if } \text{WPI}_\varepsilon^L(\delta) \text{ holds}, \\[2mm] \mathcal{E}_\varepsilon^L(u) \leq \mathfrak{E}_\varepsilon^L(u) + (C_{\mathrm{w}}+C_{\mathrm{f}})M_0\left(\frac{\varepsilon}{\delta}\right)^\omega & \text{if } \mathcal{WP}_\varepsilon^L(\delta) \text{ holds}. \end{cases} \tag{3.63}$$

Moreover, for any smooth function χ with compact support in I_{2L} we have

$$\left|\mathcal{I}_\varepsilon(\chi) - \sum_{k\in J}\chi(a_k)\mathfrak{S}_{i(k)}\right| \leq (C_{\mathrm{w}}+C_{\mathrm{f}})M_0\left(\left(\frac{\varepsilon}{\delta}\right)^\omega\|\chi\|_\infty + \varepsilon\|\chi'\|_\infty\right),$$

if $\mathcal{WP}_\varepsilon^L(\delta)$ holds, $\tag{3.64}$

where $\mathcal{I}_\varepsilon(\chi) = \int_{I_{2L}} e_\varepsilon(u)\chi(x)dx$. *The constant C_f which appears in (3.63) and (3.64) depends only on V, and the constant C_w appears in the definition of condition $\mathcal{WP}_\varepsilon^L$.*

Proof We estimate the integral of $|e_\varepsilon(u) - e_\varepsilon(\zeta_{i(k)}^{\dagger k}(\cdot - a_k))|$ on I_k as

$$\frac{\varepsilon}{2}\int_{I_k} |\dot{u}^2 - (\dot{\zeta}_{i(k),\varepsilon}^{\dagger k}(\cdot - a_k))^2|dx$$

$$\leq \varepsilon \|\dot{u} - \dot{\zeta}_{i(k),\varepsilon}^{\dagger k}(\cdot - a_k)\|_{L^\infty(I_k)} \left[\mathcal{E}_\varepsilon(u)^{\frac{1}{2}} + \mathcal{E}_\varepsilon(\zeta_{i(k),\varepsilon}^{\dagger k})^{\frac{1}{2}}\right]\sqrt{\frac{\delta}{\varepsilon}}$$

and likewise we obtain

$$\varepsilon^{-1}\int_{I_k} |V(u) - V(\zeta_{i(k),\varepsilon}^{\dagger k}(\cdot - a_k))|dx \leq C\frac{\delta}{\varepsilon}\|u - \zeta_{i(k),\varepsilon}^{\dagger k}(\cdot - a_k)\|_{L^\infty(I_k)}.$$

It suffices then to invoke $\mathrm{WPI}_\varepsilon^L(\delta)$ and $\mathrm{WPO}_\varepsilon^L(\delta)$ as well as the decay estimates (3.61) to derive (3.63), using the fact that since $\delta \geq \alpha_1\varepsilon$, negative exponentials are readily controlled by negative powers. Estimate (3.64) is derived in a very similar way, the error in $\varepsilon\|\chi'\|_\infty$ being a consequence of the approximation of $\int \chi e_\varepsilon(\zeta_{i(k),\varepsilon}^{\dagger k}(\cdot - a_k))$ by $\chi(a_k)\mathfrak{S}_{i(k)}$.   □

This result shows that, if δ is sufficiently large, the energy is close to a set of discrete values, namely the finite sums of $\mathfrak{S}_k$. We will therefore refer to this property as *the quantization of the energy*, it will play an important role later when we will obtain estimates on the dissipation rate of energy.

3.2.3 Study of the Perturbed Stationary Equation

Consider a function u defined on $\mathbb{R}$ satisfying the perturbed differential equation

$$u_{xx} = \varepsilon^{-2}V'(u) + f, \tag{3.65}$$

where $f \in L^2(\mathbb{R})$, and the energy bound (H_0). We already know, thanks to Lemma 3.21 that if $f = 0$ then u is of the form $\zeta_{i,\varepsilon}^\pm(\cdot - a)$. Our results below, summarized here in loose terms, show that if f is sufficiently small on some sufficiently large interval, then u is close to a chain of translations of the functions $\zeta_{i,\varepsilon}^\pm$ suitably glued together on that interval.

Following the approach of [4], we first recast Equation (3.65) as a system of two differential equations of first order. For that purpose, we set $w = \varepsilon u_x$ so that (3.65) is equivalent to the system

$$u_x = \frac{1}{\varepsilon}w \text{ and } w_x = \frac{1}{\varepsilon}V'(u) + \varepsilon f,$$

which we may write in a more condensed form as

$$U_x = \frac{1}{\varepsilon}G(U) + \varepsilon F \quad \text{on } \mathbb{R}, \tag{3.66}$$

where we have set $U(x) = (u(x), w(x))$ and $F(x) = (0, f(x))$, and where G denotes the vector field $G(u, w) = (w, V'(u))$. Notice that the energy bound (H_0) and assumption (A_3) together imply a global L^∞ bound on u. In turn, this L^∞ bound implies a Lipschitz bound, denoted C_0, for the non-linearity $G(u, w)$.

Lemma 3.23 *Let u_1 and u_2 satisfy (3.65) with forcing terms f_1 and f_2, and assume that both satisfy the energy bound (H_0). Denote by U_1, U_2, F_1, F_2 the corresponding solutions and forcing terms of (3.66). Then, for any x, x_0 in some arbitrary interval I,*

$$|(U_1 - U_2)(x)| \leq \left(|(U_1 - U_2)(x_0)| + \frac{\varepsilon^{\frac{3}{2}}}{\sqrt{2C_0}} \|F_1 - F_2\|_{L^2(I)} \right) \exp\left(\frac{C_0|x - x_0|}{\varepsilon} \right). \tag{3.67}$$

Proof Since $(U_1 - U_2)_x = G(U_1) - G(U_2) + \varepsilon(F_1 - F_2)$ we obtain the inequality

$$|(U_1 - U_2)_x| \leq \frac{C_0}{\varepsilon}|U_1 - U_2| + \varepsilon|F_1 - F_2|.$$

It follows from Gronwall's inequality that

$$|(U_1 - U_2)(x)| \leq \exp\left(\frac{C_0|x - x_0|}{\varepsilon} \right) |(U_1 - U_2)(x_0)|$$
$$+ \left| \int_{x_0}^{x} \varepsilon|(F_1 - F_2)(y)| \exp\left(\frac{C_0|y - x_0|}{\varepsilon} \right) dy \right|.$$

Claim (3.67) then follows from the Cauchy–Schwarz inequality.   □

We will combine the previous lemma with the following one.

Lemma 3.24 *Let u be a solution of (3.65) satisfying (H_0). Then*

$$\sup_{x,y \in I} |\xi_\varepsilon(u)(x) - \xi_\varepsilon(u)(y)| \leq \sqrt{2M_0}\varepsilon^{\frac{1}{2}} \|f\|_{L^2(I)},$$

where $I \subset \mathbb{R}$ is an arbitrary interval.

Proof This is a direct consequence of the equality $\frac{d}{dx}\xi_\varepsilon(u) = \varepsilon f \frac{d}{dx}u$, the Cauchy–Schwarz inequality, and the definition of the energy.   □

Lemma 3.25 *Let u be a solution of (3.65) satisfying* (H_0). *Let $L > 0$ and assume that*

$$\mathcal{D}(u) \cap I_{2L} \subseteq I_L.$$

There exists a constant $0 < \kappa_w < 1$, depending only on V, such that if

$$M_0 \frac{\varepsilon}{L} + M_0^{\frac{1}{2}} \varepsilon^{\frac{3}{2}} \|f\|_{L^2(I_{\frac{3}{2}L})} \leq \kappa_w, \tag{3.68}$$

then the condition $WPI_\varepsilon^L(\delta)$ holds where

$$\frac{\delta}{\varepsilon} := -\frac{2}{\rho_w} \log \left(M_0 \frac{\varepsilon}{L} + M_0^{\frac{1}{2}} \varepsilon^{\frac{3}{2}} \|f\|_{L^2(I_{\frac{3}{2}L})} \right), \tag{3.69}$$

and where the constant ρ_w depends on only M_0 and V. Moreover, κ_w is sufficiently small so that $2|\log \kappa_w|/\rho_w \geq \alpha_1$, where α_1 was defined in (3.26).

Proof If $\mathcal{D}(u) \cap I_{2L} = \emptyset$ then there is nothing to prove. If not, we first claim that there exists a point $a_1 \in I_L$ such that $u(a_1) = z_{i(1)}$ for some $i(1) \in \{1, \ldots, \mathfrak{q} - 1\}$. Indeed, if not, and since the endpoints of I_{2L} are not in the front set, the function u would have a critical point with a critical value in the complement of $\cup_j B(\sigma_j, \mu_0)$. At that point, the discrepancy would therefore be larger than C/ε for some constant $C > 0$ depending on only V (through the choice of μ_0). On the other hand, since $|\xi_\varepsilon| \leq e_\varepsilon$, by averaging there exists at least one point in $I_{\frac{3}{2}L}$ where the discrepancy of u is smaller in absolute value than $M_0/(3L)$. Combined with the estimate of Lemma 3.24 on the oscillation of the discrepancy, we hence derive our first claim, provided κ_w in (3.68) is chosen sufficiently small. Wet set $\dagger_1 = \text{sign}(u'(a_1))$, $u_1 = u$ and $u_2 = \zeta_{i(1),\varepsilon}^{\dagger_1}(\cdot - x_1))$. Since

$$V(u_1(a_1)) = V(u_2(a_1)) = V(z_{i(1)}),$$

and since

$$|\xi_\varepsilon(u_1)(a_1) - \xi_\varepsilon(u_2)(a_1)| = |\xi_\varepsilon(u_1)(a_1)| \leq M_0/(3L) + \sqrt{2M_0} \varepsilon^{\frac{1}{2}} \|f\|_{L^2(I_{\frac{3}{2}L})},$$

we obtain

$$\left| \varepsilon(u_1')^2(a_1) - \varepsilon(u_2')^2(a_1) \right| \leq M_0/(L) + 2\sqrt{2M_0} \varepsilon^{\frac{1}{2}} \|f\|_{L^2(I_{\frac{3}{2}L})}.$$

Since also

$$|u_1'(a_1) + u_2'(a_1)| \geq |u_2'(a_1)| = |\sqrt{\frac{2V(z_{i(1)})}{\varepsilon^2}}| \geq C/\varepsilon,$$

it follows that

$$\left| \varepsilon(u_1' - u_2')(a_1) \right| \leq C \left(M_0 \frac{\varepsilon}{L} + \sqrt{M_0} \varepsilon^{\frac{3}{2}} \|f\|_{L^2(I_{\frac{3}{2}L})} \right),$$

for a constant $C > 0$ which depends on only V. We may then apply Lemma 3.23 to u_1 and u_2 with the choice $x_0 = a_1$, and for which we thus have, with the notations of Lemma 3.23,

$$|(U_1 - U_2)(x_0)| \le C \left(M_0 \frac{\varepsilon}{L} + \sqrt{M_0} \varepsilon^{\frac{3}{2}} \|f\|_{L^2(I_{\frac{3}{2}L})} \right).$$

Estimate (3.67) then yields (3.24) on $I_1 = [a_1 - \delta, a_1 + \delta]$, for the choice of δ given by (3.69) with $\rho_w = 4(C_0 + 1)$, where C_0 depends on only M_0 and V and was defined above Lemma 3.23.

If $\mathcal{D}(u) \cap (I_{\frac{3}{2}L} \setminus [a_1 - \delta, a_1 + \delta]) = \emptyset$, we are done, and if not we may repeat the previous construction (the boundary points of $[a_1 - \delta, a_1 + \delta]$ are not part of the front set), until after finitely many steps we cover the whole front set. $\quad\square$

We turn to the outer condition[8] WPO_ε^L.

Lemma 3.26 *Let u be a solution of (3.65) verifying* (H_0)*, and assume that for some index $i \in \{1, \dots, \mathfrak{q}\}$*

$$u(x) \in B(\sigma_i, \mu_0) \quad \forall x \in A,$$

where A is some arbitrary bounded interval. Set $R = \text{length}(A)$, let $0 < \rho < R$, and set $B = \{x \in A \mid \text{dist}(x, A^c) > \rho\}$. Then we have the estimate

$$\mathcal{E}_\varepsilon(u, B) \le C_0 \left(\mathcal{E}_\varepsilon(u, A \setminus B)^{\frac{1}{\theta}} \left(\frac{\varepsilon}{\rho} \right)^{1 + \frac{1}{\theta}} + R^{\frac{3}{2}} M_0^{\frac{1}{2\theta}} \left(\frac{\varepsilon}{R} \right)^{1 + \frac{1}{2\theta}} \|f\|_{L^2(A)} \right),$$

where the constant C_0 depends on only V.

Proof Let $0 \le \chi \le 1$ be a smooth cut-off function with compact support in A and such that $\chi \equiv 1$ on B and $|\chi'| \le 2/\rho$ on A. We multiply (3.65) by $\varepsilon(u - \sigma_i)\chi^2$ and integrate on A. This leads to

$$\int_A \varepsilon u_x^2 \chi^2 + \frac{1}{\varepsilon} V'(u)(u - \sigma_i)\chi^2 = \int_{A \setminus B} 2\varepsilon u_x(u - \sigma_i)\chi \chi' - \int_A \varepsilon f(u - \sigma_i)\chi^2.$$

We estimate the first term on the right-hand side above by

$$\left| \int_{A \setminus B} 2\varepsilon u_x(u - \sigma_i)\chi \chi' \right| \le \left(\int_A \varepsilon u_x^2 \chi^2 \right)^{\frac{1}{2}} \left(\int_{A \setminus B} \varepsilon^\theta (u - \sigma_i)^{2\theta} \right)^{\frac{1}{2\theta}} \left(\int_{A \setminus B} |2\chi'|^{\frac{2\theta}{\theta-1}} \right)^{\frac{\theta-1}{2\theta}}$$

$$\le \frac{1}{2} \int_A \varepsilon u_x^2 \chi^2 + \frac{1}{2} \varepsilon^{1 + \frac{1}{\theta}} \left(\int_{A \setminus B} \frac{2}{\lambda_i} e_\varepsilon(u) \right)^{\frac{1}{\theta}} \left(\frac{4}{\rho} \right)^2 (2\rho)^{\frac{\theta-1}{\theta}}$$

$$\le \frac{1}{2} \int_A u_x^2 \chi^2 + 16\lambda_i^{-\frac{1}{\theta}} \left(\frac{\varepsilon}{\rho} \right)^{1 + \frac{1}{\theta}} \mathcal{E}_\varepsilon(u, A \setminus B)^{\frac{1}{\theta}},$$

---

[8] for which several adaptations have to be carried out compared to the non-degenerate case.

where we have used (3.2) and the fact that length$(A \setminus B) = 2\rho$. Similarly, we estimate

$$\left| \int_A \varepsilon f (u - \sigma_i) \chi^2 \right| \leq \varepsilon \|f\|_{L^2(A)} \left(\int_A (u - \sigma_i)^{2\theta} \right)^{\frac{1}{2\theta}} R^{\frac{\theta-1}{2\theta}}$$

$$\leq \varepsilon^{1 + \frac{1}{2\theta}} \|f\|_{L^2(A)} \left(\frac{2}{\lambda_i} \right)^{-1} M_0^{\frac{1}{2\theta}} R^{\frac{\theta-1}{2\theta}}.$$

Also, by (3.2) we have

$$\int_A \frac{1}{\varepsilon} V'(u)(u - \sigma_i) \chi^2 \geq \theta \int_B \frac{1}{\varepsilon} V(u).$$

Combining the previous inequalities, the conclusion follows. □

Combining Lemma 3.25 with Lemma 3.26 we obtain the following.

Proposition 3.27 *Let u be a solution to (3.65) satisfying assumption* (H_0), *and such that* $\mathcal{D}(u) \cap I_{3L} \subset I_L$. *There exist positive constants*[9] C_w *and* α_1, *depending on only* M_0 *and* V, *such that if* $\alpha \geq \alpha_1$ *and if*

1. $M_0 \dfrac{\varepsilon}{L} \leq \dfrac{1}{2} \exp(-\frac{\rho_w}{2}\alpha)$,

2. $\|f\|_{L^2(I_{3L})} \leq \dfrac{1}{2} M_0^{-\frac{1}{2}} \varepsilon^{-\frac{3}{2}} \exp(-\frac{\rho_w}{2}\alpha)$,

3. $\|f\|_{L^2(I_{3L})} \leq \dfrac{C_w}{2C_0} M_0^{1-\frac{1}{2\theta}} \left(\dfrac{\varepsilon}{L} \right)^{-1-\frac{1}{2\theta}} L^{-\frac{3}{2}} \alpha^{-\omega}$,

then $\mathcal{WP}_\varepsilon^L(\alpha\varepsilon)$ *holds.*

Proof Direct substitution shows that assumptions 1 and 2 imply condition (3.68), provided α_1 is choosen sufficiently large, and also imply condition WPI$_\varepsilon^L(\delta)$ for some $\delta \geq \alpha\varepsilon$ given by (3.69). It remains to consider WPO$_\varepsilon^L(\alpha\varepsilon)$. We invoke Lemma 3.26 on each of the intervals $A = (a_k + \frac{1}{2}\alpha\varepsilon, a_{k+1} - \frac{1}{2}\alpha\varepsilon)$, taking $B = (a_k + \alpha\varepsilon, a_{k+1} - \alpha\varepsilon)$. In view of WPI$_\varepsilon^L(\alpha\varepsilon)$ and (3.61), we obtain

$$\mathcal{E}_\varepsilon(u, A \setminus B) \leq C\alpha^{-\omega},$$

and therefore

$$\mathcal{E}_\varepsilon(u, A \setminus B)^{\frac{1}{\theta}} \alpha^{-1-\frac{1}{\theta}} \leq C\alpha^{-\omega},$$

---

[9] Recall that C_w enters in the definition of condition $\mathcal{WP}_\varepsilon^L$. A parameter named C_w already appears in the statement of Proposition 3.22 We impose that its updated value here is be larger than its original value in Proposition 3.22 (and Proposition 3.22 remains of course true with this updated value!).

where C depends on only V. Also, in view of assumption 3 we have

$$C_o \sum_k R^{\frac{3}{2}} M_0^{\frac{1}{2\theta}} \left(\frac{\varepsilon}{R}\right)^{1+\frac{1}{2\theta}} \|f\|_{L^2(A)} \le C_o L^{\frac{3}{2}} M_0^{\frac{1}{2\theta}} \left(\frac{\varepsilon}{L}\right)^{1+\frac{1}{2\theta}} \|f\|_{L^2(I_{3L})} \le \frac{1}{2} C_w M_0 \alpha^{-\omega},$$

provided α_1 is sufficiently large (third requirement). It remains to estimate $e_\varepsilon(u)$ on the intervals $(-2L, a_1)$ and $(a_\ell, 2L)$. We first use Lemma 3.26 with $A = (-3L, -L)$ (resp. $A = (L, 3L)$ and $B = (-\frac{5}{2}L, -\frac{3}{2}L)$ (resp. $B = (\frac{3}{2}L, \frac{5}{2}L)$). This yields, using the trivial bound $\mathcal{E}_\varepsilon(u, A \setminus B) \le M_0$, the estimate

$$\mathcal{E}_\varepsilon(u, I_{\frac{5}{2}L} \setminus I_{\frac{3}{2}L}) \le C \left(M_0^{\frac{1}{\theta}} \left(\frac{\varepsilon}{L}\right)^{1+\frac{1}{\theta}} + M_0^{\frac{1}{2\theta}} \left(\frac{\varepsilon}{L}\right)^{\frac{1}{2\theta}} \right) \le C \alpha^{-\omega}, \tag{3.70}$$

in view of 1 and provided α_1 is sufficiently large. We apply one last time Lemma 3.26, with $A = (-2L - \frac{1}{2}\alpha\varepsilon, a_1 - \frac{1}{2}\alpha\varepsilon)$ (resp. $A = (a_\ell + \frac{1}{2}\alpha\varepsilon, 2L + \frac{1}{2}\alpha\varepsilon)$) and $B = (-2L, a_1 - \alpha\varepsilon)$ (resp. $B = (a_\ell + \alpha\varepsilon, 2L)$). Since $A \setminus B \subset I_{\frac{5}{2}L} \setminus I_{\frac{3}{2}L}$, it follows from (3.70) and Lemma 3.26, combined with our previous estimates, that condition $\text{WPO}_\varepsilon^L(\alpha\varepsilon)$ is satisfied provided we choose C_w sufficiently large. $\qquad\square$

Remark 3.28 *Notice that condition 1 in Proposition 3.27 is always satisfied when $\alpha\varepsilon \le \delta_{\log}^\varepsilon$, since $L/\varepsilon \ge 1$. Also, for $\alpha = \delta_{\log}^\varepsilon/\varepsilon$, assumption 3 in Proposition 3.27 is weaker than assumption 2 We therefore deduce the following.*

Corollary 3.29 *Let u be a solution to (3.65) satisfying assumption (H_0), and such that $\mathcal{D}(u) \cap I_{3L} \subset I_L$. If*

$$\varepsilon \|f\|_{L^2(I_{3L})} \le \left(\frac{M_0}{L}\right)^{\frac{1}{2}}, \tag{3.71}$$

then $\mathcal{WP}_\varepsilon^L(\delta_{\log}^\varepsilon)$ holds.

3.3 Regularized Fronts

In this section, we assume that v_ε is a solution of $(\text{PGL})_\varepsilon$ which satisfies (H_0) and the confinement condition $\mathcal{C}_{L,S}$.

3.3.1 Finding Regularized Fronts

We provide here the proof of Proposition 3.13, which is deduced from the following.

Lemma 3.30 *Given any* $s_1 < s_2$ *in* $[0, S]$, *there exists at least one time s in* $[s_1, s_2]$ *for which* $\mathfrak{v}_\varepsilon(\cdot, s)$ *solves (3.65) with*

$$\|f\|^2_{L^2(I_{3L})} \equiv \varepsilon^{\omega-1} \|\partial_s \mathfrak{v}_\varepsilon(\cdot, s)\|^2_{L^2(I_{3L})} \leq \varepsilon^{\omega-1} \frac{\mathrm{dissip}^{3L}_\varepsilon(s_1, s_2)}{s_2 - s_1} \leq \varepsilon^{\omega-1} \frac{M_0}{s_2 - s_1}. \tag{3.72}$$

Proof It is a direct mean value argument, taking into account the rescaling of $(\mathrm{PGL})_\varepsilon$ according to our rescaling of time. $\qquad\square$

Proof of Proposition 3.13 We invoke Lemma 3.30, and from (3.72) and the assumption $s_2 - s_1 = \varepsilon^{\omega+1} L$ of Proposition 3.13, we derive exactly the assumption (3.71) in Corollary 3.29, from which the conclusion follows. $\qquad\square$

Following the same argument, but relying on Lemma 3.25 and Proposition 3.27 rather than on Corollary 3.29, we readily obtain the following.

Proposition 3.31 *For* $\alpha_1 \leq \alpha \leq \delta^\varepsilon_{\log}$:

1. *Each subinterval of* $[0, S]$ *of size* $\mathfrak{q}_0(\alpha)\varepsilon^{\omega+2}$ *contains at least one time s at which* $\mathrm{WPI}^L_\varepsilon(\alpha\varepsilon, s)$ *holds, where*

$$\mathfrak{q}_0(\alpha) = 4M_0^2 \exp(\rho_w \alpha). \tag{3.73}$$

2. *Each subinterval of* $[0, S]$ *of size* $\mathfrak{q}_0(\alpha, \beta)\varepsilon^{\omega+2}$ *contains at least one time s at which* $\mathcal{WP}^L_\varepsilon(\alpha\varepsilon, s)$ *holds, where*

$$\beta := \frac{L}{\varepsilon} \quad \text{and} \quad \mathfrak{q}_0(\alpha, \beta) = \max\left(\mathfrak{q}_0(\alpha), \left(\frac{2C_0}{C_w}\right)^2 \left(\frac{\beta}{M_0}\right)^{1-\frac{1}{\theta}} \alpha^{2\omega}\right). \tag{3.74}$$

3.3.2 Local Dissipation

For $s \in [0, S]$, set $\mathcal{E}^L_\varepsilon(s) = \mathcal{E}^L_\varepsilon(\mathfrak{v}_\varepsilon(s))$ and, when $\mathrm{WPI}^L_\varepsilon(\alpha_1\varepsilon, s)$ holds, $\mathfrak{E}^L_\varepsilon(s) = \mathfrak{E}^L_\varepsilon(\mathfrak{v}_\varepsilon(s))$, $\mathfrak{E}^L_\varepsilon$ being defined in (3.62). We assume throughout that $s_1 \leq s_2$ are contained in $[0, S]$, and in some places (in view of (3.28)) that $s_2 \geq L^2\varepsilon^\omega$.

Proposition 3.32 *If* $s_2 \geq L^2\varepsilon^\omega$, *we have*

$$\mathcal{E}^L_\varepsilon(s_2) + \mathrm{dissip}^L_\varepsilon(s_1, s_2) \leq \mathcal{E}^L_\varepsilon(s_1) + 100 C_e L^{-(\omega+2)}(s_2 - s_1) + C_e(1 + M_0)\left(\frac{L}{\varepsilon}\right)^{-\omega}. \tag{3.75}$$

Proof Let $0 \leq \varphi \leq 1$ be a smooth function with compact support in I_{2L}, such that $\varphi(x) = 1$ on $I_{\frac{5}{3}L}$, $|\varphi''| \leq 100 L^{-2}$. It follows from the properties of φ and

(3.28) that

$$\mathcal{I}_\varepsilon(s,\varphi) \le \mathcal{E}_\varepsilon^L(s) \text{ for } s \in (s_1,s_2) \text{ and } \mathcal{I}_\varepsilon(s_2,\varphi) \ge \mathcal{E}_\varepsilon^L(s_2) - C_e\left(\frac{L}{\varepsilon}\right)^{-\omega},$$

which combined with (3.36) yields

$$\mathcal{E}_\varepsilon^L(s_2) + \text{dissip}_\varepsilon^L(s_1,s_2) \le \mathcal{E}_\varepsilon^L(s_1) + C_e\left(\frac{L}{\varepsilon}\right)^{-\omega} + \varepsilon^{-\omega}\int_{s_1}^{s_2} \mathcal{F}_S(s,\varphi,\mathfrak{v}_\varepsilon)ds,$$

where $\mathcal{F}_S$ is defined in (3.37). The estimate (3.75) is then obtained invoking the inequality $|\xi_\varepsilon| \le e_\varepsilon$ to bound the term involving $\mathcal{F}_S$: combined with (3.28) for times $s \ge L^2\varepsilon^\omega$ and with assumption (H_0) for times $s \le L^2\varepsilon^\omega$. $\qquad\square$

If $\mathcal{WP}_\varepsilon^L(\delta,s_1)$ and $\text{WPI}_\varepsilon^L(\delta',s_2)$ hold, for some $\delta,\delta' \ge \alpha_1\varepsilon$ and $s_2 \ge L^2\varepsilon^\omega$, then combining inequality (3.75) with the first inequality (3.63) applied to $\mathfrak{v}_\varepsilon(s_2)$ as well as the second applied to $\mathfrak{v}_\varepsilon(s_1)$, we obtain

$$\mathfrak{E}_\varepsilon^L(s_2) + \text{dissip}_\varepsilon^L(s_1,s_2)$$
$$\le \mathcal{E}_\varepsilon^L(s_2) + C_f M_0\left(\frac{\varepsilon}{\delta'}\right)^\omega + \text{dissip}_\varepsilon^L(s_1,s_2)$$
$$\le \mathcal{E}_\varepsilon^L(s_1) + 100C_eL^{-(\omega+2)}(s_2 - s_1) + C_f M_0\left(\frac{\varepsilon}{\delta'}\right)^\omega + C_e(1+M_0)\left(\frac{\varepsilon}{L}\right)^\omega$$
$$\le \mathfrak{E}_\varepsilon^L(s_1) + (C_w + C_f)M_0\left(\frac{\varepsilon}{\delta}\right)^\omega + C_f M_0\left(\frac{\varepsilon}{\delta'}\right)^\omega + 100C_eL^{-(\omega+2)}(s_2 - s_1)$$
$$+ C_e(1+M_0)\left(\frac{\varepsilon}{L}\right)^\omega.$$

(3.76)

We deduce from this inequality an estimate for the dissipation between s_1 and s_2 and an upper bound on $\mathcal{E}_\varepsilon^L(s_2)$.

Corollary 3.33 *Assume that* $\mathcal{WP}_\varepsilon^L(\delta,s_1)$ *and* $\text{WPI}_\varepsilon^L(\delta',s_2)$ *hold, for some* $\delta,\delta' \ge \alpha_1\varepsilon$ *and* $s_2 \ge L^2\varepsilon^\omega$, *and that* $\mathfrak{E}_\varepsilon^L(s_1) = \mathfrak{E}_\varepsilon^L(s_2)$. *Then*

$$\text{dissip}_\varepsilon^L[s_1,s_2] \le (C_w + C_f)M_0\left(\frac{\varepsilon}{\delta}\right)^\omega + C_f M_0\left(\frac{\varepsilon}{\delta'}\right)^\omega + 100C_eL^{-(\omega+2)}(s_2 - s_1)$$
$$+ C_e(1+M_0)\left(\frac{\varepsilon}{L}\right)^\omega,$$

$$\mathcal{E}_\varepsilon^L(s_2) - \mathfrak{E}_\varepsilon^L(s_2) \le (C_w + C_f)M_0\left(\frac{\varepsilon}{\delta}\right)^\omega + 100C_eL^{-(\omega+2)}(s_2 - s_1)$$
$$+ C_e(1+M_0)\left(\frac{\varepsilon}{L}\right)^\omega.$$

3.3.3 Quantization of the Energy

Let $s \in [0,S]$ and $\delta \geq \alpha_1 \varepsilon$, and assume that $\mathfrak{v}_\varepsilon$ satisfies $\mathcal{WP}_\varepsilon^L(\delta,s)$. The front energy $\mathfrak{E}_\varepsilon^L(s)$, by definition, may take only a finite number of values, and is hence *quantized*. We emphasize that, at this stage, $\mathfrak{E}_\varepsilon^L(s)$ is only defined *assuming* condition $\mathrm{WPI}_\varepsilon^L(\delta,s)$ holds. However, the *value* of $\mathfrak{E}_\varepsilon^L(s)$ *does not* depend on δ, provided that $\delta \geq \alpha_1 \varepsilon$, so that it suffices *ultimately* to check that condition $\mathrm{WPI}_\varepsilon^L(\alpha_1 \varepsilon, s)$ is fulfilled.

Since $\mathfrak{E}_\varepsilon(s)$ may take only a finite number of values, let $\mu_1 > 0$ be the smallest possible difference between two distinct such values. Let $L_0 \equiv L_0(s_1,s_2) > 0$ be such that

$$100 C_e L_0^{-(\omega+2)}(s_2 - s_1) = \frac{\mu_1}{4} \tag{3.77}$$

and finally choose α_1 sufficiently large so that

$$\left((2C_f + C_w)M_0 + C_e(1 + M_0)\right)\alpha_1^{-\omega} \leq \frac{\mu_1}{4}. \tag{3.78}$$

As a direct consequence of (3.76), (3.77), (3.78) and the definition of μ_1 we obtain the following result.

Corollary 3.34 *For $s_1 \leq s_2 \in [0,S]$ with $s_2 \geq \varepsilon^\omega L^2$, assume that $\mathcal{WP}_\varepsilon^L(\alpha_1 \varepsilon, s_1)$ and $\mathrm{WPI}_\varepsilon^L(\alpha_1 \varepsilon, s_2)$ hold and that $L \geq L_0(s_1,s_2)$. Then we have $\mathfrak{E}_\varepsilon^L(s_2) \leq \mathfrak{E}_\varepsilon^L(s_1)$. Moreover, if $\mathfrak{E}_\varepsilon^L(s_2) < \mathfrak{E}_\varepsilon^L(s_1)$, then $\mathfrak{E}_\varepsilon^L(s_2) + \mu_1 \leq \mathfrak{E}_\varepsilon^L(s_1)$.*

In the opposite direction we have the following.

Lemma 3.35 *For $s_1 \leq s_2 \in [0,S]$, assume that $\mathrm{WPI}_\varepsilon^L(\alpha_1 \varepsilon, s_1)$ and $\mathrm{WPI}_\varepsilon^L(\alpha_1 \varepsilon, s_2)$ hold and that $L \geq L_0(s_1,s_2)$. Assume also that*

$$s_2 - s_1 \leq \rho_0 \left(\frac{1}{8} d_{\min}^{\varepsilon,L}(s_1)\right)^{\omega+2}. \tag{3.79}$$

Then we have $\mathfrak{E}_\varepsilon^L(s_2) \geq \mathfrak{E}_\varepsilon^L(s_1)$. In the case of equality, we have $J(s_1) = J(s_2)$ and

$$\sigma_{i(k\pm\frac{1}{2})}(s_1) = \sigma_{i(k\pm\frac{1}{2})}(s_1), \text{ for any } k \in J(s_1) \text{ and } d_{\min}^{\varepsilon,L}(s_2) \geq \frac{1}{2} d_{\min}^{\varepsilon,L}(s_1). \tag{3.80}$$

Proof It a consequence of the bound (3.7) in Theorem 3.2 on the speed of the front set combined with assumption (3.79). Indeed, this implies that for arbitrary $s \in [s_1,s_2]$, the front set at time s is contained in a neighborhood of size $d_{\min}^{\varepsilon,L}(s_1)/8$ of the front set at time s_1. In view of the definition of $d_{\min}^{\varepsilon,L}(s_1)$, and of the continuity in time of the solution, this implies that for all $k_0 \in J(s_1)$

the set

$$\mathcal{A}_{k_0} = \left\{ k \in J(s_2) \text{ such that } a_k^\varepsilon(s_2) \in \left[a_{k_0}^\varepsilon(s_1) - \frac{1}{4} d_{\min}^{\varepsilon,L}(s_1), a_{k_0}^\varepsilon(s_1) + \frac{1}{4} d_{\min}^{\varepsilon,L}(s_1) \right] \right\}$$

is non-empty, since it must contain a front connecting $\sigma_{i(k_0 - \frac{1}{2})}(s_1)$ to $\sigma_{i(k_0 + \frac{1}{2})}(s_1)$. In particular, summing over all fronts in $\mathcal{A}_{k_0}$, we obtain

$$\sum_{k \in \mathcal{A}_{k_0}} \mathfrak{S}_{i(k)}^L \geq \mathfrak{S}_{i(k_0)}^L,$$

with equality if and only if $\sharp \mathcal{A}_{k_0} = 1$. Summing over all indices k_0, we are led to the conclusion. $\qquad\square$

3.3.4 Propagating Regularized Fronts

We discuss in this subsection the case of equality $\mathfrak{E}_\varepsilon^L(s_1) = \mathfrak{E}_\varepsilon^L(s_2)$. We assume throughout that we are given $\delta_{\log}^\varepsilon \geq \delta > \alpha_1 \varepsilon$ and two times $s_1 \leq s_2 \in [\varepsilon^\omega L^2, S]$ such that

$$\mathcal{C}(\delta, L, s_1, s_2) \qquad \begin{cases} \mathcal{WP}_\varepsilon^L(\delta, s_1) \quad \text{and} \quad \text{WPI}_\varepsilon^L(\delta, s_2) \quad \text{hold}, \\ \mathfrak{E}_\varepsilon^L(s_1) = \mathfrak{E}_\varepsilon^L(s_2), \quad \text{with } L \geq L_0(s_1, s_2). \end{cases}$$

Under that assumption, our first result shows that $\mathfrak{v}_\varepsilon$ remains well-prepared on almost the whole time interval $[s_1, s_2]$, though with a smaller δ.

Proposition 3.36 *There exists $\alpha_2 \geq \alpha_1$, depending only on V, M_0 and C_w, with the following property. Assume that $\mathcal{C}(\delta, L, s_1, s_2)$ holds with $\alpha_2 \varepsilon \leq \delta \leq \delta_{\log}^\varepsilon$, then property $\mathcal{WP}_\varepsilon^L(\Lambda_{\log}(\delta), s)$ holds for any time $s \in [s_1 + \varepsilon^{2+\omega}, s_2]$, where*

$$\Lambda_{\log}(\delta) = \frac{\omega}{\rho_w} \varepsilon \left(\log \frac{\delta}{\varepsilon} \right). \tag{3.81}$$

The proof of Proposition 3.36 relies on the following.

Lemma 3.37 *Assume that $\mathcal{C}(\delta, L, s_1, s_2)$ holds with $\delta \geq \alpha_1 \varepsilon$. We have the estimate, for $s \in [s_1 + \varepsilon^{\omega+2}, s_2]$,*

$$\int_{I_{\frac{3}{2}L}} |\partial_t v_\varepsilon(x, s\varepsilon^{-\omega})|^2 dx \leq C \varepsilon^{-3} \text{dissip}_\varepsilon^L[s, s - \varepsilon^{\omega+2}].$$

Proof of Lemma 3.37 Differentiating equation (PGL_ε) with respect to time, we are led to

$$|\partial_t(\partial_t v_\varepsilon) - \partial_{xx}(\partial_t v_\varepsilon)| \leq \frac{C}{\varepsilon^2} |\partial_t v_\varepsilon|.$$

It follows from standard parabolic estimates, working for $x \in I_{2L}$ on the cylinder $\Lambda_\varepsilon(x) = [x - \varepsilon, x + \varepsilon] \times [t - \varepsilon^2, t]$, where $t := s\varepsilon^{-\omega}$, that for any point $y \in [x - \frac{\varepsilon}{2}, x + \frac{\varepsilon}{2}]$ we have

$$|\partial_t v_\varepsilon(y, t)| \le C\varepsilon^{-\frac{3}{2}} \|\partial_t v_\varepsilon\|_{L^2(\Lambda_\varepsilon(x))}.$$

Taking the square of the previous inequality, and integrating over $[x - \frac{\varepsilon}{2}, x + \frac{\varepsilon}{2}]$, we are led to

$$\int_{x-\frac{\varepsilon}{2}}^{x+\frac{\varepsilon}{2}} |\partial_t v_\varepsilon(y, t)|^2 dy \le C\varepsilon^{-2} \int_{[x-2\varepsilon, x+2\varepsilon] \times [t-\varepsilon^2, t]} |\partial_t v_\varepsilon(y, t)|^2 dy.$$

An elementary covering argument then yields

$$\int_{I_{\frac{3}{2}L}} |\partial_t v_\varepsilon(y, t)|^2 dy \le C\varepsilon^{-2} \|\partial_t v_\varepsilon\|^2_{L^2(I_{\frac{5}{3}L} \times [t-\varepsilon^2, t])} \le C\varepsilon^{-3} \text{dissip}^L_\varepsilon[s, s - \varepsilon^{\omega+2}].$$

□

Proof of Proposition 3.36 In view of Proposition 3.31, Corollary 3.34, and assumption $\mathcal{C}(\delta, L, s_1, s_2)$, we may assume, without loss of generality, that

$$s_2 - s_1 \le 2q_0(\delta/\varepsilon, L/\varepsilon). \tag{3.82}$$

Let $s \in (s_1 + \varepsilon^{\omega+2}, s_2)$, and consider once more the map $u = \mathfrak{v}_\varepsilon(\cdot, s)$, so that u is a solution to (3.65), with source term $f = \partial_t v_\varepsilon(\cdot, s\varepsilon^{-\omega})$. It follows from Lemma 3.37, combined with the first of Corollary 3.33 on the dissipation, that

$$\|f\|^2_{L^2(I_{\frac{3}{2}L})} \le C\varepsilon^{-3}\left[(C_w + 2C_f)M_0\left(\frac{\varepsilon}{\delta}\right)^\omega + 100C_e L^{-(\omega+2)}(s_2 - s_1)\right.$$
$$\left. + C_e(1 + M_0)\left(\frac{\varepsilon}{L}\right)^\omega\right].$$

Notice that (3.82) combined with the assumption $\delta \le \delta^\varepsilon_{\log}$ yields

$$100C_e L^{-(\omega+2)}(s_2 - s_1) \le C\left(\frac{\varepsilon}{\delta}\right)^\omega.$$

We deduce from Lemma 3.25, imposing on α_2 the additional condition $\frac{\omega}{\rho_w}(\log \alpha_2) \ge \alpha_1$, that $\text{WPI}^L_\varepsilon((\Lambda_{\log}(\delta), s)$ holds. It remains to show that $\text{WPO}^L_\varepsilon(\Lambda_{\log}(\delta), s)$ holds likewise. To that aim, we invoke (3.33) which we use with the choice $s_1 = s_1$ and $s_2 = s$. This yields, taking once more (3.82) into account,

$$\mathfrak{E}^L_\varepsilon(s) - \mathcal{E}^L_\varepsilon(s) \le (C + C_w)\left(\frac{\varepsilon}{\delta}\right)^\omega.$$

Combining this relation with (3.61) and the first inequality of (3.63), we deduce that

$$\int_{\Omega} e_\varepsilon(\mathfrak{v}_\varepsilon(s))ds \leq (C+C_w)\left(\frac{\varepsilon}{\delta}\right)^{\omega} + C\left(\frac{\varepsilon}{\Lambda_{\log}(\delta)}\right)^{\omega} \leq C_w M_0 \left(\frac{\varepsilon}{\Lambda_{\log}(\delta)}\right)^{\omega},$$

(3.83)

provided α_2 is chosen sufficiently large. $\qquad\square$

In view of (3.79) and (3.7), we introduce the function

$$\mathfrak{q}_1(\alpha) := \left(\frac{\mathfrak{q}_0(\alpha)}{\rho_0}\right)^{\frac{1}{\omega+2}},$$

which represents therefore the maximum displacement of the front set in the interval of time needed (at most) to find two consecutive times at which $WPI_\varepsilon^L(\alpha\varepsilon)$ holds.

From Proposition 3.36 and Lemma 3.35 we deduce the following.

Corollary 3.38 *Let* $s \in [\varepsilon^{\omega}L^2, S]$ *and* $\alpha_2 \leq \alpha \leq \delta_{\log}^{\varepsilon}$, *and assume that* $WP_\varepsilon^L(\alpha\varepsilon, s)$ *holds as well as* $d_{\min}^{\varepsilon,L}(s) \geq 16\mathfrak{q}_1(\alpha)\varepsilon$. *Then* $WP_\varepsilon^L(\Lambda_{\log}(\alpha\varepsilon), s')$ *holds for any* $s + \varepsilon^{2+\omega} \leq s' \leq T_0^{\varepsilon}(\alpha, s)$, *where*

$$T_0^{\varepsilon}(\alpha, s) = \max\left\{s + \varepsilon^{2+\omega} \leq s' \leq S \quad s.t. \quad d_{\min}^{\varepsilon,L}(s'') \geq 8\mathfrak{q}_1(\alpha)\varepsilon\right.$$
$$\left.\forall s'' \in [s + \varepsilon^{\omega+2}, s']\right\}.$$

We complete this section presenting the following proof.

Proof of Proposition 3.14 This follows directly from Corollary 3.38 with the choice $\alpha = \delta_{\log}^{\varepsilon}$, noticing that $\Lambda_{\log}(\delta_{\log}^{\varepsilon}) = \delta_{\log\log}^{\varepsilon}$. $\qquad\square$

Proof of Corollary 3.15 If we assume moreover that $s_0 \geq \varepsilon^{\omega}L^2$ and that $WP_\varepsilon^L(\delta_{\log}^{\varepsilon}, s_0)$ holds, then it is a direct consequence of the inclusion (3.7) and Proposition 3.14, taking into account the assumption (3.33). If we assume only that $s_0 \geq 0$ and that $WP_\varepsilon^L(\alpha_1\varepsilon, s_0)$ holds, then it suffices to consider the first time $s_0' \geq s_0 + \varepsilon^{\omega}L^2$ at which $WP_\varepsilon^L(\delta_{\log}^{\varepsilon}, s_0')$ holds and to rely on Proposition 3.14 likewise. Indeed, since $s_0' - s_0 \leq \varepsilon^{\omega}L^2 + \varepsilon^{\omega+1}L$ by Proposition 3.13, we may apply Corollary 3.34 and Lemma 3.35 for $s_1 = s_0$ and $s_2 = s_0'$, which yields $\mathfrak{E}_\varepsilon^L(s_0) = \mathfrak{E}_\varepsilon^L(s_0')$ and therefore also the same asymptotics for $d_{\min}^{\varepsilon,L}$ at times s_0 and s_0'. $\qquad\square$

3.4 First Compactness Results for the Front Points

The purpose of this section is to provide the proofs of Proposition 3.16 and Corollary 3.17.

Proof of Proposition 3.16 As mentioned, we choose the test functions (independently of time) so that they are affine near the front points for any $s \in I^\varepsilon(s_0)$. More precisely, for a given $k_0 \in J$ we impose the following conditions on the test functions $\chi \equiv \chi_{k_0}$ in (3.42):

$$
\begin{cases}
\chi \text{ has compact support in } [a_k^\varepsilon(s_0) - \dfrac{1}{3}d_{\min}^*(s_0), a_k^\varepsilon(s_0) + \dfrac{1}{3}d_{\min}^*(s_0)], \\[2mm]
\chi \text{ is affine on the interval } [a_k^\varepsilon(s_0) - \dfrac{1}{4}d_{\min}^*(s_0), a_k^\varepsilon(s_0) + \dfrac{1}{4}d_{\min}^*(s_0)], \\[2mm]
\qquad\qquad\qquad\qquad\qquad\qquad\qquad\qquad \text{with } \chi' = 1 \text{ there} \\[2mm]
\|\chi\|_{L^\infty(\mathbb{R})} \le C d_{\min}^*(s_0), \|\chi'\|_{L^\infty(\mathbb{R})} \le C \text{ and } \|\chi''\|_{L^\infty(\mathbb{R})} \le C d_{\min}^*(s_0)^{-1}.
\end{cases}
$$
$$(3.84)$$

It follows from Corollary 3.15 that, for ε sufficiently small, we are in position to claim (3.42) and (3.43) for arbitrary s_1 and s_2 in the full interval $I^*(s_0)$. Combined with the first estimate of Corollary 3.33, with $\delta = \delta' = \delta_{\text{loglog}}^\varepsilon$, this yields the conclusion (3.44). $\qquad\square$

Proof of Corollary 3.17 The family of functions $(\mathfrak{v}_\varepsilon)_{0<\varepsilon<1}$ is equi-continuous on every compact subset of the interval $I^*(s)$, so that the conclusion follows from the Arzela–Ascoli theorem. $\qquad\square$

3.5 Refined Asymptotics Off the Front Set

3.5.1 Relaxations Towards Stationary Solutions

Throughout this section, we assume that we are in the situation described by Corollary 3.15, in particular L is fixed and ε will tend to zero. Our main purpose is then to provide rigorous mathematical statements and proofs concerning the properties of the function $\mathfrak{W}_{\varepsilon_n} = \mathfrak{W}_{\varepsilon_n}^k$ defined in (3.45), for given $k \in J$, which have been presented, most of them in a formal way, in Subsection 3.1.5. We notice first that we may expand V' near $\sigma \equiv \sigma_{i(k+\frac{1}{2})}$ as

$$ V'(\sigma + u) = 2\theta\lambda u^{2\theta-1}\left(1 + ug(u)\right), \qquad (3.85) $$

where g is some smooth function on $\mathbb{R}$ and where we have set for the sake of simplicity $\lambda = \lambda_{i(k+\frac{1}{2})}$. We work on the sets $\mathcal{V}_k(s_0)$ defined in (3.48) and on their analogs at the ε level:

$$ \mathcal{V}_k^\varepsilon(s_0) = \bigcup_{s \in I^\varepsilon(s_0)} \mathcal{J}_\varepsilon(s) \times \{s\} \equiv \bigcup_{s \in I^\varepsilon(s_0)} \left(a_k^\varepsilon(s) + \delta_{\text{loglog}}^\varepsilon, a_{k+1}^\varepsilon(s) - \delta_{\text{loglog}}^\varepsilon\right) \times \{s\}. $$
$$(3.86)$$

We will therefore work only with *arbitrary* small values of u. Let $u_0 > 0$ be sufficiently small so that $|ug(u)| \leq 1/4$ on $(-u_0, u_0)$ and $V'(\sigma + u)$ is strictly increasing on $(-u_0, u_0)$, convex on $(0, u_0)$ and concave on $(-u_0, 0)$. For small values of ε, the value of u in (3.85), in view of (3.46) in Lemma 3.18, will not exceed u_0, and we may therefore assume for the considerations in this section that $ug(u) = u_0 g(u_0)$, if $u \geq u_0$ and $-ug(u) = u_0 g(u_0)$, if $u \leq -u_0$. Equation (PGL)$_\varepsilon$ translates into the following equation for $\mathfrak{W}_\varepsilon$:

$$L_\varepsilon(\mathfrak{W}_\varepsilon) \equiv \varepsilon^\omega \frac{\partial \mathfrak{W}_\varepsilon}{\partial s} - \frac{\partial^2 \mathfrak{W}_\varepsilon}{\partial x^2} + \lambda f_\varepsilon(\mathfrak{W}_\varepsilon) = 0, \tag{3.87}$$

where we have set

$$f_\varepsilon(w) = 2\theta w^{2\theta - 1} \left(1 + \varepsilon^{\frac{1}{\theta - 1}} wg(\varepsilon^{\frac{1}{\theta - 1}} w)\right). \tag{3.88}$$

Notice that our assumption yields in particular

$$|f_\varepsilon(w)| \geq \frac{3}{2}\theta |w|^{2\theta - 1}. \tag{3.89}$$

The analysis of the parabolic equation (3.87) is the core of this section. As mentioned, our results express convergence to stationary solutions. We first provide a few properties concerning these stationary solutions: the first lemma describes stationary solutions involved in the attractive case, whereas the second lemma is used in the repulsive case.

Lemma 3.39 *Let $r > 0$ and $0 < \varepsilon < 1$. There exist unique solutions $\overset{v+}{u}_{\varepsilon,r}$ (resp. $\overset{v-}{u}_{\varepsilon,r}$) to*

$$\begin{cases} -\dfrac{d\mathcal{U}}{dx^2} + \lambda f_\varepsilon(\mathcal{U}) = 0 \text{ on } (-r, r), \\ \mathcal{U}(-r) = +\infty \text{ (resp. } \mathcal{U}(-r) = -\infty\text{) and } \mathcal{U}(r) = +\infty \text{ (resp. } \mathcal{U}(r) = -\infty\text{).} \end{cases}$$

Moreover, we have

$$C^{-1}r^{-\frac{1}{\theta-1}} \leq \overset{v+}{u}_{\varepsilon,r} \leq C(r - |x|)^{-\frac{1}{\theta-1}} \text{ and } C^{-1}r^{-\frac{1}{\theta-1}} \leq -\overset{v-}{u}_{\varepsilon,r} \leq C(r - |x|)^{-\frac{1}{\theta-1}}, \tag{3.90}$$

for some constant $C > 0$ depending only on V.

Lemma 3.40 *Let $r > 0$ and $0 < \varepsilon < 1$ be given. There exists a unique solution $\overset{\triangleright}{u}_{\varepsilon,r}$ to*

$$-\frac{d\mathcal{U}}{dx^2} + \lambda f_\varepsilon(\mathcal{U}) = 0 \text{ on } (-r, r), \ \mathcal{U}(-r) = -\infty \text{ and } \mathcal{U}(r) = +\infty.$$

These and related results are standard and have been considered since the works of Keller [12] and Osserman [14] in the 1950s, at least regarding existence. The convexity/concavity assumptions are sufficient for uniqueness. We refer to Lemma 3.59 in Appendix A for a short discussion of the case of a pure power non-linearity.

We set $r^\varepsilon(s) = r^\varepsilon_{k+\frac{1}{2}}(s) = \frac{1}{2}(a^\varepsilon_{k+1}(s) - a^\varepsilon_k(s))$. Our aim is to provide sufficiently accurate expansions of $\mathfrak{W}_\varepsilon$ and the renormalized discrepancy $\varepsilon^{-\omega}\xi_\varepsilon$ on neighborhoods of the points $a^\varepsilon_{k+\frac{1}{2}}(s)$, for instance the intervals

$$\Theta^\varepsilon_{k+\frac{1}{2}}(s) = a^\varepsilon_{k+\frac{1}{2}}(s) + [-\frac{7}{8}r^\varepsilon(s), \frac{7}{8}r^\varepsilon(s)] = [a^\varepsilon_k(s) + \frac{1}{8}r^\varepsilon(s), a^\varepsilon_{k+1}(s) - \frac{1}{8}r^\varepsilon(s)].$$

$$(3.91)$$

We first turn to the the attractive case $\dagger_k = -\dagger_{k+1}$. We may assume additionally that

$$k \in \{1, \dots, \ell - 1\} \text{ and } \dagger_k = -\dagger_{k+1} = 1, \tag{3.92}$$

the case $\dagger_k = -\dagger_{k+1} = -1$ being handled similarly.

Proposition 3.41 *If (3.92) holds and ε is sufficiently small, then for any $s \in I^\varepsilon(s_0)$ and every $x \in \Theta^\varepsilon_{k+\frac{1}{2}}(s)$ we have the estimate*

$$|\mathfrak{W}_\varepsilon(x,s) - \lambda^{-\frac{1}{2(\theta-1)}} \overset{\vee+}{\mathfrak{u}}_{r^\varepsilon(s)}(x)| \le C\varepsilon^{\min(\frac{1}{\omega+2}, \frac{\omega-1}{2(\theta-1)})}. \tag{3.93}$$

The repulsive case corresponds to $\dagger_k = \dagger_{k+1}$ and we may assume as above that

$$k \in \{1, \dots, \ell - 1\} \text{ and } \dagger_k = \dagger_{k+1} = 1. \tag{3.94}$$

Proposition 3.42 *If (3.94) holds and ε is sufficiently small, then for any $s \in I^\varepsilon(s_0)$ and every $x \in \Theta^\varepsilon_{k+\frac{1}{2}}(s)$ we have the estimate*

$$|\mathfrak{W}_\varepsilon(x,s) - \lambda^{-\frac{1}{2(\theta-1)}} \overset{\triangleright}{\mathfrak{u}}_{r^\varepsilon(s)}(x)| \le C\varepsilon^{\min(\frac{1}{\omega+2}, \frac{\omega-1}{2(\theta-1)})}. \tag{3.95}$$

Combining these results with parabolic estimates, we obtain estimates for the discrepancy.

Proposition 3.43 *If ε is sufficiently small, then for any $s \in I^\varepsilon(s_0)$ and every $x \in \Theta^\varepsilon_{k+\frac{1}{2}}(s)$ we have the estimate*

$$|\varepsilon^{-\omega}\xi_\varepsilon(\mathfrak{v}_\varepsilon) - \lambda^{-\frac{1}{2(\theta-1)}}_{i(k+\frac{1}{2})} r^\varepsilon(s)^{-(\omega+1)}\gamma_{k+\frac{1}{2}}| \le C\varepsilon^{\frac{1}{\theta^2}}, \tag{3.96}$$

where

$$\begin{cases} \gamma_{k+\frac{1}{2}} = A_\theta \text{ if } \dagger_k = -\dagger_{k+1}, \\ \gamma_{k+\frac{1}{2}} = B_\theta \text{ if } \dagger_k = \dagger_{k+1}. \end{cases} \tag{3.97}$$

For the outer regions, corresponding to $k = 0$ and $k = \ell$, estimates for the discrepancy are directly deduced from the crude estimates provided by Proposition 3.12. Proposition 3.43 provides a rigorous ground to the formal computation (3.52) of the introduction, and hence allows us to derive the precise motion law. The proofs of Propositions 3.41 and 3.42, however, are the central part of this section. Note that by no means are the estimates provided in Propositions 3.41, 3.42 and 3.43 optimal, our goal was only to obtain convergence estimates, valid for all ε sufficiently small, uniformly on $\cup_{s \in I^\varepsilon(s_0)} \Theta^\varepsilon_{k+\frac{1}{2}}(s) \times \{s\}$.

3.5.2 Preliminary Results

We first turn to the proof of Lemma 3.18, which provides first properties of $\mathfrak{W}_\varepsilon$.

Proof of Lemma 3.18 Let $x \in (a_k(s) + \delta^\varepsilon_{\text{loglog}}, a_{k+1}(s) - \delta^\varepsilon_{\text{loglog}})$ and any $s \in I^\varepsilon(s_0)$, and recall that $d(x,s) := \text{dist}(x, \{a^\varepsilon_k(s), a^\varepsilon_{k+1}(s)\})$. In view of Proposition 3.13, and in particular of estimate (3.31), it suffices to show that

$$\mathfrak{v}_\varepsilon(y,s) \in B(\sigma_i, \mu_0) \quad \text{for all} \quad (y,s) \in \left[x - \frac{d(x,s)}{2}, x + \frac{d(x,s)}{2}\right] \times [s - \varepsilon^\omega d(x,s)^2, s].$$

By Theorem 3.2, on such a time scale the front set moves at most by a distance

$$d := \left(\frac{\varepsilon^\omega d(x,s)^2}{\rho_0}\right)^{\frac{1}{\omega+2}} \leq \rho_0^{-\frac{1}{\omega+2}} \left(\frac{\varepsilon}{\delta^\varepsilon_{\text{loglog}}}\right)^{\frac{\omega}{\omega+2}} d(x,s) \leq \frac{d(x,s)}{4},$$

provided ε/L is sufficiently small. More precisely, Theorem 3.2 only provides one inclusion, forward in time, but its combination with Corollary 3.15 provides both forward and backward inclusions (for times in the interval $I^\varepsilon(s_0)$), from which the conclusion then follows. $\square$

For the analysis of the scalar parabolic equation (3.87), we will extensively use the fact that the map f_ε is *non-decreasing* on $\mathbb{R}$, allowing comparison principles. The desired estimates for $\mathfrak{W}_\varepsilon$ will be obtained using appropriate choices of sub- and super-solutions. The construction of these functions involves a number of elementary solutions. First, we use the functions $\mathcal{W}^\pm_\varepsilon$, independent of the space variable x and solve the ordinary differential equation

$$\begin{cases} \varepsilon^\omega \dfrac{\partial \mathcal{W}^\pm_\varepsilon}{\partial s} = -\lambda f_\varepsilon(\mathcal{W}^\pm_\varepsilon), \\ \mathcal{W}_\varepsilon(0) = \pm\infty. \end{cases} \tag{3.98}$$

Using separation of variables, we may construct such a solution which verifies the bounds

$$0 < \mathcal{W}_\varepsilon^+(s) \le C\varepsilon^{\frac{\omega}{2(\theta-1)}}[\lambda s]^{-\frac{1}{2(\theta-1)}} \text{ and } 0 \ge \mathcal{W}_\varepsilon^-(s) \ge -C\varepsilon^{\frac{\omega}{2(\theta-1)}}[\lambda s]^{-\frac{1}{2(\theta-1)}},$$
(3.99)

so that it relaxes quickly to zero. We will also use solutions of the standard heat equation and rely in several places on the next remark.

Lemma 3.44 *Let Φ be a non-negative solution to the heat equation $\varepsilon^\omega \partial_s \Phi - \Phi_{xx} = 0$, and U be such that $L_\varepsilon(U) = 0$. Then $L_\varepsilon(U + \Phi) \ge 0$, and $L_\varepsilon(U - \Phi) \le 0$.*

Proof Notice that $L_\varepsilon(U \pm \Phi) = \lambda(f_\varepsilon(U \pm \Phi) - f_\varepsilon(U))$, so that the conclusion follows from the fact that f_ε is non-decreasing. $\qquad\square$

Next, let s be given in $I^\varepsilon(s_0)$. By translation invariance, we may assume without loss of generality that

$$a_{k+\frac{1}{2}}^\varepsilon(s) = 0.$$
(3.100)

We set $h_\varepsilon = (\varepsilon/2\rho_0)^{\frac{1}{\omega+2}}$, and consider the cylinders

$$\Lambda_\varepsilon^{\text{ext}}(s) = \mathcal{J}_\varepsilon^{\text{ext}}(s) \times [s - \varepsilon, s] \text{ and } \Lambda_\varepsilon^{\text{int}}(s) = \mathcal{J}_\varepsilon^{\text{int}}(s) \times [s - \varepsilon, s],$$
(3.101)

where $\mathcal{J}_\varepsilon^{\text{int}}(s) = [-r_{\text{int}}^\varepsilon(s), r_{\text{int}}^\varepsilon(s)]$, $\mathcal{J}_\varepsilon^{\text{ext}}(s) = [-r_{\text{ext}}^\varepsilon(s), r_{\text{ext}}^\varepsilon(s)]$ with

$$r_{\text{ext}}^\varepsilon(s) = r^\varepsilon(s) + 2h_\varepsilon \text{ and } r_{\text{int}}^\varepsilon(s) = r^\varepsilon(s) - 2h_\varepsilon.$$

If ε is sufficiently small, in view of (3.7) we have the inclusions, with $\mathcal{V}_k^\varepsilon(s_0)$ defined in (3.86),

$$\Lambda_\varepsilon^{\text{int}}(s) \subset \Pi_\varepsilon(s) \equiv \mathcal{V}_k^\varepsilon(s_0) \cap ([s - \varepsilon, s] \times \mathbb{R}) \subset \Lambda_\varepsilon^{\text{ext}}(s).$$

As a matter of fact, still for ε sufficiently small, we have for any $\tau \in [s - \varepsilon, s]$,

$$\begin{cases} -r_{\text{ext}}^\varepsilon(s) + h_\varepsilon \le a_k^\varepsilon(\tau) + \delta_{\text{loglog}}^\varepsilon \le -r_{\text{int}}^\varepsilon(s) - h_\varepsilon, \\ r_{\text{int}}^\varepsilon + h_\varepsilon \le a_{k+1}^\varepsilon(\tau) - \delta_{\text{loglog}}^\varepsilon \le r_{\text{ext}}^\varepsilon(s) - h_\varepsilon(s_0). \end{cases}$$
(3.102)

We also consider the parabolic boundary of $\Lambda_\varepsilon^{\text{ext}}(s)$:

$$\partial_p \Lambda_\varepsilon^{\text{ext}}(s) = [-r_{\text{ext}}^\varepsilon(s), r_{\text{ext}}^\varepsilon(s)] \times \{s - \varepsilon\} \cup \{-r_{\text{ext}}^\varepsilon\} \times [s - \varepsilon, s] \cup \{r_{\text{ext}}^\varepsilon\} \times [s - \varepsilon, s]$$

$$= \partial \Lambda_\varepsilon^{\text{ext}}(s) \setminus [-r_{\text{ext}}^\varepsilon(s), r_{\text{ext}}^\varepsilon(s)] \times \{s\},$$

and define $\partial_p \Lambda_\varepsilon^{\text{int}}(s)$ accordingly. Finally, we set

$$\partial_p \Pi_\varepsilon(s) = \partial(\Pi_\varepsilon(s)) \setminus [a_k^\varepsilon(s) + \delta_{\text{loglog}}^\varepsilon, a_{k+1}^\varepsilon(s) - \delta_{\text{loglog}}^\varepsilon] \times \{s\}.$$

A first application of the comparison principle leads to the following bounds.

Proposition 3.45 *For* $x \in \mathcal{J}_\varepsilon^{\text{int}}(s)$,

$$\begin{cases} \mathfrak{W}_\varepsilon(x,s) \le \overset{\vee+}{u}_{\varepsilon,r_{\text{int}}^\varepsilon}(x) + C\varepsilon^{\frac{\omega-1}{2(\theta-1)}}, \\ \mathfrak{W}_\varepsilon(x,s) \ge \overset{\vee-}{u}_{\varepsilon,r_{\text{int}}^\varepsilon}(x) - C\varepsilon^{\frac{\omega-1}{2(\theta-1)}}. \end{cases} \tag{3.103}$$

Proof We work on the cylinder $\Lambda_\varepsilon^{\text{int}}(s)$ and consider there the comparison map

$$W_\varepsilon^{\text{sup}}(y,\tau) = \overset{\vee+}{u}_{\varepsilon,r_{\text{int}}^\varepsilon}(y) + \mathcal{W}_\varepsilon(\tau - (s-\varepsilon)) \text{ for } (y,\tau) \in \Lambda_\varepsilon^{\text{int}}(s).$$

Since the two functions on the r.h.s. of the definition of $W_\varepsilon^{\text{sup}}$ are positive solutions to (3.87) and since f_ε is super-additive on $\mathbb{R}^+$, that is, since

$$f_\varepsilon(a+b) \ge f_\varepsilon(a) + f_\varepsilon(b) \text{ provided } a \ge 0, b \ge 0, \tag{3.104}$$

we deduce that

$$L_\varepsilon\left(W_\varepsilon^{\text{sup}}(y,\tau)\right) \ge 0 \text{ on } \Lambda_\varepsilon^{\text{int}}(s) \text{ with } W_\varepsilon^{\text{sup}}(y,\tau) = +\infty \text{ for } (y,\tau) \in \partial_p \Lambda_\varepsilon^{\text{int}},$$

so that $W^{\text{sup}\varepsilon}(x,s) \ge \mathfrak{W}_\varepsilon$ on $\partial_p \Lambda_\varepsilon^{\text{int}}$. It follows that $W_\varepsilon^{\text{sup}}(y,\tau) \ge \mathfrak{W}_\varepsilon$ on $\Lambda_\varepsilon^{\text{int}}$, which, combined with (3.99), immediately leads to the first inequality. The second is derived similarly. $\qquad\square$

At this stage, the constructions are somewhat different in the case of attractive and repulsive forces, so we need to distinguish the two cases.

3.5.3 The Attractive Case

We assume here that $\dagger_k = -\dagger_{k+1}$. Without loss of generality, we may assume that

$$\dagger_k = -\dagger_{k+1} = 1, \tag{3.105}$$

the case $\dagger_k = -\dagger_k = -1$ being handled similarly. The purpose of this subsection is to provide *the proof of Proposition 3.41*. We split the proof into separate lemmas, the main efforts being devoted to the construction of *subsolutions*. We start with the following lower bound.

Lemma 3.46 *Assume that (3.105) holds. Then, for* $x \in \mathcal{J}_\varepsilon(s - \frac{\varepsilon}{2})$, *we have the lower bound*

$$\mathfrak{W}_\varepsilon(x, s - \frac{\varepsilon}{2}) \ge -C\varepsilon^{\frac{\omega-1}{2(\theta-1)}}.$$

Proof In view of (3.47), we notice that

$$\mathfrak{W}_\varepsilon(y,\tau) \ge 0 \text{ on } \partial_p \Pi_\varepsilon(s) \setminus [a_k(s-\varepsilon) + \delta_{\text{loglog}}^\varepsilon, a_{k+1}(s-\varepsilon) - \delta_{\text{loglog}}^\varepsilon] \times \{s - \varepsilon\}.$$

We consider next the function W_ε defined for $\tau \geq s - \varepsilon$ by $W_\varepsilon(y,\tau) = W_\varepsilon^-(\tau - (s - \varepsilon))$. Since $W_\varepsilon < 0$, and since $W_\varepsilon(s - \varepsilon) = -\infty$, we obtain $W_\varepsilon \leq \mathfrak{W}_\varepsilon$ on $\partial_p \Pi_\varepsilon(s)$, so that, by the comparison principle, we are led to $W_\varepsilon \leq \mathfrak{W}_\varepsilon$ on $\Pi_\varepsilon(s)$, leading to the conclusion. $\qquad\square$

Proposition 3.47 *Assume that (3.105) holds. We have the lower bound for* $x \in \mathcal{J}_\varepsilon(s)$:

$$\mathfrak{W}_\varepsilon(x,s) \geq \overset{\vee+}{u}_{\varepsilon,r^\varepsilon_{\text{ext}}}(x) - C\varepsilon^{-\frac{1}{3\theta-1}} \exp\left(-\pi^2 \frac{\varepsilon^{-\omega+1}}{32(r^\varepsilon(s))^2}\right). \tag{3.106}$$

Proof On $\mathcal{J}_\varepsilon(s - \frac{\varepsilon}{2})$ we consider the map φ_ε defined by

$$\varphi_\varepsilon(x) = \inf\{\mathfrak{W}_\varepsilon(x, s - \frac{\varepsilon}{2}) - \overset{\vee+}{u}_{\varepsilon,r^\varepsilon_{\text{ext}}}(x), 0\} \leq 0. \tag{3.107}$$

Invoking (3.102) and estimates (3.90) for $\overset{\vee+}{u}_{\varepsilon,r^\varepsilon_{\text{ext}}}$, we obtain, for $x \in \mathcal{J}_\varepsilon(s - \frac{\varepsilon}{2})$,

$$0 \leq \overset{\vee+}{u}_{\varepsilon,r^\varepsilon_{\text{ext}}}(x) \leq Ch_\varepsilon^{-\frac{1}{\theta-1}}, \tag{3.108}$$

which, combined with Lemma 3.46, yields

$$|\varphi_\varepsilon(x)| \leq Ch_\varepsilon^{-\frac{1}{\theta-1}} \text{ for } x \in \mathcal{J}_\varepsilon(s - \frac{\varepsilon}{2}). \tag{3.109}$$

Combining (3.108), estimate (3.90) of Lemma 3.39 and estimate (3.47) of Lemma 3.18, we deduce that, if ε is sufficiently small then

$$\varphi_\varepsilon(a_k^\varepsilon(s - \frac{\varepsilon}{2}) + \delta^\varepsilon_{\text{loglog}}) = \varphi_\varepsilon(a_{k+1}^\varepsilon(s - \frac{\varepsilon}{2}) - \delta^\varepsilon_{\text{loglog}}) = 0. \tag{3.110}$$

We extend φ_ε by 0 outside the set $\mathcal{J}_\varepsilon(s - \frac{\varepsilon}{2})$, and consider the solution Φ_ε to

$$\begin{cases} \varepsilon^\omega \dfrac{\partial \Phi_\varepsilon}{\partial \tau} - \dfrac{\partial \Phi_\varepsilon}{\partial x^2} = 0 \text{ on } \Lambda_\varepsilon^{\text{ext}}(s) \cap \{\tau \geq s - \dfrac{\varepsilon}{2}\}, \\[2mm] \Phi_\varepsilon(x, s - \dfrac{\varepsilon}{2}) = \varphi_\varepsilon(x) \quad \text{for } x \in \mathcal{J}_\varepsilon^{\text{ext}}(s - \dfrac{\varepsilon}{2}), \\[2mm] \Phi_\varepsilon(\pm r^\varepsilon_{\text{ext}}(s), \tau) = 0 \quad \text{for } \tau \in (s - \dfrac{\varepsilon}{2}, s). \end{cases} \tag{3.111}$$

Notice that $\Phi_\varepsilon \leq 0$. We consider next on $\Lambda_\varepsilon^{\text{ext}}(s) \cap \{\tau \geq s - \frac{\varepsilon}{2}\}$ the function $W_\varepsilon^{\text{inf}}$ defined by

$$W_\varepsilon^{\text{inf}}(y, \tau) = \overset{\vee+}{u}_{\varepsilon,r^\varepsilon_{\text{ext}}}(y) + \Phi_\varepsilon(y, \tau).$$

It follows from Lemma 3.44 that $L_\varepsilon(W_\varepsilon^{\text{inf}}) \leq 0$, so that $W_\varepsilon^{\text{inf}}$ is a *subsolution*. Since $W_\varepsilon^{\text{inf}} \leq \mathfrak{W}_\varepsilon$ on $\partial_p \left(\Pi_\varepsilon(s) \cap \{\tau \geq s - \frac{\varepsilon}{2}\}\right)$ it follows in particular that

$$W_\varepsilon^{\text{inf}} \leq \mathfrak{W}_\varepsilon \text{ on } \mathcal{J}_\varepsilon(s). \tag{3.112}$$

To complete the proof, we rely on the next linear estimates for Φ_ε.

Lemma 3.48 *We have the bound, for $y \in \mathcal{J}_\varepsilon^{\mathrm{ext}}$ and $\tau \in (s - \frac{\varepsilon}{2}, s)$,*

$$|\Phi_\varepsilon(y, \tau)| \le C \exp\left(-\pi^2 \varepsilon^{-\omega} \frac{(\tau - (s - \frac{\varepsilon}{2}))}{16(r^\varepsilon(s))^2}\right) \|\varphi_\varepsilon\|_{L^\infty(\mathcal{J}_\varepsilon(s - \frac{\varepsilon}{2}))}.$$

We postpone the proof of Lemma 3.48 and complete the proof of Proposition 3.47.

Proof of Proposition 3.47 completed Combining Lemma 3.48 with (3.109), we are led, for $x \in \mathcal{J}_\varepsilon(s)$, to

$$|\Phi_\varepsilon(x, s)| \le C h_\varepsilon^{-\frac{1}{\theta-1}} \exp\left(-\pi^2 \frac{\varepsilon^{-\omega+1}}{32(r^\varepsilon(s))^2}\right). \tag{3.113}$$

The conclusion then follows, invoking (3.112).   ☐

Proof of Lemma 3.48 Consider on the interval $[-2r^\varepsilon(s), 2r^\varepsilon(s)]$ the function $\psi(x)$ defined by $\psi(x) = \cos\left(\frac{\pi}{4r^\varepsilon(s)} x\right)$, so that $-\ddot{\psi} = \frac{\pi^2}{16}(r^\varepsilon(s))^{-2}\psi$, $\psi \ge 0$, $\psi(-2r^\varepsilon(s)) = \psi(2r^\varepsilon(s)) = 0$ and $\psi(x) \ge 1/2$ for $x \in [-r_{\mathrm{ext}}^\varepsilon(s), r_{\mathrm{ext}}^\varepsilon(s)]$. Hence, we obtain

$$\varepsilon^\omega \Psi_\tau - \Psi_{xx} = 0 \text{ on } \Lambda_\varepsilon^{\mathrm{ext}}(s) \cap \{\tau \ge s - \frac{\varepsilon}{2}\},$$

$$\text{where } \Psi(x, \tau) = \exp\left(-\pi^2 \varepsilon^{-\omega} \frac{\tau - (s - \frac{\varepsilon}{2})}{16 r^\varepsilon(s)^2}\right) \psi(x).$$

On the other hand, for $(y, \tau) \in \partial_p\left(\Lambda_\varepsilon^{\mathrm{ext}}(s) \cap \{\tau \ge s - \frac{\varepsilon}{2}\}\right)$ we have

$$|\Phi_\varepsilon(y, \tau)| \le \|\varphi_\varepsilon\|_{L^\infty(\mathcal{J}_\varepsilon(s - \frac{\varepsilon}{2}))} 2\Psi(y, \tau)$$

and the conclusion follows therefore from the comparison principle for the heat equation.   ☐

Proof of Proposition 3.41 completed Combining the upper bound (3.103) of Proposition 3.45 with the lower bound (3.106) of Proposition 3.47, we are led, for ε sufficiently small, to

$$\overset{\vee+}{u}_{\varepsilon, r_{\mathrm{ext}}^\varepsilon}(x) - A_\varepsilon \le \mathfrak{W}_\varepsilon(x, s) \le \overset{\vee+}{u}_{\varepsilon, r_{\mathrm{int}}^\varepsilon}(x) + A_\varepsilon, \tag{3.114}$$

where we have set

$$A_\varepsilon = C \varepsilon^{\frac{\omega-1}{2(\theta-1)}}. \tag{3.115}$$

The conclusion (3.93) then follows from Proposition 3.63 of Appendix A combined with the definition of h_ε and (A.7).   ☐

3.5.4 The Repulsive Case

In this subsection, we assume throughout that $\dagger_k = \dagger_{k+1}$ and may assume moreover that

$$\dagger_k = \dagger_{k+1} = 1; \tag{3.116}$$

the case $\dagger_k = \dagger_k = -1$ is handled similarly. The main purpose of this subsection is to provide *the proof of Proposition 3.42*, the central part being the construction of accurate *supersolutions*, subsolutions being provided by the same construction. We assume as before that (3.100) holds, and use as comparison map $\mathfrak{U}_\varepsilon$ defined on $\mathcal{I}^{\text{trs}}_\varepsilon(s) \equiv (-r^\varepsilon_{\text{ext}}(s), r^\varepsilon_{\text{int}}(s))$ by

$$\mathfrak{U}_\varepsilon(\cdot) \equiv \overset{\triangleright}{u}_{\varepsilon, r^\varepsilon(s)}(\cdot + 2h_\varepsilon),$$

so that $\mathfrak{U}_\varepsilon(x) \to +\infty$ as $x \to r^\varepsilon_{\text{int}}(s)$, $\mathfrak{U}_\varepsilon(x) \to -\infty$ as $x \to -r^\varepsilon_{\text{ext}}(s)$ and $|\mathfrak{U}_\varepsilon(-r^\varepsilon(s))| \le Ch_\varepsilon^{-\frac{1}{\theta-1}}$.

Proposition 3.49 *For $x \in (a_k(s) + \delta^\varepsilon_{\text{loglog}}, r^\varepsilon_{\text{int}}(s))$ we have the inequality, where $C > 0$ denotes some constant,*

$$\mathfrak{W}_\varepsilon(x, s) \le \mathfrak{U}_\varepsilon(x) + C\varepsilon^{-\frac{1}{3\theta-1}} \exp\left(-\pi^2 \frac{\varepsilon^{-\omega+1}}{16(r^\varepsilon(s))^2}\right). \tag{3.117}$$

Proof As for (3.107), write for $x \in \mathcal{I}^{\text{trs}}_\varepsilon(s) \cap \mathcal{J}_\varepsilon(s-\varepsilon)$

$$\psi_\varepsilon(x) = \sup\{\mathfrak{W}_\varepsilon(x, s-\varepsilon) - \mathfrak{U}_\varepsilon, 0\} \ge 0.$$

We notice that

$$\psi_\varepsilon(a_k(s-\varepsilon) + \delta^\varepsilon_{\text{loglog}}) = \psi_\varepsilon(r^\varepsilon_{\text{int}}(s)) = 0.$$

Indeed, for the first relation, we argue as in (3.110), whereas for the second, we have $\mathfrak{U}_\varepsilon(r^\varepsilon_{\text{int}}(s)) = \overset{\triangleright}{u}_{\varepsilon, r^\varepsilon(s)}(r^\varepsilon(s)) = +\infty$. We extend ψ_ε by 0 outside the interval $\mathcal{I}^{\text{trs}}_\varepsilon(s) \cap \mathcal{J}_\varepsilon(s-\varepsilon)$ and derive, arguing as for (3.109),

$$|\psi_\varepsilon(x)| \le Ch_\varepsilon^{-\frac{1}{\theta-1}} \le C\varepsilon^{-\frac{1}{3\theta-1}} \text{ for } x \in \mathbb{R}. \tag{3.118}$$

We introduce the cylinder $\Lambda^{\text{trans}}_\varepsilon(s) \equiv (-r^\varepsilon_{\text{ext}}(s), r^\varepsilon_{\text{int}}(s)) \times (s-\varepsilon, s)$ and the solution Ψ_ε to

$$\begin{cases} \varepsilon^\omega \dfrac{\partial \Psi_\varepsilon}{\partial \tau} - \dfrac{\partial^2 \Psi_\varepsilon}{\partial x^2} = 0 \text{ on } \Lambda^{\text{trans}}_\varepsilon(s), \\[2mm] \Phi_\varepsilon(x, s-\varepsilon) = \psi_\varepsilon(x) \text{ for } x \in (-r^\varepsilon_{\text{ext}}(s), r^\varepsilon_{\text{int}}(s)) \text{ and} \\[2mm] \Psi_\varepsilon(-r^\varepsilon_{\text{ext}}(s), \tau) = \Psi_\varepsilon(r^\varepsilon_{\text{int}}(s), \tau) = 0 \text{ for } \tau \in (s-\varepsilon, s), \end{cases} \tag{3.119}$$

so that $\Psi_\varepsilon \geq 0$. Arguing as for (3.113), we obtain for $\tau \in (s - \varepsilon, s)$

$$|\Psi_\varepsilon(y, \tau)| \leq C\varepsilon^{-\frac{1}{3\theta-1}} \exp\left(-\pi^2 \varepsilon^{-\omega} \frac{(\tau - (s - \varepsilon))}{16(r^\varepsilon(s))^2}\right). \qquad (3.120)$$

We consider on $\Lambda_\varepsilon^{\text{trans}}(s)$ the function $W_\varepsilon^{\text{trans}}$ defined by

$$W_\varepsilon^{\text{trans}}(y, \tau) = \mathfrak{U}_\varepsilon(y) + \Psi_\varepsilon(y, \tau).$$

It follows from Lemma 3.44 that $L_\varepsilon(W_\varepsilon^{\text{trans}}) \geq 0$, that is $W_\varepsilon^{\text{trans}}$ is a *supersolution* for L_ε on $\Lambda_\varepsilon^{\text{trans}}(s)$. Consider next the subset $\Pi_\varepsilon^{\text{trans}}(s)$ of $\Lambda_\varepsilon^{\text{trans}}$ defined by

$$\Pi_\varepsilon^{\text{trans}}(s) \equiv \bigcup_{\tau \in (s - \varepsilon, s)} (a_k(\tau) + \delta_{\text{loglog}}^\varepsilon, r_{\text{int}}^\varepsilon(s)) \times \{\tau\}.$$

We claim that

$$W_\varepsilon^{\text{trans}} \geq \mathfrak{W}_\varepsilon \text{ on } \partial_p \Pi_\varepsilon^{\text{trans}}(s). \qquad (3.121)$$

Indeed, by construction, we have $W_\varepsilon^{\text{trans}} = +\infty$ on $r_{\text{int}}^\varepsilon(s) \times (s - \varepsilon, s)$ and $W_\varepsilon^{\text{trans}}(x, s - \varepsilon) \geq \mathfrak{W}_\varepsilon(x, s - \varepsilon)$ for $x \in (a_k(s - \varepsilon) + \delta_{\text{loglog}}^\varepsilon, r_{\text{int}}^\varepsilon(s))$. Finally on $\bigcup_{\tau \in (s - \varepsilon, s)} \{a_k(\tau) + \delta_{\text{loglog}}^\varepsilon\} \times \{\tau\}$, the conclusion (3.121) follows from estimate (3.47) of Lemma 3.18. Combining inequality (3.121) with the comparison principle, we are led to

$$W_\varepsilon^{\text{trans}} \geq \mathfrak{W}_\varepsilon \text{ on } \Pi_\varepsilon^{\text{trans}}(s). \qquad (3.122)$$

Combining (3.122) with (3.120), we are led to (3.117). $\qquad \square$

Our next task is to construct a *subsolution*. To that aim, we rely on the symmetries of the equation, in particular the invariance $x \to -x$ and the *almost oddness* of the non-linearity. To be more specific, we introduce the operator

$$\tilde{L}_\varepsilon(u) \equiv \varepsilon^\omega \frac{\partial u}{\partial \tau} - \frac{\partial^2 u}{\partial x^2} + \lambda \tilde{f}_\varepsilon(u) = 0, \text{ with } \tilde{f}_\varepsilon(u) = 2\theta u^{2\theta-1}\left(1 - \varepsilon^{\frac{1}{\theta-1}} u g(-\varepsilon^{\frac{1}{\theta-1}} u)\right),$$

which has the same properties as L_ε, and consider the stationary solution $\overset{\triangleleft}{u}_{\varepsilon, r^\varepsilon(s)}$ for L_ε defined on $(-r^\varepsilon(s), r^\varepsilon(s))$ by

$$-\frac{\partial^2 \overset{\triangleleft}{u}_{\varepsilon, r^\varepsilon(s)}}{\partial x^2} + \lambda f_\varepsilon(\overset{\triangleleft}{u}_{\varepsilon, r^\varepsilon(s)}) = 0, \qquad \overset{\triangleleft}{u}_{\varepsilon, r^\varepsilon(s)}(-r^\varepsilon(s)) = +\infty$$

$$\text{and } \overset{\triangleleft}{u}_{\varepsilon, r^\varepsilon(s)}(r^\varepsilon(s)) = -\infty,$$

so that $-\overset{\triangleleft}{u}_{\varepsilon, r^\varepsilon(s)}$ is a stationary solution to $\tilde{L}_\varepsilon$. Consider the function $\tilde{\mathfrak{W}}_\varepsilon$ defined by

$$\tilde{\mathfrak{W}}_\varepsilon(x, \tau) = -\mathfrak{W}_\varepsilon(-x, \tau) \qquad (3.123)$$

and observe that $\tilde{L}_\varepsilon(\tilde{\mathfrak{W}}_\varepsilon) = 0$. Finally, we define on the interval $(-r^\varepsilon_{\text{int}}(s), r^\varepsilon_{\text{ext}}(s))$ the function

$$\mathfrak{V}_\varepsilon(x) \equiv \overset{\triangleleft}{\mathfrak{u}}_{\varepsilon, r^\varepsilon(s)}(2h_\varepsilon - x),$$

so that $\mathfrak{V}_\varepsilon(x) \to -\infty$ as $x \to -r^\varepsilon_{\text{int}}(s)$ and $\mathfrak{V}_\varepsilon(x) \to +\infty$ as $x \to r^\varepsilon_{\text{ext}}(s)$.

Proposition 3.50 *For $x \in (-r^\varepsilon_{\text{int}}(s), a_{k+1}(s) - \delta^\varepsilon_{\text{loglog}})$ we have the inequality*

$$\mathfrak{W}_\varepsilon(x, s) \geq \mathfrak{V}_\varepsilon(x) - C\varepsilon^{-\frac{1}{3\theta-1}} \exp\left(-\pi^2 \frac{\varepsilon^{-\omega+1}}{16(r^\varepsilon(s))^2}\right). \tag{3.124}$$

Proof We argue as in the proof of Proposition 3.49, replacing L_ε by $\tilde{\varepsilon}$, $\mathfrak{W}_\varepsilon$ by $\tilde{\mathfrak{W}}_\varepsilon$, and $\mathfrak{U}_\varepsilon$ by $\tilde{\mathfrak{U}}_\varepsilon = -\overset{\triangleleft}{\mathfrak{u}}_{\varepsilon, r^\varepsilon(s)}(\cdot - 2h_\varepsilon(s_0))$. Inequality (3.124) for $\mathfrak{W}_\varepsilon$ is then obtained by inverting relation (3.123) and from the corresponding estimate on $\tilde{\mathfrak{W}}_\varepsilon$. $\qquad\square$

Proof of Proposition 3.42 completed Combining (3.117) with (3.124), we are led to

$$\mathfrak{U}_\varepsilon(x) - \tilde{A}_\varepsilon \leq \mathfrak{W}_\varepsilon(x, s) \leq \mathfrak{V}_\varepsilon(x) + \tilde{A}_\varepsilon, \tag{3.125}$$

where we have set $\tilde{A}_\varepsilon = C\varepsilon^{-\frac{1}{3\theta-1}} \exp\left(-\pi^2 \frac{\varepsilon^{-\omega+1}}{16(r^\varepsilon(s))^2}\right)$. The proof is then completed with the same arguments as in the proof of Proposition 3.41 $\quad\square$

3.5.5 Estimating the Discrepancy

Linear Estimates

The purpose of this section is to provide the proof of Proposition 3.43. So far Proposition 3.41 and Proposition 3.42 provide a good approximation of $\mathfrak{W}_\varepsilon$ on the level of the *uniform norm*. However, the discrepancy involves also a first-order derivative, for which we rely on the regularization property of the linear heat equation. To that aim, set

$$\begin{cases} \Lambda \equiv (-1, 1) \times [0, 1], \ \Lambda^{1/2} \equiv \left(-\dfrac{1}{2}, \dfrac{1}{2}\right) \times \left[\dfrac{3}{4}, 1\right], \\[2mm] \text{and more generally for } \varrho > 0 \\[2mm] \Lambda_\varrho \equiv (-\varrho, \varrho) \times [0, \varrho^2], \ \Lambda^{1/2}_\varrho \equiv \left(-\dfrac{1}{2}\varrho, \dfrac{1}{2}\varrho\right) \times \left[\dfrac{3}{4}\varrho^2, \varrho^2\right]. \end{cases}$$

The following standard result (see e.g. [2] Lemma A. 7 for a proof) is useful in our context.

Lemma 3.51 *Let u be a smooth real-valued function on Λ. There exists a constant $C > 0$ such that*

$$\|u_x\|_{L^\infty(\Lambda^{1/2})} \leq C(\|u_t - u_{xx}\|_{L^\infty(\Lambda)} + \|u\|_{L^\infty(\Lambda)}).$$

We deduce from this result the following scaled version.

Lemma 3.52 *Let $\varrho > 0$ and let u be defined on Λ_ϱ. Then we have for some constant $C > 0$ independent of ϱ*

$$\|u_x\|^2_{L^\infty(\Lambda_\varrho^{1/2})} \leq C\left[\|u_t - u_{xx}\|_{L^\infty(\Lambda_\varrho)}\|u\|_{L^\infty(\Lambda_\varrho)} + \varrho^{-2}\|u\|^2_{L^\infty(\Lambda_\varrho)}\right]. \quad (3.126)$$

Proof The argument is parallel to the proof of Lemma A.1 in [1], which corresponds to its elliptic version. Set $h = u_t - u_{xx}$, let (x_0, t_0) be given in $\Lambda_\varrho^{1/2}$, and let $0 < \mu \leq \frac{\varrho}{2}$ be a constant to be determined in the course of the proof. We consider the function

$$v(y, \tau) = u\left(2\mu y + x_0, 4\mu^2(\tau - 1) + t_0\right),$$

so that v is defined on Λ and satisfies there

$$v_t - v_{yy} = \mu^2 h\left((2\mu y + x_0, 4\mu^2(\tau - 1) + t_0\right) \text{ on } \Lambda.$$

Applying Lemma 3.51 to v, we are led to

$$|v_y(0, 1)| \leq C\left(\mu^2\|h\left(2\mu y + x_0, 4\mu^2(\tau - 1) + t_0\right)\|_{L^\infty(\Lambda)} + \|v\|_{L^\infty(\Lambda)}\right)$$
$$\leq C\left(\mu^2\|h\|_{L^\infty(\Lambda_\varrho)} + \|u\|_{L^\infty(\Lambda_\varrho)}\right),$$

so that, going back to u, we obtain

$$\mu|u_x(x_0, t_0)| \leq C\left(\mu^2\|h\|_{L^\infty(\Lambda_\varrho)} + \|u\|_{L^\infty(\Lambda_\varrho)}\right). \quad (3.127)$$

We distinguish two cases.

Case 1: $\|u\|_{L^\infty} \leq \varrho^2\|h\|_{L^\infty}$. In this case we apply (3.127) with $\mu = \left(\dfrac{\|u\|_{L^\infty}}{\|h\|_{L^\infty}}\right)^{\frac{1}{2}}$. This yields

$$|u_y(x_0, t_0)| \leq 2C\|u\|^{1/2}_{L^\infty}\|h\|^{1/2}_{L^\infty}.$$

Case 2: $\|u\|_{L^\infty} \geq \varrho^2\|h\|_{L^\infty}$. In this case we apply (3.127) with $\mu = \varrho$. We obtain

$$|u_x(x_0, t_0)| \leq C\left(\varrho\|h\|_{L^\infty(\Lambda_\varrho)} + \varrho^{-1}\|u\|_{L^\infty(\Lambda_\varrho)}\right)$$
$$\leq C\left(\|h\|^{1/2}_{L^\infty(\Lambda_\varrho)}\|u\|^{1/2}_{L^\infty(\Lambda_\varrho)} + r^{-1}\|u\|_{L^\infty(\Lambda_\varrho)}\right). \quad (3.128)$$

In both cases, we obtain the desired inequality. $\qquad\square$

Estimating the Derivative of $\mathfrak{W}_\varepsilon$

Consider the general situation where we are given two functions U and U_ε defined for $(x,t) \in \Lambda_\varrho$ and such that $L_0(U) = 0$ and $L_\varepsilon(U_\varepsilon) = 0$, where $s := \varepsilon^{-\omega}t$, so that, in view of (3.88),

$$|\partial_t(U - U_\varepsilon) - \partial_{xx}(U - U_\varepsilon)|$$
$$\leq C\Big[|U - U_\varepsilon|(|U|^{2\theta-2} + |U_\varepsilon|^{2\theta-2}) + \varepsilon^{\frac{1}{\theta-1}}|U_\varepsilon|^{2\theta}\Big] \text{ on } \Lambda_\varrho.$$

We deduce from (3.126) applied to the difference $U - U_\varepsilon$ that we have (we use the notation $\|\cdot\| = \|\cdot\|_{L^\infty(\Lambda_\varrho)}$ for simplicity)

$$\|(U - U_\varepsilon)_x\|^2_{L^\infty(\Lambda_\varrho^{1/2})} \leq C\,\|U - U_\varepsilon\|^2\left(\|U\|^{2\theta-2} + \|U_\varepsilon\|^{2\theta-2} + \varrho^{-2}\right)$$
$$+ C\varepsilon^{\frac{1}{\theta-1}}\|U - U_\varepsilon\|\|U_\varepsilon\|^{2\theta}.$$

Similarly, applying (3.126) to U and U_ε we obtain

$$\|(U + U_\varepsilon)_x\|^2_{L^\infty(\Lambda_\varrho^{1/2})} \leq C(\|U\|^{2\theta} + \|U_\varepsilon\|^{2\theta} + \varrho^{-2}\left(\|U\|^2 + \|U_\varepsilon\|^2\right)$$
$$+ \varepsilon^{\frac{1}{\theta-1}}(\|U_\varepsilon\|^{2\theta+1} + \|U\|\|U_\varepsilon\|^{2\theta})),$$

so that

$$\|(U^2 - U_\varepsilon^2)_x\|^2_{L^\infty(\Lambda_\varrho^{1/2})} \leq C\big[\|U - U_\varepsilon\|^2\mathcal{R}_1^\varepsilon(U,U_\varepsilon) + \|U - U_\varepsilon\|\mathcal{R}_2^\varepsilon(U,U_\varepsilon)\big],$$
$$(3.129)$$

where we have set

$$\begin{cases} \mathcal{R}_1^\varepsilon(U,U_\varepsilon) = (\|U\|^{2\theta-2} + \|U_\varepsilon\|^{2\theta-2} + \varrho^{-2})(\|U\|^{2\theta} + \|U_\varepsilon\|^{2\theta} \\ \qquad + \varrho^{-2}(\|U\|^2 + \|U_\varepsilon\|^2) + \varepsilon^{\frac{1}{\theta-1}}(\|U_\varepsilon\|^{2\theta+1} + \|U\|\|U_\varepsilon\|^{2\theta})), \\ \mathcal{R}_2^\varepsilon(U,U_\varepsilon) = \varepsilon^{\frac{1}{\theta-1}}\|U_\varepsilon\|^{2\theta}(\|U\|^{2\theta} + \|U_\varepsilon\|^{2\theta} + \varrho^{-2}(\|U\|^2 + \|U_\varepsilon\|^2) \\ \qquad + \varepsilon^{\frac{1}{\theta-1}}(\|U_\varepsilon\|^{2\theta+1} + \|U\|\|U_\varepsilon\|^{2\theta})). \end{cases}$$

We now apply the discussion to our original situation. Thanks to the general inequality (3.129), we are in position to establish the following.

Proposition 3.53 *If (3.105) holds and ε is sufficiently small, then for any $s \in I^\varepsilon(s_0)$ and every $x \in \Theta^\varepsilon_{k+\frac{1}{2}}(s)$ we have the estimate*

$$|(\mathfrak{W}_\varepsilon)_x^2(x) - \lambda^{-\frac{1}{(\theta-1)}}(\overset{\vee+}{\mathfrak{u}}_{r^\varepsilon(s)})_x^2(x)| \leq C\varepsilon^{\frac{1}{\theta^2}}.$$

Proof We apply inequality (3.129) on the cylinder Λ_ϱ with $\varrho = \frac{1}{16}d^*_{\min}(s_0)$ and to the functions $U(y,\tau) = \mathfrak{W}_\varepsilon(y+x, \varepsilon^\omega\tau + s)$ and $U_\varepsilon(y,\tau) = \overset{\vee+}{\mathcal{U}}_{r^\varepsilon(s)}(y+x)$. We first estimate $\mathcal{R}_1$ and $\mathcal{R}_2$. Since we have

$$|U(y,\tau)| + |U_\varepsilon| \leq C d_{\min}^*(s_0)^{-\frac{1}{\theta-1}}, \qquad \text{for } (y,\tau) \in \Lambda_\varrho,$$

it follows that

$$\mathcal{R}_1^\varepsilon(U, U_\varepsilon) \leq d_{\min}^*(s_0)^{-4-\frac{2}{\theta-1}} \text{ and } \mathcal{R}_2^\varepsilon(U, U_\varepsilon) \leq \varepsilon^{\frac{1}{\theta-1}} d_{\min}^*(s_0)^{-4-\frac{4}{\theta-1}}.$$

Invoking inequality (3.126) of Lemma 3.52, and combining it with (A.7) and the conclusion of Proposition 3.41, we derive the conclusion using a crude lower bound for the power of ε. □

Similarly we obtain the following.

Proposition 3.54 *If (3.94) holds and ε is sufficiently small, then for any $s \in I^\varepsilon(s_0)$ and every $x \in \Theta^\varepsilon_{k+\frac{1}{2}}(s)$ we have the estimate*

$$|(\mathfrak{W}_\varepsilon)_x^2(x) - \lambda^{-\frac{1}{(\theta-1)}} (\overset{\triangleright}{\mathfrak{u}}_{r^\varepsilon(s)})_x^2(x)| \leq C\varepsilon^{\frac{1}{\theta^2}}. \tag{3.130}$$

Proof of Proposition 3.43 completed The proof of Proposition 3.43 follows by combining Proposition 3.53 in the attractive case and Proposition 3.54 in the repulsive case with the estimates (A.10). □

3.6 The Motion Law for Prepared Data

In this section, we present the proof of Proposition 3.19.

Proof of Proposition 3.19,
  Step 1 First, by definition of L_0, assumption (H_1) and estimate (3.7), it follows that for fixed $L \geq L_0$, and for all ε sufficiently small (depending only on L),

$$\mathfrak{D}_\varepsilon(s) \cap I_{4L} \subset I_L \qquad \forall 0 \leq s \leq S,$$

so that ($\mathcal{C}_{L,S}$) holds.
  Step 2 Since the assumptions of Corollary 3.33 are met with the choice $s_0 = 0$ and $L = L_0$, we obtain that for ε sufficiently small, $\mathcal{WP}_\varepsilon^{L_0}(\delta_{\mathrm{loglog}}^\varepsilon, s)$ holds and $d_{\min}^{\varepsilon,L}(s) \geq \frac{1}{2} d_{\min}^*(0) = \frac{1}{2} \min\{a_{k+1}^0 - a_k^0, \ k = 1,\ldots,\ell_0 - 1\}$, for all $s \in I^\varepsilon(0)$, as well as the identities $J(s) = J(0)$, $\sigma_{i(k\pm\frac{1}{2})}(s) = \sigma_{i(k\pm\frac{1}{2})}(0)$ and $\dagger_k(s) = \dagger_k(0)$, for any $k \in J(0)$.
  Step 3 We claim that for any $s_1 \leq s_2 \in I^*(0)$, we have

$$\limsup_{\varepsilon \to 0} (\mathrm{dissip}_\varepsilon^L(s_1, s_2)) = 0. \tag{3.131}$$

Indeed, let $L \geq L_0$ be arbitrary. We know from Step 1 that $(C_{L,S})$ holds provided ε is sufficiently small. By Proposition 3.13, for ε sufficiently small there exist two times s_1^ε and s_2^ε such that $0 < s_1^\varepsilon \leq s_1 \leq s_2 \leq s_2^\varepsilon$, $|s_i - s_i^\varepsilon| \leq \varepsilon^{\omega+1}L$ and $\mathcal{WP}_\varepsilon^L(\delta_{\log}^\varepsilon, s_i^\varepsilon)$ holds for $i = 1, 2$. From the second step and assumption (H_1) we infer that $\mathfrak{E}_\varepsilon^L(s_1^\varepsilon) = \mathfrak{E}_\varepsilon^{L_0}(s_1^\varepsilon) = \mathfrak{E}_\varepsilon^{L_0}(s_2^\varepsilon) = \mathfrak{E}_\varepsilon^L(s_2^\varepsilon)$. Invoking Corollary 3.33 we are therefore led to the inequality

$$\text{dissip}_\varepsilon^{L_0}(s_1, s_2) \leq \text{dissip}_\varepsilon^L(s_1^\varepsilon, s_2^\varepsilon) \leq CM_0 \left(\frac{\varepsilon}{\delta_{\log}^\varepsilon}\right)^\omega + CL^{-(\omega+2)}(s_2 - s_1 + 2\varepsilon^{\omega+1}L).$$

Since $L \geq L_0$ was arbitrary the conclusion (3.131) follows by letting first $\varepsilon \to 0$ and then $L \to \infty$.

Step 4 In view of Corollary 3.17 we may find a subsequence $(\varepsilon_n)_{\in \mathbb{N}}$ tending to 0 such that the functions $a_k^{\varepsilon_n}(\cdot)_{n \in \mathbb{N}}$ converge uniformly as $n \to 0$ on compact subsets on $I^*(0)$. Consider the cylinder

$$\mathcal{C}_{k+\frac{1}{2}}^* \equiv [a_k^0 + \frac{1}{4}d_{\min}^*(0), a_{k+1}^0 - \frac{1}{4}d_{\min}^*(0)] \times I^*(0).$$

It follows from Step 2 and Proposition 3.43 that

$$\varepsilon_n^{-\omega}\xi_{\varepsilon_n}(\mathfrak{v}_{\varepsilon_n}) \to \lambda_{i(k+\frac{1}{2})}^{-\frac{1}{2(\theta-1)}} r_{k+\frac{1}{2}}(s)^{-(\omega+1)}\gamma_k \text{ as } \varepsilon_n \to 0, \text{ for } k = 1, \ldots \ell_0 - 1 \tag{3.132}$$

uniformly on every compact subset of $\mathcal{C}_{k+\frac{1}{2}}^*$, where γ_k is defined in (3.97) and where $r_{k+\frac{1}{2}}(s) = a_{k+1}(s) - a_k(s)$.

Step 5 As in (3.84), we consider a test function $\chi \equiv \chi_k$ with the following properties:

$$\begin{cases} \chi \text{ has compact support in } [a_k^0 - \frac{1}{3}d_{\min}^*(0), a_k^0 + \frac{1}{3}d_{\min}^*(0)], \\ \chi \text{ is affine on the interval } [a_k^0 - \frac{1}{4}d_{\min}^*(0), a_k^0 + \frac{1}{4}d_{\min}^*(0)], \text{ with } \chi' = 1 \text{ there} \\ \|\chi\|_{L^\infty(\mathbb{R})} \leq Cd_{\min}^*(0), \|\chi'\|_{L^\infty(\mathbb{R})} \leq C \text{ and } \|\chi''\|_{L^\infty(\mathbb{R})} \leq Cd_{\min}^*(0)^{-1}. \end{cases}$$

It follows from the definition of χ_k that $\chi_k'' = 0$ outside a, and so is $\varepsilon^{-\omega}\xi_\varepsilon(\mathfrak{v}_\varepsilon)\chi_k''$. It follows from (3.132) that for $s_1 \leq s_2 \in I^*(0)$,

$$\mathfrak{F}_{\varepsilon_n}(s_1, s_2, \chi_k) \to \left(\int_{a_k^0 - \frac{1}{3}d_{\min}^*(0)}^{a_k^0 - \frac{1}{4}d_{\min}^*(0)} \chi''(x)dx\right)\left(\int_{s_1}^{s_2} \lambda_{k-\frac{1}{2}}^{-\frac{1}{2(\theta-1)}} r_{k-\frac{1}{2}}(s)^{-\frac{1}{\theta-1}}\gamma_{k-\frac{1}{2}}ds\right)$$

$$+ \left(\int_{a_k^0 + \frac{1}{4}d_{\min}^*(0)}^{a_k^0 + \frac{1}{3}d_{\min}^*} \chi''(x)dx\right)\left(\int_{s_1}^{s_2} \lambda_{k+\frac{1}{2}}^{-\frac{1}{2(\theta-1)}} r_{k+\frac{1}{2}}(s)^{-\frac{1}{\theta-1}}\gamma_{k+\frac{1}{2}}ds\right) \tag{3.133}$$

as $\varepsilon_n \to 0$. Since the above two integrals containing χ'' are identically equal to 1 and -1 respectively, we finally deduce from (3.42) combined with (3.131) and (3.133), letting ε_n tend to 0, that for $s_1 \leq s_2 \in I^*(0)$ we have

$$[a_k(s_1) - a_k(s_2)]\mathfrak{S}_{i(k)}$$
$$= \int_{s_1}^{s_2} \left(\lambda_{i(k-\frac{1}{2})}^{-\frac{1}{2(\theta-1)}} r_{k-\frac{1}{2}}(s)^{-(\omega+1)} \gamma_{k-\frac{1}{2}} - \lambda_{i(k+\frac{1}{2})}^{-\frac{1}{2(\theta-1)}} r_{k+\frac{1}{2}}(s)^{-(\omega+1)} \gamma_{k+\frac{1}{2}} \right) ds,$$

which is nothing other than the integral formulation of the system $(\mathcal{S})$. Since the latter possesses a unique solution, the limiting points are unique and therefore convergence of the a_k^ε for $s \in I^*(s)$ holds for the full family $(\mathfrak{v}_\varepsilon)_{\varepsilon>0}$.

Step 6 We use an elementary continuation method to extend the convergence from $I^*(0)$ to the full interval $(0, S)$. Indeed, as long as $d_{\min}^*(s)$ remains bounded from below by a strictly positive constant (which holds, by definition of $S_{\max}$, as long as $s < S$) we may take s as a new origin of times (Step 2 yields $\mathcal{WP}_\varepsilon^{L_0}(\alpha_1\varepsilon, s)$) and use Steps 1 to 5 to extend the stated convergence past s. The proof is here completed. □

3.7 Clearing-out

The purpose of this section is to provide a proof of Proposition 3.20. We are led to consider the situation where for some length $L \geq 0$ we have

$$\mathfrak{D}_\varepsilon(0) \cap [-5L, 5L] \subset [-\kappa_0 L, \kappa_0 L] \tag{3.134}$$

for some (small) constant $\kappa_0 \leq \frac{1}{2}$. It follows from Theorem 3.3 that

$$\mathcal{C}_{L,S} \text{ holds}, \qquad \text{where } S = \rho_0 \left(\frac{L}{2} \right)^{\omega+2},$$

and that for $s \in [0, S]$ we have

$$\mathfrak{D}_\varepsilon(s) \cap [-4L, 4L] \subset [-\kappa_0(s)L, \kappa_0(s)L], \tag{3.135}$$

where

$$\kappa_0(s) := \kappa_0 + \left(\frac{s}{\rho_0} \right)^{\frac{1}{\omega+2}} \frac{1}{L}. \tag{3.136}$$

For those times $s \in [0, S]$ for which the preparedness assumption $\text{WPI}_\varepsilon^L(\alpha_1\varepsilon, s)$ holds, we set

$$\begin{cases} d_{\min}^{\varepsilon,+}(s) = \min\{|a_{k+1}^\varepsilon(s) - a_k^\varepsilon(s)|, \ k \in J^+(s)\}, \text{ and} \\ d_{\min}^{\varepsilon,-}(s) = \min\{|a_{k+1}^\varepsilon(s) - a_k^\varepsilon(s)|, \ k \in J^-(s)\}, \end{cases}$$

with $J^\pm(s) = \{k \in \{1,\ldots,\ell(s) - 1\}, \text{s.t.} \epsilon_{k+\frac{1}{2}} = \mp 1\}$, so that $d_{\min}^\varepsilon(s) = \min\{d_{\min}^{\varepsilon,+}(s), d_{\min}^{\varepsilon,-}(s)\}$, with the convention that the quantities are equal to L in the case when the defining set is empty.

At first, we will focus on the case $J^-(s) \neq \emptyset$. The following result provides an upper bound in terms of $d_{\min}^{\varepsilon,-}(s)$ for a dissipation time for the quantized function $\mathfrak{E}_\varepsilon^L$. This phenomenon is related to the cancellation of *a front with its anti-front*, and is the main building block for the proof of Proposition 3.20.

Proposition 3.55 *There exist $\kappa_1 > 0$, $\alpha_3 > 0$, and $\mathcal{K}_{\mathrm{col}} > 0$, all depending on only V and M_0, with the following properties. If (3.134) holds, if $s_0 \in (\varepsilon^\omega L^2, S)$ is such that $\kappa_0(s_0) \leq \kappa_1$, $\mathcal{WP}_\varepsilon^L(\alpha_3\varepsilon, s_0)$ holds, $J^-(s_0)$ is non-empty, and $s_0 + \mathcal{K}_{\mathrm{col}} d_{\min}^{\varepsilon,-}(s_0)^{\omega+2} < S$, then there exists some time $\mathcal{T}_{\mathrm{col}}^{\varepsilon,+}(s_0) \in (s_0, S)$ such that $\mathrm{WPI}_\varepsilon^L(\alpha_3\varepsilon, \mathcal{T}_{\mathrm{col}}^{\varepsilon,+}(s_0))$ holds,*

$$\mathfrak{E}_\varepsilon^L(\mathcal{T}_{\mathrm{col}}^{\varepsilon,+}(s_0)) \leq \mathfrak{E}_\varepsilon^L(s_0) - \mu_1, \tag{3.137}$$

where μ_1 is a constant introduced in Lemma 3.34, and

$$\mathcal{T}_{\mathrm{col}}^{\varepsilon,+}(s_0) - s_0 \leq \mathcal{K}_{\mathrm{col}} \left(d_{\min}^{\varepsilon,-}(s_0)\right)^{\omega+2}. \tag{3.138}$$

We postpone the proof of Proposition 3.55 until after Section 3.7.1, where we will analyze in more detail the attractive and repulsive forces at work at the ε level. We will then prove Proposition 3.55 in Section 3.7.2, and finally Proposition 3.20 in Section 3.7.3.

3.7.1 Attractive and Repulsive Forces at the ε Level

In this subsection we consider the general situation where $\mathcal{C}_{L,S}$ holds, for some length $L \geq 0$ and some $S > 0$.

In order to deal with the attractive and repulsive forces underlying annihilations or splittings, we set

$$\mathcal{F}_{k+\frac{1}{2}}(s) = -\omega^{-1} \mathcal{B}_{k+\frac{1}{2}} \left(a_{k+1}^\varepsilon(s) - a_k^\varepsilon(s)\right)^{-\omega}$$

and consider the positive functionals

$$\mathcal{F}_{\mathrm{rep}}^\varepsilon(s) = \sum_{k \in J^+(s)} \mathcal{F}_{k+\frac{1}{2}}(s), \quad \mathcal{F}_{\mathrm{att}}^\varepsilon(s) = -\sum_{k \in J^-(s)} \mathcal{F}_{k+\frac{1}{2}}(s), \tag{3.139}$$

with the convention that the quantity is equal to $+\infty$ in the case when the defining set is empty. For some constants $0 < \kappa_2 \leq \kappa_3$ depending on only M_0 and V we have

$$\begin{cases} \kappa_2 \, \mathcal{F}_{\mathrm{att}}^\varepsilon(s)^{-\frac{1}{\omega}} \leq d_{\min}^{\varepsilon,-}(s) \leq \kappa_3 \mathcal{F}_{\mathrm{att}}^\varepsilon(s)^{-\frac{1}{\omega}}, \\ \kappa_2 \, \mathcal{F}_{\mathrm{rep}}^\varepsilon(s)^{-\frac{1}{\omega}} \leq d_{\min}^{\varepsilon,+}(s) \leq \kappa_3 \mathcal{F}_{\mathrm{rep}}^\varepsilon(s)^{-\frac{1}{\omega}}. \end{cases} \tag{3.140}$$

Let $s_0 \in [\varepsilon^\omega L^2, S]$ be such that

$$\mathcal{WP}_\varepsilon^L(\alpha_2 \varepsilon, s_0) \text{ holds} \qquad \text{and} \qquad d_{min}^{\varepsilon,L}(s_0) \geq 16 q_1(\alpha_2)\varepsilon. \tag{3.141}$$

We consider as in Corollary 3.38 the stopping time

$$\mathcal{T}_0^\varepsilon(\alpha_2, s_0) = \max \left\{ s_0 + \varepsilon^{2+\omega} \leq s \leq S \quad \text{s.t.} \quad d_{min}^{\varepsilon,L}(s') \geq 8 q_1(\alpha_2)\varepsilon \atop \forall s' \in [s_0 + \varepsilon^{\omega+2}, s] \right\},$$

and for simplicity we will write $\mathcal{T}_0^\varepsilon(s_0) \equiv \mathcal{T}_0^\varepsilon(\alpha_2, s_0)$. In view of (3.141) and the statement of Corollary 3.38,

$$\mathcal{WP}_\varepsilon^L(\alpha_1 \varepsilon, s) \text{ holds} \qquad \forall s \in \mathcal{I}_0^\varepsilon(s_0) \equiv [s_0 + \varepsilon^{2+\omega}, \mathcal{T}_0^\varepsilon(s_0)].$$

The functionals $\mathcal{F}_{att}^\varepsilon$ and $\mathcal{F}_{rep}^\varepsilon$ are in particular well defined and continuous on the interval of time $\mathcal{I}_0^\varepsilon(s_0)$ with $J^+(s) = J^+(s_0)$ and $J^-(s) = J^-(s_0)$ for all s in that interval. Note that the attractive forces are dominant when $d_{min}^{\varepsilon,-}(s) \leq d_{min}^{\varepsilon,+}(s)$ and by contrast the repulsive forces are dominant when $d_{min}^{\varepsilon,+}(s) \leq d_{min}^{\varepsilon,-}(s)$.

We first focus on the attractive case, and for $s \in \mathcal{I}_0^\varepsilon(s_0)$, we introduce the new stopping times

$$\mathcal{T}_1^\varepsilon(s) = \inf\{s \leq s' \leq \mathcal{T}_0^\varepsilon(s_0), F_{att}(s') \geq \upsilon_1^\omega F_{att}(s) \text{ or } s' = \mathcal{T}_0^\varepsilon(s_0)\},$$

where $\upsilon_1 = 10 \kappa_3^2 \kappa_2^{-2}$, so that $\upsilon_1 > 10$ and $\mathcal{T}_1^\varepsilon(s) \leq \mathcal{T}_0^\varepsilon(s_0)$. In view of (3.140), we have

$$\frac{1}{10}\left(\frac{\kappa_2}{\kappa_3}\right)^3 d_{min}^{\varepsilon,-}(s) \leq d_{min}^{\varepsilon,-}(\mathcal{T}_1^\varepsilon(s)), \tag{3.142}$$

and if $\mathcal{T}_1^\varepsilon(s) < \mathcal{T}_0^\varepsilon(s_0)$ then

$$d_{min}^{\varepsilon,-}(\mathcal{T}_1^\varepsilon(s)) \leq \frac{1}{10}\frac{\kappa_2}{\kappa_3} d_{min}^{\varepsilon,-}(s) \leq \frac{1}{10} d_{min}^{\varepsilon,-}(s). \tag{3.143}$$

The next result provides an upper bound on $\mathcal{T}_1^\varepsilon(s) - s$. Central to our argument is Proposition 3.19, which we use combined with various arguments by contradiction. We have the following proposition.

Proposition 3.56 *There exists $\beta_0 > 0$, depending on only V and M_0, with the following properties. If $J^-(s_0) \neq \emptyset$, $\hat{s} \in \mathcal{I}_0^\varepsilon(s_0)$ and*

$$\beta_0 \varepsilon \leq d_{min}^{\varepsilon,-}(\hat{s}) \leq d_{min}^{\varepsilon,+}(\hat{s}), \tag{3.144}$$

then we have

$$\mathcal{T}_1^\varepsilon(\hat{s}) - \hat{s} \leq \mathcal{K}_0 \left(d_{min}^{\varepsilon,-}(\hat{s})\right)^{\omega+2}, \tag{3.145}$$

where $\mathcal{K}_0$ is defined in (3.19), and moreover if $\mathcal{T}_1^\varepsilon(\hat{s}) < S$ then

$$d_{min}^{\varepsilon,-}(\mathcal{T}_1^\varepsilon(\hat{s})) \leq d_{min}^{\varepsilon,+}(\mathcal{T}_1^\varepsilon(\hat{s})). \tag{3.146}$$

Proof Up to a translation of times we may first assume that $\hat{s} = 0$, which eases the notations somewhat. We then argue by contradiction and assume that the conclusion is false, that is, there does not exist any such constant β_0, no matter how large it is chosen, such that the conclusion holds. Taking $\beta_0 = n$, this means that given any $n \in \mathbb{N}_*$ there exists some $0 < \varepsilon_n \leq 1$, a solution v_n to $(PGL)_{\varepsilon_n}$ such that $\mathcal{E}_{\varepsilon_n}(v_n) \leq M_0$, such that $WP_{\varepsilon_n}^{L_0}(\alpha_1 \varepsilon_n, 0)$ holds, such that

$$n\varepsilon_n \leq d_{\min}^{\varepsilon_n,-}(0) = d_{\min}^{\varepsilon_n}(0) \leq d_{\min}^{\varepsilon_n,+}(0), \tag{3.147}$$

and such that one of the conclusions fails, that is such that either

$$T_1^n \equiv T_1^{\varepsilon_n}(0) > \mathcal{K}_0 (d_{\min}^{\varepsilon,-}(0))^{\omega+2}, \tag{3.148}$$

or

$$d_{\min}^{\varepsilon,-}(T_1^n) > d_{\min}^{\varepsilon,+}(T_1^n). \tag{3.149}$$

Setting $S_0^n = \mathcal{K}_0(d_{\min}^{\varepsilon_n,-}(0))^{\omega+2}$, relation (3.148) may be rephrased as

$$F_{\text{att}}^n(s) \leq v_1^\omega F_{\text{att}}^n(0) \quad \text{and} \quad d_{\min}^{\varepsilon_n}(s) \geq 8q_1(\alpha_2)\varepsilon_n \quad \text{for any } s \in [0, S_0^n], \tag{3.150}$$

where the superscripts n refer to the corresponding functionals computed for the map v_n. Passing possibly to a subsequence, we may therefore assume that at least one of the properties (3.150) or (3.149) holds for any $n \in \mathbb{N}_*$. Also, passing possibly to a further subsequence, we may assume that the total number of fronts of $v_n(0)$ inside $[-L, L]$ is constant, equal to a number ℓ, denote $a_1^n(s), \ldots, a_\ell^n(s)$ the corresponding front points, for $s \in [0, T_1^n]$, and set $d_n^-(s) = d_{\min}^{\varepsilon_n,-}(s), d_n^+(s) = d_{\min}^{\varepsilon_n,+}(s), d_n(s) = d_{\min}^{\varepsilon_n}(s)$.

In order to obtain a contradiction we shall make use of *the scale invariance* of the equation: if v_ε is a solution to $(PGL)_\varepsilon$ then the map $\tilde{v}_{\tilde{\varepsilon}}(x, t) = v_\varepsilon(rx, r^2 t)$ is a solution to $(PGL)_{\tilde{\varepsilon}}$ with $\tilde{\varepsilon} = r^{-1}\varepsilon$. As scaling factor r_n, we choose $r_n = d_{\min}^{\varepsilon_n,-}(0) \geq n\varepsilon_n$ and set

$$\tilde{v}_n(x, t) = v_n(r_n x, r_n^2 t), \quad \breve{v}_n(x, \tau) = \tilde{v}_n(x, \tilde{\varepsilon}_n^{-\omega}\tau), \tag{3.151}$$

so that $\tilde{v}_n$ is a solution to $(PGL)_{\tilde{\varepsilon}_n}$ satisfying $WP_{\tilde{\varepsilon}_n}^{L_n}(\alpha_2\tilde{\varepsilon}_n, 0)$ with $L_n = r_n^{-1}L$ and

$$\tilde{\varepsilon}_n = (r_n)^{-1}\varepsilon_n = (d_{\min}^{\varepsilon_n,-}(0))^{-1}\varepsilon_n \leq \frac{1}{n}, \quad \text{hence we have } \varepsilon_n \to 0 \text{ as } n \to +\infty.$$

The points $\tilde{a}_1^n(s) = r_n^{-1}a_1^n(r_n^{-(2+\omega)}s), \ldots, \tilde{a}_\ell^n(s) = r_n^{-1}a_\ell^n(r_n^{-(2+\omega)}s)$ are the front points of $\breve{v}_n$. Let $\tilde{d}_n^-, \tilde{d}_n^+, \tilde{d}_n$ be the quantities corresponding to $d_{\min}^{\varepsilon,-}, d_{\min}^{\varepsilon,+}, d_{\min}^\varepsilon$ for $\breve{v}_n$, so that

$$\tilde{d}_n^-(s) = r_n^{-1}d_n^-(r_n^{-(2+\omega)}s), \quad \tilde{d}_n^+(s) = r_n^{-1}d_n^+(r_n^{-(2+\omega)}s),$$

$$\text{and } \tilde{d}_n(s) = r_n^{-1}d_n(r_n^{-(2+\omega)}s),$$

and notice that $d_n^-(0) = d_n(0) = 1$. We next distinguish the following two complementary cases.

Case 1: (3.150) holds for all $n \in \mathbb{N}_*$. It follows from assumption (3.150) that $\mathrm{WP}_{\tilde{\varepsilon}_n}^{L_n}(\alpha_1 \tilde{\varepsilon}_n, \tau)$ holds for every $\tau \in (0, \tilde{S}_1^n)$, where $\tilde{S}_1^n = r_n^{-(2+\omega)} S_0^n = \mathcal{K}_0$. Let $k_0 \in \{1, \ldots, \ell\}$ be such that

$$a_{k_0+1}^n(0) - a_{k_0}^n(0) = d_{\min}^{\varepsilon n,-}(0).$$

Using a translation if necessary, we may also assume that $a_{k_0}^n(0) = 0$ so that $a_{k_0+1}^n(0) = d_{\min}^{\varepsilon n,-}(0)$. We denote by $\tilde{\mathcal{F}}_{\mathrm{att}}^n$ the functional $\mathcal{F}_{\mathrm{att}}^\varepsilon$ computed for the front points of $\tilde{\mathfrak{v}}_n$, so that

$$\tilde{\mathcal{F}}_{\mathrm{att}}^n(r_n^{-(2+\omega)}s) = r_n^{(2+\omega)} \mathcal{F}_{\mathrm{att}}^n(s).$$

By construction we have

$$\tilde{a}_{k_0}^n(0) = 0 \text{ and } \tilde{a}_{k_0+1}^n(0) = 1 = \tilde{d}_n^-(0). \tag{3.152}$$

Since $\tilde{\varepsilon}_n \to 0$ as $n \to \infty$, we may implement part of the already-established asymptotic analysis for $(\mathrm{PGL})_\varepsilon$ on the sequence $(\tilde{\mathfrak{v}}_n)_{n \in \mathbb{N}}$. First, passing possibly to a subsequence, we may assume that for some subset $\tilde{J} \subset J(0)$ the points $\{\tilde{a}_k(0)\}_{k \in \tilde{J}}$ converge to some finite limits $\{\tilde{a}_k^0\}_{k \in \tilde{J}}$, whereas the points with indices in $J(0) \setminus \tilde{J}$ diverge either to $+\infty$ or to $-\infty$. We choose $\tilde{L} \geq 1$ so that

$$\underset{k \in \tilde{J}}{\cup} \{\tilde{a}_k^0\} \subset [-\frac{\tilde{L}}{2}, \frac{\tilde{L}}{2}]. \tag{3.153}$$

In view of (3.152), we have $\tilde{a}_{k_0}(0) = 0, \tilde{a}_{k_0+1} = 1$ and $\inf\{|\tilde{a}_{k+1}(0) - \tilde{a}_k(0)|, k \in \tilde{J}\} = 1$. We are then in position to apply the convergence result stated in Proposition 3.19 to the sequence $(\tilde{\mathfrak{v}}_n(\cdot))_{n \in \mathbb{N}}$. It states that the front points $(\tilde{a}_k^n(\tau))_{k \in J_0}$ which do not escape to infinity converge to the solution $(\tilde{a}_k(\cdot))_{k \in \tilde{J}}$ of the ordinary differential equation $(\mathcal{S})$ supplemented with the corresponding initial values $(\tilde{a}_k(0))_{k \in \tilde{J}}$, uniformly in time on every compact subset of $(0, \tilde{S}_{\max})$, where $\tilde{S}_{\max}$ denotes the maximal time of existence for the solution. In particular, we have

$$\begin{cases} \tilde{d}_n^-(\tau) \to \mathfrak{d}_{\tilde{a}}^-(\tau), \text{ uniformly on every compact subset of } (0, \tilde{S}_{\max}), \\ \underset{n \to +\infty}{\limsup} \tilde{F}_{\mathrm{att}}^n(\tau) \geq F_{\mathrm{att}}(\tilde{a}(\tau)) \text{ for every } \tau \in (0, \tilde{S}_{\max}), \end{cases}$$

the presence of the lim sup being related to the fact that some points might escape to infinity so that the limiting values of the functionals are possibly smaller. We use next the properties of the differential equation $(\mathcal{S})$ established in Appendix B. We first invoke Proposition 3.7 which asserts that $\tilde{S}_{\max} \leq \mathcal{K}_0$ and that

$$F_{\text{att}}(\tilde{a}(\tau)) \to +\infty \text{ as } \tau \to \tilde{S}_{\max}.$$

Hence, there exists some $\tau_1 \in (0, \tilde{S}_{\max}) \subset (0, \mathcal{K}_0)$ such that, if n is sufficiently large, then

$$\tilde{\mathcal{F}}^n_{\text{att}}(\tau_1) > \mathfrak{v}_1^\omega \tilde{\mathcal{F}}^n_{\text{att}}(0).$$

Going back to the original time scale, this yields $\mathcal{F}^n_{\text{att}}(r_n^{\omega+2}\tau_1) > \mathfrak{v}_1^\omega \mathcal{F}^n_{\text{att}}(0)$. Since $r_n^{\omega+2}\tau_1 \in (0, r_n^{2+\omega}\mathcal{K}_0) = (0, S_0^n)$ this contradicts (3.150) and completes the proof in Case 1.

Case 2: (3.149) holds for all $n \in \mathbb{N}_*$. We consider an arbitrary index $j \in J^+$. As above, translating the origin, we may assume without loss of generality that $a_j^n(0) = 0$. We also define the map $\mathfrak{v}_n$ as in Case 1, according to the same scaling as described in (3.151), the only difference being that the origin has been shifted differently. With similar notations, we have

$$\tilde{a}_j^n(0) = 0 \text{ and } \tilde{a}_{j+1}^n(0) \geq 1 = \tilde{d}_n^-(0).$$

Passing possibly to a further subsequence, we may assume that the front points at time 0 converge to some limits in $\bar{\mathbb{R}}$ denoted $\tilde{a}_k(0)$. We are hence again in the position to apply the convergence result of Proposition 3.19, so that the front points $(\tilde{a}_k^n(s))_{k \in J_j}$ which do not escape to infinity converge to the solution $(\tilde{a}_k(\cdot))_{k \in J_j}$ of the ordinary differential equation $(\mathcal{S})$ supplemented with the corresponding initial values $(\tilde{a}_k(0))_{k \in J_j}$, uniformly in time on every compact subset of $(0, \tilde{S}'_{\max})$, where $\tilde{S}'_{\max}$ denotes the (new) maximal time of existence for the solution. It follows from assumption (3.172), Theorem 3.2 and scaling that $0 < \tilde{\mathcal{T}}_1 \equiv \liminf \tilde{\mathcal{T}}_1^n \leq \tilde{S}'_{\max}$. We claim that, for any $\tau \in (0, \tilde{\mathcal{T}}_1)$, and for sufficiently large n, we have

$$|\tilde{a}_j^n(\tau) - \tilde{a}_{j+1}^n(\tau)| \geq \frac{\kappa_2}{2\kappa_3}. \tag{3.154}$$

This is actually a property of the differential equation $(\mathcal{S})$. We have indeed, in view of Proposition 3.65, $0 < F_{\text{rep}}(\tilde{a}(\tau)) \leq F_{\text{rep}}(\tilde{a}(0))$, so that it follows from (3.140) that

$$|\tilde{a}_j(\tau) - \tilde{a}_{j+1}(\tau)| \geq \frac{\kappa_2}{\kappa_3},$$

which yields (3.154) taking the convergence into account. Since (3.154) holds for any j, we deduce that

$$d_{\min}^{\varepsilon_n,+}(\mathcal{T}_1^n) \geq \frac{\kappa_2}{2\kappa_3} d_{\min}^{\varepsilon_n,-}(0)$$

and therefore by (3.149) we have

$$d_{\min}^{\varepsilon n}(T_1^n) = d_{\min}^{\varepsilon n,-}(T_1^n) \geq d_{\min}^{\varepsilon n,+}(T_1^n) \geq \frac{\kappa_2}{2\kappa_3} d_{\min}^{\varepsilon n,-}(0) \geq \frac{\kappa_2}{2\kappa_3} n\varepsilon. \qquad (3.155)$$

For n sufficiently large, this implies that $T_1^n < T_0^n$, and therefore from (3.143) we have

$$d_{\min}^{\varepsilon n,-}(T_1^n) \leq \frac{1}{10} \frac{\kappa_2}{\kappa_3} d_{\min}^{\varepsilon n,-}(0),$$

which is in contradiction with (3.155). $\qquad\qquad\square$

We turn now to the case where $d_{\min}^{\varepsilon,+}(s) \leq d_{\min}^{\varepsilon,-}(s)$. In order to handle the repulsive forces at work, for $s \in \mathcal{I}_0^\varepsilon(s_0)$ we introduce the new stopping times

$$T_2^\varepsilon(s) = \inf\{s \leq s' \leq T_0^\varepsilon(s_0), \mathcal{F}_{\mathrm{rep}}^\varepsilon(s') \leq \upsilon_2^\omega \mathcal{F}_{\mathrm{rep}}^\varepsilon(s) \text{ or } s' = T_0^\varepsilon(s_0)\},$$

where $\upsilon_2 = \frac{\kappa_2^2}{10\kappa_3^2}$, so that $\upsilon_2 < 1$. Notice that, in view of (3.140), we have, if $T_2^\varepsilon(s) < T_0^\varepsilon(s_0)$,

$$d_{\min}^{\varepsilon,+}(T_2^\varepsilon(s)) \geq \upsilon_2^{-1} \frac{\kappa_2}{\kappa_3} d_{\min}^{\varepsilon,+}(s) \geq 10 d_{\min}^{\varepsilon,+}(s). \qquad (3.156)$$

With $\mathcal{S}_1$ introduced in Proposition 3.7, we set

$$\mathcal{K}_1 = \mathcal{S}_1^{-\omega} \left(\frac{2\kappa_3}{\kappa_2 \upsilon_2} \right)^{\omega+2}. \qquad (3.157)$$

Proposition 3.57 *There exists $\beta_1 > 0$, depending on only V and M_0, with the following properties. If $J^+(s_0) \neq \emptyset$, $\hat{s} \in \mathcal{I}_0^\varepsilon(s_0)$ and*

$$\beta_1 \varepsilon \leq d_{\min}^{\varepsilon,+}(\hat{s}) \leq d_{\min}^{\varepsilon,-}(\hat{s}), \qquad (3.158)$$

then we have

$$T_2^\varepsilon(\hat{s}) - \hat{s} \leq \mathcal{K}_1 \left(d_{\min}^{\varepsilon,+}(\hat{s}) \right)^{\omega+2}, \qquad (3.159)$$

and if $T_2^\varepsilon(\hat{s}) < S$ then $T_2^\varepsilon(\hat{s}) < T_0^\varepsilon(s_0)$ and for any $s \in [\hat{s}, T_2^\varepsilon(\hat{s})]$, we have

$$d_{\min}^\varepsilon(s) \geq \frac{1}{2} \mathcal{S}_2 d_{\min}^{\varepsilon,+}(\hat{s}) \qquad (3.160)$$

and

$$\mathcal{F}_{\mathrm{att}}^\varepsilon(s)^{-\frac{1}{\omega}} \leq \mathcal{F}_{\mathrm{att}}^\varepsilon(\hat{s})^{-\frac{1}{\omega}} + \frac{1}{\kappa_3}(d_{\min}^{\varepsilon,+}(\hat{s})), \qquad (3.161)$$

where $\mathcal{S}_2$ is defined in Proposition 3.7 and κ_3 is defined in (3.140).

Proof The argument shares strong similarities with the proof of Proposition 3.56; we therefore just sketch its main points, in particular relying implicitly on the notations introduced there, as far as this is possible. By translation in time we also assume that $\hat{s} = 0$ and argue by contradiction assuming that for any $n \in \mathbb{N}_*$ there exists some $0 < \varepsilon_n \leq 1$, a solution v_n to $(PGL)_{\varepsilon_n}$ such that $\mathcal{E}_{\varepsilon_n}(v_n) \leq M_0$, $\mathcal{WP}^L_{\varepsilon_n}(\alpha_1 \varepsilon_n, 0)$ holds, such that $n \varepsilon_n \leq d_n^+(0)$, and such that either we have for any $s \in (0, S_1^n)$, where $S_1^n = \mathcal{K}_1 d_n^+(0)^{\omega+2}$, $d_n(s) \geq 8q(\alpha_2)\varepsilon_n$ and

$$\kappa_3^\omega (d_n^+(s'))^{-\omega} \geq F_{\mathrm{rep}}^n(s') \geq \upsilon_2^\omega F_{\mathrm{rep}}^n(0) \geq \upsilon_2^\omega \kappa_2^\omega (d_n^+(0))^{-\omega} \qquad (3.162)$$

or there is some $\tau_n \in (0, \mathcal{T}_2^n)$ such that

$$d_{\min}^{\varepsilon,+}(\tau_n) < \frac{1}{2} \mathcal{S}_2 d_{\min}^{\varepsilon,+}(s) \qquad (3.163)$$

or

$$\mathcal{F}_{\mathrm{att}}^n(\tau_n)^{-\frac{1}{\omega}} < \mathcal{F}_{\mathrm{att}}^n(0)^{-\frac{1}{\omega}} + \frac{1}{32\kappa_3}(d_n^+(s)). \qquad (3.164)$$

As in (3.151), but with a different scaling r_n, we set

$$r_n = d_{\min}^{\varepsilon_n,+}(0) \geq n\varepsilon_n, \ \tilde{v}_n(x,t) = v_n(r_n x, r_n^2 t), \ \text{and} \ \tilde{\mathfrak{v}}_n(x,s) = \tilde{v}_n(x, \tilde{\varepsilon}_n^{-\omega} s). \ (3.165)$$

We verify that $\tilde{v}_n$ is a solution to $(PGL)_{\tilde{\varepsilon}_n}$ with $\tilde{\varepsilon}_n = (r_n)^{-1}\varepsilon_n \to 0$ as $n \to \infty$ and that the points $\tilde{a}_k^n(\tau) = r_n^{-1} a_k^n(r_n^{-(2+\omega)}\tau)$ for $k \in J$ are the front points of $\tilde{\mathfrak{v}}_n$. We distinguish three cases, which together were all possible cases, taking subsequences if necessary.

Case 1: (3.162) holds, for any $n \in \mathbb{N}$. It follows that $\mathcal{WP}^{L_n}_{\tilde{\varepsilon}_n}(\alpha_1 \tilde{\varepsilon}_n, \tau)$ holds for every $\tau \in (0, \tilde{S}_1^n)$, where $\tilde{S}_1^n = r_n^{-(2+\omega)} S_1^n = \mathcal{K}_1$. Let j be an arbitrary index in J^+. Translating the origin if necessary, we may assume that $a_j^n(0) = 0$ so that $a_{j+1}^n(0) \geq d_n^+(0) \geq n\varepsilon_n$ and hence $\tilde{a}_{j+1}^n(0) - \tilde{a}_j^n(0) \geq 1$. Since $\tilde{\varepsilon}_n \to 0$ as $n \to \infty$, we may implement part of the already-established asymptotic analysis for $(PGL)_\varepsilon$ on the sequence $(\tilde{v}_n)_{n \in \mathbb{N}}$. First, passing possibly to a subsequence, we may assume that for some subset $\tilde{J} \subset J(0)$ the points $\{\tilde{a}_k(0)\}_{k \in \tilde{J}}$ converge to some finite limits $\{\tilde{a}_k^0\}_{k \in \tilde{J}}$, whereas the points with indices in $J(0) \setminus \tilde{J}$ diverge either to $+\infty$ or to $-\infty$. We choose $\tilde{L} \geq 1$ so that (3.153) holds. It follows from Proposition 3.19 that for $\tau \in (0, \mathcal{K}_1)$, we have

$$|\tilde{a}_{j+1}^n(\tau) - \tilde{a}_j^n(\tau)| \to |\tilde{a}_{j+1}(\tau) - \tilde{a}_j(\tau)| \geq \left(\mathcal{S}_1 \tau + \mathcal{S}_2 \mathfrak{d}_{\tilde{a}}^+(0)^{\omega+2}\right)^{\frac{1}{\omega+2}}$$

$$= (\mathcal{S}_1 \tau + \mathcal{S}_2)^{\frac{1}{\omega+2}}$$

as $n \to \infty$, where the last inequality is a consequence of Proposition 3.7. Taking the infimum over J^+, we obtain, for n sufficiently large,

$$\tilde{d}_n^+(\tau) = \inf_{j \in J^+} |\tilde{a}_{j+1}^n(\tau) - \tilde{a}_j^n(\tau)| \geq \frac{1}{2}(\mathcal{S}_1\tau + \mathcal{S}_2)^{\frac{1}{\omega+2}} \geq \frac{1}{2}(\mathcal{S}_1\tau)^{\frac{1}{\omega+2}}, \forall \tau \in (0, \mathcal{K}_1).$$

(3.166)

On the other hand, going back to (3.162), with the same notation as in Proposition 3.56, we are led to the inequality

$$\tilde{d}_n^+(\tau) \leq \kappa_3 \kappa_2^{-1} \upsilon_2^{-1} \text{ for } \tau \in (0, \mathcal{K}_1).$$

(3.167)

In view of our choice (3.157) of $\mathcal{K}_1$, relations (3.166) and (3.167) are contradictory for τ close to $\mathcal{K}_1$, thus yielding a contradiction in Case 1.

Case 2: (3.162) does not hold, but (3.163) holds for any $n \in \mathbb{N}$. The argument is almost identical, we conclude again thanks to (3.166) but keeping $\mathcal{S}_2$ instead of $\mathcal{S}_1\tau$ in its last inequality.

Case 3: (3.162) does not hold but (3.164) holds for any $n \in \mathbb{N}$. As in the proof of Proposition 3.56, we conclude that $0 < \tilde{\mathcal{T}}_2 \equiv \liminf_{n \to +\infty} \tilde{\mathcal{T}}_2^n$. This situation is slightly more delicate than the ones analyzed so far, and we have *also to track the fronts escaping possibly to infinity*. Up to a subsequence, we may assume that the set J is decomposed as a disjoint union of clusters $J = \bigcup_{i=1}^{q} J_p$, where each of the sets J_p is an ordered set of $m_p + 1$ consecutive points, that is $J_p = \{k_p, k_p + 1, \ldots k_p + m_p\}$ and such that the two following properties hold.

- There exists a constant $C > 0$ independent of n such that

$$|\tilde{a}_{k_p}^n(0) - \tilde{a}_{k_p+r}^n(0)| \leq C \text{ for any } p \in \{1, \ldots, q\} \text{ and any } r \in \{k_p, \ldots, m_p\}.$$

(3.168)

- For $1 \leq p_1 < p_2 \leq q$, we have $\tilde{a}_{k_{p_2}}^n - \tilde{a}_{k_{p_1}}^n \to +\infty$.

For a given $p \in \{1, \ldots, q\}$, translating the origin if necessary, we may assume that $\tilde{a}_{k_p}^n(0) = 0$, and passing possibly to a further subsequence, that the front points at time 0 converge as $n \to +\infty$ to some limits denoted $\tilde{a}_{p,k}(0)$, for $k \in \{k_p, \ldots, k_p + m_p\}$. Notice that, as an effect of the scaling, all other front points diverge to infinity, in the chosen frame. We now apply Proposition 3.19 to this cluster of points: it yields uniform convergence, for $k \in \{k_p, \ldots, k_p + m_p\}$ of the front points $\tilde{a}_k^n(\cdot)$ to the solution $\tilde{a}_{p,k}(\cdot)$ of the differential equation $(\mathcal{S})$ supplemented with the initial time conditions $\tilde{a}_{p,k}(0)$ defined above. If F_{att}^p denotes the functional F_{att} defined in (3.139) restricted to the points of the cluster J_p, we have in view of (B.17)

$$\frac{d}{d\tau} F_{\text{att}}^p(\tau) \geq 0, \qquad \text{for any } p = 1, \ldots, q, \quad \text{for any } \tau \in (0, \tilde{\mathcal{T}}_2).$$

On the other hand, since the mutual distances between the distinct clusters diverge towards infinity, and hence their mutual interaction energies tend to zero, one obtains, in view of the uniform convergence for each separate cluster, that

$$\lim_{n \to +\infty} \tilde{\mathcal{F}}_{\text{att}}^n(\tau) = \sum_{p=1}^{q} F_{\text{att}}^p(\tau) \geq \sum_{p=1}^{q} F_{\text{att}}^p(0) = \lim_{n \to +\infty} \tilde{\mathcal{F}}_{\text{att}}^n(0), \qquad \text{for } \tau \in (0, \tilde{\mathcal{T}}_2).$$

Therefore, for n sufficiently large we are led to

$$\tilde{\mathcal{F}}_{\text{att}}^n(\tilde{\mathcal{T}}_2)^{-\frac{1}{\omega}} \leq \tilde{\mathcal{F}}_{\text{att}}^n(0)^{-\frac{1}{\omega}} + \frac{1}{2\kappa_3}.$$

Scaling back to the original variables, this contradicts (3.164) and hence completes the proof. $\qquad\square$

From Propositions 3.56 and 3.57 we obtain the following.

Proposition 3.58 *There exists $\mathcal{K}_2 > 0$, depending on only V and M_0, with the following properties. Assume that $J^-(s_0) \neq \emptyset$ and that $s \in \mathcal{I}_0^\varepsilon(s_0)$ satisfies*

$$d_{\min}^{\varepsilon,L}(s) \geq \max(\beta_0, \beta_1)\varepsilon, \qquad \text{and} \qquad s + \mathcal{K}_2 d_{\min}^{\varepsilon,-}(s)^{\omega+2} < S. \tag{3.169}$$

Then there exists some time $\mathcal{T}_{\text{col}}^-(s) \in \mathcal{I}_0^\varepsilon(s_0)$ such that

$$\mathcal{T}_{\text{col}}^{\varepsilon,-}(s) - s \leq \mathcal{K}_2 d_{\min}^{\varepsilon,-}(s)^{\omega+2} \tag{3.170}$$

and

$$d_{\min}^{\varepsilon,L}(\mathcal{T}_{\text{col}}^{\varepsilon,-}(s)) \leq \max(\beta_0, 8\mathfrak{q}_1(\alpha_2))\,\varepsilon. \tag{3.171}$$

Proof We distinguish two cases.

Case I:

$$d_{\min}^{\varepsilon,L}(s) = d_{\min}^{\varepsilon,-}(s) \leq d_{\min}^{\varepsilon,+}(s). \tag{3.172}$$

In that case we will make use of Proposition 3.56 in an iterative argument. In view of (3.169), we are in a position to invoke Proposition 3.56 at time $\hat{s} = s$ and set $s_1 = \mathcal{T}_1^\varepsilon(s)$, so that in particular

$$s_1 - s \leq \mathcal{K}_0 d_{\min}^{\varepsilon,-}(s)^{\omega+2} \qquad \text{and} \qquad d_{\min}^{\varepsilon,-}(s_1) \leq d_{\min}^{\varepsilon,+}(s_1). \tag{3.173}$$

Notice that by (3.169) and (3.173) we have $s_1 < S$.

We distinguish two sub-cases.

Case I.1: $s_1 = \mathcal{T}_0^\varepsilon(s_0)$ or $d_{\min}^{\varepsilon,-}(s_1) < \beta_0\varepsilon$. In that case, we simply set $\mathcal{T}_{\text{col}}^{\varepsilon,-}(s) = s_1$ and we are done if we require $\mathcal{K}_2 \geq \mathcal{K}_0$, by (3.173) and the definition of $\mathcal{T}_0^\varepsilon(s_0)$.

Case I.2: $s_1 < T_0^\varepsilon(s_0)$ and $d_{\min}^{\varepsilon,-}(s_1) \geq \beta_0\varepsilon$. In that case, we may apply Proposition 3.56 at time $\hat{s} = s_1$ and set $s_2 = T_1^\varepsilon(s_1)$, so that in particular

$$s_2 - s_1 \leq K_0 d_{\min}^{\varepsilon,-}(s_1)^{\omega+2} \quad \text{and} \quad d_{\min}^{\varepsilon,-}(s_2) \leq d_{\min}^{\varepsilon,+}(s_2). \tag{3.174}$$

Moreover, since in that case $s_1 = T_1^\varepsilon(s) < T_0^\varepsilon(s_0)$, it follows from (3.143) that

$$d_{\min}^{\varepsilon,-}(s_1) \leq \frac{1}{10} d_{\min}^{\varepsilon,-}(s), \tag{3.175}$$

and therefore from (3.174) we actually have

$$s_2 - s_1 \leq K_0 10^{-(\omega+2)} d_{\min}^{\varepsilon,-}(s)^{\omega+2} \quad \text{and} \quad d_{\min}^{\varepsilon,-}(s_2) \leq d_{\min}^{\varepsilon,+}(s_2). \tag{3.176}$$

We then iterate the process until we fall into Case I.1. If we have not reached that stage up to step m, then thanks to Proposition 3.56 applied at time $\hat{s} = s_m$ we obtain, with $s_{m+1} := T_1^\varepsilon(s_m)$,

$$s_{m+1} - s_m \leq K_0 d_{\min}^{\varepsilon,-}(s_m)^{\omega+2} \quad \text{and} \quad d_{\min}^{\varepsilon,-}(s_{m+1}) \leq d_{\min}^{\varepsilon,+}(s_{m+1}). \tag{3.177}$$

Moreover, since Case I.1 was not reached before step m, we have $s_p = T_1^\varepsilon(s_{p-1}) < T_0^\varepsilon(s_0)$ for all $p \leq m$, so that repeated use of (3.143) yields

$$d_{\min}^{\varepsilon,-}(s_p) \leq \left(\frac{1}{10}\right)^p d_{\min}^{\varepsilon,-}(s), \qquad \forall p \leq m. \tag{3.178}$$

From (3.177) we thus also have

$$s_{p+1} - s_p \leq K_0 10^{-p(\omega+2)} d_{\min}^{\varepsilon,-}(s)^{\omega+2}, \qquad \forall p \leq m, \tag{3.179}$$

and therefore by summation

$$s_{m+1} - s \leq K_0 \left(\sum_{p=0}^{m} 10^{-p(\omega+2)}\right) d_{\min}^{\varepsilon,-}(s)^{\omega+2}, \tag{3.180}$$

so that in particular from (3.169) it holds that $s_{m+1} < S$ if we choose $K_2 \geq 2K_0$. It follows from (3.178) that Case I.1 is necessarily reached in a finite number of steps, thus defining $T_{\mathrm{col}}^{\varepsilon,-}(s)$, and from (3.180) we obtain the upper bound

$$T_{\mathrm{col}}^{\varepsilon,-}(s) - s \leq K_0 \left(\sum_{p=0}^{\infty} 10^{-p(\omega+2)}\right) d_{\min}^{\varepsilon,-}(s)^{\omega+2} \leq 2K_0 d_{\min}^{\varepsilon,-}(s)^{\omega+2}, \tag{3.181}$$

from which (3.170) follows.

Case II:

$$d_{\min}^{\varepsilon,L}(s) = d_{\min}^{\varepsilon,+}(s) < d_{\min}^{\varepsilon,-}(s). \tag{3.182}$$

Note that this implies that $J^+(s_0) \neq \emptyset$. We will show that Case II can be reduced to Case I after some controlled interval of time necessary for the repulsive forces to push $d_{\min}^{\varepsilon,+}$ above $d_{\min}^{\varepsilon,-}$. More precisely, we define the stopping time

$$\mathcal{T}_{\mathrm{cros}}^{\varepsilon}(s) = \inf\{\mathcal{T}_0^{\varepsilon}(s) \geq s' \geq s, \, d_{\min}^{\varepsilon,-}(s') \leq d_{\min}^{\varepsilon,+}(s')\}.$$

As in Case I, we implement an iterative argument, but based this time on Proposition 3.57. In view of (3.182) and (3.169), we may apply Proposition 3.57 at time $\hat{s} = s$ and set $s_1 = \mathcal{T}_2^{\varepsilon}(s)$, so that in particular

$$s_1 - s \leq \mathcal{K}_1 d_{\min}^{\varepsilon,+}(s)^{\omega+2} \leq \mathcal{K}_1 d_{\min}^{\varepsilon,-}(s)^{\omega+2}. \tag{3.183}$$

Notice that by (3.169) and (3.183) we have $s_1 < S$ and therefore $d_{\min}^{\varepsilon,+}(s_1) \geq 10 d_{\min}^{\varepsilon,+}(s) \geq \beta_1$, and by (3.161)

$$
\begin{aligned}
\mathcal{F}_{\mathrm{att}}^{\varepsilon}(s_1)^{-\frac{1}{\omega}} &\leq \mathcal{F}_{\mathrm{att}}^{\varepsilon}(s)^{-\frac{1}{\omega}} + \frac{1}{\kappa_3} d_{\min}^{\varepsilon,+}(s) \\
&\leq \mathcal{F}_{\mathrm{att}}^{\varepsilon}(s)^{-\frac{1}{\omega}} + \frac{1}{10\kappa_3} d_{\min}^{\varepsilon,+}(s_1).
\end{aligned}
\tag{3.184}
$$

We distinguish two sub-cases.

Case II.1: $s_1 \geq \mathcal{T}_{\mathrm{cros}}^{\varepsilon}(s)$. In that case we proceed to Case I which we will apply starting at s_1 instead of s and we set $\mathcal{T}_{\mathrm{col}}^{\varepsilon,-}(s) := \mathcal{T}_{\mathrm{col}}^{\varepsilon,-}(s_1)$. Since, combining the first inequality of (3.184) with (3.140), we deduce that

$$d_{\min}^{\varepsilon,-}(s_1) \leq \kappa_3 \kappa_2^{-1} d_{\min}^{\varepsilon,-}(s) + d_{\min}^{\varepsilon,+}(s) \leq \left(\kappa_3 \kappa_2^{-1} + 1\right) d_{\min}^{\varepsilon,-}(s), \tag{3.185}$$

the equivalent of (3.181) becomes

$$
\begin{aligned}
\mathcal{T}_{\mathrm{col}}^{\varepsilon,-}(s_1) - s_1 &\leq \mathcal{K}_0 \Big(\sum_{p=0}^{\infty} 10^{-p(\omega+2)} \Big) d_{\min}^{\varepsilon,-}(s_1)^{\omega+2} \\
&\leq 2\mathcal{K}_0 d_{\min}^{\varepsilon,-}(s_1)^{\omega+2} \\
&\leq 2\mathcal{K}_0 \left(\kappa_3 \kappa_2^{-1} + 1\right)^{\omega+2} d_{\min}^{\varepsilon,-}(s)^{\omega+2},
\end{aligned}
\tag{3.186}
$$

and therefore it follows from (3.183) that

$$\mathcal{T}_{\mathrm{col}}^{\varepsilon,-}(s) - s \leq \mathcal{T}_{\mathrm{col}}^{\varepsilon,-}(s_1) - s_1 + (s_1 - s) \leq \left(\mathcal{K}_1 + 2\mathcal{K}_0 \left(\kappa_3 \kappa_2^{-1} + 1\right)^{\omega+2}\right) d_{\min}^{\varepsilon,-}(s)^{\omega+2}, \tag{3.187}$$

and (3.170) follows if $\mathcal{K}_2 \geq \mathcal{K}_1 + 2\mathcal{K}_0 \left(\kappa_3 \kappa_2^{-1} + 1\right)^{\omega+2}$.

Case II.2: $s_1 < \mathcal{T}_{\mathrm{cros}}^{\varepsilon}(s)$. In that case we proceed to construct $s_2 = \mathcal{T}_2^{\varepsilon}(s_1)$. Notice that combining the second inequality of (3.184) with (3.140), we deduce that

$$d^{\varepsilon,-}_{\min}(s_1) \leq \kappa_3\kappa_2^{-1}d^{\varepsilon,-}_{\min}(s) + \frac{1}{5}d^{\varepsilon,+}_{\min}(s_1) \leq \kappa_3\kappa_2^{-1}d^{\varepsilon,-}_{\min}(s) + \frac{1}{5}d^{\varepsilon,-}_{\min}(s_1), \quad (3.188)$$

so that

$$d^{\varepsilon,+}_{\min}(s_1) \leq d^{\varepsilon,-}_{\min}(s_1) \leq \frac{5}{4}\kappa_3\kappa_2^{-1}d^{\varepsilon,-}_{\min}(s). \quad (3.189)$$

We now explain the iterative argument. Assume that for some $m \geq 1$ we have already constructed $s_1,\dots,s_m$, such that for $2 \leq p \leq m$

$$s_p < S, \qquad \beta_1\varepsilon \leq d^{\varepsilon,+}_{\min}(s_p) \leq d^{\varepsilon,-}_{\min}(s_p), \qquad s_p = T^{\varepsilon}_2(s_{p-1}).$$

First, repeated use of (3.156) yields

$$d^{\varepsilon,+}_{\min}(s_p) \geq 10^p d^{\varepsilon,+}_{\min}(s), \qquad \forall 1 \leq p \leq m, \quad (3.190)$$

and actually

$$d^{\varepsilon,+}_{\min}(s_p) \geq 10^{p-q} d^{\varepsilon,+}_{\min}(s), \qquad \forall 1 \leq q \leq p \leq m. \quad (3.191)$$

Hence, by repeated use of (3.161), we obtain

$$\mathcal{F}^{\varepsilon}_{\mathrm{att}}(s_m)^{-\frac{1}{\omega}} \leq \mathcal{F}^{\varepsilon}_{\mathrm{att}}(s)^{-\frac{1}{\omega}} + \frac{1}{\kappa_3}\left(d^{\varepsilon,+}_{\min}(s) + \sum_{p=1}^{m-1}d^{\varepsilon,+}_{\min}(s_p)\right)$$

$$\leq \mathcal{F}^{\varepsilon}_{\mathrm{att}}(s)^{-\frac{1}{\omega}} + \frac{1}{\kappa_3}\sum_{p=0}^{m-1}10^{-p}d^{\varepsilon,+}_{\min}(s_{m-1})$$

$$\leq \mathcal{F}^{\varepsilon}_{\mathrm{att}}(s)^{-\frac{1}{\omega}} + \frac{2}{\kappa_3}d^{\varepsilon,+}_{\min}(s_{m-1})$$

$$\leq \mathcal{F}^{\varepsilon}_{\mathrm{att}}(s)^{-\frac{1}{\omega}} + \frac{1}{5\kappa_3}d^{\varepsilon,+}_{\min}(s_m).$$

Combining the latter with (3.140), we deduce that

$$d^{\varepsilon,-}_{\min}(s_m) \leq \kappa_3\kappa_2^{-1}d^{\varepsilon,-}_{\min}(s) + \frac{1}{5}d^{\varepsilon,+}_{\min}(s_m) \leq \kappa_3\kappa_2^{-1}d^{\varepsilon,-}_{\min}(s) + \frac{1}{5}d^{\varepsilon,-}_{\min}(s_m),$$

so that

$$d^{\varepsilon,+}_{\min}(s_m) \leq d^{\varepsilon,-}_{\min}(s_m) \leq \frac{5}{4}\kappa_3\kappa_2^{-1}d^{\varepsilon,-}_{\min}(s). \quad (3.192)$$

Let $s_{m+1} := \mathcal{T}_2^\varepsilon(s_m)$. Then by (3.159) and (3.190)

$$s_{m+1} - s \leq \mathcal{K}_1 \left(d_{\min}^{\varepsilon,+}(s)^{\omega+2} + \sum_{p=1}^m d_{\min}^{\varepsilon,+}(s_p)^{\omega+2} \right)$$

$$\leq \mathcal{K}_1 \sum_{p=0}^{m-1} 10^{-(\omega+2)(m-p)} d_{\min}^{\varepsilon,+}(s_m)^{\omega+2} \qquad (3.193)$$

$$\leq 2\mathcal{K}_1 d_{\min}^{\varepsilon,+}(s_m)^{\omega+2}.$$

Combining (3.193) with (3.192), we are led to

$$s_{m+1} - s \leq 2 \left(\frac{5\kappa_3}{4\kappa_2} \right)^{\omega+2} \mathcal{K}_1 d_{\min}^{\varepsilon,-}(s)^{\omega+2}$$

and therefore by (3.169) we have $s_{m+1} < S$.

Combining (3.192) with (3.190), we obtain

$$0 \leq d_{\min}^{\varepsilon,-}(s_m) - d_{\min}^{\varepsilon,+}(s_m) \leq \frac{\kappa_3}{\kappa_2} d_{\min}^{\varepsilon,-}(s) - 10^m (d_{\min}^{\varepsilon,+}(s)),$$

and therefore necessarily

$$m \leq \log_{10} \left(\frac{\kappa_3 d_{\min}^{\varepsilon,-}(s)}{\kappa_2 d_{\min}^{\varepsilon,+}(s)} \right).$$

It follows that the number $m_0 = \sup\{m \in \mathbb{N}_*, d_{\min}^{\varepsilon,-}(s_m) \geq d_{\min}^{\varepsilon,+}(s_m)\}$ is finite, and at that stage we proceed to Case I as in Case II.1 above, and the conclusion follows likewise, replacing (3.185) by

$$d_{\min}^{\varepsilon,+}(s_{m_0+1}) \leq \frac{9}{4} \kappa_3 \kappa_2^{-1} d_{\min}^{\varepsilon,-}(s)$$

which is obtained by combining

$$d_{\min}^{\varepsilon,+}(s_{m_0+1}) \leq \kappa_3 \kappa_2^{-1} d_{\min}^{\varepsilon,-}(s) + d_{\min}^{\varepsilon,+}(s_{m_0}),$$

with

$$d_{\min}^{\varepsilon,+}(s_{m_0}) \leq d_{\min}^{\varepsilon,-}(s_{m_0}) \leq \frac{5}{4} \kappa_3 \kappa_2^{-1} d_{\min}^{\varepsilon,-}(s).$$

$\square$

3.7.2 Proof of Proposition 3.55

We will fix the value of the constants κ_1, α_3 and $\mathcal{K}_{col}$ in the course of the proof. Let s_0 be as in the statement. We first require that

$$\alpha_3 \geq \alpha_2 \qquad \text{and that} \qquad \alpha_3 \geq 16 q_1(\alpha_2),$$

so that assumption $\mathcal{WP}_\varepsilon^L(\alpha_3, s_0)$ implies assumption 3.141 of Subsection 3.7.1.

Next, we set $s = s_0 + \varepsilon^{\omega+2}$ and we wish to make sure that the assumptions of Proposition 3.55 are satisfied at time s. In view of the upper bound (3.7) on the velocity of the front set, we deduce that

$$d_{\min}^{\varepsilon,L}(s) \geq d_{\min}^{\varepsilon,L}(s_0) - C\rho_0^{\frac{1}{\omega+2}}\varepsilon \geq \alpha_3\varepsilon - C\rho_0^{\frac{1}{\omega+2}}\varepsilon \geq \max(\beta_0, \beta_1)\varepsilon$$

provided we choose α_3 sufficiently large. Also,

$$\frac{1}{2}d_{\min}^{\varepsilon,-}(s_0) \leq d_{\min}^{\varepsilon,L}(s_0) - C\rho_0^{\frac{1}{\omega+2}}\varepsilon \leq d_{\min}^{\varepsilon,L}(s) \leq d_{\min}^{\varepsilon,L}(s_0) + C\rho_0^{\frac{1}{\omega+2}}\varepsilon \leq 2d_{\min}^{\varepsilon,L}(s_0),$$
$$(3.194)$$

and therefore provided we choose

$$\mathcal{K}_{\mathrm{col}} \geq 2^{\omega+2}\mathcal{K}_2$$

it follows from the assumption $s_0 + \mathcal{K}_{\mathrm{col}}d_{\min}^{\varepsilon,L}(s_0)^{\omega+2} < S$ that $s + \mathcal{K}_2 d_{\min}^{\varepsilon,L}(s)^{\omega+2} < S$. Therefore we may apply Proposition 3.58. Let $\mathcal{T}_{\mathrm{col}}^{\varepsilon,-}(s) \in \mathcal{I}_0^\varepsilon(s_0)$ be given by its statement, so that by (3.194)

$$\mathcal{T}_{\mathrm{col}}^{\varepsilon,-}(s) - s \leq 2^{\omega+2}\mathcal{K}_2 d_{\min}^{\varepsilon,-}(s_0)^{\omega+2},$$

and

$$d_{\min}^{\varepsilon,L}(\mathcal{T}_{\mathrm{col}}^{\varepsilon,-}(s)) \leq \max\left(\beta_0, 8\mathfrak{q}_1(\alpha_2)\right)\varepsilon. \tag{3.195}$$

By Proposition 3.31, there exists some time $\mathcal{T}_{\mathrm{col}}^{\varepsilon,+}(s_0) \in [\mathcal{T}_{\mathrm{col}}^{\varepsilon,-}(s), \mathcal{T}_{\mathrm{col}}^{\varepsilon,-}(s) + \mathfrak{q}_0(\alpha_3)\varepsilon^{\omega+2}]$ such that $\mathrm{WPI}_\varepsilon^L(\alpha_3\varepsilon, \mathcal{T}_{\mathrm{col}}^{\varepsilon,+}(s_0))$ holds. In view of (3.194) and since $d_{\min}^{\varepsilon,-}(s_0) \geq \alpha_3\varepsilon$, it follows that

$$0 \leq \mathcal{T}_{\mathrm{col}}^{\varepsilon,+}(s_0) - s_0 \leq \varepsilon^{\omega+2} + 2^{\omega+2}\mathcal{K}_2 d_{\min}^{\varepsilon,-}(s_0)^{\omega+2} + \mathfrak{q}_0(\alpha_3)\varepsilon^{\omega+2}$$
$$\leq \left(2^{\omega+2}\mathcal{K}_2 + \frac{1+\mathfrak{q}_0(\alpha_3)}{\alpha_3^{\omega+2}}\right)d_{\min}^{\varepsilon,-}(s_0)^{\omega+2} \tag{3.196}$$
$$\leq \mathcal{K}_{\mathrm{col}}d_{\min}^{\varepsilon,-}(s_0)^{\omega+2}$$

provided we finally *fix* the value of $\mathcal{K}_{\mathrm{col}}$ as

$$\mathcal{K}_{\mathrm{col}} = \left(2^{\omega+2}\mathcal{K}_2 + \frac{1+\mathfrak{q}_0(\alpha_3)}{\alpha_3^{\omega+2}}\right).$$

[Note that at this stage $\mathcal{K}_{\mathrm{col}}$ is fixed but its definition depends on α_3 which has not yet been fixed. Of course when we fix α_3 below we shall do it without any reference to $\mathcal{K}_{\mathrm{col}}$, in order to avoid impossible loops.]

Next, we first claim that

$$\mathfrak{E}_\varepsilon^L(s_0) \geq \mathfrak{E}_\varepsilon^L(s) \geq \mathfrak{E}_\varepsilon^L(\mathcal{T}_{\mathrm{col}}^{\varepsilon,+}(s_0)).$$

In view of Corollary 3.34, it suffices to check that $L \geq L_0(s_0, \mathcal{T}_{col}^{\varepsilon,+}(s_0))$, where we recall that the function $L_0(\cdot)$ was defined in (3.77). In view of (3.196), this reduces to

$$100 C_e L^{-(\omega+2)} \mathcal{K}_{col} d_{min}^{\varepsilon,-}(s_0)^{\omega+2} \leq \frac{\mu_1}{4}.$$

Since by (3.135) we have $d_{min}^{\varepsilon,-}(s_0) \leq 2\kappa_0(s_0)L$, it therefore suffices that

$$\kappa_0(s_0) \leq \frac{1}{2}\left(\frac{\mu_1}{400 C_e \mathcal{K}_{col}}\right)^{\frac{1}{\omega+2}} \equiv \kappa_1,$$

and we have now *fixed* the value of κ_1.

Next, we claim that actually

$$\mathcal{E}_\varepsilon^L(\mathcal{T}_{col}^{\varepsilon,+}(s_0)) \leq \mathcal{E}_\varepsilon^L(s_0) - \mu_1.$$

Indeed this is so, otherwise by Corollary 3.34 we would have $\mathcal{E}_\varepsilon^L(\mathcal{T}_{col}^{\varepsilon,+}(s_0)) = \mathcal{E}_\varepsilon^L(s_0)$, and therefore condition $\mathcal{C}(\alpha_3\varepsilon, L, s_0, \mathcal{T}_{col}^{\varepsilon,+}(s_0))$ of Subsection 3.3.4 would hold. Invoking Proposition 3.36, this would imply that condition $\mathcal{WP}^L(\Lambda_{log}(\alpha_3\varepsilon), \tau)$ holds for $\tau \in (s_0 + \varepsilon^{\omega+2}, \mathcal{T}_{col}^{\varepsilon,+}(s_0))$, so that in particular

$$d_{min}^\varepsilon(\mathcal{T}_{col}^{\varepsilon,-}(s_0)) \geq \Lambda_{log}(\alpha_3\varepsilon).$$

It suffices thus to choose α_3 sufficiently big so that

$$\Lambda_{log}(\alpha_3\varepsilon) > \max(\beta_0, 8q_1(\alpha_2))\varepsilon,$$

and the contradiction then follows from (3.195). $\qquad\qquad\square$

3.7.3 Proof of Proposition 3.20

We will fix the values of κ_* and ρ_* in the course of the proofs, as the smallest numbers which satisfy a finite number of lower bound inequalities.

First, recall that it follows from (3.53) and (3.7) that if $0 \leq s \leq \rho_0(R-r)^{\omega+2}$ then

$$\mathfrak{D}_\varepsilon(s) \cap I_{4L} \subset \bigcup_{k \in J_0} (b_k^\varepsilon - R, b_k^\varepsilon + R) \subset I_{2\kappa_0 L}, \text{ where the union is disjoint,} \quad (3.197)$$

and in particular $\mathcal{C}_{L,s}$ holds, where

$$S := \rho_0(R-r)^{\omega+2} \geq \rho_0 \left(\frac{R}{2}\right)^{\omega+2}.$$

Having (3.77) in mind, and in view of (3.197) and (3.54), we estimate

$$100 C_e L^{-(\omega+2)} S \leq 100 C_e \left(\frac{R}{2L}\right)^{\omega+2} S \leq 100 C_e \alpha_*^{-(\omega+2)} \leq \frac{\mu_1}{4},$$

where the last inequality follows provided we choose α_* sufficiently large. As a consequence, the function $\mathfrak{E}^L_\varepsilon$ is non-increasing on the set of times s in the interval $[\varepsilon^\omega L^2, S]$ where $\mathcal{WP}^L_\varepsilon(\alpha_1 \varepsilon, s)$ holds.

For such times s, the front points $\{a^\varepsilon_k(s)\}_{k \in J(s)}$ are well defined, and for $q \in J_0$, we have defined in the introduction $J_q(s) = \{k \in J(s), a^\varepsilon_k(s) \in [b^\varepsilon_q - R, b^\varepsilon_q + R]\}$, and we have set $\ell_q = \sharp J_q$ and $J_q(s) = \{k_q, k_{q+1}, \ldots, k_{q+\ell_q-1}\}$, where $k_1 = 1$, and $k_q = \ell_1 + \cdots + \ell_{q-1} + 1$, for $q \geq 2$.

Step 1. Annihilations of all the pairs of fronts–anti-fronts We claim that there exists some time $\tilde{s} \in (\varepsilon^\omega L^2, \frac{1}{2} S)$ such that $\mathcal{WP}^L_\varepsilon(\delta^\varepsilon_{\log}, \tilde{s})$ holds and such that for any $q \in J_0$, $\epsilon_{k+\frac{1}{2}}(\tilde{s}) = +1$, for $k \in J_q(\tilde{s}) \setminus \{k_q(\tilde{s}) + \ell_q(\tilde{s}) - 1\}$ or $\sharp J_q(\tilde{s}) \leq 1$, or equivalently that $\dagger_k(\tilde{s}) = \dagger_{k'}(\tilde{s})$ for k and k' in the same $J_q(\tilde{s})$. In particular, $d^{\varepsilon,-}_{\min}(\tilde{s}) \geq 2R$.

Proof of the claim If we require α_* to be sufficiently large, then by (3.54) we have that $\varepsilon^\omega L^2 + \varepsilon^{\omega+1} L \leq S/2$, and therefore by Proposition 3.13 we may choose a first time

$$s_0 \in [\varepsilon^\omega L^2, \varepsilon^\omega L^2 + \varepsilon^{\omega+1} L] \quad \text{such that} \quad \mathcal{WP}^L_\varepsilon(\delta^\varepsilon_{\log}, s_0) \text{ holds.}$$

Actually, we have

$$s_0 \leq \varepsilon^\omega L^2 + \varepsilon^{\omega+1} L \leq 2\varepsilon^\omega L^2 \leq 2\alpha_*^{-(\omega+2)} r^{\omega+2} \leq 2\alpha_*^{-2(\omega+2)} R^{\omega+2}$$

$$\leq \frac{1}{\rho_0} 2^{\omega+3} \alpha_*^{-2(\omega+2)} S. \tag{3.198}$$

Note that by (3.7) we have the inclusion

$$\mathfrak{D}_\varepsilon(s_0) \cap I_L \subset \mathcal{N}(b, r_0),$$

where

$$r_0 = r + \left(\frac{s_0}{\rho_0} \right)^{\frac{1}{\omega+2}} \leq 2r,$$

provided once more that α_* is sufficiently large, and where for $\rho > 0$ we have set $\mathcal{N}(b, \rho) = \cup_{q \in J_0} [b^\varepsilon_j - \rho, b^\varepsilon_j + \rho]$. In view of the confinement condition (3.53) only two cases can occur:

(i) $\quad d^{\varepsilon,-}_{\min}(s_0) \geq 3R - 2r_0 \quad$ or $\quad$ (ii) $\quad d^{\varepsilon,-}_{\min}(s_0) \leq 2r_0$.

If case (i) occurs, then, for any $q \in J_0$, we have $\epsilon_{k+\frac{1}{2}} = +1$, for any $k \in J_q(\tau_1) \setminus \{k_q(s_0) + \ell_q(s_0) - 1\}$. Choosing $\tilde{s} = s_0$, Step 1 is completed in the case considered.

If instead case (ii) occurs, then we will make use of Proposition 3.55 to remove the small pairs of fronts–anti-fronts present at small scales. More precisely, assume that for some $j \geq 0$ we have constructed $0 \leq s_j \leq S$ and $r_j > 0$ such that $\mathcal{WP}_\varepsilon^L(\delta_{\log}^\varepsilon, s_j)$ holds, such that we have

$$\mathfrak{D}_\varepsilon(s_j) \cap I_L \subset \mathcal{N}(b, r_j), \qquad \mathfrak{E}_\varepsilon^L(s_j) \leq \mathfrak{E}_\varepsilon^L(s_0) - j\mu_1, \tag{3.199}$$

as well as the estimates

$$r_0 \leq r_j \leq \gamma^j r_0 \leq \frac{R}{2}, \qquad s_j \leq s_0 + (2^{\omega+2}\mathcal{K}_{\mathrm{col}} + 1)\frac{\gamma^{j(\omega+2)} - 1}{\gamma^{\omega+2} - 1}r_0^{\omega+2} \leq \frac{S}{2}, \tag{3.200}$$

where $\gamma := \left(2 + 2\left(\frac{\mathcal{K}_{\mathrm{col}}}{\rho_0}\right)^{\frac{1}{\omega+2}}\right)$, and moreover that case (ii) holds at step j, that is

$$d_{\min}^{\varepsilon,-}(s_j) \leq 2r_j \leq R. \tag{3.201}$$

Let $\tilde{s}_j := \mathcal{T}_{\mathrm{col}}^{\varepsilon,+}(s_j)$ be given by Proposition 3.55 (the confinement condition holds in view of (3.197) and we have $\delta_{\log}^\varepsilon \geq \alpha_3\varepsilon$ provided α_* is sufficiently large), and then let $s_{j+1} \in [\tilde{s}_j, \tilde{s}_j + \varepsilon^{\omega+1}L]$ satisfying $\mathcal{WP}_\varepsilon^L(\delta_{\log}^\varepsilon, s_{j+1})$ be given by Proposition 3.13. In particular, we have

$$\mathfrak{E}_\varepsilon^L(s_{j+1}) \leq \mathfrak{E}_\varepsilon^L(\tilde{s}_j) \leq \mathfrak{E}_\varepsilon^L(s_j) - \mu_1 \leq \mathfrak{E}_\varepsilon^L(s_0) - (j+1)\mu_1. \tag{3.202}$$

Since

$$s_{j+1} - s_j \leq \mathcal{K}_{\mathrm{col}}(2r_j)^{\omega+2} + \varepsilon^{\omega+1}L \leq (2^{\omega+2}\mathcal{K}_{\mathrm{col}} + 1)r_j^{\omega+2},$$

we have, in view of (3.200),

$$s_{j+1} \leq s_0 + (2^{\omega+2}\mathcal{K}_{\mathrm{col}} + 1)\left[\frac{\gamma^{j(\omega+2)} - 1}{\gamma^{\omega+2} - 1} + \gamma^{j(\omega+2)}\right]r_0^{\omega+2}$$

$$\leq s_0 + (2^{\omega+2}\mathcal{K}_{\mathrm{col}} + 1)\frac{\gamma^{(j+1)(\omega+2)} - 1}{\gamma^{\omega+2} - 1}r_0^{\omega+2}, \tag{3.203}$$

and by (3.7) $\mathfrak{D}_\varepsilon(s_{j+1}) \cap I_L \subset \mathcal{N}(b, r_{j+1})$, where

$$r_{j+1} = r_j + 2\left(\frac{\mathcal{K}_{\mathrm{col}}}{\rho_0}\right)^{\frac{1}{\omega+2}}r_j + \frac{1}{\rho_0}\varepsilon\left(\frac{L}{\varepsilon}\right)^{\frac{1}{\omega+2}} \leq \left(2 + 2\left(\frac{\mathcal{K}_{\mathrm{col}}}{\rho_0}\right)^{\frac{1}{\omega+2}}\right)r_j = \gamma r_j. \tag{3.204}$$

In view of (3.198) and (3.54), we also have

$$\gamma^{j+1}r_0 \leq 2\gamma^{j+1}\alpha_*^{-1}R \tag{3.205}$$

and

$$s_0 + (2^{\omega+2}\mathcal{K}_{\mathrm{col}} + 1)\frac{\gamma^{(j+1)(\omega+2)} - 1}{\gamma^{\omega+2} - 1}r_0^{\omega+2}$$

$$\leq \left[\frac{2^{\omega+3}}{\rho_0}\alpha_*^{-(\omega+2)} + \frac{2^{2\omega+4}}{\rho_0}(2^{\omega+2}\mathcal{K}_{\mathrm{col}} + 1)\frac{\gamma^{(j+1)(\omega+2)} - 1}{\gamma^{\omega+2} - 1}\alpha_*^{-(\omega+2)}\right]S. \tag{3.206}$$

It follows from (3.203), (3.204), (3.205) and (3.206) that if α_* is sufficiently large (depending on only M_0, V and j), then (3.200) holds also for s_{j+1}. As above we distinguish two cases:

(i) $\quad d_{\min}^{\varepsilon,-}(s_{j+1}) \geq 3R - 2r_{j+1}$ $\quad$ or $\quad$ (ii) $\quad d_{\min}^{\varepsilon,-}(s_{j+1}) \leq 2r_{j+1}$.

If case (i) holds then by (3.200) we have $d_{\min}^{\varepsilon,-}(s_{j+1}) \geq 2R$, we set $\tilde{s} = s_{j+1}$ which therefore satisfies the requirements of the claim, and we proceed to Step 2.

If case (ii) occurs then we proceed to construct s_{j+2} as above. The key fact in this recurrence construction is the second inequality in (3.199), which, since $\mathcal{E}_\varepsilon^L(s_j) \geq 0$ independently of j, implies that *the process has to reach case (i) in a number of steps less than or equal to M_0/μ_1*. In particular, choosing the constant α_* sufficiently big so that the right-hand side of (3.205) is smaller than $R/2$ for all $0 \leq j \leq M_0/\mu_1$ and so that the right-hand side of (3.206) is smaller than $S/2$ for all $0 \leq j \leq M_0/\mu_1$ ensures that the construction was licit and that the process necessarily reaches case (i) before it could reach $j = M_0/\mu_1 + 1$, so defining $\tilde{s}$ as above. $\qquad\square$

Step 2: Divergence of the remaining repulsing fronts at small scale At this stage we have constructed $\tilde{s} \in [\varepsilon^\omega L^2, \frac{1}{2}S]$ which satisfies the requirements of the claim in Step 1. Moreover, note that in view of (3.198) and (3.200) we have the upper bound

$$\tilde{s} \leq \left(2\alpha_*^{-(\omega+2)} + 2^{\omega+2}(2^{\omega+2}\mathcal{K}_{\text{col}} + 1) \frac{\gamma^{\frac{M_0}{\mu_1}(\omega+2)} - 1}{\gamma^{\omega+2} - 1} \right) r^{\omega+2}. \qquad (3.207)$$

In order to complete the proof, we next distinguish two cases:

(i) $\quad \sharp J_q(\tilde{s}) \leq 1$, for any $q \in J_0$. $\qquad$ (ii) $\quad \sharp J_{q_0}(\tilde{s}) > 1$, for some $q_0 \in J_0$.

If case (i) holds, then we actually have

$$d_{\min}^{\varepsilon,L}(\tilde{s}) \geq 2R. \qquad (3.208)$$

Since $2R \geq 16\mathfrak{q}_1(\delta_{\log}^\varepsilon)\varepsilon$ when α_* is sufficiently large, it follows from Corollary 3.38 that $\mathcal{WP}_\varepsilon^L(\delta_{\log\log}^\varepsilon, s)$ holds for any $\tilde{s} + \varepsilon^{2+\omega} \leq s \leq \mathcal{T}_0^\varepsilon(\delta_{\log}^\varepsilon, \tilde{s})$, where

$$\mathcal{T}_0^\varepsilon(\delta_{\log}^\varepsilon, \tilde{s}) = \max \left\{ \tilde{s} + \varepsilon^{2+\omega} \leq s \leq S \text{ s.t. } d_{\min}^{\varepsilon,L}(s') \geq 8\mathfrak{q}_1(\delta_{\log}^\varepsilon)\varepsilon \right.$$
$$\left. \forall s' \in [\tilde{s} + \varepsilon^{\omega+2}, s] \right\}.$$

In particular, $\mathcal{WP}_\varepsilon^L(\delta_{\log\log}^\varepsilon, s)$ holds for any s in $\tilde{s} + \varepsilon^{2+\omega} \leq s \leq \mathcal{T}_3^\varepsilon(\delta_{\log}^\varepsilon, \tilde{s})$, where

$$\mathcal{T}_3^\varepsilon(\delta_{\log}^\varepsilon, \tilde{s}) = \max \left\{ \tilde{s} + \varepsilon^{2+\omega} \leq s \leq S \text{ s.t. } d_{\min}^{\varepsilon,L}(s') \geq R \ \forall s' \in [\tilde{s} + \varepsilon^{\omega+2}, s] \right\}.$$

In view of (3.208) and estimate (3.7), we obtain the lower bound

$$T_3^\varepsilon(\delta_{\log}^\varepsilon, \tilde{s}) \geq \tilde{s} + \rho_0 R^{\omega+2}. \qquad (3.209)$$

Note that (3.209) and (3.54) also yield

$$T_3^\varepsilon(\delta_{\log}^\varepsilon, \tilde{s}) \geq \tilde{s} + \rho_0 \alpha_*^{-1} r^{\omega+2} \geq \rho_0 \alpha_*^{-1} r^{\omega+2}. \qquad (3.210)$$

Combining (3.207) and (3.210), we deduce in particular that

$$\mathcal{WP}_\varepsilon^L(\alpha_1 \varepsilon, s_r) \text{ holds} \qquad \text{and} \qquad d_{\min}^{\varepsilon,L}(s_r) \geq R \geq r,$$

which is the claim of Proposition 3.20, provided that

$$\rho_* \geq \left(3 + 2^{\omega+2}(2^{\omega+2}\mathcal{K}_{\mathrm{col}} + 1)\frac{\gamma^{\frac{M_0}{\mu_1}(\omega+2)} - 1}{\gamma^{\omega+2} - 1}\right) \qquad \text{and} \qquad \rho_* \leq \rho_0 \alpha_*^{-1}. \qquad (3.211)$$

It remains to consider the situation where case (ii) holds. In that case, we have

$$d_{\min}^{\varepsilon,-}(\tilde{s}) \geq 2R \qquad \text{and} \qquad d_{\min}^{\varepsilon,+}(\tilde{s}) \leq 2\gamma^{M_0/\mu_1} r \leq R,$$

so that we are in a situation suited for Proposition 3.57. We may actually apply Proposition 3.57 recursively with $s_0 := \tilde{s}$ and $\hat{s} \equiv \hat{s}_k = (T_2^\varepsilon)^k(\hat{s}_0)$, where $\hat{s}_0 = \tilde{s} + \varepsilon^{\omega+2}$, as long as $d_{\min}^{\varepsilon,+}(\hat{s}_k)$ remains sufficiently small with respect to R (say e.g. $d_{\min}^{\varepsilon,+}(\hat{s}_k) \leq \alpha_*^{-\frac{1}{2}} R$ provided α_* is chosen sufficiently large), the details are completely similar to the ones in Case II of Proposition 3.58 and are therefore not repeated here. If we denote by k_0 the first index for which $d_{\min}^{\varepsilon,+}(\hat{s}_{k_0})$ becomes larger than $\frac{2}{S_2} r$ (in view of (3.160)) and k_1 the last index before $d_{\min}^{\varepsilon,+}$ reaches $\alpha_*^{-\frac{1}{2}} R$, then we have

$$\hat{s}_{k_0} \leq C r^{\omega+2} \qquad \text{and} \qquad \hat{s}_{k_1} \geq \frac{1}{C}\alpha_*^{-(\omega+2)/2} R^{\omega+2} \geq \frac{1}{C}\alpha_*^{(\omega+2)/2} r^{\omega+2},$$

for some constant $C > 0$ depending on only M_0 and V, and the conclusion that $\mathcal{WP}_\varepsilon^L(\alpha_1 \varepsilon, s_r)$ holds follows as in case (i) above, choosing first ρ_* sufficiently large (independently of α_*) and then α_* sufficiently large (given ρ_*). $\qquad \square$

3.8 Proofs of Theorems 3.3, 3.5 and 3.9

3.8.1 Proof of Theorem 3.3

Theorem 3.3 being essentially a special case of Theorem 3.5, we go directly to the proof of Theorem 3.5. Notice, however, that in Theorem 3.3 the solution to the limiting system is unique, so that the result is not constrained by the need to pass to a subsequence.

3.8.2 Proof of Theorem 3.5

We fix $S < S_{\max}$ and let $L \geq \kappa_*^{-1} L_0$, where L_0 is defined in the statement of Proposition 3.19 and κ_* in the statement of Proposition 3.20. We set $R = \frac{1}{2}\min\{a_{k+1}^0 - a_k^0, k = 1, \ldots, \ell_0 - 1\}$ and consider an arbitrary $0 < r < R/\alpha_*$. Since (H_1) holds, there exists some constant $\varepsilon_r > 0$ such that, if $0 < \varepsilon \leq \varepsilon_r$, then (3.53) holds with $b_k \equiv a_k^0$ for any $k \in \{1, \ldots, \ell_0\}$. We are therefore in position to make use of Proposition 3.20 and assert that for all such ε condition $\mathcal{WP}_\varepsilon^L(\alpha_1 \varepsilon, s_r)$ holds as well as (3.55) and (3.56). It follows in particular from (3.55) and (3.56) that for every $k \in 1, \ldots, \ell_0$ we have $\sharp J_k(s_r) = |m_k^0|$, where m_k^0 is defined in (3.12), and therefore $\sharp J(s_r) = \sum_{k=1}^{\ell_0} |m_k^0| \equiv \ell_1$, in other words the number of fronts as well as their properties do not depend on ε nor on r.

We construct next the limiting splitting solution to the ordinary differential equation and the corresponding subsequence proceeding backwards in time and using a diagonal argument. For that purpose, we introduce an arbitrary decreasing sequence $\{r_m\}_{m\in\mathbb{N}_*}$ such that $0 < r_1 \leq R/\alpha_*$, and such that $r_m \to 0$ as $m \to +\infty$. For instance, we may take $r_m = \frac{1}{m}R/\alpha_*$, and we set $s_m = s_{r_m}$. Taking first $m = 1$, we find a subsequence $\{\varepsilon_{n,1}\}_{n\in\mathbb{N}_*}$ such that $\varepsilon_{n,1} \to 0$ as $n \to \infty$, and such that all points $\{a_k^{\varepsilon_{n,1}}(s_1)\}_{k\in J}$ converge to some limits $\{a_k^1(s_1)\}_{k\in J}$ as $n \to +\infty$. It follows from (3.56), passing to the limit $n \to +\infty$, that

$$d_{\min}^*(s_1) \geq r_1. \tag{3.212}$$

We are therefore in position to apply the convergence result of Proposition 3.19, which yields in particular that

$$\mathfrak{D}_{\varepsilon_{n,1}}(s) \cap I_{4L} \longrightarrow \cup_{k=1}^{\ell_1}\{a_k^1(s)\} \qquad \forall s_1 < s < S_{\max}^1, \tag{3.213}$$

as $n \to +\infty$, where $\{a_k^1(\cdot)\}_{k\in J}$ is the unique solution of $(\mathcal{S})$ with initial data $\{a_k^1(s_1)\}_{k\in J}$ on its maximal time of existence $(s_1, S_{\max}^1)$.

Taking next $m = 2$, we may extract a subsequence $\{\varepsilon_{n,2}\}_{n\in\mathbb{N}_*}$ from the sequence $\{\varepsilon_{n,1}\}_{n\in\mathbb{N}_*}$ such that all the points $\{a_k^{\varepsilon_{n,2}}(s_2)\}_{k\in J}$ converge to some limits $\{a_k^2(s_2)\}_{k\in J}$ as $n \to +\infty$. Arguing as above, we may assert that

$$\mathfrak{D}_{\varepsilon_{n,2}}(s) \cap I_{4L} \longrightarrow \cup_{k=1}^{\ell_1}\{a_k^2(s)\} \qquad \forall s_2 < s < S_{\max}^2, \tag{3.214}$$

as $n \to +\infty$, where $\{a_k^2(\cdot)\}_{k\in J}$ is the unique solution of $(\mathcal{S})$ with initial data $\{a_k^2(s_2)\}_{k\in J}$ on its maximal time of existence $(s_2, S_{\max}^2)$. It follows from (3.55), namely that only repulsive forces are present at scale smaller than R, that $S_{\max}^2 \geq s_1$. Therefore, since we have extracted a subsequence, it follows from (3.213) and (3.214) that $a_k^2(s_1) = a_k^1(s_1)$ for all $k \in J$, and therefore also that $S_{\max}^2 = S_{\max}^1 \equiv S_{\max}$ and $a_k^2(\cdot) = a_k^1(\cdot) = a_k(\cdot)$ on $(s_2, S_{\max})$.

We proceed similarly at each step $m \in \mathbb{N}_*$, extracting a subsequence $\{\varepsilon_{n,m}\}_{n \in \mathbb{N}_*}$ from the sequence $\{\varepsilon_{n,m-1}\}_{n \in \mathbb{N}^*}$ such that all the points $\{a_k^{\varepsilon_{n,m}}(s_m)\}_{k \in J}$ converge to some limits $\{a_k^m(s_m)\}_{k \in J}$. Finally setting, for $n \in \mathbb{N}_*$, $\varepsilon_n = \varepsilon_{n,n}$, we obtain that

$$\mathfrak{D}_{\varepsilon_n}(s) \cap I_{4L} \longrightarrow \cup_{k=1}^{\ell_1} \{a_k(s)\} \qquad \forall 0 < s < S_{max}^m,$$

where $\{a_k(\cdot)\}_{k \in J}$ is a splitting solution of $(\mathcal{S})$ with initial data $\{a_k^0\}_{k \in J_0}$, on its maximal time of existence $(0, S_{max})$. Since $L \geq L_0$ was arbitrary, it follows that (3.15) holds.

It remains to prove that (3.14) holds. This is actually a direct consequence of (3.15), the continuity of the trajectories $a_k(\cdot)$ and the regularizing effect of the front set stated in Proposition 3.12 (e.g. (3.31) for the L^∞ norm). As a matter of fact, it is standard to deduce from this the fact that the convergence towards the equilibria σ_q, locally outside the trajectory set, holds in any C^m norm, since the potential V was assumed to be smooth. □

3.8.3 Proof of Theorem 3.9

As underlined in the introduction, Theorem 3.9 follows rather directly from Theorem 3.5 and more importantly its consequence, Corollary 3.8 (whose proof is elementary and explained after Proposition 3.7), combined with Theorem 3.2 and Proposition 3.12.

Let therefore $L > L_0$ and $\delta > 0$ be arbitrarily given; we shall prove that, at least for $\varepsilon \equiv \varepsilon_n$ sufficiently small,

$$\mathcal{D}_\varepsilon(S_{max}) \cap I_L \subset \cup_{j \in \{1,\dots,\ell_2\}}[b_j - \delta, b_j + \delta] \tag{3.215}$$

and

$$|\mathfrak{v}_\varepsilon(x, S_{max}) - \sigma_{i(j+\frac{1}{2})}| \leq C(\delta, L)\varepsilon^{\frac{1}{\theta-1}} \tag{3.216}$$

for all $j \in \{0, \dots, \ell_2\}$ and for all $x \in (b_j + \delta, b_{j+1} - \delta)$, where we have used the convention that $b_0 = -L$ and $b_{\ell_2+1} = L$. Since L can be arbitrarily big and δ arbitrarily small, this will imply that assumption (H_1) is verified at times S_{max}, which is the claim of Theorem 3.9.

Concerning (3.215), by Corollary 3.8 there exists

$$s^- \in \left[S_{max} - \rho_0 \left(\frac{\delta}{4} \right)^{\omega+2}, S_{max} \right) \tag{3.217}$$

such that

$$\cup_{k \in \{1,\dots,\ell_1\}} \{a_k(s^-)\} \subset \cup_{j \in \{1,\dots,\ell_2\}}[b_j - \frac{1}{4}\delta, b_j + \frac{1}{4}\delta].$$

This, together with Theorem 3.5 implies that, for ε sufficiently small,

$$\mathcal{D}_\varepsilon(s^-) \cap I_{2L} \cap \left(\cup_{j \in \{1,\dots,\ell_2\}} [b_j - \frac{1}{2}\delta, b_j + \frac{1}{2}\delta] \right)^c = \emptyset. \tag{3.218}$$

In turn, Theorem 3.2 (including (3.7)) and (3.218), combined with the upper bound (3.217) on $S_{\max} - s^-$, imply that

$$\mathcal{D}_\varepsilon(s) \cap I_{\frac{3}{2}L} \cap \left(\cup_{j \in \{1,\dots,\ell_2\}} [b_j - \frac{3}{4}\delta, b_j + \frac{3}{4}\delta] \right)^c = \emptyset, \qquad \forall s \in [s^-, S_{\max}]. \tag{3.219}$$

For $s = S_{\max}$ this is stronger than (3.215).

We proceed to (3.216). In view of (3.219), for any $x_0 \in I_L \setminus \left(\cup_{j \in \{1,\dots,\ell_2\}} [b_j - \delta, b_j + \delta] \right)$ we have, for ε sufficiently small,

$$\mathfrak{v}_\varepsilon(y,s) \in B(\sigma_i, \mu_0) \qquad \forall (y,s) \in [x_0 - \frac{1}{8}\delta, x_0 + \frac{1}{8}\delta] \times [s^-, S_{\max}].$$

The latter is nothing but (3.29) for $r = \frac{1}{8}\delta$, $s_0 = s^-$ and $S = S_{\max}$, and therefore the conclusion (3.216) follows from (3.31) of Proposition 3.12, with $C(\delta, L) = \frac{1}{5}C_e(8/\delta)^{\frac{1}{\theta-1}}$ as soon as $\varepsilon^\omega/(S_{\max} - s^-) \le \delta^2/64$. $\qquad\qquad \square$

Appendix A

In this appendix we establish properties concerning the stationary solutions $\overset{\triangledown}{u}^+$, $\overset{\triangleright}{u}$, $\overset{\triangledown}{u}^+_{\varepsilon,r}$, etc., which we have used in the course of the previous discussion, mainly in Section 3.5.

A.1 The Operator $\mathcal{L}_\mu$

Consider for $\mu > 0$ the non-linear operator $\mathcal{L}_\mu$, defined for a smooth function U on $\mathbb{R}$ by

$$\mathcal{L}_\mu(U) = -\frac{d^2}{dx^2}U + 2\mu\theta\, U^{2\theta-1},$$

and set for simplicity $\mathcal{L} \equiv \mathcal{L}_1$. Most results in this section are deduced from the comparison principle: if u_1 and u_2 are two functions defined on some non-empty interval I, such that

$$\mathcal{L}_\mu(u_1) \ge 0, \ \mathcal{L}_\mu(u_2) \le 0, \text{ and } u_1 \ge u_2 \text{ on } \partial I, \tag{A.1}$$

then $u_1(x) \ge u_2(x)$ for $x \in I$. Scaling arguments are also used extensively. Given $r > 0$ and $\eta > 0$ we provide a rescaling of a given smooth function

U as follows:

$$\begin{cases} U_{\eta,R} = \eta\, U\left(\dfrac{U}{r}\right), \text{ and we verify that} \\[3mm] \mathcal{L}_\mu(U_{\eta,r})(x) = \dfrac{\eta}{r^2}\mathcal{L}_\gamma(U)\left(\dfrac{x}{r}\right), \text{ where } \gamma = \mu\eta^{2(\theta-1)}r^2. \end{cases} \tag{A.2}$$

In particular, if $\mathcal{L}_\mu(U) = 0$, then we have

$$\mathcal{L}_\mu(r^{-\frac{1}{(\theta-1)}}U(\tfrac{\cdot}{r})) = 0 \text{ and } \mathcal{L}_{\lambda\mu}(\lambda^{-\frac{1}{2(\theta-1)}}U) = 0, \text{ for any } r > 0 \text{ and any } \lambda > 0.$$

Notice also that U^* defined on $(0,+\infty)$ by $U^*(x) = [\sqrt{2}(\theta-1)x]^{-\frac{1}{\theta-1}}$ solves $\mathcal{L}(U^*) = 0$.

Lemma 3.59 *There exists a unique smooth map $\overset{\vee+}{u}_r$ on $(-r,r)$ such that $\mathcal{L}(\overset{\vee+}{u}_r) = 0$ and $\overset{\vee+}{u}_r(\pm r) = +\infty$, and a unique solution $\overset{\triangleright}{u}_r$ such that $\mathcal{L}(\overset{\triangleright}{u}_r) = 0$ and $\overset{\triangleright}{u}_r(\pm r) = \pm\infty$. Moreover, $\overset{\vee+}{u}_r$ is even, $\overset{\triangleright}{u}_r$ is odd, and, setting $\overset{\vee+}{u} \equiv \overset{\vee+}{u}_1$ and $\overset{\triangleright}{u} \equiv \overset{\triangleright}{u}_1$, we have*

$$\overset{\vee+}{u}_r(x) = r^{-\frac{1}{\theta-1}}\overset{\vee+}{u}\left(\tfrac{x}{r}\right) \text{ and } \overset{\triangleright}{u}_r(x) = r^{-\frac{1}{\theta-1}}\overset{\triangleright}{u}(\tfrac{x}{r}). \tag{A.3}$$

Proof For $n \in \mathbb{N}^*$, we construct on $(-r,r)$ a unique solution $\overset{\vee+}{u}_{r,n}$ that solves $\mathcal{L}(\overset{\vee+}{u}_{r,n}) = 0$ and $\overset{\vee+}{u}_{r,n}(\pm r) = n$, minimizing the corresponding convex energy. By the comparison principle, $\overset{\vee+}{u}_{r,n}$ is non-negative, increasing with n and uniformly bounded on compact subsets of $(-r,r)$ in view of (A.5). Hence a unique limit $\overset{\vee+}{u}_r$ exists, solution to $\mathcal{L}(\overset{\vee+}{u}_r) = 0$. We observe that $\overset{\vee+}{u}_{r,n}(\cdot) \geq U^*(r_n - \cdot)$, where $r_n = r + [\sqrt{2}(\theta-1)]^{-1}n^{-(\theta-1)}$, so that we obtain the required boundary conditions for $\overset{\vee+}{u}_r$. Uniqueness may again be established thanks to the comparison principle. We construct similarly a unique solution $\overset{\triangleright}{u}_{r,n}$ that solves $\mathcal{L}(\overset{\triangleright}{u}_{r,n}) = 0$ and $\overset{\vee+}{u}_{r,n}(\pm r) = \pm n$. We notice that $\overset{\triangleright}{u}_{r,n}$ is odd, its restriction on $(0,r)$ non-negative and increasing with n. Moreover, on some interval (a,r), where $0 < a < r$ does not depend on n, we have $\overset{\vee+}{u}_{r,n}(\cdot) \geq U^*(\tilde{r}_n - \cdot)$, where $\tilde{r}_n = r + [(\theta-1)]^{-1}n^{-(\theta-1)}$, and we conclude as for the first assertion. $\square$

Remark 3.60 Given $r > 0$ and $\lambda > 0$ we notice that the functions $\overset{\vee}{U}_r^\lambda$ and $\overset{\triangleright}{U}_r^\lambda$ defined by

$$\overset{\vee}{U}_r^\lambda(x) = \lambda^{-\frac{1}{2(\theta-1)}}\overset{\vee+}{u}_r(x) \text{ and } \overset{\triangleright}{U}_r^\lambda(x) = \lambda^{-\frac{1}{2(\theta-1)}}\overset{\triangleright}{u}_r(x) \tag{A.4}$$

solve $\mathcal{L}_\lambda(\overset{\vee}{U}{}_r^\lambda) = 0$ and $\mathcal{L}_\lambda(\overset{\triangleright}{U}{}_r^\lambda) = 0$ with the same boundary conditions as $\overset{\vee+}{u}_r$ and $\overset{\triangleright}{u}_r$.

Lemma 3.61 *(i) Assume that $\mathcal{L}(u) \le 0$ on $(-r, r)$. Then, we have, for $x \in (-r, r)$,*

$$u(x) \le \left(\sqrt{2}(\theta - 1)\right)^{-\frac{1}{\theta-1}} \left[(x+r)^{-\frac{1}{\theta-1}} + (x-r)^{-\frac{1}{\theta-1}}\right]. \tag{A.5}$$

(ii) Assume that $\mathcal{L}(u) \ge 0$ on $(-r, r)$ and that $u(-r) = u(r) = +\infty$. Then we have

$$u(x) \ge \left(\sqrt{2}(\theta - 1)\right)^{-\frac{1}{\theta-1}} \max\{(x+r)^{-\frac{1}{\theta-1}}, (r-x)^{-\frac{1}{\theta-1}}\}. \tag{A.6}$$

Proof Set $\tilde{U} = U^*(\cdot + r) + U^*(r - \cdot)$. By subaddivity and translation invariance, we have $\mathcal{L}(\tilde{U}) \ge 0$ on $(-r, r)$ with $\tilde{U}(\pm r) = +\infty$, so that (A.5) follows from the comparison principle (A.1) with $u_1 = \tilde{U}$ and $u_2 = u$. Similarly, (A.6) follows from (A.1) with $u_1 = u$ and $u_2 = U^*(\cdot + r)$ or $u_2 = U^*(r - \cdot)$. □

Combining estimate (ii) of Lemma 3.61 with the scaling law of Lemma 3.59, we are led to

$$\left|\frac{d}{dr}\overset{\vee+}{u}_r(x)\right| + \left|\frac{d}{dr}\overset{\triangleright}{u}_r(x)\right| \le Cr^{-\frac{\theta}{\theta-1}}, \quad \text{for } x \in \left(-\frac{7}{8}r, \frac{7}{8}r\right). \tag{A.7}$$

A.2 The Discrepancy for $\mathcal{L}_\mu$

The discrepancy Ξ_μ for $\mathcal{L}_\mu$ relates to a given function u the function $\Xi_\mu(u)$ defined by

$$\Xi_\mu(u) = \frac{\dot{u}^2}{2} - \mu u^{2\theta}. \tag{A.8}$$

This function is *constant* if u solves $\mathcal{L}_\mu(u) = 0$. We set $\Xi = \Xi_1$,

$$A_\theta \equiv \Xi(\overset{\vee+}{u}) = -(\overset{\vee+}{u}(0))^{2\theta} < 0 \text{ and } B_\theta \equiv \Xi(\overset{\triangleright}{u}) = \frac{(\overset{\triangleright}{u}(0))_x^2}{2} > 0. \tag{A.9}$$

In view of the scaling relations (A.3) and Remark 3.60, we are led to the identities

$$\begin{cases} \Xi_\lambda(\overset{\vee}{U}{}_r^\lambda) = \lambda^{-\frac{1}{\theta-1)}} r^{-\frac{2\theta}{\theta-1}} A_\theta = \lambda^{-\frac{1}{\theta-1)}} r^{-(\omega+1)} A_\theta, \\ \Xi_\lambda(\overset{\triangleright}{U}{}_r^\lambda) = \lambda^{-\frac{1}{\theta-1}} r^{-\frac{2\theta}{\theta-1}} B_\theta = \lambda^{-\frac{1}{\theta-1}} r^{-(\omega+1)} B_\theta. \end{cases} \tag{A.10}$$

A.3 The Operator $\mathcal{L}^{\varepsilon}$

In this subsection, we consider more generally, for given $\lambda > 0$, the operator $\mathcal{L}^{\varepsilon}$ given by

$$\mathcal{L}^{\varepsilon}(U) = -\frac{d^2}{dx^2}U + 2\lambda f_{\varepsilon}(U),$$

with f_{ε} defined in (3.88), and the solutions $\overset{\vee+}{u}_{\varepsilon,r}$, $\overset{\vee-}{u}_{\varepsilon,r}$, and $\overset{\triangleright}{u}_{\varepsilon,r}$ of $\mathcal{L}^{\varepsilon}(U) = 0$ on $(-r, r)$ with corresponding infinite boundary conditions, whose existence and uniqueness are proved as for Lemma 3.59.

Lemma 3.62 *We have the estimates*

$$|\overset{\vee+}{u}_{\varepsilon,r}(x)| + |\overset{\triangleright}{u}_{\varepsilon,r}(x)| \le C\left(\lambda^2(\theta-1)\right)^{-\frac{1}{\theta-1}}\left[(x+r)^{-\frac{1}{\theta-1}} + (x-r)^{-\frac{1}{\theta-1}}\right].$$

Proof It follows from (3.89) that $\mathcal{L}_{\frac{3}{4}\lambda}(\overset{\vee+}{u}_{\varepsilon,r}) \le 0$, so that, invoking the comparison principle as well as the scaling law (A.2) we deduce that $\overset{\vee+}{u}_{\varepsilon,r} \le (3\lambda/4)^{-2(\theta-1)}\overset{\vee+}{u}_{r}$. A similar estimate holds for $\overset{\triangleright}{u}_{\varepsilon,r}$ and the conclusion follows from Lemma 3.61. $\qquad\square$

We complete this appendix by comparing the solutions $\overset{\vee+}{u}_{r}$ and $\overset{\vee+}{u}_{\varepsilon,r}$, as well as $\overset{\triangleright}{u}_{r}$ and $\overset{\triangleright}{u}_{\varepsilon,r}$.

Proposition 3.63 *In the interval $(-\frac{7}{8}r, \frac{7}{8}r)$ we have the estimate*

$$|\overset{\vee+}{u}_{\varepsilon,r} - \overset{\vee+}{u}_{r}| \le C\varepsilon^{\frac{1}{\theta}}r^{-\frac{2\theta-1}{\theta(\theta-1)}}.$$

Proof Let $\varepsilon < \delta < r/16$ to be fixed. It follows from Lemma 3.62 that, for $x \in (-r+\delta, r-\delta)$, we have

$$0 \le \varepsilon^{\frac{1}{\theta-1}}\overset{\vee+}{u}_{\varepsilon,r}(x) \le C\left(\frac{\varepsilon}{\delta}\right)^{\frac{1}{\theta-1}},$$

and therefore also

$$\left|\varepsilon^{\frac{1}{\theta-1}}\overset{\vee+}{u}_{\varepsilon,r}(x)g\left(\varepsilon^{\frac{1}{\theta-1}}\overset{\vee+}{u}_{\varepsilon,r}(x)\right)\right| \le C\left(\frac{\varepsilon}{\delta}\right)^{\frac{1}{\theta-1}}. \tag{A.11}$$

It follows from (A.11) and the fact that $\mathcal{L}^{\varepsilon}(\overset{\vee+}{u}_{\varepsilon,r}) = 0$, that $\mathcal{L}_{\lambda_{\varepsilon}^-}(\overset{\vee+}{u}_{\varepsilon,r}) \le 0$, where $\lambda_{\varepsilon}^{\pm} = \lambda(1 \pm C(\frac{\varepsilon}{\delta})^{\frac{1}{\theta-1}})$. On the other hand, by the scaling law (A.2), we have

$$\mathcal{L}_{\lambda_{\varepsilon}^-}\left(\left(\frac{\lambda_{\varepsilon}^-}{\lambda}\right)^{-\frac{1}{2(\theta-1)}}\overset{\vee+}{u}_{r-\delta}\right) = 0.$$

It follows from the comparison principle, since the second function is infinite on the boundary of the interval $[-r+\delta, r-\delta]$, that

$$\overset{v+}{u}_{\varepsilon,r} \leq \left(\frac{\lambda_\varepsilon^-}{\lambda}\right)^{-\frac{1}{2(\theta-1)}} \overset{v+}{u}_{r-\delta} \quad \text{on } [-r+\delta, r-\delta].$$

Integrating the inequality (A.7) between $r-\delta$ and r, we deduce that for $x \in (-\frac{7}{8}r, \frac{7}{8}r)$, we have the inequality

$$|\overset{v+}{u}_{r-\delta}(x) - \overset{v+}{u}_r(x)| \leq C\delta r^{-\frac{\theta}{\theta-1}}. \tag{A.12}$$

On the other hand, it follows from estimate (A.5) of Lemma 3.61 that for $x \in (-\frac{7}{8}r, \frac{7}{8}r)$,

$$\left|\left(\frac{\lambda_\varepsilon^-}{\lambda}\right)^{-\frac{1}{2(\theta-1)}} \overset{v+}{u}_{r-\delta}(x) - \overset{v+}{u}_{r-\delta}(x)\right| \leq C\left(\frac{\varepsilon}{\delta}\right)^{\frac{1}{\theta-1}} r^{-\frac{1}{\theta-1}}. \tag{A.13}$$

We optimize the outcome of (A.12) and (A.13) by choosing $\delta := \varepsilon^{\frac{1}{\theta}} r^{\frac{\theta-1}{\theta}}$ and we therefore obtain

$$\overset{v+}{u}_{\varepsilon,r} \leq \overset{v+}{u}_r + C\varepsilon^{\frac{1}{\theta}} r^{-\frac{2\theta-1}{\theta(\theta-1)}} \quad \text{on } (-\frac{7}{8}r, \frac{7}{8}r).$$

The lower bound for $\overset{v+}{u}_{\varepsilon,r}$ is obtained in a similar way but reversing the role of super- and subsolutions: the function $\left(\frac{\lambda_\varepsilon^+}{\lambda}\right)^{-\frac{1}{2(\theta-1)}} \overset{v+}{u}_{r+\delta}$ yields a subsolution for $\mathcal{L}^\varepsilon$ on $[-r,r]$ whereas $\overset{v+}{u}_{\varepsilon,r}$ is a solution. The conclusion then follows. $\square$

Similarly, we have the following.

Proposition 3.64 *In the interval* $(-\frac{7}{8}r, \frac{7}{8}r)$ *we have the estimate*

$$|\overset{\triangleright}{u}_{\varepsilon,r} - \overset{\triangleright}{u}_r| \leq C\varepsilon^{\frac{1}{\theta}} r^{-\frac{2\theta-1}{\theta(\theta-1)}}.$$

Proof We only sketch the necessary adaptations since the argument is closely parallel to the proof of Proposition 3.63. First, by the maximum principle $\overset{\triangleright}{u}_{\varepsilon,r}$ can only have negative maximae and positive minimae, so that actually $\overset{\triangleright}{u}_{\varepsilon,r}$ has no critical point and a single zero, which we call a_ε. Arguing as in Proposition 3.63, one first obtains

$$\left(\lambda_\varepsilon^-/\lambda\right)^{-\frac{1}{2(\theta-1)}} \overset{\triangleright}{u}_{r-\delta-a_\varepsilon}(\cdot - a_\varepsilon) \geq \overset{\triangleright}{u}_{\varepsilon,r} \geq \left(\lambda_\varepsilon^+/\lambda\right)^{-\frac{1}{2(\theta-1)}} \overset{\triangleright}{u}_{r+\delta-a_\varepsilon}(\cdot - a_\varepsilon)$$
$$\text{on } [a_\varepsilon, r-\delta],$$

and

$$-\left(\lambda_\varepsilon^-/\lambda\right)^{-\frac{1}{2(\theta-1)}} \overset{\triangleright}{u}_{r+a_\varepsilon-\delta}(\cdot - a_\varepsilon) \geq -\overset{\triangleright}{u}_{\varepsilon,r} \geq -\left(\lambda_\varepsilon^+/\lambda\right)^{-\frac{1}{2(\theta-1)}} \overset{\triangleright}{u}_{r+a_\varepsilon+\delta}(\cdot - a_\varepsilon)$$
$$\text{on } [-r+\delta, a_\varepsilon].$$

Since $\overset{\triangleright}{u}_{\varepsilon,r}$ is continuously differentiable at the point a_ε (indeed it solves $\mathcal{L}^\varepsilon(\overset{\triangleright}{u}_{\varepsilon,r}) = 0$), and since the derivative at zero of the function $\overset{\triangleright}{u}_r$ is a decreasing function of r, it first follows from the last two sets of inequalities that $|a_\varepsilon| \leq \delta$, and the conclusion then follows as in Proposition 3.63.   □

Appendix B

B.1 Some Properties of the Ordinary Differential Equation ($\mathcal{S}$)

This appendix is devoted to properties of the system of ordinary differential equations ($\mathcal{S}$), the result being somewhat parallel to the results in Section 2 of [5]. We assume that we are given $\ell \in \mathbb{N}^*$, and a solution, for $k \in J = \{1, \ldots, \ell\}$ $t \mapsto a(t) = (a_1(t), \ldots, a_\ell(t))$ to the system

$$\mathfrak{q}_k \frac{\mathrm{d}}{\mathrm{d}s} a_k(s) = -\mathcal{B}_{(k-\frac{1}{2})}[(a_k(s) - a_{k-1}(s)]^{-(\omega+1)} + \mathcal{B}_{(k+\frac{1}{2})}[(a_{k+1}(s) - a_k(s)]^{-(\omega+1)},$$
                                                                    (B.1)

where the numbers $\mathfrak{q}_k$ are supposed to be positive, and are actually taken in ($\mathcal{S}$) equal to $\mathfrak{S}_{i(k)}$, whereas the numbers $\mathcal{B}_{k+1/2}$, which may have positive or negative signs, are taken in ($\mathcal{S}$) to be equal to $\Gamma_{i(k-1/2)}$. We also define $\mathfrak{q}_{min} = \min\{\mathfrak{q}_i\}$ and $\mathfrak{q}_{max} = \max\{\mathfrak{q}_i\}$. We consider a solution on its maximal interval of existence, which we call $[0, T_{max})$. An important feature of the equation (B.1) is its gradient flow structure. The behavior of this system is indeed related to the function F defined on $\mathbb{R}^\ell$ by

$$\begin{cases} F(U) = \sum_{k=0}^{\ell-1} F_{k+1/2}(U), \text{ where, for } k = 0, \ldots, \ell-1, \text{ and } U = (u_1, \ldots, u_\ell), \\ \\ \text{we set} \\ \\ F_{k+\frac{1}{2}}(U) = -\omega^{-1} \mathcal{B}_{k+1/2}(u_{k+1} - u_k)^{-\omega} \text{ with the convention that } u_0 = -\infty. \end{cases}$$

If $u_1 < u_2 < \cdots < u_\ell$, for $k = 1, \ldots, \ell$, then we have

$$\frac{\partial F}{\partial u_k}(U) = \mathcal{B}_{k-1/2}(u_k - u_{k-1})^{-(\omega+1)} - \mathcal{B}_{k+1/2}(u_{k+1} - u_k)^{-(\omega+1)}, \quad (B.2)$$

so that ($\mathcal{S}$) becomes $\dfrac{\mathrm{d}}{\mathrm{d}s} a_k(s) = -\mathfrak{q}_k^{-1} \dfrac{\partial F}{\partial u_k}(a(s))$. Hence, we have

$$\frac{\mathrm{d}}{\mathrm{d}t} F(a(t)) = \sum_{k=1}^{\ell} \frac{\partial F}{\partial u_k}(a(t)) \frac{\mathrm{d}a_k}{\mathrm{d}t}(t) = -\sum_{k=1}^{\ell} \mathfrak{q}_k^{-1} \left(\frac{\partial F}{\partial u_k}(a(t))\right)^2$$

$$\leq -\mathfrak{q}_{max}^{-1} |\nabla F(a(t))|^2, \qquad (B.3)$$

hence F decreases along the flow. We also consider the positive functionals defined by

$$F_{\text{rep}}(U) = \sum_{k \in J^+} F_{k+1/2}(U), \; F_{\text{att}} = -\sum_{k \in J^-} F_{k+1/2}(U), \text{ for } U = (u_1, \ldots, u_\ell),$$

where $J^\pm = \{k \in \{0, \ell - 1\}$ such that $\epsilon_{k+1/2} \equiv \text{sgn}(\mathcal{B}_{k+1/2}) = \pm 1\}$.

Proposition 3.65 *Let* $a = (a_1, \ldots, a_\ell)$ *be a solution to (B.1) on its maximal interval of existence* $[0, T_{\max}]$. *Then, we have, for any time* $t \in [0, T_{\max})$,

$$
\begin{cases}
\left(F_{\text{rep}}(a(t))\right)^{-\frac{\omega+2}{\omega}} \geq \left(F_{\text{rep}}(a(0))\right)^{-\frac{\omega+2}{\omega}} + \mathcal{S}_0 t, \\
\\
\delta_a^+(t) \geq \left(\mathcal{S}_1 t + \mathcal{S}_2 \delta_a^+(0)^{\omega+2}\right)^{\frac{1}{\omega+2}}, \\
\\
(F_{\text{att}}(a(t)))^{-\frac{\omega+2}{\omega}} \leq (F_{\text{att}}(a(0)))^{-\frac{\omega+2}{\omega}} - \mathcal{S}_0 t, \\
\\
\delta_a^-(t) \leq \left(\mathcal{S}_3 \delta_a^-(0)^{\omega+2} - \mathcal{S}_4 t\right)^{\frac{1}{\omega+2}},
\end{cases}
\tag{B.4}
$$

where $\mathcal{S}_0 > 0, \mathcal{S}_1 > 0, \mathcal{S}_2 > 0, \mathcal{S}_3 > 0$ *and* $\mathcal{S}_4 > 0$ *depend on only the coefficients of (B.1).*

Since $\delta_a^-(s) \geq 0$, an immediate consequence of (B.4) is that

$$T_{\max} \leq \frac{\mathcal{S}_3}{\mathcal{S}_4} \delta_a^-(0). \tag{B.5}$$

Since the system (B.1) involves both attractive and repulsive forces, for the proof of Proposition 3.65 it is convenient to divide the collection $\{a_1(t), a_2(t), \ldots, a_\ell(t)\}$ into repulsive and attractive chains. We say that a subset A of J is a *chain* if $A = \{k, k+1, k+2, \ldots, k+m\}$ is an ordered subset of m consecutive elements in J, with $m \geq 1$.

Definition 3.66 *A chain* A *is said to be* repulsive *(resp. attractive) if and only if* $\epsilon_{j+1/2} = -1$ *(resp. +1) for* $j = k, \ldots, k+m$. *It is said to be a* maximal repulsive chain *(resp. maximal attractive chain), if there exists any repulsive (resp. attractive) chain which contains* A *strictly.*

It follows from our definition that a repulsive or attractive chain contains at least two elements. We may decompose J, in increasing order, as

$$J = B_0 \cup A_1 \cup B_1 \cup A_2 \cup B_2 \cup \ldots \cup B_{p-1} \cup A_p \cup B_p, \tag{B.6}$$

where the chains A_i are maximal repulsive chains, the chains B_i are maximal attractive for $i = 1, \ldots, p-1$, and the sets B_0 and B_p being possibly empty or maximal attractive chains. Moreover, for $i = 1 \ldots, p$ the sets $A_i \cap B_i$ and $B_i \cap A_{i+1}$ contain one element.

B.2 Maximal Repulsive Chains

In this section, we restrict ourselves to the study of the behavior of a *maximal repulsive chain* $A = \{j, j+1, \ldots, j+m\}$ of $m+1$ consecutive points, $m \leq \ell - 2$ *within the general system (B.1).* Setting, for $k = 0, \ldots, m$, $\mathfrak{u}_k(\cdot) = a_{k+j}(\cdot)$, we are led to study $\mathfrak{U}(\cdot) = (\mathfrak{u}_0(\cdot), \mathfrak{u}_1(\cdot), \ldots, \mathfrak{u}_m(\cdot))$. Since $\mathcal{B}_{k+1/2} < 0$ in the repulsive case the chain $\mathfrak{U}$ is moved through a system of $m-1$ ODEs,

$$\mathfrak{q}_k \frac{d}{ds} \mathfrak{u}_k(s) = -|\mathcal{B}_{(k-1/2)}| [(\mathfrak{u}_k(s) - \mathfrak{u}_{k-1}(s)]^{-(\omega+1)}$$

$$+ |\mathcal{B}_{(k+1/2)}| [(\mathfrak{u}_{k+1}(s) - \mathfrak{u}_k(s)]^{-(\omega+1)} \qquad (B.7)$$

for $k = 1, \ldots, m-1$, whereas the end points satisfy two differential inequalities

$$\frac{d}{ds} \mathfrak{u}_m(s) \geq -\mathfrak{q}_m^{-1} \frac{\partial \tilde{F}_{\text{rep}}}{\partial u_m}(\mathfrak{u}_m(s)), \quad \frac{d}{ds} \mathfrak{u}_0(s) \leq -\mathfrak{q}_0^{-1} \frac{\partial \tilde{F}_{\text{rep}}}{\partial u_0}(a(s)), \qquad (B.8)$$

where we have set $\tilde{F}_{\text{rep}}(U) = \sum_{k=0}^{m-1} F_{k+1/2}(U)$. We assume that at initial time we have

$$\mathfrak{u}_0(0) < \mathfrak{u}_1(0) < \cdots < \mathfrak{u}_m(0). \qquad (B.9)$$

Set $\delta_{\mathfrak{u}}(t) = \min\{\mathfrak{u}_{k+1}(t) - \mathfrak{u}_k(t), \ k = 0, \ldots, m-1\}$. We prove the following.

Proposition 3.67 *Assume that the function $\mathfrak{U}$ satisfies the system (B.7) and (B.8) on $[0, T_{\max}]$, and that (B.9) holds. Then, we have, for any $t \in [0, T_{\max})$,*

$$\left(\tilde{F}_{\text{rep}}(\mathfrak{U}(t))\right)^{-\frac{\omega+2}{\omega}} - \left(\tilde{F}_{\text{rep}}(\mathfrak{U}(0))\right)^{-\frac{\omega+2}{\omega}} \geq \frac{\omega+1}{4\omega} \mathfrak{q}_{\max}^{-1} (\omega \mathcal{B}_{\max})^{-\frac{2(\omega+1)}{\omega}} t, \quad \text{so that}$$
$$(B.10)$$

$$\delta_{\mathfrak{u}}(t) \geq \left(\mathcal{S}_1 t + \mathcal{S}_2 \delta_{\mathfrak{u}}(0)^{\omega+2}\right)^{\frac{1}{\omega+2}}, \qquad (B.11)$$

where $\mathcal{S}_1 > 0$ and $\mathcal{S}_2 > 0$ depend on only the coefficients of the equation (B.1).

The proof relies on several elementary observations.

Lemma 3.68 *Let $\mathfrak{U}$ be a solution to (B.7), (B.8) and (B.9). Then, we have*

$$\frac{d}{dt} \tilde{F}_{\text{rep}}(\mathfrak{U}(t)) \leq -\mathfrak{q}_{\max}^{-1} \left|\nabla \tilde{F}_{\text{rep}}(\mathfrak{U}(t))\right|^2, \text{ for every } t \in [0, T_{\max}]. \qquad (B.12)$$

The proof is similar to (B.3) and we omit it. For $U = (u_0, u_1, \ldots, u_m) \in \mathbb{R}^{m+1}$ with $u_0 < \cdots < u_m$ set $\rho_{\min}(U) = \inf\{|u_{k+1} - u_k|, \ k = 0, \ldots, m-1\}$ and $\mathcal{B}_{\min} = \inf\{|\mathcal{B}_{k+1/2}|\}$, $\mathcal{B}_{\max} = \sup\{|\mathcal{B}_{k+1/2}|\}$.

Lemma 3.69 *Let* $U = (u_0, \ldots, u_m)$ *be such that* $u_0 < u_1 < \cdots < u_m$. *We have*

$$\mathcal{B}_{\min} \omega^{-1} \rho_{\min}(U)^{-\omega} \leq \tilde{F}_{\mathrm{rep}}(U) \leq \mathcal{B}_{\max}(m+1)(\omega)^{-1} \rho_{\min}(U)^{-\omega}. \qquad (\mathrm{B.13})$$

$$|\nabla \tilde{F}_{\mathrm{rep}}(U)| \leq (m+2)\mathcal{B}_{\max}((\omega-1)\mathcal{B}_{\min})^{-\frac{\omega+1}{\omega}} (\tilde{F}_{\mathrm{rep}}(U))^{\frac{\omega+1}{\omega}}, \qquad (\mathrm{B.14})$$

$$\left| \frac{\partial \tilde{F}_{\mathrm{rep}}(U)}{\partial u_k} \right| \geq \frac{1}{2} (\omega \mathcal{B}_{\max})^{-\frac{\omega+1}{\omega}} \left(\tilde{F}_{\mathrm{rep}}(U) \right)^{\frac{\omega+1}{\omega}}, \; \textit{for every } k = 0, \ldots, m.$$

$$(\mathrm{B.15})$$

Proof Inequalities (B.13) and (B.14) are direct consequences of the definition of $\tilde{F}_{\mathrm{rep}}$. We turn therefore to estimate (B.15). In view of formula (B.2), the cases $k = 0$ and $k = m+1$ are straightforward. Next, let $k = 1, \ldots, m$ and set $T_{k+1/2} = \mathcal{B}_{k+1/2}(u_{k+1} - u_k)^{-(\omega+1)}$. We distinguish two cases.

Case 1: $T_{k-\frac{1}{2}} \leq \frac{1}{2} T_{k+\frac{1}{2}}$. Then, we have, in view of (B.2), $T_{k+1/2} \leq 2$ $\left| \frac{\partial F}{\partial u_k}(U) \right| \leq 2 |\nabla F(U)|$, and we are done.

Case 2: $T_{k-1/2} \geq \frac{1}{2} T_{k+1/2}$. In that case, we repeat the argument with k replaced by $k - 1$. Then either $T_{k-3/2} \leq \frac{1}{2} T_{k-1/2}$, which yields as above $T_{k-1/2} \leq 2|\nabla F(U)|$, so that we are done, or $T_{k-3/2} \geq \frac{1}{2} T_{k+1/2}$, and we go on. Since we have to stop at $k = 0$, this leads to the desired inequality (B.15). $\qquad \square$

Proof of Proposition 3.67 Combining (B.12) with (B.15), we are led to

$$\frac{\mathrm{d}}{\mathrm{d}t} \tilde{F}_{\mathrm{rep}}(\mathfrak{U}(t)) \leq -\frac{1}{4} \mathfrak{q}_{\max}^{-1} (\omega \mathcal{B}_{\max})^{-\frac{2(\omega+1)}{\omega}} \left(\tilde{F}_{\mathrm{rep}}(\mathfrak{u}(t)) \right)^{\frac{2(\omega+1)}{\omega}}.$$

Integrating this differential equation, we obtain (B.10). Combining the last inequality of Lemma 3.68 with inequality (B.13), inequality (B.11) follows. $\qquad \square$

B.3 Maximal Attractive Chains

Maximal attractive chains $B = \{j, j+1, \ldots j+m\}$, with $m \leq \ell - 1$ within the general system (B.1) are handled similarly. Defining $\mathfrak{U}$ as above, the function $\mathfrak{U}$ still satisfies (B.7), but the inequalities (B.8) are now replaced by

$$\frac{\mathrm{d}}{\mathrm{d}s} \mathfrak{u}_m(s) \leq -\mathfrak{q}_m^{-1} \frac{\partial \tilde{F}_{\mathrm{att}}}{\partial u_m}(\mathfrak{u}_m(s)), \quad \frac{\mathrm{d}}{\mathrm{d}s} \mathfrak{u}_0(s) \geq -\mathfrak{q}_0^{-1} \frac{\partial \tilde{F}_{\mathrm{att}}}{\partial u_0}(a(s)). \qquad (\mathrm{B.16})$$

$\tilde{F}_{\mathrm{att}}(U)$ is defined by $\tilde{F}_{\mathrm{att}} = -\tilde{F}_{\mathrm{rep}}$, so that we have in the attractive case $\tilde{F}_{\mathrm{att}} \geq 0$. Up to a change of sign, the function $\tilde{F}_{\mathrm{att}}$ verifies the properties (B.13), (B.14) and (B.15) stated in Proposition 3.67. However, the differential inequality

(B.12) is now turned into

$$\frac{d}{dt}\tilde{F}_{\text{att}}(\mathcal{U}(t)) \geq q_{\max}^{-1} \left| \nabla \tilde{F}_{\text{att}}(\mathcal{U}(t)) \right|^2 \geq C\tilde{F}_{\text{att}}(\mathcal{U}(t))^{\frac{2(\omega+1)}{\omega}}, \tag{B.17}$$

where the last inequality follows from (B.15) and where C is some constant depending on only the coefficients in (B.1). Integrating (B.17) and invoking (B.13) once more, we obtain the following lemma.

Lemma 3.70 *Assume that* $\mathcal{U}$ *satisfies the system (B.7) and (B.16) on* $[0, T_{\max}]$ *with (B.9). Then for constants* $\mathcal{S}_3 > 0$ *and* $\mathcal{S}_4 > 0$ *depending on only the coefficients of (B.1), we have*

$$\delta_{\mathfrak{u}}(t) \leq \left(\mathcal{S}_3 \delta_{\mathfrak{u}}(0)^{\omega+2} - \mathcal{S}_4 t \right)^{\frac{1}{\omega+2}}.$$

B.4 Proof of Proposition 3.65 Completed

Inequalities (B.4) of Proposition 3.65 follow immediately from Proposition 3.67 and Lemma 3.70 applied to each separate maximal chain provided by the decomposition (B.6). □

References

[1] F. Bethuel, H. Brezis and F. Hélein *Ginzburg-Landau vortices*, Birkhaüser (1994).

[2] F. Bethuel, G. Orlandi and D. Smets, *Collisions and phase-vortex interactions in dissipative Ginzburg-Landau dynamics*, Duke Math. J. **130** (2005) 523–614.

[3] F. Bethuel, G. Orlandi and D. Smets, *Dynamics of multiple degree Ginzburg-Landau vortices*, Comm. Math. Phys. **272** (2007) 229–261.

[4] F. Bethuel, G. Orlandi and D. Smets, *Slow motion for gradient systems with equal depth multiple-well potentials*, J. Diff. Equations **250** (2011), 53–94.

[5] F. Bethuel and D. Smets, *Slow motion for equal depth multiple-well gradient systems: the degenerate case*, Discrete Contin. Dyn. Syst. **33** (2013), 67–87.

[6] F. Bethuel and D. Smets, *On the motion law of fronts for scalar reaction-diffusion equations with equal depth multiple-well potentials*, Chin. Ann. Math. **38 B(1)** (2007), 83–148.

[7] L. Bronsard and R.V. Kohn, *On the slowness of phase boundary motion in one space dimension*, Comm. Pure Appl. Math. **43** (1990), 983–997.

[8] X. Chen, *Generation, propagation, and annihilation of metastable patterns.* J. Diff. Equations **206** (2004), no. 2, 399–437.

[9] J. Carr and R. L. Pego, *Metastable patterns in solutions of* $u_t = \epsilon^2 u_{xx} - f(u)$. Comm. Pure Appl. Math. **42** (1989), 523–576.

[10] G. Fusco and J. K. Hale, *Slow-motion manifolds, dormant instability, and singular perturbations* J. Dynam. Differential Equations **1** (1989), 75–94.

[11] R. L. Jerrard and H. M. Soner, *Dynamics of Ginzburg-Landau vortices,* Arch. Rational Mech. Anal. **142** (1998), 99–125.

[12] J. B. Keller, *On solutions of* $\Delta u = f(u)$, Comm. Pure Appl. Math. **10** (1957), 503–510.

[13] F. H. Lin, *Some dynamical properties of Ginzburg-Landau vortices*, Comm. Pure Appl. Math. **49** (1996), 323–359.

[14] R. Osserman, *On the inequality* $\Delta u \geq f(u)$, Pacific J. Math. **7** (1957), 1641–1647.

[15] E. Sandier and S. Serfaty, *Gamma-convergence of gradient flows with applications to Ginzburg-Landau*, Comm. Pure Appl. Math. **57** (2004), 1627–1672.

[*,†] Sorbonne Universités, UPMC Université Paris 06, UMR #7598, Laboratoire Jacques-Louis Lions, F-75005, Paris, France

4

Finite-time Blowup for some Nonlinear Complex Ginzburg–Landau Equations

Thierry Cazenave[*] and Seifeddine Snoussi[†]

In this article, we review finite-time blowup criteria for the family of complex Ginzburg–Landau equations $u_t = e^{i\theta}[\Delta u + |u|^\alpha u] + \gamma u$ on $\mathbb{R}^N$, where $0 \le \theta \le \frac{\pi}{2}$, $\alpha > 0$ and $\gamma \in \mathbb{R}$. We study in particular the effect of the parameters θ and γ, and the dependence of the blowup time on these parameters.

In memory of our friend Abbas Bahri

4.1 Introduction

In this paper, we review certain known results, and present some new ones, on the problem of finite-time blowup for the family of complex Ginzburg–Landau equations on $\mathbb{R}^N$

$$\begin{cases} u_t = e^{i\theta}[\Delta u + |u|^\alpha u] + \gamma u, \\ u(0) = u_0, \end{cases} \tag{4.1}$$

where $0 \le \theta \le \frac{\pi}{2}$[1], $\alpha > 0$ and $\gamma \in \mathbb{R}$[2]. The case $\theta = 0$ of equation (4.1) is the well-known nonlinear heat equation

$$\begin{cases} u_t = \Delta u + |u|^\alpha u + \gamma u, \\ u(0) = u_0, \end{cases} \tag{4.2}$$

2010 *Mathematics Subject Classification.* Primary 35Q56; secondary 35B44, 35K91, 35Q55
Key words and phrases. Complex Ginzburg–Landau equation, finite-time blowup, energy, variance
* Université Pierre et Marie Curie & CNRS, Laboratoire Jacques-Louis Lions
† UR Analyse Non-Linéaire et Géométrie, UR13ES32, Department of Mathematics, Faculty of Sciences of Tunis, University of Tunis El-Manar
[1] One might consider $-\frac{\pi}{2} \le \theta \le 0$, which is equivalent, by changing u to $\bar{u}$.
[2] For a general $\gamma \in \mathbb{C}$, the imaginary part is eliminated by changing $u(t,x)$ to $e^{-it\Im\gamma} u(t,x)$.

which arises in particular in chemistry and biology. See e.g. [10]. The case $\theta = \frac{\pi}{2}$ of (4.1) is the equally well-known nonlinear Schrödinger equation

$$\begin{cases} u_t = \mathrm{i}[\Delta u + |u|^\alpha u] + \gamma u, \\ u(0) = u_0, \end{cases} \tag{4.3}$$

which is an ubiquitous model for weakly nonlinear dispersive waves and nonlinear optics. See e.g. [43]. Therefore, equation (4.1) can be considered as "intermediate" between the nonlinear heat and Schrödinger equations. Equation (4.1) is itself a particular case of the more general complex Ginzburg–Landau equations on $\mathbb{R}^N$:

$$\begin{cases} u_t = \mathrm{e}^{\mathrm{i}\theta} \Delta u + \mathrm{e}^{\mathrm{i}\phi} |u|^\alpha u + \gamma u, \\ u(0) = u_0, \end{cases} \tag{4.4}$$

where $0 \le \theta \le \frac{\pi}{2}$, $\phi \in \mathbb{R}$, $\alpha > 0$ and $\gamma \in \mathbb{R}$, which is a generic modulation equation describing the nonlinear evolution of patterns at near-critical conditions. See e.g. [42, 8, 27].

Two strategies have been developed for studying finite-time blowup. The first one consists of deriving conditions on the initial value, as general as possible, which ensure that the corresponding solution of (4.1) blows up in finite time. The proofs often use a differential inequality which is satisfied by some quantity related to the solution, and one shows that this differential inequality can hold only on a finite-time interval. The major difficulty is to guess the appropriate quantity to calculate. However, when such a method can be applied, it usually provides a simple proof of blowup, under explicit conditions on the initial value. On the other hand, this strategy does not give any information on how the blowup occurs, nor on the mechanism that leads to blowup. Concerning the family (4.1), this is the type of approach used in [16, 20, 2] for the nonlinear heat equation; in [48, 13, 18, 44, 31, 32] for the nonlinear Schrödinger equation; and in [41, 6, 5] for the intermediate case of (4.1).

Another strategy consists of looking for an ansatz of an approximate blowing-up solution, and then showing that the remainder remains bounded, or becomes small with respect to the approximate solution, as time tends to the blowup time of the approximate solution. The first difficulty is to find the appropriate ansatz. Then, proving the boundedness of the remainder is often quite involved technically. When this method is successful, it provides a precise description of how the corresponding solutions blow up. It may also explain the mechanism that makes these solutions blow up. For the family (4.1), this is the strategy employed in particular in [26] for the nonlinear heat equation; in [22, 23, 24, 38, 39, 25] for the nonlinear Schrödinger equation; and in [37, 47, 36, 21] for the intermediate case of (4.1) (and even (4.4)).

Equation (4.1) enjoys certain properties which the general equation (4.4) does not. in particular its solutions satisfy certain energy identities (see Section 4.2), which make it possible to study blowup by the first approach described above. We review sufficient conditions for finite-time blowup (obtained using this approach), and we study the influence of the parameters θ and γ. In Sections 4.3 and 4.4, we recall the standard results for the heat equation (4.2) and the Schrödinger equation (4.3), respectively. We are not aware of any previous reference for Theorem 4.10, nor for the case $\gamma > 0$ of Theorem 4.13, although the proofs use standard arguments. The case $\gamma > 0$ of Theorem 4.14 seems to be new. Section 4.5 is devoted to the complex Ginzburg–Landau equation (4.1). In Subsection 4.5.1, we review sufficient conditions for blowup, and the case $\gamma < 0$ of Theorem 4.17 is partially new. Finally, we study in Section 4.5.2 the behavior of the blowup time as the parameter θ approaches $\frac{\pi}{2}$, i.e. as the equation gets close to the nonlinear Schrödinger equation (4.3). We consider separately the cases $\alpha < \frac{4}{N}$ (Subsection 4.5.2) and $\alpha \geq \frac{4}{N}$ (Subsection 4.5.2). The case $\gamma > 0$ of Theorem 4.22, and Theorem 4.25, are new. A few open questions are collected in Section 4.6.

Notation. We denote by $L^p(\mathbb{R}^N)$, for $1 \leq p \leq \infty$, the usual (complex valued) Lebesgue spaces. $H^1(\mathbb{R}^N)$ and $H^{-1}(\mathbb{R}^N)$ are the usual (complex valued) Sobolev spaces. (See e.g. [1] for the definitions and properties of these spaces.) We denote by $C_c^\infty(\mathbb{R}^N)$ the set of (complex valued) functions that have compact support and are of class C^∞. We denote by $C_0(\mathbb{R}^N)$ the closure of $C_c^\infty(\mathbb{R}^N)$ in $L^\infty(\mathbb{R}^N)$. In particular, $C_0(\mathbb{R}^N)$ is the space of functions u that are continuous $\mathbb{R}^N \to \mathbb{C}$ and such that $u(x) \to 0$ as $|x| \to \infty$. $C_0(\mathbb{R}^N)$ is endowed with the sup norm.

4.2 The Cauchy Problem and Energy Identities

For the study of the local well-posedness of (4.1), it is convenient to consider the equivalent integral formulation, given by Duhamel's formula,

$$u(t) = \mathcal{T}_\theta(t)u_0 + \int_0^t \mathcal{T}_\theta(t-s)[e^{i\theta}|u(s)|^\alpha u(s) + \gamma u(s)]\,\mathrm{d}s, \qquad (4.5)$$

where $(\mathcal{T}_\theta(t))_{t \geq 0}$ is the semigroup of contractions on $L^2(\mathbb{R}^N)$ generated by the operator $e^{i\theta}\Delta$ with domain $H^2(\mathbb{R}^N)$. Moreover, $\mathcal{T}_\theta(t)\psi = G_\theta(t) \star \psi$, where the kernel $G_\theta(t)$ is defined by

$$G_\theta(t)(x) \equiv (4\pi t e^{i\theta})^{-\frac{N}{2}} e^{-\frac{|x|^2}{4t e^{i\theta}}}.$$

If $0 \leq \theta < \frac{\pi}{2}$, it is not difficult to show that $(\mathcal{T}_\theta(t))_{t \geq 0}$ is an analytic semigroup on $L^2(\mathbb{R}^N)$, and a bounded C_0 semigroup on $L^p(\mathbb{R}^N)$ for $1 \leq p < \infty$, and on $C_0(\mathbb{R}^N)$. In particular (see [6, formula (2.2)])

$$\|\mathcal{T}_\theta(t)u_0\|_{L^\infty} \leq (\cos\theta)^{-\frac{N}{2}} \|u_0\|_{L^\infty}. \tag{4.6}$$

It is immediate by a contraction mapping argument (see [40, Theorem 1]) that the Cauchy problem (4.1) is locally well-posed in $C_0(\mathbb{R}^N)$. Moreover, it is easy to see using the analyticity of the semigroup that $C_0(\mathbb{R}^N) \cap H^1(\mathbb{R}^N)$ is preserved under the action of (4.1). The following result is established in [6, Proposition 2.1 and Remark 2.2] in the case $\gamma = 0$, and the argument equally applies when $\gamma \neq 0$.

Proposition 4.1 *Suppose* $0 \leq \theta < \frac{\pi}{2}$. *Given any* $u_0 \in C_0(\mathbb{R}^N)$, *there exist* $T > 0$ *and a unique solution* $u \in C([0,T], C_0(\mathbb{R}^N))$ *of (4.5) on* $(0,T)$. *Moreover, u can be extended to a maximal interval* $[0, T_{\max})$, *and if* $T_{\max} < \infty$, *then* $\|u(t)\|_{L^\infty} \to \infty$ *as* $t \uparrow T_{\max}$. *If, in addition,* $u_0 \in H^1(\mathbb{R}^N)$, *then* $u \in C([0,T], H^1(\mathbb{R}^N)) \cap C((0,T), H^2(\mathbb{R}^N)) \cap C^1((0,T), L^2(\mathbb{R}^N))$ *and u satisfies (4.1) in* $L^2(\mathbb{R}^N)$ *for all* $t \in (0,T)$. *Furthermore, if* $\alpha < \frac{4}{N}$ *and* $T_{\max} < \infty$, *then* $\|u(t)\|_{L^2} \to \infty$ *as* $t \uparrow T_{\max}$.

Remark 4.2 Whether the solution given by Proposition 4.1 is global or not is discussed throughout this paper, but we can observe that, given $0 \leq \theta < \frac{\pi}{2}$ and $u_0 \in C_0(\mathbb{R}^N)$, the corresponding solution of (4.1) is global if γ is sufficiently negative. More precisely, if $\gamma < -\frac{1}{\alpha}[2(\cos\theta)^{-\frac{N}{2}}\|u_0\|_{L^\infty}]^{\alpha+1}$, then the corresponding solution u of (4.1) is global and satisfies $\|u(t)\|_{L^\infty} \leq 2(\cos\theta)^{-\frac{N}{2}} e^{\gamma t} \|u_0\|_{L^\infty}$ for all $t \geq 0$. Indeed, $v(t) = e^{-\gamma t}u(t)$ satisfies $v_t = e^{i\theta}(\Delta v + e^{\gamma \alpha t}|v|^\alpha v)$, so that

$$v(t) = \mathcal{T}_\theta(t)u_0 + \int_0^t e^{\gamma \alpha s}\mathcal{T}_\theta(t-s)[|v(s)|^\alpha v(s)]\,ds.$$

Setting $\phi(t) = \sup\{\|v(s)\|_{L^\infty}; 0 \leq s \leq t\}$, it follows from (4.6) that $\phi(t) \leq c\|u_0\|_{L^\infty} + \frac{c}{-\gamma\alpha}\phi(t)^{\alpha+1}$ with $c = (\cos\theta)^{-\frac{N}{2}}$. Therefore, if $\gamma < -\frac{1}{\alpha}(2c)^{\alpha+1}\|u_0\|_{L^\infty}^\alpha$, then $\phi(t) \leq 2c\|u_0\|_{L^\infty}$ for all $0 \leq t < T_{\max}$, and the desired conclusion follows.

If $\theta = \frac{\pi}{2}$, then (4.1) is the nonlinear Schrödinger equation, and $(\mathcal{T}_\theta(t))_{t \geq 0}$ is a group of isometries (which is not analytic). More restrictive conditions are needed for the local solvability of (4.1), and the proofs make use of Strichartz's estimates. The following result is proved in [17, Theorem I] (except for the blowup alternatives, which follow from [4, Theorems 4.4.1 and 4.6.1]).

Proposition 4.3 *Suppose* $\theta = \frac{\pi}{2}$ *and* $(N-2)\alpha < 4$. *Given any* $u_0 \in H^1(\mathbb{R}^N)$, *there exist* $T > 0$ *and a unique* $u \in C([0,T], H^1(\mathbb{R}^N)) \cap C^1((0,T), H^{-1}(\mathbb{R}^N))$

which satisfies (4.1) for all $t \in [0,T]$ and such that $u(0) = u_0$. Moreover, u can be extended to a maximal interval $[0, T_{\max})$, and if $T_{\max} < \infty$, then $\|u(t)\|_{H^1} \to \infty$ as $t \uparrow T_{\max}$. In addition, if $\alpha < \frac{4}{N}$ and $T_{\max} < \infty$, then $\|u(t)\|_{L^2} \to \infty$ as $t \uparrow T_{\max}$.

As observed above, an essential feature of equation (4.1) is the energy identities satisfied by its solutions. Set

$$I(w) = \int_{\mathbb{R}^N} |\nabla w|^2 - \int_{\mathbb{R}^N} |w|^{\alpha+2}, \tag{4.7}$$

$$E(w) = \frac{1}{2} \int_{\mathbb{R}^N} |\nabla w|^2 - \frac{1}{\alpha+2} \int_{\mathbb{R}^N} |w|^{\alpha+2}. \tag{4.8}$$

The functionals I and E are well defined on $C_0(\mathbb{R}^N) \cap H^1(\mathbb{R}^N)$; and if $(N-2)\alpha \leq 4$, they are well defined on $H^1(\mathbb{R}^N)$.

Suppose $0 \leq \theta < \frac{\pi}{2}$, let $u_0 \in C_0(\mathbb{R}^N) \cap H^1(\mathbb{R}^N)$ and let u be the corresponding solution of (4.1) defined on the maximal interval $[0, T_{\max})$, given by Proposition 4.1. Multiplying the equation by $\bar{u}$, we obtain

$$\int_{\mathbb{R}^N} \bar{u} u_t = \gamma \int_{\mathbb{R}^N} |u|^2 - e^{i\theta} I(u) \tag{4.9}$$

and in particular, taking the real part,

$$\frac{d}{dt} \int_{\mathbb{R}^N} |u|^2 = 2\gamma \int_{\mathbb{R}^N} |u|^2 - 2\cos\theta I(u) \tag{4.10}$$

for all $0 < t < T_{\max}$. Multiplying the equation by $e^{-i\theta} u_t$, taking the real part and using (4.9) yields

$$\frac{d}{dt} E(u(t)) = -\cos\theta \int_{\mathbb{R}^N} |u_t|^2 + \gamma^2 \cos\theta \int_{\mathbb{R}^N} |u|^2 - \gamma \cos(2\theta) I(u) \tag{4.11}$$

for all $0 < t < T_{\max}$. Applying (4.10), we see that this is equivalent to

$$\frac{d}{dt} \left[E(u(t)) - \frac{\gamma}{2} \cos\theta \int_{\mathbb{R}^N} |u|^2 \right] = -\cos\theta \int_{\mathbb{R}^N} |u_t|^2 + \gamma \sin^2\theta I(u). \tag{4.12}$$

Suppose now $\theta = \frac{\pi}{2}$ and $(N-2)\alpha < 4$, let $u_0 \in H^1(\mathbb{R}^N)$ and let u be the corresponding solution of (4.1) defined on the maximal interval $[0, T_{\max})$, given by Proposition 4.3. Identities corresponding to (4.9), (4.10) and (4.11) hold. More precisely, the functions $t \mapsto \|u(t)\|_{L^2}^2$ and $t \mapsto E(u(t))$ are C^1 on $[0, T_{\max})$

and

$$\int_{\mathbb{R}^N} \overline{u} u_t = \gamma \int_{\mathbb{R}^N} |u|^2 - iI(u), \tag{4.13}$$

$$\frac{d}{dt} \int_{\mathbb{R}^N} |u|^2 = 2\gamma \int_{\mathbb{R}^N} |u|^2, \tag{4.14}$$

$$\frac{d}{dt} E(u(t)) = \gamma I(u) \tag{4.15}$$

for all $0 \le t < T_{\max}$. Identity (4.13) is obtained by taking the $H^{-1} - H^1$ duality product of the equation with $\overline{u}$ (the term $\int_{\mathbb{R}^N} \overline{u} u_t$ is understood as the duality bracket $\langle u_t, u \rangle_{H^1, H^{-1}}$). (4.14) follows, by taking the real part. Identity (4.15) is formally obtained by multiplying the equation by $e^{-i\theta} u_t$ and taking the real part. However, the solution is not smooth enough to do so, thus a regularization process is necessary. See [34] for a simple justification. Still in the case of the Schrödinger equation $\theta = \frac{\pi}{2}$, an essential tool in the blowup arguments is the variance identity. It concerns the variance

$$V(w) = \int_{\mathbb{R}^N} |x|^2 |w|^2, \tag{4.16}$$

which is not defined on $L^2(\mathbb{R}^N)$, but on the weighted space $L^2(\mathbb{R}^N, |x|^2 dx)$. It can be proved that if $u_0 \in H^1(\mathbb{R}^N) \cap L^2(\mathbb{R}^N, |x|^2 dx)$, then the corresponding solution u of (4.1) satisfies $u \in C([0, T_{\max}), L^2(\mathbb{R}^N, |x|^2 dx))$. Moreover, the map $t \mapsto V(u(t))$ is C^2 on $[0, T_{\max})$ and

$$\frac{d}{dt} V(u(t)) = -4J(u(t)) + 2\gamma V(u(t)), \tag{4.17}$$

$$\frac{d}{dt} J(u(t)) = -2 \int_{\mathbb{R}^N} |\nabla u|^2 + \frac{N\alpha}{\alpha + 2} \int_{\mathbb{R}^N} |u|^{\alpha+2} + 2\gamma J(u(t)) \tag{4.18}$$

for all $0 \le t < T_{\max}$, where the functional J is defined by

$$J(w) = \Im \int_{\mathbb{R}^N} (x \cdot \nabla \overline{w}) w \tag{4.19}$$

for $w \in H^1(\mathbb{R}^N) \cap L^2(\mathbb{R}^N, |x|^2 dx)$. The proofs of these properties require appropriate regularizations and multiplications, see [4, Section 6.5].

4.3 The Nonlinear Heat Equation

In this section, we consider the nonlinear heat equation (4.2). The first blowup result was obtained by Kaplan [16, Theorem 8]. Its argument applies to positive solutions of the equation set on a bounded domain, and is based on a differential inequality satisfied by the scalar product of the solution with the first eigenfunction. It is easy to extend the argument to the equation set on $\mathbb{R}^N$. Let $w(x) \equiv e^{-\sqrt{N^2+|x|^2}}$, so that $\Delta w \geq -w$ by elementary calculations. If $\psi = \|w\|_{L^1}^{-1} w$ and $\psi_\lambda(x) = \lambda^N \psi(\lambda x)$ for $\lambda > 0$, then $\|\psi_\lambda\|_{L^1} = 1$ and $\Delta \psi_\lambda \geq -\lambda^2 \psi_\lambda$. Let now $u_0 \in C_0(\mathbb{R}^N) \cap H^1(\mathbb{R}^N)$, $u_0 \geq 0$, $u_0 \not\equiv 0$, and let u be the corresponding solution of (4.2) defined on the maximal interval $[0, T_{\max})$. The maximum principle implies that $u(t) \geq 0$ for all $0 \leq t < T_{\max}$. Multiplying the equation by ψ_λ, integrating by parts on $\mathbb{R}^N$ and using Jensen's inequality, we obtain

$$\frac{\mathrm{d}}{\mathrm{d}t} \int_{\mathbb{R}^N} u \psi_\lambda = \int_{\mathbb{R}^N} u \Delta \psi_\lambda + \int_{\mathbb{R}^N} u^{\alpha+1} \psi_\lambda + \gamma \int_{\mathbb{R}^N} u \psi_\lambda$$

$$\geq (\gamma - \lambda^2) \int_{\mathbb{R}^N} u \psi_\lambda + \left(\int_{\mathbb{R}^N} u \psi_\lambda \right)^{\alpha+1}.$$

It follows that $f(t) = \int_{\mathbb{R}^N} u \psi_\lambda$ satisfies

$$\frac{\mathrm{d}f}{\mathrm{d}t} \geq (\gamma - \lambda^2 + f^\alpha) f \tag{4.20}$$

on $[0, T_{\max})$. It is not difficult to show that if $f(0)^\alpha > \lambda^2 - \gamma$, then (4.20) can hold on only a finite interval, so that $T_{\max} < \infty$. Therefore, we can distinguish two cases. If $\gamma \leq 0$, we choose for instance $\lambda = 1$ and we see that if u_0 is sufficiently "large" so that $\int_{\mathbb{R}^N} u_0 \psi_1 > (1 - \gamma)^{\frac{1}{\alpha}}$, then the solution blows up in finite time. If $\gamma > 0$, then we let $\lambda = \sqrt{\gamma}$, so that the condition $f(0)^\alpha > \lambda^2 - \gamma$ is always satisfied if $u_0 \not\equiv 0$. In this case, we see that every nonnegative, nonzero initial value produces a solution of (4.2) which blows up in finite time.

Levine [20] established blowup by a different argument. It is based on a differential inequality satisfied by the L^2 norm of the solution, derived from the energy identities. This argument applies to sign-changing solutions and, more generally, to complex valued solutions, and to the equation set on any domain, bounded or not. Strangely enough, even though Kaplan's argument seems to indicate that blowup is more likely to happen if $\gamma > 0$, it turns out that Levine's result only applies to the case $\gamma \leq 0$, which we consider first.

4.3.1 The Case $\gamma \leq 0$

It is convenient to set

$$I_\gamma(w) = \int_{\mathbb{R}^N} |\nabla w|^2 - \int_{\mathbb{R}^N} |w|^{\alpha+2} - \gamma \int_{\mathbb{R}^N} |u|^2, \tag{4.21}$$

$$E_\gamma(w) = \frac{1}{2} \int_{\mathbb{R}^N} |\nabla w|^2 - \frac{1}{\alpha+2} \int_{\mathbb{R}^N} |w|^{\alpha+2} - \frac{\gamma}{2} \int_{\mathbb{R}^N} |u|^2 \tag{4.22}$$

for $w \in C_0(\mathbb{R}^N) \cap H^1(\mathbb{R}^N)$.

Theorem 4.4 ([20], Theorem I) *Let $u_0 \in C_0(\mathbb{R}^N) \cap H^1(\mathbb{R}^N)$ and let u be the corresponding solution of (4.2) defined on the maximal interval $[0, T_{\max})$, given by Proposition 4.1. If $\gamma \leq 0$ and $E_\gamma(u_0) < 0$, where E_γ is defined by (4.22), then u blows up in finite time, i.e. $T_{\max} < \infty$.*

Proof We obtain a differential inequality on the quantity

$$M(t) = \frac{1}{2} \int_0^t \|u(s)\|_{L^2}^2 \, ds. \tag{4.23}$$

Formulas (4.9) and (4.12) (with $\theta = 0$) yield

$$\int_{\mathbb{R}^N} \bar{u} u_t = -I_\gamma(u), \tag{4.24}$$

$$\frac{d}{dt} E_\gamma(u(t)) = -\int_{\mathbb{R}^N} |u_t|^2 \leq 0. \tag{4.25}$$

Identity (4.25) implies

$$E_\gamma(u(t)) + \int_0^t \|u_t\|_{L^2}^2 = E_\gamma(u_0). \tag{4.26}$$

Moreover, it follows from (4.21) and (4.22) that

$$I_\gamma(u(t)) \leq (\alpha + 2)E_\gamma(u(t)) - \frac{(-\gamma)\alpha}{2} \|u\|_{L^2}^2 \tag{4.27}$$

so that by (4.26),

$$I_\gamma(u(t)) \leq (\alpha + 2)E_\gamma(u_0) - \frac{(-\gamma)\alpha}{2} \|u\|_{L^2}^2 - (\alpha + 2) \int_0^t \|u_t\|_{L^2}^2 < 0. \tag{4.28}$$

We deduce from (4.23), (4.24) and (4.28) that

$$M''(t) = \Re \int_{\mathbb{R}^N} \bar{u} u_t = -I_\gamma(u) \geq -(\alpha + 2)E_\gamma(u_0) + (\alpha + 2) \int_0^t \|u_t\|_{L^2}^2 > 0. \tag{4.29}$$

All the above formulas hold for $0 \leq t < T_{\max}$. Assume now by contradiction that $T_{\max} = \infty$. We deduce in particular from (4.29) that

$$M'(t) \underset{t \to \infty}{\longrightarrow} \infty, \quad M(t) \underset{t \to \infty}{\longrightarrow} \infty. \tag{4.30}$$

It follows from (4.23), (4.29), and Cauchy–Schwarz's inequality (in time and space) that

$$
\begin{aligned}
M(t)M''(t) &\geq \frac{\alpha+2}{2}\left(\int_0^t \|u\|_{L^2}^2\right)\left(\int_0^t \|u_t\|_{L^2}^2\right) \\
&\geq \frac{\alpha+2}{2}\left(\int_0^t \left|\int_{\mathbb{R}^N} \bar{u}u_t\right|\right)^2 \geq \frac{\alpha+2}{2}\left(\int_0^t \left|\Re\int_{\mathbb{R}^N} \bar{u}u_t\right|\right)^2 \\
&= \frac{\alpha+2}{2}\left(\int_0^t M''(s)\right)^2 = \frac{\alpha+2}{2}(M'(t)-M'(0))^2.
\end{aligned}
\tag{4.31}
$$

We deduce from (4.30) that $\frac{\alpha+2}{2}(M'(t)-M'(0))^2 \geq \frac{\alpha+4}{4}M'(t)^2$ for t sufficiently large. Therefore (4.31) yields

$$
M(t)M''(t) \geq \frac{\alpha+4}{4}M'(t)^2
$$

which means that $M(t)^{-\frac{\alpha}{4}}$ is concave for t large. Since $M(t)^{-\frac{\alpha}{4}} \to 0$ as $t \to \infty$ by (4.30), we obtain a contradiction. $\square$

The proof of Theorem 4.4 does not immediately provide an estimate of $T_{\max}$ in terms of u_0. It turns out that a variant of that proof, given in [14, Proposition 5.1] yields such an estimate.

Theorem 4.5 *Under the assumptions of Theorem 4.4, we have*

$$
T_{\max} \leq
\begin{cases}
\dfrac{\|u_0\|_{L^2}^2}{\alpha(\alpha+2)(-E_\gamma(u_0))}, & \gamma = 0, \\[3mm]
\dfrac{1}{-\gamma\alpha}\log\left(1 + \dfrac{-2\gamma\|u_0\|_{L^2}^2}{2(\alpha+2)(-E_\gamma(u_0)) - \gamma\alpha\|u_0\|_{L^2}^2}\right), & \gamma < 0.
\end{cases}
\tag{4.32}
$$

Proof Set

$$
f(t) = \|u(t)\|_{L^2}^2, \quad e(t) = E_\gamma(u(t)).
\tag{4.33}
$$

We first obtain an upper bound on e in terms of f, then a differential inequality on f. It follows from (4.24) and (4.27) that

$$
\frac{df}{dt} \geq 2(\alpha+2)(-e) + (-\gamma)\alpha f.
\tag{4.34}
$$

Since $\frac{df}{dt} > 0$ by (4.29), we deduce from (4.25), Cauchy–Schwarz's inequality, (4.24) and (4.34) that

$$
\begin{aligned}
-f\frac{de}{dt} &= \int |u|^2 \int |u_t|^2 \geq \left|\int \bar{u}u_t\right|^2 = \frac{1}{4}\left(\frac{df}{dt}\right)^2 \\
&\geq \frac{1}{2}\left(-(\alpha+2)e + \frac{(-\gamma)\alpha}{2}f\right)\frac{df}{dt}.
\end{aligned}
\tag{4.35}
$$

This means that

$$\frac{d}{dt}\left(-ef^{-\frac{\alpha+2}{2}} + \frac{-\gamma}{2}f^{-\frac{\alpha}{2}}\right) \geq 0 \qquad (4.36)$$

so that

$$-e + \frac{-\gamma}{2}f \geq \eta f^{\frac{\alpha+2}{2}} \qquad (4.37)$$

with

$$\eta = -E_\gamma(u_0)\|u_0\|_{L^2}^{-(\alpha+2)} + \frac{-\gamma}{2}\|u_0\|_{L^2}^{-\alpha} > 0. \qquad (4.38)$$

It follows from (4.34) and (4.37) that

$$\frac{df}{dt} \geq -2(-\gamma)f + 2(\alpha+2)\eta f^{\frac{\alpha+2}{2}}.$$

Therefore,

$$\frac{d}{dt}\left[(e^{-2\gamma t}f)^{-\frac{\alpha}{2}}\right] + \alpha(\alpha+2)\eta e^{\gamma\alpha t} \leq 0. \qquad (4.39)$$

After integration, then letting $t \uparrow T_{\max}$, we deduce that

$$\alpha(\alpha+2)\eta \int_0^{T_{\max}} e^{\gamma\alpha t}dt \leq f(0)^{-\frac{\alpha}{2}}. \qquad (4.40)$$

Expressing η and $f(0)$ in terms of u_0, estimate (4.32) easily follows in both the cases $\gamma = 0$ and $\gamma < 0$. $\qquad\square$

Remark 4.6 Here are some comments on Theorems 4.4 and 4.5.

1. Suppose $u_0 \neq 0$ and $E_\gamma(u_0) = 0$. In particular, $I_\gamma(u_0) < 0$. Therefore, u_0 is not a stationary solution of (4.2), and it follows from (4.25) that $E_\gamma(u(t)) < 0$ for all $0 < t < T_{\max}$. Thus we can apply Theorems 4.4 and 4.5 with u_0 replaced by $u(\varepsilon)$, and let $\varepsilon \downarrow 0$. In particular, we see that $T_{\max} < \infty$. Moreover, estimate (4.32) holds if $\gamma < 0$.
2. Given any nonzero $\varphi \in C_0(\mathbb{R}^N) \cap H^1(\mathbb{R}^N)$, we have $E_\gamma(\lambda\varphi)) < 0$ if $|\lambda|$ is sufficiently large. Thus we see that the sufficient condition $E_\gamma(u_0) < 0$ can indeed be achieved by certain initial values, for any $\alpha > 0$ and $\gamma \leq 0$.
3. Let $\alpha > 0$ and fix $u_0 \in C_0(\mathbb{R}^N) \cap H^1(\mathbb{R}^N)$. It is clear that if γ is sufficiently negative, then $E_\gamma(u_0) \geq 0$, so that one cannot apply Theorem 4.4. This is not surprising, since the corresponding solution of (4.2) is global if γ is sufficiently negative. (See Remark 4.2.)
4. Suppose $\gamma < 0$. It follows from (4.32) and the assumption $E_\gamma(u_0) < 0$ that $T_{\max} \leq \frac{1}{-\gamma\alpha}\log(1 + \frac{2}{\alpha})$. In particular, we see that for u_0 as in Theorem 4.5, the blowup time is bounded in terms of α and γ only, independently of u_0.

As observed in Remark 4.6 (4), in the case $\gamma < 0$, Theorem 4.5 does not apply to solutions for which the blowup time would be arbitrarily large. When

$$(N-2)\alpha < 4 \tag{4.41}$$

this can be improved by using the potential well argument of Payne and Sattinger [35]. To this end, we introduce some notation. Assuming (4.41) and $\gamma < 0$, we denote by Q_γ the unique positive, radially symmetric, H^1 solution of the equation

$$-\Delta Q - \gamma Q = |Q|^\alpha Q, \tag{4.42}$$

and we recall below the following well-known properties of Q_γ.

Proposition 4.7 *Assume (4.41) and $\gamma < 0$, and let $Q_\gamma \in H^1(\mathbb{R}^N)$ be the unique positive, radially symmetric solution of (4.42).*

1. $E_\gamma(Q_\gamma) > 0$ and $I_\gamma(Q_\gamma) = 0$.
2. $E_\gamma(Q_\gamma) = \inf\left\{E_\gamma(v); v \in H^1(\mathbb{R}^N), v \neq 0, I_\gamma(v) = 0\right\}$.
3. *If $u \in H^1(\mathbb{R}^N)$, $E_\gamma(u) < E_\gamma(Q_\gamma)$ and $I_\gamma(u) < 0$, then $I_\gamma(u) \leq -(E_\gamma(Q_\gamma) - E_\gamma(u))$.*

Proof The first two properties are classical, see for instance [46, Chapter 3]. Next, let $u \in H^1(\mathbb{R}^N)$ with $I_\gamma(u) < 0$, and set $h(t) = E_\gamma(tu)$ for $t > 0$. It follows easily that $h'(t) = \frac{1}{t}I_\gamma(tu)$, so that there exists a unique $t^* > 0$ such that h is increasing on $[0, t^*]$ and decreasing and concave on $[t^*, \infty)$. In particular, $I_\gamma(t^*u) = 0$, thus $h(t^*) = E_\gamma(t^*u) \geq E_\gamma(Q_\gamma)$. Moreover, since $I_\gamma(u) < 0$ we have $t^* < 1$ and by the concavity of h on $[t^*, 1]$,

$$E_\gamma(u) = h(1) \geq h(t^*) + (1 - t^*)h'(1) = h(t^*) + (1 - t^*)I_\gamma(u)$$
$$\geq E_\gamma(Q_\gamma) + (1 - t^*)I_\gamma(u) \geq E_\gamma(Q_\gamma) + I_\gamma(u)$$

from which (3) follows. □

We have the following result.

Theorem 4.8 *Assume (4.41), $\gamma < 0$, and let $Q_\gamma \in H^1(\mathbb{R}^N)$ be the unique positive, radially symmetric solution of (4.42). Let $u_0 \in C_0(\mathbb{R}^N) \cap H^1(\mathbb{R}^N)$ and let u be the corresponding solution of (4.2) defined on the maximal interval $[0, T_{\max})$. If $E_\gamma(u_0) < E_\gamma(Q_\gamma)$ and $I_\gamma(u_0) < 0$, then u blows up in finite time, and*

$$T_{\max} \leq \frac{1}{-\gamma\alpha}\left[\frac{(\alpha+4)[E_\gamma(u_0)]^+}{E_\gamma(Q_\gamma) - E_\gamma(u_0)} + \log\left(\frac{2(\alpha+2)}{\alpha}\right)\right]. \tag{4.43}$$

Proof The key observation is that

$$I_\gamma(u(t)) \leq -(E_\gamma(Q_\gamma) - E_\gamma(u_0)) < 0 \tag{4.44}$$

for $0 \le t < T_{\max}$. Indeed, it follows from (4.25) that $E_\gamma(u(t)) \le E_\gamma(u_0) < E_\gamma(Q_\gamma)$. Therefore, Proposition 4.7 (3) implies that (4.44) holds as long as $I_\gamma(u(t)) < 0$. Since the right-hand side of (4.44) is a negative constant, we see by continuity and Proposition 4.7 (2) that $I_\gamma(u(t))$ must remain negative; and so (4.44) holds for all $0 \le t < T_{\max}$.

If $E_\gamma(u_0) \le 0$, then the result follows from Remark 4.6 (2) and (4), so we suppose

$$0 < E_\gamma(u_0) < E_\gamma(Q_\gamma). \tag{4.45}$$

We use the notation (4.33) introduced in the proof of Theorem 4.5. It follows from (4.24) and (4.44) that $\frac{df}{dt} \ge 2(E_\gamma(Q_\gamma) - E_\gamma(u_0))$, so that

$$f(t) \ge 2(E_\gamma(Q_\gamma) - E_\gamma(u_0))t. \tag{4.46}$$

In particular, we see that

$$\sigma(t) \overset{\text{def}}{=} -e(t) + \frac{-\gamma}{2}f(t) \ge -E_\gamma(u_0) - \gamma(E_\gamma(Q_\gamma) - E_\gamma(u_0))t. \tag{4.47}$$

We set

$$\tau = \frac{(\alpha+4)E_\gamma(u_0)}{-\gamma\alpha(E_\gamma(Q_\gamma) - E_\gamma(u_0))} > \frac{E_\gamma(u_0)}{-\gamma(E_\gamma(Q_\gamma) - E_\gamma(u_0))}. \tag{4.48}$$

If $T_{\max} \le \tau$ then (4.43) follows from (4.48). We now suppose

$$T_{\max} > \tau. \tag{4.49}$$

Setting $v_0 = u(\tau)$, we see that the solution v of (4.2) with the initial condition $v(0) = v_0$ is $v(t) = u(t+\tau)$ and that its maximal existence time $S_{\max}$ is $S_{\max} = T_{\max} - \tau$. Since $\sigma(\tau) > 0$ by (4.47) and (4.48) we can argue as in the proof of Theorem 4.5 (note that $\eta > 0$, where η is given by (4.38) with u_0 replaced by v_0), and we deduce (see (4.40)) that

$$\frac{2(\alpha+2)(-E_\gamma(v_0)) - \gamma\alpha\|v_0\|_{L^2}^2}{2(\alpha+2)(-E_\gamma(v_0)) - \gamma(\alpha+2)\|v_0\|_{L^2}^2} \le e^{\gamma\alpha S_{\max}}. \tag{4.50}$$

(Observe that $\sigma(\tau) > 0$, so that the denominator on the left-hand side of (4.50) is positive.) Note that by (4.46), (4.48), and the fact that $e(t)$ is nonincreasing

$$\|v_0\|_{L^2}^2 = \|u(\tau)\|_{L^2}^2 \ge 2(E_\gamma(Q_\gamma) - E_\gamma(u_0))\tau$$
$$\ge \frac{2(\alpha+4)}{-\gamma\alpha}E_\gamma(u_0) \ge \frac{2(\alpha+4)}{-\gamma\alpha}E_\gamma(v_0)$$

from which it follows that

$$\frac{2(\alpha+2)(-E_\gamma(v_0)) - \gamma\alpha\|v_0\|_{L^2}^2}{2(\alpha+2)(-E_\gamma(v_0)) - \gamma(\alpha+2)\|v_0\|_{L^2}^2} \ge \frac{\alpha}{2(\alpha+2)}. \tag{4.51}$$

(4.50) and (4.51) yield $S_{\max} \le \frac{1}{\alpha} \log\left(\frac{2(\alpha+2)}{\alpha}\right)$. Since $T_{\max} = \tau + S_{\max}$, the result follows by applying (4.48).   □

Remark 4.9 Theorem 4.8 applies to solutions for which the maximal existence time is arbitrary large. Indeed, given $\varepsilon > 0$, let $u_0^\varepsilon = (1 + \varepsilon)Q_\gamma$ and u^ε the corresponding solution of (4.2). It is straightforward to verify that for all $\varepsilon > 0$, $E_\gamma(u_0^\varepsilon) < E_\gamma(Q_\gamma)$ and $I_\gamma(u_0^\varepsilon) < 0$. Indeed, the function $\varepsilon \mapsto E_\gamma((1+\varepsilon)Q_\gamma)$ is decreasing on $[1, +\infty)$, $I_\gamma(u_0^\varepsilon) < (1+\varepsilon)^2 I_\gamma(Q_\gamma)$ and $I_\gamma(Q_\gamma) = 0$. Hence u_0^ε satisfies the assumptions of Theorem 4.8. On the other hand, Q_γ is a stationary (hence global) solution of (4.2), so that the blowup time of v^ε goes to infinity as $\varepsilon \downarrow 0$, by continuous dependence.

4.3.2 The Case $\gamma > 0$

Levine's method (Subsection 4.3.1) does not immediately apply when $\gamma > 0$, but it can easily be adapted, after a suitable change of variable.

Theorem 4.10 *Suppose $\gamma > 0$. Let $u_0 \in C_0(\mathbb{R}^N) \cap H^1(\mathbb{R}^N)$ and let u be the corresponding solution of (4.2) defined on the maximal interval $[0, T_{\max})$. If $E(u_0) < 0$, where E is defined by (4.8), then u blows up in finite time, i.e., $T_{\max} < \infty$. Moreover,*

$$T_{\max} \le \frac{1}{\alpha\gamma} \log\left(1 + \frac{\gamma \|u_0\|_{L^2}^2}{(\alpha+2)(-E(u_0))}\right) < \infty. \tag{4.52}$$

Proof We set $v(t) = e^{-\gamma t} u(t)$, so that

$$\begin{cases} v_t = \Delta v + e^{\alpha\gamma t} |v|^\alpha v, \\ v(0) = u_0, \end{cases} \tag{4.53}$$

and we use the arguments in the proof of Theorem 4.5. Setting

$$\tilde{f} = \|v\|_{L^2}^2, \quad \tilde{j} = \|\nabla v\|_{L^2}^2 - e^{\alpha\gamma t}\|v\|_{L^{\alpha+2}}^{\alpha+2}, \quad \tilde{e} = \frac{1}{2}\|\nabla v\|_{L^2}^2 - \frac{e^{\alpha\gamma t}}{\alpha+2}\|v\|_{L^{\alpha+2}}^{\alpha+2}, \tag{4.54}$$

it follows from (4.53) that

$$\int_{\mathbb{R}^N} \overline{v} v_t = -\tilde{j}(t) \tag{4.55}$$

and $\frac{d\tilde{e}}{dt} = -\int_{\mathbb{R}^N} |v_t|^2 + \alpha\gamma\tilde{e} - \frac{\alpha\gamma}{2}\int_{\mathbb{R}^N} |\nabla u|^2$, so that

$$\frac{d\tilde{e}}{dt} - \alpha\gamma\tilde{e} \le -\int_{\mathbb{R}^N} |v_t|^2. \tag{4.56}$$

In particular,

$$\tilde{e}(t) \le e^{\alpha\gamma t}\tilde{e}(0) = e^{\alpha\gamma t}E(u_0) < 0. \tag{4.57}$$

Applying (4.56), Cauchy–Schwarz's inequality and (4.55), we obtain

$$-\tilde{f}\left(\frac{d\tilde{e}}{dt} - \alpha\gamma\tilde{e}\right) \ge \int |v|^2 \int |v_t|^2 \ge \left|\int \overline{v}v_t\right|^2 = \tilde{j}^2 = \frac{1}{2}(-\tilde{j})\frac{d\tilde{f}}{dt}. \tag{4.58}$$

Note that

$$\tilde{j} = (\alpha + 2)\tilde{e} - \frac{\alpha}{2}\int_{\mathbb{R}^N}|\nabla v|^2 \le (\alpha + 2)\tilde{e}. \tag{4.59}$$

Since $\frac{d\tilde{f}}{dt} > 0$ by (4.55), (4.59) and (4.57), we deduce from (4.58) and (4.59) that $-\tilde{f}(\frac{d\tilde{e}}{dt} - \alpha\gamma\tilde{e}) \ge -\frac{\alpha+2}{2}\tilde{e}\frac{d\tilde{f}}{dt}$. Therefore $\frac{d}{dt}[e^{-\alpha\gamma t}(-\tilde{e})\tilde{f}^{-\frac{\alpha+2}{2}}] \ge 0$, and so

$$e^{-\alpha\gamma t}(-\tilde{e}(t)) \ge [-\tilde{e}(0)]\tilde{f}(0)^{-\frac{\alpha+2}{2}}\tilde{f}(t)^{\frac{\alpha+2}{2}} = (-E(u_0))\|u_0\|_{L^2}^{-(\alpha+2)}\tilde{f}(t)^{\frac{\alpha+2}{2}}.$$

Thus we see that

$$\frac{d\tilde{f}}{dt} = -2\tilde{j} \ge -2(\alpha+2)\tilde{e} \ge 2(\alpha+2)(-E(u_0))\|u_0\|_{L^2}^{-(\alpha+2)}\tilde{f}^{\frac{\alpha+2}{2}}e^{\alpha\gamma t}.$$

This shows that $\alpha(\alpha + 2)(-E(u_0))\|u_0\|_{L^2}^{-(\alpha+2)}e^{\alpha\gamma t} + \frac{d}{dt}(\tilde{f}^{-\frac{\alpha}{2}}) \le 0$, and (4.52) easily follows. $\qquad\square$

Remark 4.11 Below are a few comments on Theorem 4.10.

1. Kaplan's calculations at the beginning of Section 4.3 show that if $\gamma > 0$, every nonnegative, nonzero initial value produces finite-time blowup. On the other hand, in dimension $N \ge 2$, there exist nontrivial stationary solutions in $C_0(\mathbb{R}^N)$, which are global solutions of (4.2). Indeed, it is not difficult to prove that for every $\eta > 0$, the solution u of the ODE $u'' + \frac{N-1}{r}u' + \gamma u + |u|^\alpha u = 0$ with the initial conditions $u(0) = \eta$ and $u'(0) = 0$ oscillates indefinitely and converges to 0 as $r \to \infty$. This yields a solution $u \in C^2(\mathbb{R}^N) \cap C_0(\mathbb{R}^N)$ of the equation $\Delta u + \gamma u + |u|^\alpha u = 0$, hence a stationary solution of (4.2). Note that if $(N-2)\alpha \le 4$, Pohozaev's identity implies that there is no nontrivial stationary solution in $H^1(\mathbb{R}^N) \cap C_0(\mathbb{R}^N)$. When $N \ge 3$ and $(N-2)\alpha > 4$, whether or not there exist nontrivial stationary solutions in $H^1(\mathbb{R}^N) \cap C_0(\mathbb{R}^N)$ seems to be an open problem. Note also that in dimension $N = 1$, there is no stationary solution in $C_0(\mathbb{R}^N)$, this can be easily deduced from the resulting ODE.

2. In the case $\gamma = 0$, $\alpha = \frac{2}{N}$ is the Fujita critical exponent. If $\alpha > \frac{2}{N}$, then small initial values in an appropriate sense give rise to global solutions of (4.2). On the other hand, if $\alpha \le \frac{2}{N}$, then every nonnegative, nonzero initial value produces finite-time blowup. (See [11, Theorem 1], [15], [19,

Theorem 2.1], [45, Theorem 1], [18, Corollary 1.12].) However, given any $\alpha \leq \frac{2}{N}$, there exist nonzero initial values producing global solutions. In the one-dimensional case, they can be initial values that change sign sufficiently many times and are sufficiently small [29, Theorem 1.1], [30, Theorem 1.2]. In any dimension, they can be self-similar solutions [14, Theorem 3]. If $\gamma > 0$, then equation (4.2) is not scaling-invariant, so that one cannot expect self-similar solutions.

4.4 The Nonlinear Schrödinger Equation

In this section, we consider the nonlinear Schrödinger equation (4.3). We assume $\alpha < \frac{4}{N-2}$, and it follows from Proposition 4.3 that the Cauchy problem is locally well-posed in $H^1(\mathbb{R}^N)$. In contrast with the nonlinear heat equation, for which blowup may occur no matter how small α is, blowup for (4.3) cannot occur if α is too small.

Proposition 4.12 ([12], Theorem 3.1) *Suppose $0 < \alpha < \frac{4}{N}$ and let $\gamma \in \mathbb{R}$. It follows that for every $u_0 \in H^1(\mathbb{R}^N)$, the corresponding solution of (4.3) is global, i.e. $T_{\max} = \infty$.*

Proof Let $u_0 \in H^1(\mathbb{R}^N)$ and u be the corresponding solution of (4.3) defined on the maximal interval $[0, T_{\max})$. Formula (4.14) yields

$$\|u(t)\|_{L^2} = e^{\gamma t}\|u_0\|_{L^2} \tag{4.60}$$

for all $0 \leq t < T_{\max}$. Applying the blowup alternative on the L^2 norm of Proposition 4.3, we conclude that $T_{\max} = \infty$. □

When $\alpha \geq \frac{4}{N}$, finite-time blowup may occur. This was proved in [48, p. 911] in the three-dimensional cubic, radial case with $\gamma = 0$, then in [13, p. 1795] in the general case (still with $\gamma = 0$). Note that all solutions have locally bounded L^2-norm by (4.60), so that Levine's method used in Section 4.3 cannot be applied. Instead, the proof in [48, 13] is based on the variance identity (4.17)–(4.18). This argument can easily be applied to the case $\gamma \geq 0$, which we consider first.

4.4.1 The Case $\gamma \geq 0$

The following result is proved in [48, 13] when $\gamma = 0$.

Theorem 4.13 *Suppose $\frac{4}{N} \leq \alpha < \frac{4}{N-2}$ and $\gamma \geq 0$. Let $u_0 \in H^1(\mathbb{R}^N)$ and u be the corresponding solution of (4.3) defined on the maximal interval $[0, T_{\max})$.*

If $E(u_0) < 0$ and $u_0 \in L^2(\mathbb{R}^N, |x|^2 dx)$, where E is defined by (4.8), then u blows up in finite time, i.e., $T_{\max} < \infty$.

Proof The proof is based on a differential inequality for the variance. More precisely, it follows from (4.17) that

$$\frac{d}{dt}(e^{-2\gamma t}V(u(t))) = -4e^{-2\gamma t}J(u) \tag{4.61}$$

and from (4.18) that

$$\frac{d}{dt}(e^{-2\gamma t}J(u(t))) = e^{-2\gamma t}\left[-4E(u(t)) + \frac{N\alpha - 4}{\alpha + 2}\|u\|_{L^{\alpha+2}}^{\alpha+2}\right] \geq -4e^{-2\gamma t}E(u(t)), \tag{4.62}$$

where we used the assumption $N\alpha \geq 4$ in the last inequality. (4.61) and (4.62) yield

$$\frac{d^2}{dt^2}(e^{-2\gamma t}V(u(t))) = -4\frac{d}{dt}(e^{-2\gamma t}J(u(t))) \leq 16e^{-2\gamma t}E(u(t)). \tag{4.63}$$

Since

$$I(w) = (\alpha + 2)E(w) - \frac{\alpha}{2}\int_{\mathbb{R}^N}|\nabla u|^2 \leq (\alpha + 2)E(w)$$

and $\gamma \geq 0$, it follows from (4.15) that

$$\frac{d}{dt}E(u(t)) \leq \gamma(\alpha + 2)E(u(t)) \tag{4.64}$$

so that

$$E(u(t)) \leq e^{\gamma(\alpha+2)t}E(u_0) < 0. \tag{4.65}$$

Applying (4.63) and (4.65) we obtain

$$\frac{d^2}{dt^2}(e^{-2\gamma t}V(u(t))) \leq 16e^{\alpha\gamma t}E(u_0) \leq 16E(u_0). \tag{4.66}$$

Note that by (4.61)

$$\frac{d}{dt}(e^{-2\gamma t}V(u(t)))_{|t=0} = -4J(u_0). \tag{4.67}$$

Integrating (4.66) twice and applying (4.67) yields

$$e^{-2\gamma t}V(u(t)) \leq V(u_0) - 4tJ(u_0) + 16E(u_0)\int_0^t\int_0^s e^{\alpha\gamma\sigma}\,d\sigma\,ds \tag{4.68}$$

for all $0 \leq t < T_{\max}$. The right-hand side of (4.68), considered as a function of $t \geq 0$, is negative for t large (because $E(u_0) < 0$). Since $e^{-2\gamma t}V(u(t)) \geq 0$, we conclude that $T_{\max} < \infty$. $\qquad\square$

The "natural" condition in Theorem 4.13 is $E(u_0) < 0$. However, we require that $u_0 \in L^2(\mathbb{R}^N, |x|^2 dx)$ because we calculate the variance $V(u)$. Whether the finite variance assumption is necessary or not in Theorem 4.13 seems to be an open question. A partial answer is known in the case $\alpha = \frac{4}{N}$ and $\gamma = 0$: if $E(u_0) < 0$ and $\|u_0\|_{L^2}$ is not too large, then $T_{\max} < \infty$. (See [24, Theorem 1.1].) Another partial answer is given by Ogawa and Tsutsumi [31, p. 318] for radially symmetric solutions in the case $\gamma = 0$ and $N \geq 2$. The proof can be adapted to the case $\gamma \geq 0$, under the additional restriction $N \geq 3$. (The case $N = 1$, $\gamma = 0$ and $\alpha = 4$ is considered in [32, p. 488], but we do not study its extension to $\gamma > 0$ here.)

Theorem 4.14 *Suppose $\frac{4}{N} \leq \alpha < \frac{4}{N-2}$ and $\gamma \geq 0$. Assume further $N \geq 2$ and $\alpha \leq 4$ if $\gamma = 0$, and $N \geq 3$ if $\gamma > 0$. Let $u_0 \in H^1(\mathbb{R}^N)$ and u be the corresponding solution of (4.3) defined on the maximal interval $[0, T_{\max})$. If $E(u_0) < 0$, where E is defined by (4.8), and if u_0 is radially symmetric, then u blows up in finite time, i.e., $T_{\max} < \infty$.*

Proof The proof uses calculations similar to those in the proof of Theorem 4.13, but for a truncated variance. It is convenient to set $v(t) = e^{-\gamma t} u(t)$, so that v satisfies the equation $v_t = i(\Delta v + e^{\alpha \gamma t} |v|^\alpha v)$. Moreover,

$$\|v(t)\|_{L^2} = \|u_0\|_{L^2} \tag{4.69}$$

by formula (4.60). Let $\Psi \in C^\infty(\mathbb{R}^N) \cap W^{4,\infty}(\mathbb{R}^N)$ be spherically symmetric and set

$$\zeta(t) = \int_{\mathbb{R}^N} \Psi |v|^2 dx. \tag{4.70}$$

It follows from (4.150) and (4.152) that

$$\frac{1}{2} \zeta'(t) = \Im \int_{\mathbb{R}^N} \overline{v}(\nabla v \cdot \nabla \Psi) \tag{4.71}$$

and

$$\frac{1}{2} \zeta'' = \int_{\mathbb{R}^N} \left(-\frac{1}{2} |v|^2 \Delta^2 \Psi - \frac{\alpha e^{\alpha \gamma t}}{\alpha + 2} |v|^{\alpha+2} \Delta \Psi + 2|\nabla v|^2 \Psi'' \right). \tag{4.72}$$

Note that the calculations in Proposition 4.31 are formal in the case $\theta = \frac{\pi}{2}$. However, they are easily justified for H^2 solutions, and then by continuous dependence for H^1 solutions. We observe that

$$\int_{\mathbb{R}^N} \left(-\frac{\alpha e^{\alpha \gamma t}}{\alpha + 2} |v|^{\alpha+2} \Delta \Psi + 2|\nabla v|^2 \Psi'' \right) = 2N\alpha e^{-2\gamma t} E(u) - (N\alpha - 4) \int_{\mathbb{R}^N} |\nabla v|^2$$

$$+ 2 \int_{\mathbb{R}^N} (\Psi'' - 2)|\nabla v|^2 + \frac{\alpha e^{\alpha \gamma t}}{\alpha + 2} \int_{\mathbb{R}^N} (2N - \Delta \Psi)|v|^{\alpha+2}.$$

Since $N\alpha \geq 4$ and $E(u(t)) \leq e^{\gamma(\alpha+2)t}E(u_0)$ by (4.65), we deduce that

$$\int_{\mathbb{R}^N}\left(-\frac{\alpha e^{\alpha\gamma t}}{\alpha+2}|v|^{\alpha+2}\Delta\Psi + 2|\nabla v|^2\Psi''\right) \leq 2N\alpha e^{\alpha\gamma t}E(u_0)$$

$$+ 2\int_{\mathbb{R}^N}(\Psi''-2)|\nabla v|^2 + \frac{\alpha e^{\alpha\gamma t}}{\alpha+2}\int_{\mathbb{R}^N}(2N-\Delta\Psi)|v|^{\alpha+2}$$

so that (4.72) yields

$$\frac{1}{2}\zeta'' \leq 2N\alpha e^{\alpha\gamma t}E(u_0)$$

$$+ \int_{\mathbb{R}^N}\left(-\frac{1}{2}|v|^2\Delta^2\Psi + \frac{\alpha e^{\alpha\gamma t}}{\alpha+2}|v|^{\alpha+2}(2N-\Delta\Psi) + 2|\nabla v|^2(\Psi''-2)\right).$$

$$(4.73)$$

We first consider the case $\gamma = 0$. We apply Lemma 4.32 with $A = \|u_0\|_{L^2}$, $\mu = \frac{\alpha}{\alpha+2}$ and $\varepsilon > 0$ sufficiently small so that $\chi\mu\varepsilon^{2(N-1)} < 1$ and $\kappa(\mu,\varepsilon) \leq -N\alpha E(u_0)$. With $\Psi = \Psi_\varepsilon$ given by Lemma 4.32, it follows from (4.69), (4.73) and (4.160) that $\zeta'' \leq 2N\alpha E(u_0) < 0$. Since $\zeta(t) \geq 0$ for all $0 \leq t < T_{\max}$, we conclude as in the proof of Theorem 4.13 that $T_{\max} < \infty$.

We next consider the case $\gamma > 0$ and $N \geq 3$. Let $0 < \tau < T_{\max}$. We set

$$\mu_\tau = e^{\alpha\gamma\tau} \geq 1, \tag{4.74}$$

we fix

$$\frac{1}{2} > \lambda > \frac{1}{2(N-1)} \tag{4.75}$$

(here we use $N \geq 3$) and we set

$$\varepsilon_\tau = a\mu_\tau^{-\lambda} \leq a. \tag{4.76}$$

Here, the constant $0 < a \leq 1$ is chosen sufficiently small so that $\chi a^{2(N-1)} < 1$, where χ is the constant in Lemma 4.32. Since $\mu_\tau \geq 1$ and $1 - 2(N-1)\lambda < 0$ by (4.75), it follows in particular that $\chi\mu_\tau\varepsilon_\tau^{2(N-1)} = \chi\mu_\tau^{1-2(N-1)\lambda}a^{2(N-1)} \leq \chi a^{2(N-1)} < 1$. Moreover, we deduce from (4.75) that κ defined by (4.161) satisfies $\kappa(\mu,\varepsilon) \leq C\mu_\tau^{1-\delta}$, where C is independent of τ, and

$$\delta = \alpha \min\left\{\frac{\lambda N}{2}, \frac{2(N-1)\lambda-1}{4-\alpha}\right\} > 0. \tag{4.77}$$

We now let $\Psi = \Psi_{\varepsilon_\tau}$ where Ψ_ε is given by Lemma 4.32 for this choice of ε. It follows from (4.159), (4.160) and the inequality $\kappa(\mu,\varepsilon) \leq C\mu_\tau^{1-\delta}$ that

$$-2\int_{\mathbb{R}^N}(2-\Psi''_{\varepsilon_\tau})|\nabla v|^2 + \frac{\alpha}{\alpha+2}e^{\alpha\gamma t}\int_{\mathbb{R}^N}(2N-\Delta\Psi_{\varepsilon_\tau})|v|^{\alpha+2}$$

$$-\frac{1}{2}\int_{\mathbb{R}^N}|v|^2\Delta^2\Psi_{\varepsilon_\tau} \leq C\mu_\tau^{1-\delta}. \tag{4.78}$$

Estimates (4.73) and (4.78) yield

$$\frac{1}{2}\zeta'' \leq 2N\alpha e^{\alpha\gamma t}E(u_0) + C\mu_\tau^{1-\delta} \tag{4.79}$$

for all $0 \leq t \leq \tau$. Integrating (4.79) twice and applying (4.71), we deduce that

$$\begin{aligned}
\frac{1}{2}\zeta(\tau) \leq &\frac{1}{2}\|\Psi_{\varepsilon_\tau}\|_{L^\infty}\|u_0\|_{L^2}^2 + \tau\|\nabla\Psi_{\varepsilon_\tau}\|_{L^\infty}\|u_0\|_{H^1}^2 \\
&+ \frac{2N}{\alpha\gamma^2}E(u_0)(e^{\alpha\gamma\tau} - \tau - \alpha\gamma\tau) + C\mu_\tau^{1-\delta}\tau.
\end{aligned} \tag{4.80}$$

Using (4.158) to estimate Ψ in the above inequality, applying (4.74) and (4.76) to express μ_τ and ε_τ in terms of τ, and since $\zeta(\tau) \geq 0$, we obtain

$$0 \leq Ce^{2\lambda\alpha\gamma\tau} + C\tau e^{\lambda\alpha\gamma\tau} + \frac{2N}{\alpha\gamma^2}E(u_0)(e^{\alpha\gamma\tau} - \tau - \alpha\gamma\tau) + C\tau e^{(1-\delta)\alpha\gamma\tau}. \tag{4.81}$$

Since $E(u_0) < 0$ and $\max\{2\lambda, 1 - \delta\} < 1$, the right-hand side of (4.81) is negative for τ large. Since $\tau < T_{\max}$ is arbitrary, we conclude that $T_{\max} < \infty$. $\quad\square$

4.4.2 The Case $\gamma < 0$

If $\gamma < 0$, the argument used in the proof of Theorem 4.13 breaks down because (4.64) does not hold. Yet blowup occurs when $\alpha > \frac{4}{N}$, as is shown by the following result of Tsutsumi [44, Theorem 1].

Theorem 4.15 *Suppose $\frac{4}{N} < \alpha < \frac{4}{N-2}$ and $\gamma < 0$. Let $u_0 \in H^1(\mathbb{R}^N)$ and let u be the corresponding solution of (4.3) defined on the maximal interval $[0, T_{\max})$. If*

$$V(u_0) + \frac{N\alpha - 4}{\gamma\alpha}J(u_0) + \frac{(N\alpha - 4)^2}{\gamma^2\alpha^2}E(u_0) < 0, \tag{4.82}$$

where the functionals E, V and J are defined by (4.8), (4.16) and (4.19), respectively, then u blows up in finite time, i.e., $T_{\max} < \infty$.

Proof We follow the simplified argument given in [33]. We define

$$W(w) = \frac{1}{2}\int_{\mathbb{R}^N}|\nabla w|^2 - \frac{N\alpha}{4(\alpha+2)}\int_{\mathbb{R}^N}|w|^{\alpha+2} \tag{4.83}$$

and we set $e(t) = E(u(t))$, $v(t) = V(u(t))$, $\iota(t) = I(u(t))$, $\jmath(t) = J(u(t))$, $w(t) = W(u(t))$, where the functionals E, V, I, J and W are defined by (4.8), (4.16), (4.7), (4.19) and (4.83), respectively. It follows from (4.15), (4.17) and (4.18) that

$$\frac{de}{dt} = \gamma\iota(t), \quad \frac{dv}{dt} = 2\gamma v(t) - 4\jmath(t), \quad \frac{d\jmath}{dt} = 2\gamma\jmath(t) - 4w(t). \tag{4.84}$$

It is convenient to define

$$b = -2\gamma \frac{4 - (N-2)\alpha}{N\alpha - 4} > 0, \quad \eta = -2\gamma + b = \frac{-4\gamma\alpha}{N\alpha - 4} > 0.$$

Using the identity $\gamma\iota - be = -\eta w$, we deduce from (4.84) that

$$\frac{d}{dt}(e^{-bt}e(t)) = e^{-bt}(\gamma\iota - be) = -\eta e^{-bt}w(t), \tag{4.85}$$

$$\frac{d}{dt}(e^{-bt}v(t)) = -\eta e^{-bt}v(t) - 4e^{-bt}J(t), \tag{4.86}$$

$$\frac{d}{dt}(e^{-bt}J(t)) = -\eta e^{-bt}J(t) - 4e^{-bt}w(t). \tag{4.87}$$

Integrating (4.85) on $(0,t)$, we obtain

$$e^{-bt}e(t) + \eta \int_0^t e^{-bs}w(t) = E(u_0).$$

Since $\alpha \geq \frac{4}{N}$, we have $e \geq w$ so that

$$e^{-bt}w(t) + \eta \int_0^t e^{-bs}w(s)\,ds \leq E(u_0). \tag{4.88}$$

We now set

$$\widetilde{w}(t) = \int_0^t e^{-bs}\widetilde{w}(s)\,ds, \quad \widetilde{J}(t) = \int_0^t e^{-bs}\widetilde{J}(s)\,ds$$

so that (4.88) becomes $\frac{d\widetilde{w}}{dt} + \eta\widetilde{w} \leq E(u_0)$. Therefore, $e^{\eta t}\widetilde{w}(t) \leq \frac{e^{\eta t}-1}{\eta}E(u_0)$, which implies

$$\int_0^t e^{\eta s}\widetilde{w}(s)\,ds \leq \frac{e^{\eta t} - 1 - \eta t}{\eta^2}E(u_0). \tag{4.89}$$

Integrating now (4.87) on $(0,t)$, we obtain $\frac{d\widetilde{J}}{dt} + \eta\widetilde{J} = J(u_0) - 4\widetilde{w}$, so that

$$e^{\eta t}\widetilde{J}(t) = \int_0^t e^{\eta s}[J(u_0) - 4\widetilde{w}(s)]\,ds. \tag{4.90}$$

We deduce from (4.90) and (4.89) that

$$e^{\eta t}\widetilde{J}(t) \geq \frac{e^{\eta t} - 1}{\eta}J(u_0) - 4\frac{e^{\eta t} - 1 - \eta t}{\eta^2}E(u_0). \tag{4.91}$$

Finally, since $v(t) \geq 0$ we deduce from (4.86) that $\frac{d}{dt}(e^{-bt}v(t)) \leq -4e^{-bt}J(t)$, so that

$$e^{-bt}v(t) \leq V(u_0) - 4\widetilde{J}(t). \tag{4.92}$$

It now follows from (4.92) and (4.91) that

$$e^{-bt}v(t) \leq V(u_0) - 4\frac{1 - e^{-\eta t}}{\eta}J(u_0) + 16\frac{1 - (1 - \eta t)e^{-\eta t}}{\eta^2}E(u_0). \tag{4.93}$$

Assumption (4.82) means that $V(u_0) - \frac{4}{\eta} J(u_0) + \frac{16}{\eta^2} E(u_0) < 0$. Therefore, the right-hand side of (4.93) becomes negative for t large, which implies that $T_{\max} < \infty$. □

Remark 4.16 Here are a few comments on Theorem 4.15.

1. The condition (4.82) is satisfied if $u_0 = c\varphi$ with $\varphi \in H^1(\mathbb{R}^N)$, $\varphi \neq 0$ and c is large.
2. The condition $\alpha > \frac{4}{N}$ is essential in the proof, for the definition of η and b. If $\alpha = 4/N$, then finite-time blowup occurs for some initial data [28, 7], but the proof follows a very different argument.
3. We are not aware of a result similar to Theorem 4.15 for initial values of infinite variance (in the spirit of Theorem 4.14).

4.5 The Complex Ginzburg–Landau Equation

4.5.1 Sufficient Condition for Finite-time Blowup

In this section, we derive sufficient conditions for finite-time blowup in equation (4.1), and upper estimates of the blowup time. Such conditions are obtained in [41, Theorem 1.2] in the case $\gamma \leq 0$, in [6, Theorem 1.1] in the case $\gamma = 0$, and in [41, Theorem 1.3] and [5, Theorem 1.1] in the case $\gamma > 0$. The upper bound is established in [6, Theorem 1.1] in the case $\gamma = 0$.

Theorem 4.17 *Let* $\gamma \in \mathbb{R}$, $\alpha > 0$, $0 \leq \theta < \frac{\pi}{2}$, $u_0 \in C_0(\mathbb{R}^N) \cap H^1(\mathbb{R}^N)$, *and let* u *be the corresponding solution of (4.1) defined on the maximal interval* $[0, T_{\max})$. *If*

$$
\begin{cases}
E(u_0) < 0, & \gamma \geq 0, \\
E_{\frac{\gamma}{\cos\theta}}(u_0) < 0, & \gamma \leq 0
\end{cases}
\tag{4.94}
$$

(with the definitions (4.8) and (4.22)) then u *blows up in finite time, i.e.,* $T_{\max} < \infty$. *Moreover,*

$$
T_{\max} \leq
\begin{cases}
\dfrac{1}{\gamma\alpha} \log\left(1 + \dfrac{\gamma \|u_0\|_{L^2}^2}{(\alpha+2)(-E(u_0))\cos\theta}\right), & \gamma > 0, \\[3ex]
\dfrac{\|u_0\|_{L^2}^2}{\alpha(\alpha+2)(-E(u_0))\cos\theta}, & \gamma = 0, \\[3ex]
\dfrac{1}{-\gamma\alpha} \log\left(1 + \dfrac{-2\gamma \|u_0\|_{L^2}^2}{2(\alpha+2)(-E_{\frac{\gamma}{\cos\theta}}(u_0))\cos\theta - \gamma\alpha\|u_0\|_{L^2}^2}\right), & \gamma < 0.
\end{cases}
\tag{4.95}
$$

Proof We consider separately the cases $\gamma \geq 0$ and $\gamma < 0$.

The case $\gamma \geq 0$ We follow the argument of the proof of Theorem 4.10, and in particular we use the same notation (4.54). We indicate only the minor changes that are necessary. The function $v(t) = e^{-\gamma t} u(t)$ now satisfies the equation

$$
\begin{cases}
v_t = e^{i\theta} [\Delta v + e^{\alpha \gamma t} |v|^\alpha v], \\
v(0) = u_0.
\end{cases}
\tag{4.96}
$$

Identity (4.55) becomes

$$
\int_{\mathbb{R}^N} \overline{v} v_t = -e^{i\theta} \widetilde{f}(t),
\tag{4.97}
$$

so that

$$
\frac{d\widetilde{f}}{dt} = 2\Re \int_{\mathbb{R}^N} \overline{v} v_t = -2\widetilde{f}(t) \cos\theta.
\tag{4.98}
$$

Moreover, $\frac{d\widetilde{e}}{dt} = -\cos\theta \int_{\mathbb{R}^N} |v_t|^2 + \alpha\gamma\widetilde{e} - \frac{\alpha\gamma}{2} \int_{\mathbb{R}^N} |\nabla u|^2$, so that inequality (4.56) becomes

$$
\frac{d\widetilde{e}}{dt} - \alpha\gamma\widetilde{e} \leq -\cos\theta \int_{\mathbb{R}^N} |v_t|^2.
\tag{4.99}
$$

Applying (4.99), Cauchy–Schwarz, (4.97) and (4.98), we obtain

$$
-\widetilde{f}\left(\frac{d\widetilde{e}}{dt} - \alpha\gamma\widetilde{e}\right) \geq \cos\theta \int |v|^2 \int |v_t|^2 \geq \cos\theta \left|\int \overline{v} v_t\right|^2 = \cos\theta \widetilde{f}^2 = \frac{1}{2}(-\widetilde{f})\frac{d\widetilde{f}}{dt}.
\tag{4.100}
$$

The crux is that the factor $\cos\theta$ in the first inequalities in (4.100) has been cancelled in the last one by using (4.98). In particular, the left-hand and the right-hand terms in (4.100) are the same as in (4.58). Therefore, we may now continue the argument as in the proof of Theorem 4.10. Using (4.98) instead of (4.55), we arrive at the inequality

$$
\alpha(\alpha + 2)(-E(u_0)) \|u_0\|_{L^2}^{-(\alpha+2)} e^{\alpha\gamma t} \cos\theta + \frac{d}{dt}(\widetilde{f}^{-\frac{\alpha}{2}}) \leq 0
$$

and estimate (4.95) easily follows in both the cases $\gamma > 0$ and $\gamma = 0$.

The case $\gamma < 0$ Since the result in the case $\gamma \geq 0$ is obtained by the argument of the proof of Theorem 4.10, one could try now to follow the proof of Theorem 4.5. It turns out that this strategy leads to intricate calculations and unnecessary conditions. (See [5].) Instead, we follow the strategy of [41] and we set

$$
\mu = (-\gamma)^{-\frac{1}{2}} (\cos\theta)^{\frac{1}{2}},
\tag{4.101}
$$

$$
v(t,x) = e^{-it\sin\theta} \mu^{\frac{2}{\alpha}} u(\mu^2 t, \mu x),
\tag{4.102}
$$

$$
v_0(x) = \mu^{\frac{2}{\alpha}} u_0(\mu x),
\tag{4.103}
$$

so that

$$\begin{cases} v_t = e^{i\theta}[\Delta v + |v|^\alpha v - v], \\ v(0) = v_0. \end{cases} \quad (4.104)$$

Since u is defined on $[0, T_{\max})$, v is defined on $[0, S_{\max})$ with

$$S_{\max} = \frac{-\gamma T_{\max}}{\cos \theta}. \quad (4.105)$$

We introduce the notation

$$\widetilde{f} = \|v\|_{L^2}^2, \quad \widetilde{j} = I_{-1}(v(t)), \quad \widetilde{e} = E_{-1}(v(t)), \quad (4.106)$$

where I_{-1} and E_{-1} are defined by (4.21) and (4.22), and we observe that

$$\|v_0\|_{L^2}^2 = \mu^{\frac{4}{\alpha}-N}\|u_0\|_{L^2}^2, \quad E_{-1}(v_0) = \mu^{2+\frac{4}{\alpha}-N}E_{\frac{\gamma}{\cos\theta}}(u_0). \quad (4.107)$$

We now follow the proof of Theorem 4.5. Equation (4.104) yields

$$\int_{\mathbb{R}^N} \overline{v}v_t = -e^{i\theta}\widetilde{j}(t), \quad (4.108)$$

$$\frac{d\widetilde{f}}{dt} = -2\widetilde{j}(t)\cos\theta, \quad (4.109)$$

$$\frac{d\widetilde{e}}{dt} = -\cos\theta\int_{\mathbb{R}^N}|v_t|^2. \quad (4.110)$$

Since $E_{\frac{\gamma}{\cos\theta}}(u_0) < 0$, we deduce from (4.107) that $\widetilde{e}(0) < 0$. Therefore, $\widetilde{e}(t) < 0$ by (4.110) (hence $\widetilde{j}(t) < 0$) and $\frac{d\widetilde{f}}{dt} > 0$ by (4.109). Applying (4.110), Cauchy–Schwarz, (4.108) and (4.109), we obtain

$$-\widetilde{f}\frac{d\widetilde{e}}{dt} = \cos\theta\int|v|^2\int|v_t|^2 \geq \cos\theta\left|\int\overline{v}v_t\right|^2 = \cos\theta\widetilde{j}^2 = \frac{1}{2}(-\widetilde{j})\frac{d\widetilde{f}}{dt}. \quad (4.111)$$

At this point, we use the property

$$\widetilde{j}(t) = (\alpha+2)\widetilde{e}(t) - \frac{\alpha}{2}\widetilde{f}(t) - \frac{\alpha}{2}\int_{\mathbb{R}^N}|\nabla v(t)|^2 \leq (\alpha+2)\widetilde{e}(t) - \frac{\alpha}{2}\widetilde{f}(t) < 0, \quad (4.112)$$

so that (4.111) yields $-\widetilde{f}\frac{d\widetilde{e}}{dt} \leq \frac{1}{2}(-(\alpha+2)\widetilde{e} + \frac{\alpha}{2}\widetilde{f})\frac{d\widetilde{f}}{dt}$. Therefore, $\frac{d}{dt}(-\widetilde{e}\widetilde{f}^{-\frac{\alpha+2}{2}} + \frac{1}{2}\widetilde{f}^{-\frac{\alpha}{2}}) \geq 0$, and so

$$-\widetilde{e} + \frac{1}{2}\widetilde{f} \geq \eta\widetilde{f}^{\frac{\alpha+2}{2}}, \quad (4.113)$$

with

$$\eta = (-\widetilde{e}(0))\widetilde{f}(0)^{-\frac{\alpha+2}{2}} + \frac{1}{2}\widetilde{f}(0)^{-\frac{\alpha}{2}} > 0. \quad (4.114)$$

It follows from (4.109), (4.112) and (4.113) that

$$\frac{d\widetilde{f}}{dt} \geq [2(\alpha+2)(-\widetilde{e}) + \alpha\widetilde{f}]\cos\theta \geq (-2\widetilde{f} + 2(\alpha+2)\eta\widetilde{f}^{\frac{\alpha+2}{2}})\cos\theta.$$

Therefore, $\frac{d}{dt}[(e^{2t\cos\theta}\widetilde{f})^{-\frac{\alpha}{2}} - (\alpha+2)\eta e^{-\alpha t\cos\theta}] \leq 0$, so that $(\alpha+2)\eta(1 - e^{-\alpha t\cos\theta}) \leq \widetilde{f}(0)^{-\frac{\alpha}{2}}$ for all $0 \leq t < S_{max}$. It follows easily that

$$S_{max} \leq \frac{1}{\alpha\cos\theta} \log\left(1 + \frac{2\|v_0\|_{L^2}^2}{2(\alpha+2)(-E_{-1}(v_0) + \alpha\|v_0\|_{L^2}^2)}\right). \tag{4.115}$$

Applying (4.105) (4.107), and (4.101), estimate (4.95) follows. $\qquad\square$

Remark 4.18 One can study equation (4.104) for its own sake. The proof of Theorem 4.17 shows that if $v_0 \in C_0(\mathbb{R}^N) \cap H^1(\mathbb{R}^N)$ satisfies $E_{-1}(v_0) \leq 0$ and $v_0 \not\equiv 0$, then the corresponding solution of (4.104) defined on the maximal interval $[0, S_{max})$ blows up in finite time, and (4.115) holds. It follows from (4.115) that

$$S_{max} \leq \frac{1}{\alpha\cos\theta} \log\left(\frac{\alpha+2}{\alpha}\right). \tag{4.116}$$

In particular, the bound in (4.116) is independent of v_0, so that this result does not apply to solutions for which the blowup time would be large. When $\alpha < \frac{4}{N-2}$, this restriction can be improved by the potential well argument we used in Theorem 4.8 for the heat equation. More precisely, if $v_0 \in C_0(\mathbb{R}^N) \cap H^1(\mathbb{R}^N)$ satisfies $E_{-1}(v_0) < E_{-1}(Q_{-1})$ and $I_{-1}(v_0) < 0$, where Q_{-1} is as in Theorem 4.8, then the corresponding solution of (4.104) defined on the maximal interval $[0, S_{max})$. blows up in finite time, i.e., $S_{max} < \infty$, and

$$S_{max} \leq \frac{1}{\alpha\cos\theta}\left[\frac{(\alpha+4)[E_{-1}(v_0)]^+}{E_{-1}(Q_{-1}) - E_{-1}(v_0)} + \log\left(\frac{2(\alpha+2)}{\alpha}\right)\right]. \tag{4.117}$$

The proof is easily adapted from the proof of Theorem 4.8, in the same way as the proof of Theorem 4.17 (case $\gamma \leq 0$) is adapted from the proof of Theorem 4.5. Note that this last result applies to solutions for which the maximal existence time is arbitrary large. Indeed, given $\varepsilon > 0$, $v_0^\varepsilon = (1+\varepsilon)Q_{-1}$ satisfies $E_{-1}(v_0^\varepsilon) < E_{-1}(Q_{-1})$ and $I_{-1}(v_0^\varepsilon) < 0$, while the blowup time of the corresponding solution of (4.104) goes to infinity as $\varepsilon \downarrow 0$. (See Remark 4.9.)

Remark 4.19 Here are some comments on Theorem 4.17.

1. If $\gamma \leq 0$, one can replace assumption (4.94) by the slightly weaker assumption $E_{\frac{\gamma}{\cos\theta}}(u_0) \leq 0$ and $u_0 \not\equiv 0$. See Remark 4.6 (1).
2. Let $\alpha > 0$, $0 \leq \theta < \frac{\pi}{2}$, and fix $u_0 \in C_0(\mathbb{R}^N) \cap H^1(\mathbb{R}^N)$ such that $E(u_0) < 0$. It follows from (4.95) that $T_{max} \to 0$ as $\gamma \to \infty$. On the other hand, it is clear that if γ is sufficiently negative, then $E_{\frac{\gamma}{\cos\theta}}(u_0) \geq 0$, so that one cannot apply Theorem 4.17. This is not surprising, since the corresponding solution of (4.1) is global if γ is sufficiently negative. (See Remark 4.2.)

3. Let $\alpha > 0$ and $\gamma < 0$. Given $u_0 \in C_0(\mathbb{R}^N) \cap H^1(\mathbb{R}^N)$ such that $E_{\frac{\gamma}{\cos\theta}}(u_0) < 0$, it follows from (4.95) that $T_{\max} < \frac{1}{-\gamma\alpha} \log(\frac{\alpha+2}{\alpha})$. In particular, we see that Theorem 4.17 does not apply to solutions for which the blowup time would be large. However, one can use the result presented in Remark 4.18. Using the transformation (4.101)–(4.103), and formulas (4.105) and (4.107), we deduce from (4.117) that if $E_{\frac{\gamma}{\cos\theta}}(u_0) < \mu^{-2-\frac{4}{\alpha}+N}E_{-1}(Q_{-1})$ and $I_{\frac{\gamma}{\cos\theta}}(u_0) < 0$, then the corresponding solution of (4.1) blows up in finite time and

$$
T_{\max} \leq \frac{1}{-\gamma\alpha}\left[\frac{(\alpha+4)[E_{\frac{\gamma}{\cos\theta}}(u_0)]^+}{\mu^{-2-\frac{4}{\alpha}+N}E_{-1}(Q_{-1}) - E_{\frac{\gamma}{\cos\theta}}(u_0)} + \log\left(\frac{2(\alpha+2)}{\alpha}\right)\right].
$$

Moreover, this property applies to solutions for which the maximal existence time is arbitrary large. (The stationary solution Q_{-1} of (4.104) corresponds to the standing wave $u(t,x) = \mu^{-\frac{2}{\alpha}} e^{\mathrm{i}t\mu^{-2}\sin\theta} Q_{-1}(\mu^{-1}x)$ of (4.1).)

4.5.2 Behavior of the Blowup Time as a Function of θ

Fix $\alpha > 0$ and $\gamma \in \mathbb{R}$. Given $u_0 \in C_0(\mathbb{R}^N) \cap H^1(\mathbb{R}^N)$, we let u^θ, for $0 \leq \theta < \frac{\pi}{2}$, be the solution of (4.1) defined on the maximal interval $[0, T^\theta_{\max})$. If $\gamma \geq 0$ and $E(u_0) < 0$, or if $\gamma < 0$ and $E_{\frac{\gamma}{\cos\theta}}(u_0) < 0$, then we know that $T^\theta_{\max} < \infty$. (See Theorem 4.17.) We now study the behavior of $T^\theta_{\max}$ as $\theta \to \frac{\pi}{2}$, i.e. as the equation (4.1) approaches the nonlinear Schrödinger equation (4.3). We consider separately the cases $\alpha < \frac{4}{N}$ and $\alpha \geq \frac{4}{N}$.

The Case $\alpha < \frac{4}{N}$

If $\alpha < \frac{4}{N}$, then all solutions of the limiting equation (4.3) are global by Proposition 4.12, and so we may expect that $T^\theta_{\max} \to \infty$ as $\theta \to \frac{\pi}{2}$. This is indeed what happens. Indeed, one possible proof of global existence for the nonlinear Schrödinger equation (4.3) is based on the Gagliardo–Nirenberg inequality

$$
\|w\|_{L^{\alpha+2}}^{\alpha+2} \leq \|\nabla w\|_{L^2}^2 + A\|w\|_{L^2}^{2+\frac{4\alpha}{4-N\alpha}}, \tag{4.118}
$$

where A is a constant independent of $w \in H^1(\mathbb{R}^N)$. (See e.g. [1].) Similarly, using (4.118) one can prove the following result. (For $\gamma = 0$, this is [6, Theorem 1.2].)

Theorem 4.20 *Let* $0 < \alpha < \frac{4}{N}$ *and* $\gamma \in \mathbb{R}$. *Given* $u_0 \in C_0(\mathbb{R}^N) \cap H^1(\mathbb{R}^N)$, *let* u^θ, *for* $0 \leq \theta < \frac{\pi}{2}$, *be the solution of (4.1) defined on the maximal interval*

$[0, T^\theta_{\max})$. *If $\gamma \geq 0$, then*

$$
T^\theta_{\max} \geq
\begin{cases}
\dfrac{4 - N\alpha}{4\alpha\gamma} \log\left(1 + \dfrac{\gamma}{A\|u_0\|_{L^2}^{\frac{4\alpha}{4-N\alpha}} \cos\theta}\right), & \gamma > 0, \\[4mm]
\dfrac{4 - N\alpha}{4\alpha A\|u_0\|_{L^2}^{\frac{4\alpha}{4-N\alpha}} \cos\theta}, & \gamma = 0,
\end{cases}
\tag{4.119}
$$

and if $\gamma < 0$, then

$$
T^\theta_{\max} \geq
\begin{cases}
\infty, & \cos\theta \leq \dfrac{-\gamma}{A\|u_0\|_{L^2}^{\frac{4\alpha}{4-N\alpha}}}, \\[4mm]
\dfrac{4 - N\alpha}{4\alpha\gamma} \log\left(1 + \dfrac{\gamma}{A\|u_0\|_{L^2}^{\frac{4\alpha}{4-N\alpha}} \cos\theta}\right), & \cos\theta > \dfrac{-\gamma}{A\|u_0\|_{L^2}^{\frac{4\alpha}{4-N\alpha}}},
\end{cases}
\tag{4.120}
$$

where A is the constant in (4.118).

Proof We combine (4.10) and (4.118) to obtain the desired conclusion. Setting $f(t) = \|u^\theta(t)\|_{L^2}^2$, we deduce from (4.10) and (4.118) that

$$
\frac{df}{dt} = 2\gamma f + \cos\theta(-2\|\nabla u\|_{L^2}^2 + 2\|w\|_{L^{\alpha+2}}^{\alpha+2}) \leq 2\gamma f + 2Af^{1 + \frac{2\alpha}{4-N\alpha}} \cos\theta
$$

so that

$$
\frac{d}{dt}(e^{-2\gamma t} f) \leq 2Ae^{\frac{4\alpha\gamma t}{4-N\alpha}} (e^{-2\gamma t} f)^{1 + \frac{2\alpha}{4-N\alpha}} \cos\theta.
$$

This means that $\frac{d}{dt}(-(e^{-2\gamma t} f)^{-\frac{2\alpha}{4-N\alpha}}) \leq \frac{4A\alpha \cos\theta}{4-N\alpha} e^{\frac{4\alpha\gamma t}{4-N\alpha}}$, from which we deduce that

$$
(e^{-2\gamma t} f)^{-\frac{2\alpha}{4-N\alpha}} \geq \|u_0\|^{-\frac{4\alpha}{4-N\alpha}} - \frac{4A\alpha \cos\theta}{4-N\alpha} \int_0^t e^{\frac{4\alpha\gamma s}{4-N\alpha}} ds.
$$

The above inequality yields a control on $\|u(t)\|_{L^2}$ for all $0 \leq t < T^\theta_{\max}$ such that $\frac{4A\alpha \cos\theta}{4-N\alpha} \int_0^t e^{\frac{4\alpha\gamma s}{4-N\alpha}} ds < \|u_0\|^{-\frac{4\alpha}{4-N\alpha}}$. Therefore, if we set

$$
\tau = \sup\left\{0 \leq t < \infty; \frac{4A\alpha \cos\theta}{4-N\alpha} \int_0^t e^{\frac{4\alpha\gamma s}{4-N\alpha}} ds < \|u_0\|^{-\frac{4\alpha}{4-N\alpha}}\right\}
\tag{4.121}
$$

then it follows from the blowup alternative on the L^2-norm in Proposition 4.1 that $T^\theta_{\max} \geq \tau$. The result follows by calculating the integral in (4.121) in the various cases $\gamma > 0$, $\gamma = 0$ and $\gamma < 0$. $\qquad\square$

Remark 4.21 Here are some comments on Theorem 4.20.

1. Suppose $\gamma \geq 0$. It follows from (4.120) that the upper estimate (4.95) of $T^\theta_{\max}$ established in Theorem 4.17 is optimal with respect to the dependence

in θ. Indeed, if $E(u_0) < 0$, then it follows from (4.120) and (4.95) that

$$0 < \liminf_{\theta \to \frac{\pi}{2}} \phi(\theta) T^{\theta}_{\max} \leq \limsup_{\theta \to \frac{\pi}{2}} \phi(\theta) T^{\theta}_{\max} < \infty,$$

with $\phi(\theta) = \cos\theta$ for $\gamma = 0$ and $\phi(\theta) = [\log((\cos\theta)^{-1})]^{-1}$ for $\gamma > 0$.

2. Suppose $\gamma < 0$. It follows from (4.120) that, given any initial value $u_0 \in C_0(\mathbb{R}^N) \cap H^1(\mathbb{R}^N)$, the corresponding solution of (4.1) is global for all θ sufficiently close to $\frac{\pi}{2}$.

3. We can apply Theorem 4.20 to equation (4.104). The upper estimate (4.120), together with formulas (4.107) and (4.105), shows that if $v_0 \in C_0(\mathbb{R}^N) \cap H^1(\mathbb{R}^N)$ and v is the corresponding solution of (4.104) defined on the maximal interval $[0, S_{\max})$, then $S_{\max} = \infty$ if $A\|v_0\|_{L^2}^{\frac{4\alpha}{4-N\alpha}} \leq 1$, and

$$S_{\max} \geq -\frac{4 - N\alpha}{4\alpha\cos\theta} \log\left(1 - \frac{1}{A\|v_0\|_{L^2}^{\frac{4\alpha}{4-N\alpha}}}\right) \tag{4.122}$$

if $A\|v_0\|_{L^2}^{\frac{4\alpha}{4-N\alpha}} > 1$. Suppose now that either $E_{-1}(v_0) \leq 0$ and $v_0 \neq 0$ or else $0 < E_{-1}(v_0) < E_{-1}(Q_{-1})$ and $I_{-1}(v_0) < 0$. In both cases $I_{-1}(v_0) < 0$, and it follows from (4.118) that

$$\|\nabla v_0\|_{L^2}^2 + \|v_0\|_{L^2}^2 < \|v_0\|_{L^{\alpha+2}}^{\alpha+2} \leq \|\nabla v_0\|_{L^2}^2 + A\|v_0\|_{L^2}^{2+\frac{4\alpha}{4-N\alpha}}.$$

In particular, $A\|v_0\|_{L^2}^{\frac{4\alpha}{4-N\alpha}} > 1$ so that

$$0 < \liminf_{\theta \to \frac{\pi}{2}} (\cos\theta) S^{\theta}_{\max} \leq \limsup_{\theta \to \frac{\pi}{2}} (\cos\theta) S^{\theta}_{\max} < \infty$$

by (4.122) and either (4.115) or (4.117).

The Case $\alpha > \frac{4}{N}$

If $\frac{4}{N} \leq \alpha < \frac{4}{N-2}$, then the solution of the limiting nonlinear Schrödinger equation (4.3) blows up in finite time, under appropriate assumptions on the initial value u_0. See Theorems 4.13, 4.14 and 4.15. Under these assumptions, one might expect that $T^{\theta}_{\max}(u_0)$, which is finite (under suitable assumptions) by Theorem 4.17, remains bounded as $\theta \to \frac{\pi}{2}$. It appears that no complete answer is known to this problem.

We first consider the case $\gamma \geq 0$, for which one can give a partial answer. If $E(u_0) < 0$, then $T^{\theta}_{\max} < \infty$ for all $0 \leq \theta < \frac{\pi}{2}$ by Theorem 4.17. However, the bound in (4.95) blows up as $\theta \to \frac{\pi}{2}$. The proof of (4.95) is based on Levine's argument for blowup in the nonlinear heat equation (4.2). Since, as observed before, Levine's argument does not apply to the limiting nonlinear Schrödinger equation, it is not surprising that the bound in (4.95) becomes inaccurate as $\theta \to \frac{\pi}{2}$. This observation suggests we should adapt the proof of blowup for (4.3) to equation (4.1), in order to obtain a bound on $T_{\max}^{\theta}$ as $\theta \to \frac{\pi}{2}$. This strategy proved to be successful in [6, Theorem 1.5] for $\gamma = 0$, and it can be extended to the case $\gamma \geq 0$. More precisely, we have the following result.

Theorem 4.22 *Let* $\frac{4}{N} \leq \alpha \leq 4$, $\gamma \geq 0$ *and* $N \geq 2$. *Assume that* $N \geq 3$ *if* $\gamma > 0$. *Let* $u_0 \in H^1(\mathbb{R}^N) \cap C_0(\mathbb{R}^N)$ *be radially symmetric and, given any* $0 \leq \theta < \frac{\pi}{2}$, *let* u^{θ} *be the corresponding solution of (4.1) defined on the maximal interval* $[0, T_{\max}^{\theta})$. *If* $E(u_0) < 0$, *then* $\sup_{0 \leq \theta < \frac{\pi}{2}} T_{\max}^{\theta} < \infty$.

We prove Theorem 4.22 by following the strategy of [6]. We adapt the proof of Theorem 4.14, and in particular we consider $v(t) = \mathrm{e}^{-\gamma t} u(t)$, which satisfies equation (4.96). The corresponding identities for the truncated variance are given by Proposition 4.31; and the Caffarelli–Kohn–Nirenberg estimate by Lemma 4.32. The terms involving $\cos \theta$ in (4.152) can be controlled using (4.98). Yet there is a major difference between equations (4.3) and (4.1) that must be taken care of. For (4.3), the L^2-norm of the solutions is controlled by formula (4.14). The resulting estimate for v is essential when applying Lemma 4.32. For (4.1), there is no such *a priori* estimate. However, one can estimate v on an interval $[0, T]$, where T is proportional to $T_{\max}^{\theta}$. More precisely, we have the following result, similar to [6, Lemma 5.2].

Lemma 4.23 *Fix* $0 \leq \theta < \frac{\pi}{2}$. *Let* $u_0 \in C_0(\mathbb{R}^N) \cap H^1(\mathbb{R}^N)$, *and consider the corresponding solution* v *of (4.96) defined on the maximal interval* $[0, T_{\max})$. *Set*

$$\tau = \sup\{t \in [0, T_{\max}); \|v(s)\|_{L^2}^2 \leq K\|u_0\|_{L^2}^2 \text{ for } 0 \leq s \leq t\}, \qquad (4.123)$$

where

$$K = \left[1 - \left(\frac{\alpha + 4}{2\alpha + 4}\right)^{\frac{1}{2}}\right]^{-1} > 1 \qquad (4.124)$$

so that $0 < \tau \leq T_{\max}$. *If* $E(u_0) < 0$, *then* $T_{\max} \leq \frac{\alpha + 4}{\alpha} \tau$.

Proof We use the notation of the proof of Theorem 4.17, and in particular (4.54). The proof is based Levine's argument used in the proof of Theorem 4.4, which shows that if $\|v\|_{L^2}^2$ achieves the value $K\|u_0\|_{L^2}^2$ at a certain time t, then v must blow up within a lapse of time which is controlled by t. More precisely, let τ be given by (4.123). If $\tau = T_{\max}$, there is nothing to prove, so we assume $\tau < T_{\max}$. It follows that $\|v(\tau)\|_{L^2}^2 = K\|u_0\|_{L^2}^2$, so that

$$\widetilde{f}(t) \leq \widetilde{f}(\tau) = K\widetilde{f}(0), \quad 0 \leq t \leq \tau. \qquad (4.125)$$

Since $E(u_0) < 0$, it follows (see the proof of Theorem 4.17) that $\widetilde{f}$ is nondecreasing on $[0, T_{\max})$; and so, using (4.125),

$$\widetilde{f}(t) \geq K\widetilde{f}(0), \quad \tau \leq t < T_{\max}. \tag{4.126}$$

We deduce from (4.99) that

$$e^{-\alpha\gamma t}\widetilde{e}(t) \leq E(u_0) - \cos\theta \int_0^t e^{-\alpha\gamma s} \int_{\mathbb{R}^N} |v_t(s)|^2,$$

so that

$$\widetilde{e}(t) \leq -\cos\theta \int_0^t \int_{\mathbb{R}^N} |v_t|^2. \tag{4.127}$$

Since $\frac{d\widetilde{f}}{dt} \geq -2\cos\theta\,\widetilde{J} \geq -2(\alpha+2)\cos\theta\,\widetilde{e}$ by (4.98), we deduce from (4.127) that

$$\frac{d\widetilde{f}}{dt} \geq 2(\alpha+2)\cos^2\theta \int_0^t \int_{\mathbb{R}^N} |v_t|^2. \tag{4.128}$$

Set

$$\widetilde{M}(t) = \frac{1}{2}\int_0^t \widetilde{f}(s)\,ds. \tag{4.129}$$

It follows from (4.128) and Cauchy–Schwarz's inequality that (see the proof of Theorem 4.4)

$$\widetilde{M}\widetilde{M}'' \geq \frac{\alpha+2}{2}\cos^2\theta\left(\int_0^t\left|\int_{\mathbb{R}^N} v_t\overline{v}\right|\right)^2. \tag{4.130}$$

Since $\widetilde{J} \leq (\alpha+2)\widetilde{e} \leq 0$, identities (4.97) and (4.98) yield

$$\left|\int_{\mathbb{R}^N} v_t\overline{v}\right| = -\widetilde{J} = \frac{1}{2\cos\theta}\frac{d\widetilde{f}}{dt} = \frac{1}{\cos\theta}\widetilde{M}''(t)$$

so that (4.130) becomes

$$\widetilde{M}\widetilde{M}'' \geq \frac{\alpha+2}{2}(\widetilde{M}'(t) - \widetilde{M}'(0))^2 = \frac{\alpha+2}{8}(\widetilde{f}(t) - \widetilde{f}(0))^2. \tag{4.131}$$

It follows from (4.131), (4.126) and (4.124) that

$$\widetilde{M}\widetilde{M}'' \geq \frac{\alpha+2}{8}\left(\frac{K-1}{K}\right)^2\widetilde{f}(t)^2 = \frac{\alpha+4}{16}\widetilde{f}(t)^2 = \frac{\alpha+4}{4}[\widetilde{M}'(t)]^2 \tag{4.132}$$

for all $\tau \leq t < T_{\max}$. This means that $(\widetilde{M}^{-\frac{\alpha}{4}})'' \leq 0$ on $[\tau, T_{\max})$; and so

$$\widetilde{M}(t)^{-\frac{\alpha}{4}} \leq \widetilde{M}(\tau)^{-\frac{\alpha}{4}} + (t-\tau)(\widetilde{M}^{-\frac{\alpha}{4}})'(\tau) = \widetilde{M}(\tau)^{-\frac{\alpha}{4}}\left[1 - \frac{\alpha}{4}(t-\tau)\widetilde{M}(\tau)^{-1}\widetilde{M}'(\tau)\right]$$

for $\tau \leq t < T_{\max}$. Since $\tilde{M}(t)^{-\frac{\alpha}{4}} > 0$, we deduce that for every $\tau \leq t < T_{\max}$, $\frac{\alpha}{4}(t-\tau)\tilde{M}(\tau)^{-1}\tilde{M}'(\tau) \leq 1$, i.e.,

$$(t-\tau)\tilde{f}(\tau) \leq \frac{4}{\alpha}\int_0^\tau \tilde{f}(s)\,\mathrm{d}s \leq \frac{4}{\alpha}\tau\tilde{f}(\tau), \tag{4.133}$$

where we used (4.125) in the last inequality. Thus $t \leq \frac{\alpha+4}{\alpha}\tau$ for all $\tau \leq t < T_{\max}$, which proves the desired inequality. $\qquad\square$

Proof of Theorem 4.22 We set $v^\theta(t) = \mathrm{e}^{-\gamma t}u^\theta(t)$, thus v^θ is the solution of (4.96) on $[0, T_{\max}^\theta)$. We let K be defined by (4.124) and we set

$$\tau_\theta = \sup\{t \in [0, T_{\max}^\theta); \|v^\theta(s)\|_{L^2}^2 \leq K\|u_0\|_{L^2}^2 \text{ for } 0 \leq s \leq t\}. \tag{4.134}$$

Therefore,

$$\sup_{0\leq\theta<\frac{\pi}{2}} \sup_{0\leq t<\tau_\theta} \|v^\theta(t)\|_{L^2}^2 \leq K\|u_0\|_{L^2}^2 \tag{4.135}$$

and, by Lemma 4.23,

$$T_{\max}^\theta \leq \frac{\alpha+4}{\alpha}\tau_\theta \tag{4.136}$$

so that we only need a bound on τ_θ. We first derive an inequality (4.141) by calculating a truncated variance. Let $\Psi \in C^\infty(\mathbb{R}^N) \cap W^{4,\infty}(\mathbb{R}^N)$ be real-valued, nonnegative, and radially symmetric. We set

$$\zeta_\theta(t) = \int_{\mathbb{R}^N} \Psi(x)|v^\theta(t,x)|^2\mathrm{d}x,$$

$$\mathcal{H}^\theta(t) = \int_{\mathbb{R}^N}\left\{-2(2-\Psi'')|v_r^\theta|^2 + \frac{\alpha}{\alpha+2}\mathrm{e}^{\alpha\gamma t}(2N-\Delta\Psi)|v^\theta|^{\alpha+2} - \frac{1}{2}|v^\theta|^2\Delta^2\Psi\right\},$$

$$\mathcal{K}^\theta(t) = \int_{\mathbb{R}^N}\left\{-2\Psi|v_r^\theta|^2 + \frac{\alpha+4}{\alpha+2}\mathrm{e}^{\alpha\gamma t}\Psi|v^\theta|^{\alpha+2} + |v^\theta|^2\Delta\Psi\right\}$$

and we observe that

$$-\mathcal{K}^\theta(0) \leq C(\|\Psi\|_{L^\infty} + \|\Delta\Psi\|_{L^\infty})\|u_0\|_{H^1}^2. \tag{4.137}$$

We apply Proposition 4.31 with $f(t) \equiv \mathrm{e}^{\alpha\gamma t}$. It follows from (4.150) that

$$\frac{1}{2}\zeta_\theta'(0) \leq C(\|\Psi\|_{L^\infty} + \|\nabla\Psi\|_{L^\infty} + \|\Delta\Psi\|_{L^\infty})(1 + \|u_0\|_{H^1}^{\alpha+2}) \tag{4.138}$$

and from (4.152) that

$$\frac{1}{2}\zeta_\theta'' \leq -\frac{1}{2}\int_{\mathbb{R}^N}|v^\theta|^2\Delta^2\Psi - \frac{\alpha\mathrm{e}^{\alpha\gamma t}}{\alpha+2}\int_{\mathbb{R}^N}|v^\theta|^{\alpha+2}\Delta\Psi + 2\int_{\mathbb{R}^N}\Psi''|v_r^\theta|^2$$
$$+ \cos\theta\frac{\mathrm{d}}{\mathrm{d}t}\mathcal{K}^\theta. \tag{4.139}$$

Using the identity

$$-\frac{1}{2}\int_{\mathbb{R}^N}|v^\theta|^2\Delta^2\Psi - \frac{\alpha e^{\alpha\gamma t}}{\alpha+2}\int_{\mathbb{R}^N}|v^\theta|^{\alpha+2}\Delta\Psi + 2\int_{\mathbb{R}^N}\Psi''|v_r^\theta|^2,$$

$$= 2N\alpha\widetilde{e}(t) + \mathcal{H}^\theta(t) - (N\alpha-4)\int_{\mathbb{R}^N}|v_r^\theta|^2$$

where $\widetilde{e}(t)$ is defined by (4.54), the estimate $\widetilde{e}(t) \le e^{\alpha\gamma t}E(u_0)$ by (4.99), and the assumption $N\alpha \ge 4$, we deduce from (4.139) that

$$\frac{1}{2}\zeta_\theta'' \le 2N\alpha e^{\alpha\gamma t}E(u_0) + \mathcal{H}^\theta(t) + \cos\theta\frac{d}{dt}\mathcal{K}^\theta. \tag{4.140}$$

Integrating the above inequality twice, and since $\zeta_\theta \ge 0$, we obtain

$$0 \le \tau_\theta\left(\frac{1}{2}\zeta_\theta'(0) - \cos\theta\mathcal{K}^\theta(0)\right) + 2N\alpha E(u_0)\int_0^{\tau_\theta}\int_0^s e^{\alpha\gamma\sigma}\,d\sigma\,ds$$

$$+\int_0^{\tau_\theta}\int_0^s \mathcal{H}^\theta(\sigma)\,d\sigma\,ds + \cos\theta\int_0^{\tau_\theta}\mathcal{K}^\theta(s)\,ds. \tag{4.141}$$

We derive a bound on τ_θ from (4.141). In order to do so we show that, for large time, the dominating term in the right-hand side is the middle one, which is negative.

We first obtain an estimate of the last term in (4.141), for which the factor $\cos\theta$ is essential. Indeed, it follows from (4.98) that

$$\frac{d}{dt}\int_{\mathbb{R}^N}|v^\theta|^2 = 2\cos\theta\left(-2\widetilde{e}(t) + \frac{\alpha}{\alpha+2}e^{\alpha\gamma t}\int_{\mathbb{R}^N}|v^\theta|^{\alpha+2}\right),$$

where $\widetilde{e}(t)$ is defined by (4.54). Since $\widetilde{e}(t) \le 0$ by (4.99), we deduce by integrating on $(0,\tau_\theta)$ and applying (4.135) that

$$\frac{2\alpha}{\alpha+2}\cos\theta\int_0^t e^{\alpha\gamma s}\int_{\mathbb{R}^N}|v^\theta|^{\alpha+2} \le (K-1)\|u_0\|_{L^2}^2$$

for all $0 \le t \le \tau_\theta$. It follows that

$$\cos\theta\int_0^{\tau_\theta}\mathcal{K}^\theta(s)\,ds \le C(\|\Psi\|_{L^\infty} + \|\Delta\Psi\|_{L^\infty})\|u_0\|_{L^2}^2, \tag{4.142}$$

where we used (4.135) again to estimate the factor of $\Delta\Psi$.

We conclude by estimating the term involving $\mathcal{H}^\theta$ in (4.141) with Lemma 4.32. We first consider the case $\gamma = 0$. We apply Lemma 4.32 with $A = \|u_0\|_{L^2}$, $\mu = \frac{\alpha}{\alpha+2}$ and $\varepsilon > 0$ chosen sufficiently small so that $\chi\mu\varepsilon^{2(N-1)} < 1$ and $\kappa(\mu,\varepsilon) \le -N\alpha E(u_0)$. With $\Psi = \Psi_\varepsilon$ given by Lemma 4.32, it follows from (4.160) that $\mathcal{H}^\theta(t) \le -N\alpha E(u_0)$ for all $0 \le t < \tau_\theta$. Therefore, we deduce from (4.141) and (4.142) that

$$0 \leq \tau_\theta \left(\frac{1}{2} \zeta_\theta'(0) - \cos\theta \mathcal{K}^\theta(0) \right) + N\alpha \frac{(\tau_\theta)^2}{2} E(u_0).$$

Since $E(u_0) < 0$, we conclude that $\sup_{0 \leq \theta < \frac{\pi}{2}} \tau_\theta < \infty$.

We next consider the case $\gamma > 0$ (and so $N \geq 3$). We apply Lemma 4.32, this time with with $A = \|u_0\|_{L^2}$ and

$$\mu = \mu_\theta = e^{\alpha\gamma\tau_\theta}. \tag{4.143}$$

The additional difficulty with respect to the case $\gamma = 0$ is that μ_θ may, in principle, be large. We fix λ satisfying (4.75) (we use the assumption $N \geq 3$) and we set

$$\varepsilon_\theta = a\mu_\theta^{-\lambda} \leq a. \tag{4.144}$$

Here, $0 < a \leq 1$ is chosen sufficiently small so that $\chi a^{2(N-1)} < 1$, where χ is the constant in Lemma 4.32. Since $\mu_\theta \geq 1$ and $1 - 2(N-1)\lambda < 0$ by (4.75), it follows in particular that $\chi \mu_\theta \varepsilon_\theta^{2(N-1)} = \chi \mu_\theta^{1-2(N-1)\lambda} a^{2(N-1)} \leq \chi a^{2(N-1)} < 1$. Moreover, we deduce from (4.75) that κ defined by (4.161) satisfies $\kappa(\mu_\theta, \varepsilon_\theta) \leq C\mu_\theta^{1-\delta}$, where C is independent of θ, and $\delta > 0$ is given by (4.77). We now let $\Psi = \Psi_{\varepsilon_\theta}$ where Ψ_ε is given by Lemma 4.32 for this choice of ε, and it follows from (4.159) and (4.160) that

$$\mathcal{H}^\theta(t) \leq Ce^{(1-\delta)\alpha\gamma\tau_\theta}, \tag{4.145}$$

and from (4.158), (4.144) and (4.143) that

$$\|\Psi_{\varepsilon_\theta}\|_{L^\infty} + \|\nabla\Psi_{\varepsilon_\theta}\|_{L^\infty} + \|\Delta\Psi_{\varepsilon_\theta}\|_{L^\infty} \leq Ce^{2\lambda\alpha\gamma\tau_\theta}. \tag{4.146}$$

Finally, we estimate the first term in the right-hand side of (4.141) and we deduce from (4.138), (4.137), (4.146) and (4.143) that

$$\frac{1}{2}\zeta_\theta'(0) - \cos\theta \mathcal{K}^\theta(0) \leq Ce^{2\lambda\alpha\gamma\tau_\theta}. \tag{4.147}$$

Estimates (4.141), (4.147), (4.145), (4.142), (4.146) and (4.143) now yield

$$0 \leq C(1+\tau_\theta)e^{2\lambda\alpha\gamma\tau_\theta} + \frac{2N}{\alpha\gamma^2}E(u_0)(e^{\alpha\gamma\tau_\theta} - 1 - \alpha\gamma\tau_\theta) + C\tau_\theta^2 e^{(1-\delta)\alpha\gamma\tau_\theta}. \tag{4.148}$$

Since $E(u_0) < 0$, and $\max\{2\lambda, 1-\delta\} < 1$, the right-hand side of the above inequality is negative if τ_θ is large. Thus $\sup_{0 \leq \theta < \frac{\pi}{2}} \tau_\theta < \infty$, which completes the proof. $\square$

Remark 4.24 Under the assumptions of Theorem 4.22, we know that T_{max}^θ remains bounded. On the other hand, we do not know if T_{max}^θ has a limit as

$\theta \to \frac{\pi}{2}$, and if it does, if this limit is the blowup time of the solution of the limiting Schrödinger equation.

We end this section by considering the case $\gamma < 0$. The condition for blowup in Theorem 4.17 in this case is $E_{\frac{\gamma}{\cos\theta}}(u_0) < 0$. Given $u_0 \in C_0(\mathbb{R}^N) \cap H^1(\mathbb{R}^N)$, $u_0 \neq 0$, it is clear that $E_{\frac{\gamma}{\cos\theta}}(u_0) > 0$ for all θ sufficiently close to $\frac{\pi}{2}$, and we do not know if there exists an initial value u_0 such that the corresponding solution of (4.1) blows up in finite time for all θ close to $\frac{\pi}{2}$. (See Open Problem 4.28.)

Another point of view concerning the case $\gamma < 0$ is to apply the transformation (4.101)–(4.103) and study the resulting equation (4.104). Let $v_0 \in C_0(\mathbb{R}^N) \cap H^1(\mathbb{R}^N)$ and, given $0 \leq \theta < \frac{\pi}{2}$, let v^θ the corresponding solution of (4.104) defined on the maximal interval $[0, S_{\max}^\theta)$. If $E_{-1}(v_0) < 0$, then it follows from (4.116) that $S_{\max}^\theta < \infty$ for all $0 \leq \theta < \frac{\pi}{2}$. Therefore, it makes sense to study the behavior of $S_{\max}^\theta$ as $\theta \to \frac{\pi}{2}$, and we have the following result.

Theorem 4.25 *Suppose $N \geq 2$, $\frac{4}{N} \leq \alpha \leq 4$, and fix a radially symmetric initial value $v_0 \in H^1(\mathbb{R}^N) \cap C_0(\mathbb{R}^N)$. Given any $0 \leq \theta < \frac{\pi}{2}$, let v^θ be the corresponding solution of (4.104) defined on the maximal interval $[0, S_{\max}^\theta)$. If $E_{-1}(v_0) < 0$, then $\sup_{0 \leq \theta < \frac{\pi}{2}} T_{\max}^\theta < \infty$.*

The proof of Theorem 4.25 is very similar to the proof of [6, Theorem 1.5], with minor modifications only. More precisely, it is not difficult to adapt the proof of Lemma 4.23 to show that if

$$\tau_\theta = \sup\{t \in [0, S_{\max}^\theta); \|v^\theta(s)\|_{L^2}^2 \leq K \|v_0\|_{L^2}^2 \text{ for } 0 \leq s \leq t\},$$

where K is defined by (4.124), then $S_{\max}^\theta \leq \frac{\alpha+4}{\alpha} \tau_\theta$. Moreover, given a real-valued, radially symmetric function $\Psi \in C^\infty(\mathbb{R}^N) \cap W^{4,\infty}(\mathbb{R}^N)$, and setting

$$\zeta_\theta(t) = \int_{\mathbb{R}^N} \Psi(x)|v^\theta(t,x)|^2 dx,$$

it is not difficult to deduce from Proposition 4.31 the variance identities

$$\frac{1}{2}\zeta_\theta'(t) = \cos\theta \int_{\mathbb{R}^N} \left\{ -\Psi|v_r^\theta|^2 + \Psi|v^\theta|^{\alpha+2} - \Psi|v^\theta|^2 + \frac{1}{2}|v^\theta|^2 \Delta\Psi \right\}$$
$$+ \sin\theta \Im \int_{\mathbb{R}^N} \overline{v^\theta}(\nabla v^\theta \cdot \nabla\Psi)$$

and

$$\frac{1}{2}\zeta_\theta''(t) \leq 2N\alpha E_{-1}(v^\theta)$$

$$+ \int_{\mathbb{R}^N} \left\{ -2(2 - \Psi'')|v_r^\theta|^2 + \frac{\alpha}{\alpha+2}(2N - \Delta\Psi)|v^\theta|^{\alpha+2} - \frac{1}{2}(2N\alpha + \Delta^2\Psi)|v^\theta|^2 \right\}$$

$$+ \cos\theta \frac{d}{dt} \int_{\mathbb{R}^N} \left\{ -2\Psi|v_r^\theta|^2 + \frac{\alpha+4}{\alpha+2}\Psi|v^\theta|^{\alpha+2} + (\Delta\Psi - 2\Psi)|v^\theta|^2 \right\}.$$

One can then conclude exactly as in the case $\gamma = 0$ of the proof of Theorem 4.22.

4.6 Some Open Problems

Open problem 4.26 Suppose $\frac{4}{N} < \alpha < \frac{4}{N-2}$. Let $u_0 \in H^1(\mathbb{R}^N)$, $u_0 \neq 0$ and let u be the corresponding solution of (4.3). It follows from [33, Theorem 1] that, if γ is sufficiently negative, then u is global. Does u blow up in finite time for $\gamma > 0$ sufficiently large (this is true if $E(u_0) < 0$ and $u_0 \in L^2(\mathbb{R}^N, |x|^2 dx)$, by Theorem 4.13), or does there exist $u_0 \neq 0$ such that u is global for all $\gamma > 0$?

Open problem 4.27 Let $0 \leq \theta < \frac{\pi}{2}$, $u_0 \in C_0(\mathbb{R}^N) \cap H^1(\mathbb{R}^N)$, $u_0 \neq 0$, and let u be the corresponding solution of (4.1). If γ is sufficiently negative, then u is global, by Remark 4.2. Does u blow up in finite time for all sufficiently large $\gamma > 0$ (this is true if $E(u_0) < 0$, by Theorem 4.17), or does there exist $u_0 \neq 0$ such that u is global for all $\gamma > 0$? (The question is open even for the nonlinear heat equation (4.2).)

Open problem 4.28 Let $\gamma < 0$. Does there exist an initial value $u_0 \in C_0(\mathbb{R}^N) \cap H^1(\mathbb{R}^N)$, $u_0 \neq 0$ such that the corresponding solution of (4.1) blows up in finite time for all θ close to $\frac{\pi}{2}$? One possible strategy for constructing such initial values when $\frac{4}{N} < \alpha < \frac{4}{N-2}$ would be to adapt the proof of Theorem 4.15 to equation (4.1).

Open problem 4.29 Suppose $\gamma \geq 0$ and $0 \leq \theta < \frac{\pi}{2}$. The sufficient condition for blowup in Theorem 4.17 is $E(u_0) < 0$. Does there exist a constant $\kappa > \frac{2}{\alpha+2}$ such that the (weaker) condition $\int_{\mathbb{R}^N} |\nabla u_0|^2 - \kappa \int_{\mathbb{R}^N} |u_0|^{\alpha+2} < 0$ implies finite-time blowup? Note that for the equation with $\gamma = 0$ set on a bounded domain with Dirichlet boundary conditions, $\kappa = 1$ is not admissible. Indeed, there exist initial values for which $I(u_0) < 0$ and $T_{max} = \infty$. (See [9, Theorem 1].)

Open problem 4.30 Theorems 4.20 and 4.25 require that $\alpha \leq 4$ and the solution is radially symmetric. Are these assumptions necessary? Note that they are necessary in Lemma 4.32 (see Section 6 in [6]) which is an essential tool in our proof. Could these assumptions be replaced by stronger decay conditions on the initial value, such as $u_0 \in L^2(\mathbb{R}^N, |x|^2 dx)$? In particular, one could think of adapting the proof of Theorem 4.13 (instead of the proof of Theorem 4.14), but this does not seem to be simple, see Section 7 in [6].

4.7 A Truncated Variance Identity

We prove the following result, which is a slightly more general form of [6, Lemma 5.1].

Proposition 4.31 *Fix $\alpha > 0$, $0 \le \theta < \frac{\pi}{2}$, and a real-valued function $\Psi \in C^\infty(\mathbb{R}^N) \cap W^{4,\infty}(\mathbb{R}^N)$. Let $u_0 \in C_0(\mathbb{R}^N) \cap H^1(\mathbb{R}^N)$, $f \in C^1(\mathbb{R}, \mathbb{R})$, and consider the corresponding solution v of*

$$\begin{cases} v_t = e^{i\theta}[\Delta v + f(t)|v|^\alpha v], \\ v(0) = u_0, \end{cases} \tag{4.149}$$

defined on the maximal interval $[0, T_{\max})$. If ζ is defined by

$$\zeta(t) = \int_{\mathbb{R}^N} \Psi(x)|v(t,x)|^2 dx$$

then $\zeta \in C^2([0, T_{\max}))$,

$$\frac{1}{2}\zeta'(t) = \cos\theta \int_{\mathbb{R}^N} \left\{ -\Psi|\nabla v|^2 + f(t)\Psi|v|^{\alpha+2} + \frac{1}{2}|v|^2 \Delta \Psi \right\} \\ + \sin\theta \, \Im \int_{\mathbb{R}^N} \overline{v}(\nabla v \cdot \nabla \Psi) \tag{4.150}$$

and

$$\frac{1}{2}\zeta''(t) = \int_{\mathbb{R}^N} \left\{ -\frac{1}{2}|v|^2 \Delta^2 \Psi - \frac{\alpha f(t)}{\alpha+2}|v|^{\alpha+2}\Delta\Psi + 2\Re\langle H(\Psi)\nabla\overline{v}, \nabla v\rangle \right\} \\ + \cos\theta \frac{d}{dt} \int_{\mathbb{R}^N} \left\{ -2\Psi|\nabla v|^2 + \frac{\alpha+4}{\alpha+2}f(t)\Psi|v|^{\alpha+2} + |v|^2 \Delta\Psi \right\} \\ - 2\cos^2\theta \int_{\mathbb{R}^N} \Psi|v_t|^2 - \frac{2f'(t)}{\alpha+2}\cos\theta \int_{\mathbb{R}^N} \Psi|v|^{\alpha+2} \tag{4.151}$$

for all $0 \leq t < T_{\max}$, where $H(\Psi)$ is the Hessian matrix $(\partial_{ij}^2 \Psi)_{i,j}$. In particular, if both Ψ and u_0 (hence, v) are radially symmetric, then

$$
\begin{aligned}
\frac{1}{2}\zeta''(t) = & \int_{\mathbb{R}^N} \left\{ -\frac{1}{2}|v|^2 \Delta^2 \Psi - \frac{\alpha f(t)}{\alpha + 2}|v|^{\alpha+2}\Delta\Psi + 2\Psi''|v_r|^2 \right\} \\
& + \cos\theta \frac{d}{dt} \int_{\mathbb{R}^N} \left\{ -2\Psi|v_r|^2 + \frac{\alpha+4}{\alpha+2}f(t)\Psi|v|^{\alpha+2} + |v|^2\Delta\Psi \right\} \quad (4.152) \\
& - 2\cos^2\theta \int_{\mathbb{R}^N} \Psi|v_t|^2 - \frac{2f'(t)}{\alpha+2}\cos\theta \int_{\mathbb{R}^N} \Psi|v|^{\alpha+2}
\end{aligned}
$$

for all $0 \leq t < T_{\max}$.

Proof Multiplying the equation (4.149) by $\Psi(x)\overline{v}$, taking the real part, and using the identity

$$
2\Re[\overline{v}(\nabla v \cdot \nabla\Psi)] = \nabla \cdot (|v|^2\nabla\Psi) - |v|^2\Delta\Psi
$$

we obtain (4.150). Next, the identity

$$
\overline{v}(\nabla v_t \cdot \nabla\Psi) = \nabla \cdot (v_t\overline{v}\nabla\Psi) - (\nabla\Psi \cdot \nabla\overline{v})v_t - \overline{v}v_t\Delta\Psi
$$

and integration by parts yield

$$
\frac{d}{dt}\left(\sin\theta\Im \int_{\mathbb{R}^N} \overline{v}(\nabla v \cdot \nabla\Psi) \right) = -\sin\theta\Im \int_{\mathbb{R}^N} [\overline{v}\Delta\Psi + 2\nabla\Psi \cdot \nabla\overline{v}]v_t.
$$

We rewrite this last identity in the form

$$
\begin{aligned}
\frac{d}{dt}\left(\sin\theta\Im \int_{\mathbb{R}^N} \overline{v}(\nabla v \cdot \nabla\Psi) \right) = & \cos\theta \Re \int_{\mathbb{R}^N} [\overline{v}\Delta\Psi + 2\nabla\Psi \cdot \nabla\overline{v}]v_t \\
& - \Re \int_{\mathbb{R}^N} [\overline{v}\Delta\Psi + 2\nabla\Psi \cdot \nabla\overline{v}]e^{-i\theta}v_t.
\end{aligned}
\quad (4.153)
$$

Using (4.149) and the identities

$$
\Re[(\nabla\Psi \cdot \nabla\overline{v})|v|^\alpha v] = \frac{1}{\alpha+2}\nabla \cdot (|v|^{\alpha+2}\nabla\Psi) - \frac{1}{\alpha+2}|v|^{\alpha+2}\Delta\Psi,
$$

$$
\begin{aligned}
\Re[\Delta v(\overline{v}\Delta\Psi + 2\nabla\overline{v} \cdot \nabla\Psi)] = & \Re\nabla \cdot \left[\nabla v(\overline{v}\Delta\Psi + 2\nabla\overline{v} \cdot \nabla\Psi) - |\nabla v|^2\nabla\Psi \right. \\
& \left. - \frac{1}{2}|v|^2\nabla(\Delta\Psi) \right] - 2\Re\langle H(\Psi)\nabla\overline{v}, \nabla v\rangle \\
& + \frac{1}{2}|v|^2\Delta^2\Psi,
\end{aligned}
$$

we see that

$$-\Re \int_{\mathbb{R}^N} [\bar{v}\Delta\Psi + 2\nabla\Psi \cdot \nabla\bar{v}]e^{-i\theta}v_t$$

$$= -\Re \int_{\mathbb{R}^N} [\bar{v}\Delta\Psi + 2\nabla\Psi \cdot \nabla\bar{v}](\Delta v + f(t)|v|^\alpha v)$$

$$= -\frac{1}{2}\int_{\mathbb{R}^N} |v|^2 \Delta^2\Psi - \frac{\alpha f(t)}{\alpha+2}\int_{\mathbb{R}^N}|v|^{\alpha+2}\Delta\Psi + 2\Re\int_{\mathbb{R}^N}\langle H(\Psi)\nabla\bar{v},\nabla v\rangle. \tag{4.154}$$

Next, we observe that

$$\Re \int_{\mathbb{R}^N} [\bar{v}\Delta\Psi + 2\nabla\Psi \cdot \nabla\bar{v}]v_t = \frac{1}{2}\frac{\mathrm{d}}{\mathrm{d}t}\int_{\mathbb{R}^N}|v|^2\Delta\Psi + 2\Re\int_{\mathbb{R}^N}(\nabla\bar{v}\cdot\nabla\Psi)v_t. \tag{4.155}$$

On the other hand,

$$\frac{f'}{\alpha+2}\int_{\mathbb{R}^N}\Psi|v|^{\alpha+2} + \frac{\mathrm{d}}{\mathrm{d}t}\int_{\mathbb{R}^N}\Psi\left(\frac{|\nabla v|^2}{2} - \frac{f(t)}{\alpha+2}|v|^{\alpha+2}\right)$$

$$= \Re \int_{\mathbb{R}^N} \Psi(\nabla\bar{v}\cdot\nabla v_t - f|v|^\alpha\bar{v}v_t) = -\Re\int_{\mathbb{R}^N}[\Psi(\Delta\bar{v}+f|v|^\alpha\bar{v})v_t + (\nabla\Psi\cdot\nabla\bar{v})v_t]$$

$$= -\cos\theta\int_{\mathbb{R}^N}\Psi|v_t|^2 - \Re\int_{\mathbb{R}^N}(\nabla\Psi\cdot\nabla\bar{v})v_t$$

so that

$$2\Re\int_{\mathbb{R}^N}(\nabla\Psi\cdot\nabla\bar{v})v_t = -2\cos\theta\int_{\mathbb{R}^N}\Psi|v_t|^2 - \frac{2f'}{\alpha+2}\int_{\mathbb{R}^N}\Psi|v|^{\alpha+2} \tag{4.156}$$

$$- \frac{\mathrm{d}}{\mathrm{d}t}\int_{\mathbb{R}^N}\Psi\left(|\nabla v|^2 - \frac{2f(t)}{\alpha+2}|v|^{\alpha+2}\right).$$

Applying (4.153), (4.154), (4.155) and (4.156), we deduce that

$$\frac{\mathrm{d}}{\mathrm{d}t}\left(\sin\theta\,\Im\int_{\mathbb{R}^N}\bar{v}(\nabla\Psi\cdot\nabla v)\right)$$

$$= \int_{\mathbb{R}^N}\left\{-\frac{1}{2}|v|^2\Delta^2\Psi - \frac{\alpha f(t)}{\alpha+2}|v|^{\alpha+2}\Delta\Psi + 2\Re\langle H(\Psi)\nabla\bar{v},\nabla v\rangle\right\} \tag{4.157}$$

$$+ \cos\theta\frac{\mathrm{d}}{\mathrm{d}t}\int_{\mathbb{R}^N}\left(-\Psi|\nabla v|^2 + \frac{2f(t)}{\alpha+2}\Psi|v|^{\alpha+2} + \frac{1}{2}|v|^2\Delta\Psi\right)$$

$$- 2\cos^2\theta\int_{\mathbb{R}^N}\Psi|v_t|^2 - \frac{2f'(t)}{\alpha+2}\cos\theta\int_{\mathbb{R}^N}\Psi|v|^{\alpha+2}.$$

Taking now the time derivative of (4.150) and applying (4.157), we obtain (4.151). Finally, if both Ψ and u_0 (hence v) are radially symmetric, then $\Re\langle H(\Psi)\nabla\bar{v},\nabla v\rangle = \Psi''|v_r|^2$, so that (4.152) follows from (4.151). $\square$

4.8 A Caffarelli–Kohn–Nirenberg Inequality

We use the following form of Caffarelli–Kohn–Nirenberg inequality [3]. It extends an inequality which was established in [31] and generalized in [6, Lemma 5.3].

Lemma 4.32 *Suppose $N \geq 2$ and $\alpha \leq 4$ and let $A > 0$. There exist a constant χ and a family $(\Psi_\varepsilon)_{\varepsilon>0} \subset C^\infty(\mathbb{R}^N) \cap W^{4,\infty}(\mathbb{R}^N)$ of radially symmetric functions such that $\Psi_\varepsilon(x) > 0$ for $x \neq 0$,*

$$\sup_{\varepsilon>0}[\varepsilon^2\|\Psi_\varepsilon\|_{L^\infty} + \varepsilon\|\partial_r\Psi_\varepsilon\|_{L^\infty} + \|\Delta\Psi_\varepsilon\|_{L^\infty} + \varepsilon^{-2}\|\Delta^2\Psi_\varepsilon\|_{L^\infty}] < \infty, \quad (4.158)$$

$$2N - \Delta\Psi_\varepsilon \geq 0 \quad (4.159)$$

and

$$-2\int_{\mathbb{R}^N}(2-\Psi_\varepsilon'')|u_r|^2 + \mu\int_{\mathbb{R}^N}(2N-\Delta\Psi_\varepsilon)|u|^{\alpha+2}$$
$$-\frac{1}{2}\int_{\mathbb{R}^N}|u|^2\Delta^2\Psi_\varepsilon \leq \kappa(\mu,\varepsilon) \quad (4.160)$$

for all radially symmetric $u \in H^1(\mathbb{R}^N)$ such that $\|u\|_{L^2} \leq A$ and all $\mu,\varepsilon > 0$ such that $\chi\mu\varepsilon^{2(N-1)} < 1$, where

$$\kappa(\mu,\varepsilon) = \begin{cases} \chi\mu\left(\varepsilon^{\frac{N\alpha}{2}} + [\chi\mu\varepsilon^{2(N-1)}]^{\frac{\alpha}{4-\alpha}}\right) + \chi\varepsilon^2 & \text{if } 0 < \alpha < 4, \\ \chi\mu\varepsilon^{\frac{N\alpha}{2}} + \chi\varepsilon^2 & \text{if } \alpha = 4. \end{cases} \quad (4.161)$$

Proof We follow the method of [31], and we construct a family $(\Psi_\varepsilon)_{\varepsilon>0}$ such that, given A, the estimate (4.160) holds with $\Psi = \Psi_\varepsilon$ provided $\varepsilon > 0$ is sufficiently small. Fix a function $h \in C^\infty([0,\infty))$ such that

$$h \geq 0, \quad \operatorname{supp} h \subset [1,2], \quad \int_0^\infty h(s)\,\mathrm{d}s = 1$$

and let

$$\zeta(t) = t - \int_0^t (t-s)h(s)\,\mathrm{d}s = t - \int_0^t\int_0^s h(\sigma)\,\mathrm{d}\sigma\,\mathrm{d}s$$

for $t \geq 0$. It follows that $\zeta \in C^\infty([0,\infty)) \cap W^{4,\infty}((0,\infty))$, $\zeta' \geq 0$, $\zeta'' \leq 0$, $\zeta(t) = t$ for $t \leq 1$, and $\zeta(t) = M$ for $t \geq 2$ with $M = \int_0^2 sh(s)\,\mathrm{d}s$. Set

$$\Phi(x) = \zeta(|x|^2).$$

It follows in particular that $\Phi \in C^\infty(\mathbb{R}^N) \cap W^{4,\infty}(\mathbb{R}^N)$. Given any $\varepsilon > 0$, set

$$\Psi_\varepsilon(x) = \varepsilon^{-2}\Phi(\varepsilon x),$$

so that

$$\|D^\beta \Psi_\varepsilon\|_{L^\infty} = \varepsilon^{|\beta|-2}\|D^\beta \Phi\|_{L^\infty}, \qquad (4.162)$$

where β is any multi-index such that $0 \le |\beta| \le 4$. Next, set

$$\xi(t) = \sqrt{2(1-\zeta'(t)) - 4t\zeta''(t)} = \sqrt{2\int_0^t h(s)\,ds + 4th(t)}. \qquad (4.163)$$

It is not difficult to check that $\xi \in C^1([0,\infty)) \cap W^{1,\infty}(0,\infty)$. Let

$$\gamma(r) = \xi(r^2)$$

and, given $\varepsilon > 0$, let

$$\gamma_\varepsilon(r) = \gamma(\varepsilon r).$$

It easily follows that γ_ε is supported in $[\varepsilon^{-1}, \infty)$, so that

$$\|r^{-(N-1)}\gamma_\varepsilon'\|_{L^\infty} \le \varepsilon^{N-1}\|\gamma_\varepsilon'\|_{L^\infty} = \varepsilon^N\|\gamma'\|_{L^\infty}, \qquad (4.164)$$

$$\|r^{-(N-1)}\gamma_\varepsilon u_r\|_{L^2} \le \varepsilon^{N-1}\|\gamma_\varepsilon u_r\|_{L^2}. \qquad (4.165)$$

Set

$$I_\varepsilon(u) = -2\int_{\mathbb{R}^N}(2-\Psi_\varepsilon'')|u_r|^2 + \mu\int_{\mathbb{R}^N}(2N-\Delta\Psi_\varepsilon)|u|^{\alpha+2}$$
$$-\frac{1}{2}\int_{\mathbb{R}^N}|u|^2\Delta^2\Psi_\varepsilon. \qquad (4.166)$$

Elementary but long calculations using in particular (4.163) show that

$$2-\Psi_\varepsilon''(x) = \gamma_\varepsilon(|x|)^2, \qquad (4.167)$$

$$2N-\Delta\Psi_\varepsilon(x) = N[\gamma_\varepsilon(|x|)]^2 + 4(N-1)(\varepsilon|x|)^2\zeta''(\varepsilon^2|x|^2) \le N[\gamma_\varepsilon(|x|)]^2. \qquad (4.168)$$

We deduce from (4.166), (4.167), (4.168), and (4.162) that

$$I_\varepsilon(u) \le -2\int_{\mathbb{R}^N}\gamma_\varepsilon^2|u_r|^2 + N\mu\int_{\mathbb{R}^N}\gamma_\varepsilon^2|u|^{\alpha+2} + \frac{\varepsilon^2}{2}\|\Delta^2\Phi\|_{L^\infty}\|u\|_{L^2}^2. \qquad (4.169)$$

We next claim that

$$\|\gamma_\varepsilon^{\frac{1}{2}}u\|_{L^\infty}^2 \le \varepsilon^N\|\gamma'\|_{L^\infty}\|u\|_{L^2}^2 + 2\varepsilon^{N-1}\|u\|_{L^2}\|\gamma_\varepsilon u_r\|_{L^2}. \qquad (4.170)$$

Indeed,

$$\gamma_\varepsilon(r)|u(r)|^2 = -\int_r^\infty \frac{d}{ds}[\gamma_\varepsilon(s)|u(s)|^2] \le \int_0^\infty |\gamma_\varepsilon'||u|^2 + 2\int_0^\infty \gamma_\varepsilon|u|\,|u_r|$$
$$\le \|r^{-(N-1)}\gamma_\varepsilon'\|_{L^\infty}\|u\|_{L^2}^2 + 2\|u\|_{L^2}\|r^{-(N-1)}\gamma_\varepsilon u_r\|_{L^2}. \qquad (4.171)$$

(The above calculation is valid for a smooth function u and is easily justified for a general u by density.) The estimate (4.170) follows from (4.171), (4.164), and (4.165).

In what follows, χ denotes a constant that may depend on N, γ, Φ and A and change from line to line, but is independent of $0 < \alpha \leq 4$ and $\varepsilon > 0$. We assume $\|u\|_{L^2} \leq A$, and we observe that

$$\int_{\mathbb{R}^N} \gamma_\varepsilon^2 |u|^{\alpha+2} = \int_{\mathbb{R}^N} \gamma_\varepsilon^{\frac{4-\alpha}{2}} [\gamma_\varepsilon^{\frac{1}{2}} |u|]^\alpha |u|^2 \leq \|\gamma\|_{L^\infty}^{\frac{4-\alpha}{2}} \|\gamma_\varepsilon^{\frac{1}{2}} u\|_{L^\infty}^\alpha A^2$$

$$\leq \chi \|\gamma_\varepsilon^{\frac{1}{2}} u\|_{L^\infty}^\alpha. \tag{4.172}$$

Applying (4.170) and the inequality $(x + y)^{\frac{\alpha}{2}} \leq \chi (x^{\frac{\alpha}{2}} + y^{\frac{\alpha}{2}})$, we deduce from (4.172) that

$$\int_{\mathbb{R}^N} \gamma_\varepsilon^2 |u|^{\alpha+2} \leq \chi \varepsilon^{\frac{N\alpha}{2}} + \chi \varepsilon^{-\frac{(N-1)\alpha}{2}} \|\gamma_\varepsilon u_r\|_{L^2}^{\frac{\alpha}{2}}. \tag{4.173}$$

We first consider the case $\alpha < 4$. Applying the inequality $xy \leq \frac{x^p}{p\delta^p} + \frac{\delta^{p'} y^{p'}}{p'}$ with $\delta > 0$ and $p = \frac{4}{4-\alpha}, p' = \frac{4}{\alpha}$, we see that

$$\varepsilon^{-\frac{(N-1)\alpha}{2}} \|\gamma_\varepsilon u_r\|_{L^2}^{\frac{\alpha}{2}} \leq \chi \delta^{-\frac{4}{4-\alpha}} \varepsilon^{\frac{2(N-1)\alpha}{4-\alpha}} + \chi \delta^{\frac{4}{\alpha}} \|\gamma_\varepsilon u_r\|_{L^2}^2$$

so that (4.173) yields

$$\int_{\mathbb{R}^N} \gamma_\varepsilon^2 |u|^{\alpha+2} \leq \chi \delta^{\frac{4}{\alpha}} \|\gamma_\varepsilon u_r\|_{L^2}^2 + \chi \left(\varepsilon^{\frac{N\alpha}{2}} + \delta^{-\frac{4}{4-\alpha}} \varepsilon^{\frac{2(N-1)\alpha}{4-\alpha}} \right). \tag{4.174}$$

Estimates (4.169) and (4.174) now yield

$$I_\varepsilon(u) \leq -\left(2 - \chi \mu \delta^{\frac{4}{\alpha}} \right) \|\gamma_\varepsilon u_r\|_{L^2}^2 + \chi \mu \left(\varepsilon^{\frac{N\alpha}{2}} + [\delta^{-4} \varepsilon^{2(N-1)\alpha}]^{\frac{1}{4-\alpha}} \right) + \chi \varepsilon^2. \tag{4.175}$$

We choose $\delta > 0$ so that the first term in the right-hand side of (4.175) vanishes, i.e. $\chi \mu \delta^{\frac{4}{\alpha}} = 2$. For this choice of δ, it follows from (4.175) that if $\|u\|_{L^2} \leq A$, then $I_\varepsilon(u) \leq \kappa(\mu, \varepsilon)$, where κ is defined by (4.161). This proves inequality (4.160) for $\alpha < 4$. The case $\alpha = 4$ follows by letting $\alpha \uparrow 4$ and observing that $[\chi \mu \varepsilon^{2(N-1)}]^{\frac{\alpha}{4-\alpha}} \to 0$ as $\alpha \uparrow 4$ when $\chi \mu \varepsilon^{2(N-1)} < 1$. $\qquad \square$

References

[1] Adams R. A. and Fournier John J. F.: *Sobolev spaces. Second edition.* Pure and Applied Mathematics (Amsterdam) **140**. Elsevier/Academic Press, Amsterdam, 2003. (MR2424078)

[2] Ball J.M.: Remarks on blow-up and nonexistence theorems for nonlinear evolution equations. Quart. J. Math. Oxford Ser. (2) **28** (1977), no. 112, 473–486. (MR0473484) (doi: 10.1093/qmath/28.4.473)

[3] Caffarelli L., Kohn R. V. and Nirenberg L.: First order interpolation inequalities with weights. Compositio Math. **53** (1984), no. 3, 259–275. (MR0768824) (link: http://www.numdam.org/item?id=CM_1984__53_3_259_0)

[4] Cazenave T.: *Semilinear Schrödinger equations*. Courant Lecture Notes in Mathematics, **10**. New York University, Courant Institute of Mathematical Sciences, New York; American Mathematical Society, Providence, RI, 2003. (MR2002047)

[5] Cazenave T., Dias J. P. and Figueira M.: Finite-time blowup for a complex Ginzburg–Landau equation with linear driving. J. Evol. Equ. **14** (2014), no. 2, 403–415. (MR3207620) (doi: 10.1007/s00028-014-0220-z)

[6] Cazenave T., Dickstein F. and Weissler F. B.: Finite-time blowup for a complex Ginzburg-Landau equation. SIAM J. Math. Anal. **45** (2013), no. 1, 244–266. (MR3032976) (doi: 10.1137/120878690)

[7] Correia S.: Blowup for the nonlinear Schrödinger equation with an inhomogeneous damping term in the L^2-critical case. Commun. Contemp. Math. **17** (2015), no. 3, 1450030, 16 pp. (MR3325044) (doi: 10.1142/S0219199714500308)

[8] Cross M.C. and Hohenberg P. C.: Pattern formation outside of equilibrium. Rev. Mod. Phys. **65** (1993), no. 3, 851–1112. (doi: 10.1103/RevModPhys.65.851)

[9] Dickstein F., Mizoguchi N., Souplet P. and Weissler F. B.: Transversality of stable and Nehari manifolds for a semilinear heat equation. Calc. Var. Partial Differential Equations, **42** (2011), no. 3–4, 547–562. (MR2846266) (doi: 10.1007/s00526-011-0397-8)

[10] Fife P. C.: *Mathematical aspects of reacting and diffusing systems*. Lecture Notes in Biomathematics **28**, Springer, New York, 1979. (MR0527914) (doi: 10.1007/978-3-642-93111-6)

[11] Fujita H.: On the blowing-up of solutions of the Cauchy problem for $u_t = \triangle u + u^{\alpha+1}$, J. Fac. Sci. Univ. Tokyo Sect. IA Math. **13** (1966), 109–124. (MR0214914) (link: http://hdl.handle.net/2261/6061)

[12] Ginibre J. and Velo G.: On a class of nonlinear Schrödinger equations. I. The Cauchy problem, general case, J. Funct. Anal. **32**, no. 1 (1979), 1–32. (MR0533218) (doi: 10.1016/0022-1236(79)90076-4)

[13] Glassey R. T.: On the blowing up of solutions to the Cauchy problem for nonlinear Schrödinger equations. J. Math. Phys. **18** (1977), no. 9, 1794–1797. (MR0460850) (doi: 10.1063/1.523491)

[14] Haraux A. and Weissler F. B.: Non uniqueness for a semilinear initial value problem. Indiana Univ. Math. J. **31** (1982), no. 2, 167–189. (MR0648169) (doi: 10.1512/iumj.1982.31.31016)

[15] Hayakawa K.: On nonexistence of global solutions of some semilinear parabolic equations, Proc. Japan Acad. Ser. A Math. Sci. **49** (1973), 503–505. (MR0338569) (doi: 10.3792/pja/1195519254)

[16] Kaplan S.: On the growth of solutions of quasilinear parabolic equations. Comm. Pure Appl. Math. **16** (1963), 305–330. (MR0160044) (doi: 10.1002/cpa.3160160307)

[17] Kato T.: On nonlinear Schrödinger equations. Ann. Inst. H. Poincaré Phys. Théor. **46** (1987), no. 1, 113–129. (MR0877998) (link: http://www.numdam.org/item?id=AIHPA_1987__46_1_113_0)

[18] Kavian O.: A remark on the blowing-up of solutions to the Cauchy problem for nonlinear Schrödinger equations. Trans. Amer. Math. Soc. **299** (1987), no. 1, 193–205. (MR0869407) (doi: 10.2307/2000489)

[19] Kobayashi K., Sirao T. and Tanaka H.: On the growing up problem for semilinear heat equations, J. Math. Soc. Japan **29** (1977), no. 3, 407–424. (MR0450783) (doi: 10.2969/jmsj/02930407)

[20] Levine H. A.: Some nonexistence and instability theorems for formally parabolic equations of the form $Pu_t = -Au + f(u)$. Arch. Ration. Mech. Anal. **51** (1973), pp. 371–386. (MR0348216) (doi: 10.1007/BF00263041)

[21] Masmoudi N. and Zaag H.: Blow-up profile for the complex Ginzburg–Landau equation. J. Funct. Anal. **255** (2008), no. 7, 1613–1666. (MR2442077) (doi: 10.1016/j.jfa.2008.03.008)

[22] Merle F. and Raphaël P.: Sharp upper bound on the blowup rate for the critical nonlinear Schrödinger equation. Geom. Funct. Anal. **13** (2003), 591–642. (MR1995801) (doi: 10.1007/s00039-003-0424-9)

[23] Merle F. and Raphaël P.: On universality of blow-up profile for L^2 critical nonlinear Schrödinger equation. Invent. Math. **156** (2004), no. 3, 565–672. (MR2061329) (doi: 10.1007/s00222-003-0346-z)

[24] Merle F. and Raphaël P.: The blow-up dynamic and upper bound on the blow-up rate for critical nonlinear Schrödinger equation. Ann. of Math. (2) **161** (2005), no. 1, 157–222. (MR2150386) (doi: 10.4007/annals.2005.161.157)

[25] Merle F., Raphaël P. and Szeftel J.: Stable self-similar blow-up dynamics for slightly L^2 super-critical NLS equations. Geom. Funct. Anal. **20** (2010), no. 4, 1028–1071. (MR2729284) (doi: 10.1007/s00039-010-0081-8)

[26] Merle F. and Zaag H.: Stability of the blow-up profile for equations of the type $u_t = \triangle u + |u|^{p-1}u$. Duke Math. J. **86** (1997), no. 1, 143–195. (MR1427848) (doi: 10.1215/S0012-7094-97-08605-1)

[27] Mielke A.: The Ginzburg–Landau equation in its role as a modulation equation. in *Handbook of dynamical systems. Vol. 2*, 759–834, North-Holland, Amsterdam, 2002. (MR1901066) (doi: 10.1016/S1874-575X(02)80036-4)

[28] Mohamad D.: Blow-up for the damped L^2-critical nonlinear Schrödinger equation. Adv. Differential Equations **17** (2012), no. 3–4, 337–367. (MR2919105) (link: http://projecteuclid.org/euclid.ade/1355703089)

[29] Mizoguchi N. and Yanagida E.: Critical exponents for the blow-up of solutions with sign changes in a semilinear parabolic equation, Math. Ann. **307** (1997), no. 4, 663–675. (MR1464136) (doi: 10.1007/s002080050055)

[30] Mizoguchi N. and Yanagida E.: Critical exponents for the blowup of solutions with sign changes in a semilinear parabolic equation. II, J. Differential Equations **145** (1998), no. 2, 295–331. (MR1621030) (doi: 10.1006/jdeq.1997.3387)

[31] Ogawa T. and Tsutsumi Y.: Blow-up of H^1 solutions for the nonlinear Schrödinger equation. J. Differential Equations **92** (1991), pp. 317–330. (MR1120908) (doi: 10.1016/0022-0396(91)90052-B)

[32] Ogawa T. and Tsutsumi Y.: Blow-up of H^1 solutions for the one dimensional nonlinear Schrödinger equation with critical power nonlinearity. Proc. Amer. Math. Soc. **111** (1991), 487–496. (MR1045145) (doi: 10.2307/2048340)

[33] Ohta M. and Todorova G.: Remarks on global existence and blowup for damped nonlinear Schrödinger equations. Discrete Contin. Dyn. Syst. **23** (2009), no. 4, 1313–1325. (MR2461853) (doi: 10.3934/dcds.2009.23.1313)

[34] Ozawa T.: Remarks on proofs of conservation laws for nonlinear Schrödinger equations. Calc. Var. Partial Differential Equations **25** (2006), no. 3, 403–408. (MR2201679) (doi: 10.1007/s00526-005-0349-2)

[35] Payne L. and Sattinger D. H.: Saddle points and instability of nonlinear hyperbolic equations. Israel J. Math. **22** (1975), 273–303. (MR0402291) (doi: 10.1007/BF02761595)

[36] Plecháč P. and Šverák V.: On self-similar singular solutions of the complex Ginzburg–Landau equation, Comm. Pure Appl. Math. **54** (2001), no. 10, 1215–1242. (MR1843986) (doi: 10.1002/cpa.3006)

[37] Popp S., Stiller O., Kuznetsov E. and Kramer L. The cubic complex Ginzburg–Landau equation for a backward bifurcation. Phys. D **114** (1998), no. 1–2, 81–107. (MR1612047)(doi: 10.1016/S0167-2789(97)00170-X)

[38] Raphaël P.: Existence and stability of a solution blowing up on a sphere for an L^2-supercritical non linear Schrödinger equation. Duke Math. J. **134** (2006), no. 2, 199–258. (MR2248831) (doi: 10.1215/S0012-7094-06-13421-X)

[39] Raphaël P. and Szeftel J.: Standing ring blow up solutions to the N-dimensional quintic nonlinear Schrödinger equation. Comm. Math. Phys. **290** (2009), no. 3, 973–996. (MR2525647) (doi: 10.1007/s00220-009-0796-2)

[40] Segal I. E.: Nonlinear semi-groups. Ann. of Math. (2), **78** (1963), no. 2, 339–364. (doi: 10.2307/1970347)

[41] Snoussi S and Tayachi S.: Nonglobal existence of solutions for a generalized Ginzburg–Landau equation coupled with a Poisson equation. J. Math. Anal. Appl. **254** (2001), 558–570. (MR1805524) (doi: 10.1006/jmaa.2000.7235)

[42] Stewartson K. and Stuart J. T.: A non-linear instability theory for a wave system in plane Poiseuille flow. J. Fluid Mech. **48** (1971), 529–545. (MR0309420) (doi: 10.1017/S0022112071001733)

[43] Sulem C. and Sulem P.-L.: *The nonlinear Schrödinger equation. Self-focusing and wave collapse.* Applied Mathematical Sciences. **139**, Springer-Verlag, New York, 1999. (MR1696311) (doi: 10.1007/b98958)

[44] Tsutsumi M.: Nonexistence of global solutions to the Cauchy problem for the damped nonlinear Schrödinger equation. SIAM J. Math. Anal. **15** (1984), no. 2, 357–366. (MR0731873) (doi: 10.1137/0515028)

[45] Weissler F. B.: Existence and nonexistence of global solutions for a semilinear heat equation, Israel J. Math. **38** (1981), no. 1–2, 29–40. (MR0599472) (doi: 10.1007/BF02761845)

[46] Willem M.: *Minimax theorems.* Progress in Nonlinear Differential Equations and their Applications, **24**. Birkhauser Boston, Inc., Boston, MA, 1996. (MR1400007) (doi: 10.1007/978-1-4612-4146-1)

[47] Zaag H.: Blow-up results for vector-valued nonlinear heat equation with no gradient structure. Ann. Inst. H. Poincaré Anal. Non Linéaire **15** (1998), 581–622. (MR1643389) (doi: 10.1016/S0294-1449(98)80002-4)

[48] Zakharov V. E.: Collapse of Langmuir waves. Soviet Phys. JETP **35** (1972), 908–914.

* B.C. 187, 4 place Jussieu, 75252 Paris Cedex 05, France
  E-mail address: thierry.cazenave@upmc.fr
† UR Analyse Non-Linéaire et Géométrie, UR13ES32, Department of Mathematics, Faculty of Sciences of Tunis, University of Tunis El-Manar

5

Asymptotic Analysis for the Lane–Emden Problem in Dimension Two

Francesca De Marchis[*], Isabella Ianni[†] and Filomena Pacella[‡]

Introduction

We consider the Lane–Emden Dirichlet problem

$$\begin{cases} -\Delta u = |u|^{p-1}u & \text{in } \Omega, \\ u = 0 & \text{on } \partial\Omega, \end{cases} \tag{5.1}$$

when $p > 1$ and $\Omega \subset \mathbb{R}^2$ is a smooth bounded domain. The aim of the paper is to survey some recent results on the asymptotic behavior of solutions of (5.1) as the exponent $p \to \infty$.

We will start in Section 5.1 with a summary of some basic and well-known facts about the solutions of (5.1). We will also describe a recent result about the existence, for p large, of a special class of sign-changing solutions of (5.1) in symmetric domains (see [19]) and we will provide, for p large, the exact computation of the Morse index of least energy nodal radial solutions of (5.1) in the ball, as obtained in [23].

The asymptotic behavior as $p \to \infty$ will be described in Sections 5.2–5.3. In Section 5.2 a general "profile decomposition" theorem obtained in [20] and holding for both positive and sign-changing solutions will be presented with a detailed proof, together with some additional new results for positive solutions recently obtained in [22]. Finally, in Section 5.3 we will describe the precise limit profile of the symmetric nodal solutions found in [19] and then studied in [20]. In particular, the result of this section will show that, asymptotically, as

2010 *Mathematics Subject classification*: 35B05, 35B06, 35J91.
Keywords: semilinear elliptic equations, superlinear elliptic boundary value problems, asymptotic analysis, concentration of solutions.
Research supported by: PRIN 201274FYK7_005 grant, INDAM - GNAMPA and Sapienza Research Funds: "Avvio alla ricerca 2015" and "Awards Project 2014"
* Francesca De Marchis, University of Roma *Sapienza*
† Isabella Ianni, Second University of Napoli
‡ Filomena Pacella, University of Roma *Sapienza*

$p \to \infty$, these solutions look like a superposition of two bubbles with different sign corresponding to radial solutions of the regular and singular Liouville problem in $\mathbb{R}^2$.

Acknowledgments

This paper originates from a short course given by F. Pacella at a Conference-School held in Hammamet in March 2015 in honor of Abbas Bahri. She would like to thank all the organizers for the wonderful and warm hospitality.

5.1 Various Results for Solutions of the Lane–Emden Problem

We consider the Lane–Emden Dirichlet problem

$$\begin{cases} -\Delta u = |u|^{p-1}u & \text{in } \Omega, \\ u = 0 & \text{on } \partial\Omega, \end{cases} \tag{5.2}$$

where $p > 1$ and $\Omega \subset \mathbb{R}^2$ is a smooth bounded domain.

Since in two dimensions any exponent $p > 1$ is subcritical (with respect to the Sobolev embedding), it is well known, by standard variational methods, that (5.2) has at least one positive solution. Moreover, exploiting the oddness of the nonlinearity $f(u) = |u|^{p-1}u$ and using topological tools it can be proved that (5.2) admits infinitely many solutions.

It was first proved in [6], and later in [5] for more general nonlinearities, that there exists at least one solution which changes sign, so it makes sense to study the properties of both positive and sign-changing solutions. The latter will be often referred to as nodal solutions. Among these solutions one can select those which have the least energy, therefore named "least energy" (or "least energy nodal") solutions. More precisely, considering the energy functional:

$$E_p(u) = \frac{1}{2}\int_\Omega |\nabla u|^2 \,dx - \frac{1}{p+1}\int_\Omega |u|^{p+1}\,dx, \qquad u \in H_0^1(\Omega)$$

and the Nehari manifold

$$\mathcal{N} = \{u \in H_0^1(\Omega) : \langle E_p'(u), u \rangle = 0\}$$

or the nodal Nehari set

$$\mathcal{N}^\pm = \{u \in H_0^1(\Omega) : \langle E_p'(u), u^\pm \rangle = 0\},$$

where $u^\pm$ are the positive and negative part of u, it is possible to prove that the $\inf_{\mathcal{N}} E_p$ (resp. $\inf_{\mathcal{N}^\pm} E_p$) is achieved. The corresponding minimizers are the least energy positive (resp. nodal) solutions (see [46], [5]). Note that any minimizer on $\mathcal{N}$ cannot change sign and we will assume that it is positive (rather than negative).

Let us observe that $\mathcal{N}$ is a codimension one manifold in $H_0^1(\Omega)$ while $\mathcal{N}^\pm$ is a C^1-manifold of codimension two in $H_0^1(\Omega) \cap H^2(\Omega)$ (but not in $H_0^1(\Omega)$, see [5]).

For the least energy solutions several qualitative properties can be obtained. We start by considering the case of positive solutions.

Let us first define the Morse index of a solution of (5.2).

Definition 5.1 *The Morse index $m(u)$ of a solution u of (5.2) is the maximal dimension of a subspace of $C_0^1(\Omega)$ on which the quadratic form*

$$Q(\varphi) = \int_\Omega |\nabla \varphi|^2 \, dx - p \int_\Omega |u|^{p-1} \varphi^2 \, dx$$

is negative definite.

In the case when Ω is a bounded domain, $m(u)$ can be equivalently defined as the number of the negative Dirichlet eigenvalue of the linearized operator at u:

$$L_u = -\Delta - p|u|^{p-1}$$

in the domain Ω.

It is easy to see, just multiplying the equation by u and integrating, that $Q(u) < 0$, so that there is at least one negative direction for $Q(u)$, i.e. $m(u) \geq 1$. This holds for any solution of (5.2), either positive or sign-changing. For the least energy solution u, since it minimizes the energy on a codimension one manifold, one could guess that $m(u) = 1$. This is what was indeed proved in [44] (see also [46]), for more general nonlinearities. Another important property of a solution, both for theoretical reasons and for applications, is its symmetry in symmetric domains.

For positive solutions u of (5.2), as a consequence of the famous result by Gidas, Ni and Nirenberg [29] it holds that if Ω is symmetric and convex with respect to a line, then u is invariant by reflection with respect to that line. In particular a positive solution of (5.2) in a ball is radial and strictly radially decreasing.

This result allows us to prove that if Ω is a ball there exists only one positive solution of (5.2) (this holds also in higher dimensions, when p is a subcritical exponent) ([29], [45], [3], [14]).

The question of the uniqueness of the positive solution in more general bounded domains is a very difficult one, still open, and will be addressed in the subsequent paper [16]. Here we recall that it has been conjectured ([29]) that the uniqueness should hold in convex domains (also in higher dimensions) and that, so far, in the case of planar domains it has only been proved when the domains are symmetric and convex with respect to two orthogonal lines passing through the origin ([14], [38]). If one restricts the question to the least energy solutions (or more generally to solutions of Morse index one) then the uniqueness, in convex planar domains, has been proved in [34]. On the other side it is easy to see that there are nonconvex domains for which multiple positive solutions exist; examples of such domains are annular domains or dumbbell domains ([15], [38]).

More properties of positive solutions and, actually, a good description of their profile, can be obtained, for large exponents p, by the asymptotic analysis of the solutions of (5.2), as $p \to \infty$.

This study started in [41] and [42] where the authors considered families (u_p) of least energy (hence positive) solutions and, for some domains, proved concentration at a single point, as well as asymptotic estimates, as $p \to \infty$. Later, inspired by the paper [2], Adimurthi and Grossi in [1] identified a "limit problem" by showing that suitable rescalings of u_p converge, in $C^2_{loc}(\mathbb{R}^2)$ to a regular solution U of the Liouville problem

$$\begin{cases} -\Delta U = e^U & \text{in } \mathbb{R}^2, \\ \int_{\mathbb{R}^2} e^U dx = 8\pi. \end{cases} \qquad (5.3)$$

They also considered general bounded domains and showed that $\|u_p\|_\infty$ converges to $\sqrt{e}$ as $p \to \infty$, as it had been previously conjectured.

So the asymptotic profile of the least energy solutions is clear, as well as their energy.

Concerning general positive solutions, a first asymptotic analysis (actually holding for general families of solutions, both positive and sign-changing) under the following energy condition:

$$p \int_\Omega |\nabla u_p|^2 dx \le C \qquad (5.4)$$

for some positive constant $C \ge 8\pi e$ and independent of p, was carried out in [20]. Then, recently in [22], starting from this, a complete description of the asymptotic profile of any family of positive solutions u_p satisfying (5.4) has been obtained, showing that (u_p) concentrates at a finite number of distinct points in Ω, having the limit profile of the solution U of (5.3) when a suitable rescaling around each of the concentration points is made. These results have been recently improved in [17] where it has been proved that $\|u_p\|_\infty$ converges to $\sqrt{e}$ as $p \to \infty$, generalizing the result in [1], and that a quantization of the energy holds (see the end of Section 5.2 for the complete statement). Positive solutions with the described profile have been found in [27].

Now let us analyze the case of sign-changing solutions of (5.2).

Since any such solution u has at least two nodal regions (i.e. connected components of the set where u does not vanish), multiplying the equation by u and integrating on each nodal domain, we get that the Morse index $m(u)$ is at least two. For the least energy nodal solution, since it minimizes the energy functional E_p on $\mathcal{N}^\pm$, it is proved in [5] that its Morse index is exactly two.

Concerning symmetry properties of sign-changing solutions, a general result as the one of Gidas, Ni and Nirenberg for positive solutions cannot hold. This is easily understood by just thinking of the Dirichlet eigenfunctions of the Laplacian in a ball.

Nevertheless, by using maximum principles, properties of the linearized operator and bounds on the Morse index, partial symmetry results can be obtained also for nodal solutions. This direction of research started in [37] and continued in [39] and [30]. In particular in these papers, semilinear elliptic equations with nonlinear terms $f(u)$ either convex or with a convex derivative were studied in rotationally symmetric domains, showing the foliated Schwarz symmetry of solutions (of any sign) having Morse index $m(u) \leq N$, where N is the dimension of the domain. We recall the definition of foliated Schwarz symmetry.

Definition 5.2 *Let $B \subseteq \mathbb{R}^N, N \geq 2$, be a ball or an annulus. A continuous function v in B is said to be foliated Schwarz symmetric if there exists a unit vector $p \in \mathbb{R}^N$ such that $v(x)$ depends only on $|x|$ and $\vartheta = \arccos(\frac{x}{|x|} \cdot p)$ and is nonincreasing in ϑ.*

In other words a foliated Schwarz symmetric function is axially symmetric and monotone with respect to the angular coordinate.

In particular, in dimension two, the results of [39] allow us to claim that, in a ball or in an annulus, any solution u of (5.2) with Morse index $m(u) \leq 2$ is foliated Schwarz symmetric. Thus, in such domains, the least energy nodal solutions are foliated Schwarz symmetric.

Since radial functions are, obviously, foliated Schwarz symmetric, one may ask whether the least energy nodal solutions are radial or not. The answer to this question was provided by [4] where it was proved that any sign-changing solution u of a semilinear elliptic equation with a general autonomous nonlinearity $f(u)$ in a ball or an annulus must have Morse index $m(u) \geq N + 2$ (again N denotes the dimension of the domain).

An immediate corollary of this theorem is that, since a least energy nodal solution of (5.2) has Morse index two, it cannot be radial.

Another interesting consequence of the result of [4] is that the nodal set of a least energy nodal solution of (5.2) in a ball or an annulus must intersect the boundary of Ω. We recall that the nodal set $N(u)$ of a function u defined in the domain Ω is:

$$N(u) = \overline{\{x \in \Omega : u(x) = 0\}}.$$

To understand the property of the nodal line is important while studying sign-changing functions. It is an old question related to the study of the nodal eigenfunctions of the Laplacian, in particular of the second eigenfunction. In [36] it has been proved that in convex planar domains the nodal set of a second eigenfunction touches the boundary, but the question is still open in higher dimension, except for the case of some symmetric domains ([33], [13]).

Coming back to nodal solutions of (5.2) we observe that if Ω is a ball or an annulus, it is easy to see that there exist both nodal solutions with an interior nodal line and solutions whose nodal line intersects the boundary. Examples of solutions of the first type are the radial ones while those of the second type are antisymmetric with respect to a line passing through the center. It is natural to ask whether both kinds of solution exist in more general domains. While it is not difficult to provide examples of symmetric domains where there are nodal solutions whose nodal line intersects the boundary (rectangles, regular polygons etc.) it is not obvious at all that solutions with an interior nodal line exist. In the paper [19] we have succeeded in proving the existence of this type of solution in some symmetric planar domains, for large exponents p. The precise statement is the following.

Theorem 5.3 *Assume that Ω is simply connected, invariant under the action of a finite group G of orthogonal transformations of $\mathbb{R}^2$. If $|G| \geq 4$ ($|G|$ is the order of the group) then, for p sufficiently large (5.2) admits a sign-changing G-symmetric solution u_p, with two nodal domains, whose nodal line neither touches $\partial\Omega$, nor passes through the origin. Moreover*

$$p \int_\Omega |\nabla u_p|^2 dx \leq \alpha \, 8\pi e \quad \text{for some } \alpha < 5 \text{ and } p \text{ large.}$$

Let us now come back to the question of the Morse index of nodal solutions of (5.2). As recalled before, the result of [4] allows us to give an estimate from below in the radial case:

$$m(u) \geq 4$$

for any radial sign-changing solution u of (5.2) in a ball or an annulus. In the recent paper [23] we have been able to compute the Morse index exactly for these solutions when the exponent p is large and u has the least energy among the radial nodal solutions. The result is the following.

Theorem 5.4 *Let u_p be the least energy sign-changing radial solution of (5.2). Then*

$$m(u_p) = 12$$

for p sufficiently large.

The proof of this theorem is based on a decomposition of the spectrum of the linearized operator at u_p, as well as on fine estimates of the radial solution u_p obtained in [32].

As in the case of positive solutions, a better description of nodal solutions and of their profile can be obtained, for large exponents p, by performing an asymptotic analysis, as $p \to \infty$. This study started in [31] by considering a family (u_p) of solutions of (5.2) satisfying the condition

$$p \int_\Omega |\nabla u|^2 \, dx \to 16\pi e \qquad \text{as } p \to +\infty,$$

where $16\pi e$ is the "least-asymptotic" energy for nodal solutions. Under some additional conditions it was proved in [31] that these low-energy solutions concentrate at two distinct points of Ω and suitable scalings of u_p^+ and u_p^- converge to a regular solution U of (5.3).

Next, the case of least energy radial nodal solutions was considered in [32] where the new phenomenon of u_p^+ and u_p^- concentrating at the same point but with different profiles was shown. The precise result is the following.

Theorem 5.5 *Let (u_p) be a family of least energy radial nodal solutions of (5.2) in the ball with u_p positive at the center. Then*

(i) *a suitable scaling of u_p^+ converges in $C^2_{loc}(\mathbb{R}^2)$ to a regular solution U of (5.3),*

(ii) *a suitable scaling and translation of u_p^- converges in $C^2_{loc}(\mathbb{R}^2 \setminus \{O\})$ to a singular radial solution V of*

$$\begin{cases} -\Delta V = e^V + H\delta_0 & \text{in } \mathbb{R}^2, \\ \int_{\mathbb{R}^2} e^V dx < +\infty, \end{cases} \tag{5.5}$$

where H is a suitable negative constant and δ_0 is the Dirac measure centered at O.

Moreover:

$$p \int_\Omega |\nabla u_p|^2 \, dx \to C > 16\pi e \qquad \text{as } p \to +\infty.$$

So the theorem shows the existence of solutions which asymptotically look like a tower of two bubbles corresponding to solutions of two different Liouville problems in $\mathbb{R}^2$, namely (5.3) and (5.5).

In Section 5.3 of this paper we show that the same phenomenon appears in other symmetric domains different from the balls. We obtain this through the asymptotic analysis of the sign-changing solutions found in Theorem 5.3.

The starting point for this result is an asymptotic analysis of a general family (u_p) of solutions of (5.2) satisfying the condition (5.4). This first result, inspired by the paper [25] (see also [26]) can be viewed as a first step towards the complete classification of the asymptotic behavior of general sign-changing solutions of (5.2).

The hardest part of the proof of this result relies on showing that the rescaling about the minimum point x_p^- converges to a radial singular solution of a singular Liouville problem in $\mathbb{R}^2$. Indeed, while the rescaling of u_p about the maximum point x_p^+ can be studied in a "canonical" way, the analysis of the rescaling about x_p^- requires additional arguments. In particular the presence of the nodal line, with an unknown geometry, gives difficulties which, obviously, are not present when dealing with positive solutions or with radial sign-changing solutions. Also the proofs of the results for nodal radial solutions of [32] cannot be of any help since they depend strongly on one-dimensional estimates.

We would like to point out that the analysis carried out in [20] also allows us to get the same asymptotic result by substituting the bound on the energy with a bound on the Morse index of the solutions (see [21]).

Finally, we observe that the bubble-tower solutions of (5.1) are also interesting in the study of the associated heat flow because they induce a peculiar blow-up phenomenon (see [7, 24, 35] and in particular [18]).

We conclude by remarking that the phenomenon of nodal solutions of (5.2) with positive and negative parts concentrating at the same point and having different asymptotic profiles does not seem to appear in higher dimensions as p approaches the critical Sobolev exponent.

Finally, nodal solutions to (5.2) concentrating at a finite number of point without exhibiting the bubble tower phenomenon, i.e. only simple concentration points, also exist (see [28]).

5.2 General Asymptotic Analysis

This section is mostly devoted to the study of the asymptotic behavior of a general family $(u_p)_{p>1}$ of nontrivial solutions of (5.2) satisfying the uniform upper bound

$$p \int_{\Omega} |\nabla u_p|^2 dx \le C, \quad \text{for some } C > 0 \text{ independent of } p. \tag{5.6}$$

At the end of the section we also exhibit some recent results related to families of positive solutions. The material presented is based mainly on some of the results contained in [20] plus smaller additions or minor improvements. We also refer to [22] (see also [16]) for the complete analysis in the case of positive solutions.

Recall that in [41] it has been proved that for any family $(u_p)_{p>1}$ of nontrivial solutions of (5.1) the following lower bound holds:

$$\liminf_{p \to +\infty} p \int_{\Omega} |\nabla u_p|^2 dx \ge 8\pi e, \tag{5.7}$$

so the constant C in (5.6) is intended to satisfy $C \ge 8\pi e$. Moreover, if u_p is sign-changing then we also know that (see again [41])

$$\liminf_{p \to +\infty} p \int_{\Omega} |\nabla u_p^{\pm}|^2 dx \ge 8\pi e. \tag{5.8}$$

If we denote by E_p the energy functional associated with (5.2), i.e.

$$E_p(u) := \frac{1}{2} \|\nabla u\|_2^2 - \frac{1}{p+1} \|u\|_{p+1}^{p+1}, \quad u \in H_0^1(\Omega),$$

since for a solution u of (5.2)

$$E_p(u) = \left(\frac{1}{2} - \frac{1}{p+1} \right) \|\nabla u\|_2^2 = \left(\frac{1}{2} - \frac{1}{p+1} \right) \|u\|_{p+1}^{p+1}, \tag{5.9}$$

then (5.6), (5.7) and (5.8) are equivalent to uniform upper and lower bounds for the energy E_p or for the L^{p+1}-norm, indeed

$$\limsup_{p \to +\infty} 2pE_p(u_p) = \limsup_{p \to +\infty} p \int_{\Omega} |u_p|^{p+1} dx = \limsup_{p \to +\infty} p \int_{\Omega} |\nabla u_p|^2 dx \le C,$$

$$\liminf_{p \to +\infty} 2pE_p(u_p) = \liminf_{p \to +\infty} p \int_{\Omega} |u_p|^{p+1} dx = \liminf_{p \to +\infty} p \int_{\Omega} |\nabla u_p|^2 dx \ge 8\pi e,$$

and if u_p is sign-changing, also

$$\liminf_{p \to +\infty} 2pE_p(u_p^{\pm}) = \liminf_{p \to +\infty} p \int_{\Omega} |u_p^{\pm}|^{p+1} dx = \liminf_{p \to +\infty} p \int_{\Omega} |\nabla u_p^{\pm}|^2 dx \ge 8\pi e.$$

We will use all these equivalent formulations throughout the paper.

Observe that by the assumption in (5.6) we have that

$$E_p(u_p) \to 0, \quad \|\nabla u_p\|_2 \to 0, \quad \text{as } p \to +\infty,$$
$$E_p(u_p^{\pm}) \to 0, \quad \|\nabla u_p^{\pm}\|_2 \to 0, \quad \text{as } p \to +\infty \qquad \text{(if } u_p \text{ is sign-changing),}$$

so in particular $u_p^\pm \to 0$ a.e. as $p \to +\infty$.

In this section we will show that the solutions u_p do not vanish as $p \to +\infty$ (both $u_p^\pm$ do not vanish if u_p is sign-changing) and that moreover, differently from what happens in higher dimension, they do not blow up (see Theorem 5.6). Moreover, we will show that they concentrate at a finite number of points and we will also describe the asymptotic behavior of suitable rescalings of u_p ("bubbles") around suitable "concentrating" sequences of points (see Theorem 5.8).

Our first result is the following.

Theorem 5.6 *Let (u_p) be a family of solutions to (5.2) satisfying (5.6). Then the following held.*

(i) *(No vanishing)*

$$\|u_p\|_\infty^{p-1} \geq \lambda_1,$$

where $\lambda_1 = \lambda_1(\Omega)(> 0)$ *is the first eigenvalue of the operator $-\Delta$ in $H_0^1(\Omega)$.*
If u_p is sign-changing then also $\|u_p^\pm\|_\infty^{p-1} \geq \lambda_1$.

(ii) *(Existence of the first bubble) Let $(x_p^+)_p \subset \Omega$ such that $|u_p(x_p^+)| = \|u_p\|_\infty$. Let us set*

$$\mu_p^+ := \left(p|u_p(x_p^+)|^{p-1}\right)^{-\frac{1}{2}} \tag{5.10}$$

and for $x \in \tilde{\Omega}_p^+ := \{x \in \mathbb{R}^2 : x_p^+ + \mu_p^+ x \in \Omega\}$

$$v_p^+(x) := \frac{p}{u_p(x_p^+)}(u_p(x_p^+ + \mu_p^+ x) - u_p(x_{1,p})). \tag{5.11}$$

Then $\mu_p^+ \to 0$ as $p \to +\infty$ and

$$v_p^+ \longrightarrow U \text{ in } C_{\text{loc}}^2(\mathbb{R}^2) \text{ as } p \to +\infty,$$

where

$$U(x) = \log\left(\frac{1}{1 + \frac{1}{8}|x|^2}\right)^2 \tag{5.12}$$

is the solution of $-\Delta U = e^U$ in $\mathbb{R}^2$, $U \leq 0$, $U(0) = 0$ and $\int_{\mathbb{R}^2} e^U = 8\pi$. Moreover

$$\liminf_{p \to +\infty} \|u_p\|_\infty \geq 1. \tag{5.13}$$

(iii) *(No blow-up) There exists $C > 0$ such that*

$$\|u_p\|_\infty \leq C, \text{ for } p \text{ large.} \tag{5.14}$$

(iv) There exist constants $c, C > 0$, such that for all p sufficiently large we have

$$c \le p \int_\Omega |u_p|^p dx \le C. \tag{5.15}$$

(v) $\sqrt{p}\, u_p \rightharpoonup 0$ in $H_0^1(\Omega)$ as $p \to +\infty$.

Proof Point *(i)* was first proved for positive solutions in [41], here we follows the proof in [31, Proposition 2.5]. If u_p is sign-changing, just observe that $u_p^\pm \in H_0^1(\Omega)$, where we know that

$$0 < 8\pi e - \varepsilon \overset{(5.7)/(5.8)}{\le} \int_\Omega |\nabla u_p^\pm|^2 dx \overset{(5.6)}{\le} C < +\infty$$

and that also by the Poincaré inequality

$$\int_\Omega |\nabla u_p^\pm|^2 dx = \int_\Omega |u_p^\pm|^{p+1} dx \le \|u_p^\pm\|_\infty^{p-1} \int_\Omega |u_p^\pm|^2 dx \le \frac{\|u_p^\pm\|_{L^\infty(\Omega)}^{p-1}}{\lambda_1(\Omega)} \int_\Omega |\nabla u_p^\pm|^2 dx.$$

If u_p is not sign-changing just observe that either $u_p = u_p^+$ or $u_p = u_p^-$ and the same proof as before applies.

The proof of *(ii)* follows the same ideas as in [1] where the same result has been proved for least energy (positive) solutions. We let x_p^+ be a point in Ω where $|u_p|$ achieves its maximum. Without loss of generality we can assume that

$$u_p(x_p^+) = \max_\Omega u_p > 0. \tag{5.16}$$

By *(i)* we have that $p u_p(x_p^+)^{p-1} \to +\infty$ as $p \to +\infty$, so (5.13) holds and moreover $\mu_{1,p} \to 0$, where μ_p^+ is defined in (5.10). Let $\widetilde{\Omega}_p^+$ and v_p^+ be defined as in (5.11), then by (5.16) we have

$$v_p^+(0) = 0 \quad \text{and} \quad v_p^+ \le 0 \text{ in } \widetilde{\Omega}_p^+. \tag{5.17}$$

Moreover, v_p^+ solves

$$-\Delta v_p^+ = \left|1 + \frac{v_p^+}{p}\right|^p \left(1 + \frac{v_p^+}{p}\right) \quad \text{in } \widetilde{\Omega}_p^+, \tag{5.18}$$

with

$$\left|1 + \frac{v_p^+}{p}\right| \le 1 \quad \text{and} \quad v_p^+ = -p \quad \text{on } \partial\widetilde{\Omega}_p^+.$$

Then

$$|-\Delta v_p^+| \le 1 \quad \text{in } \widetilde{\Omega}_p^+. \tag{5.19}$$

Using (5.17) and (5.19) we prove that

$$\widetilde{\Omega}_p^+ \to \mathbb{R}^2 \quad \text{as } p \to +\infty. \tag{5.20}$$

Indeed, since $\mu_p^+ \to 0$ as $p \to +\infty$, either $\widetilde{\Omega}_p^+ \to \mathbb{R}^2$ or $\widetilde{\Omega}_p^+ \to \mathbb{R}\times]-\infty, R[$ as $p \to +\infty$ for some $R \geq 0$ (up to a rotation). In the second case we let

$$v_p^+ = \varphi_p + \psi_p \quad \text{in } \widetilde{\Omega}_p^+ \cap B_{2R+1}(0)$$

with $-\Delta\varphi_p = -\Delta v_p^+$ in $\widetilde{\Omega}_p^+ \cap B_{2R+1}(0)$ and $\psi_p = v_p^+$ in $\partial\left(\widetilde{\Omega}_p^+ \cap B_{2R+1}(0)\right)$.

Thanks to (5.19) we have, by standard elliptic theory, that φ_p is uniformly bounded in $\widetilde{\Omega}_1^+ \cap B_{2R+1}(0)$. So the function ψ_p is harmonic in $\widetilde{\Omega}_p^+ \cap B_{2R+1}(0)$, bounded from above by (5.17) and satisfies $\psi_p = -p \to -\infty$ on $\partial\widetilde{\Omega}_p^+ \cap B_{2R+1}(0)$. Since $\partial\widetilde{\Omega}_p^+ \cap B_{2R+1}(0) \to (\mathbb{R} \times \{R\}) \cap B_{2R+1}(0)$ as $p \to +\infty$ one easily gets that $\psi_p(0) \to -\infty$ as $p \to +\infty$ (if $R = 0$ this is trivial, if $R > 0$ it follows by the Harnack inequality). This is a contradiction since $\psi_p(0) = -\varphi_p(0)$ and φ_p is bounded, hence (5.20) is proved.

Then for any $R > 0$, $B_R(0) \subset \widetilde{\Omega}_{1,p}$ for p sufficiently large. So (v_p^+) is a family of nonpositive functions with uniformly bounded Laplacian in $B_R(0)$ and with $v_{p^+}(0) = 0$.

Thus, arguing as before, we write $v_p^+ = \varphi_p + \psi_p$, where φ_p is uniformly bounded in $B_R(0)$ and ψ_p is an harmonic function which is uniformly bounded from above. By the Harnack inequality, either ψ_p is uniformly bounded in $B_R(0)$ or it tends to $-\infty$ on each compact set of $B_R(0)$. The second alternative cannot happen because, by definition, $\psi_p(0) = v_p^+(0) - \varphi_p(0) = -\varphi_p(0) \geq -C$. Hence we obtain that v_p^+ is uniformly bounded in $B_R(0)$, for all $R > 0$. By standard elliptic regularity theory one has that v_p^+ is bounded in $C_{loc}^{2,\alpha}(\mathbb{R}^2)$. Thus, by the Arzela–Ascoli theorem and a diagonal process on $R \to +\infty$, after passing to a subsequence

$$v_p^+ \to U \quad \text{in } C_{loc}^2(\mathbb{R}^2) \text{ as } p \to +\infty, \tag{5.21}$$

with $U \in C^2(\mathbb{R}^2)$, $U \leq 0$ and $U(0) = 0$. Thanks to (5.18) (on each ball also $1 + \frac{v_p^+}{p} > 0$ for p large) and (5.21) we get that U is a solution of $-\Delta U = e^U$ in $\mathbb{R}^2$. Moreover for any $R > 0$, by (5.24), we have

$$\int_{B_R(0)} e^{U(x)}dx \overset{(5.21)+\text{Fatou}}{\leq} \int_{B_R(0)} \frac{|u_p(x_p^+ + \mu_p^+ x)|^{p+1}}{u_p(x_p^+)^{p+1}}dx + o_p(1)$$

$$= \frac{p}{\|u_p\|_\infty^2} \int_{B_{R\mu_{1,p}}(x_{1,p})} |u_p(y)|^{p+1}dy + o_p(1)$$

$$\overset{(5.13)}{\leq} \frac{p}{(1-\varepsilon)^2} \int_\Omega |u_p(y)|^{p+1}dy + o_p(1) \overset{(5.6)}{\leq} C < +\infty,$$

so that $e^U \in L^1(\mathbb{R}^2)$. Thus, since $U(0) = 0$, by virtue of the classification due to Chen and Li [8] we obtain (5.12). Lastly an easy computation shows that $\int_{\mathbb{R}^2} e^U = 8\pi$.

Point (*iii*) was first proved in [41], here we write a simpler proof which follows directly from (*ii*) by applying Fatou's lemma. An analogous argument can be found in [1, Lemma 3.1]. Indeed,

$$C \overset{(5.6)}{\geq} p \int_\Omega |u_p(y)|^{p+1} dy = \|u_p\|_\infty^2 \int_{\tilde\Omega_p^+} \left|1 + \frac{v_p^+(x)}{p}\right|^{p+1} dx$$

$$\overset{(ii)\text{-Fatou}}{\geq} \|u_p\|_\infty^2 \int_{\mathbb{R}^2} e^{U(x)} dx = \|u_p\|_\infty^2 8\pi.$$

(*iv*) follows directly from (*iii*). Indeed, on the one hand

$$0 < C \overset{(5.7)-(5.9)}{\leq} p \int_\Omega |u_p|^{p+1} dx \leq \|u_p\|_\infty p \int_\Omega |u_p|^p dx \overset{(iii)}{\leq} Cp \int_\Omega |u_p|^p dx.$$

On the other hand, by the Hölder inequality

$$p \int_\Omega |u_p|^p dx \leq |\Omega|^{\frac{1}{p+1}} p \left(\int_\Omega |u_p|^{p+1} dx\right)^{\frac{p}{p+1}} \overset{(5.6)}{\leq} C.$$

To prove (*v*) we need (*iv*). Indeed, let us note that, since (5.6) holds, there exists $w \in H_0^1(\Omega)$ such that, up to a subsequence, $\sqrt{p}u_p \rightharpoonup w$ in $H_0^1(\Omega)$. We want to show that $w = 0$ a.e. in Ω.

Using the equation (5.2), for any test function $\varphi \in C_0^\infty(\Omega)$, we have

$$\int_\Omega \nabla(\sqrt{p}u_p)\nabla\varphi \, dx = \sqrt{p} \int_\Omega |u_p|^{p-1} u_p \varphi \, dx \leq \frac{\|\varphi\|_\infty}{\sqrt{p}} p \int_\Omega |u_p|^p dx \overset{(iv)}{\leq} \frac{\|\varphi\|_\infty}{\sqrt{p}} C$$

for p large. Hence

$$\int_\Omega \nabla w \nabla\varphi \, dx = 0 \quad \forall \varphi \in C_0^\infty(\Omega),$$

which implies that $w = 0$ a.e. in Ω. $\qquad\square$

In order to show our next results we need to introduce some notation. Given a family (u_p) of solutions of (5.2) and assuming that there exists $n \in \mathbb{N} \setminus \{0\}$ families of points $(x_{i,p})$, $i = 1,\dots,n$ in Ω such that

$$p|u_p(x_{i,p})|^{p-1} \to +\infty \quad \text{as } p \to +\infty, \tag{5.22}$$

we define the parameters $\mu_{i,p}$ by

$$\mu_{i,p}^{-2} = p|u_p(x_{i,p})|^{p-1}, \quad \text{for all } i = 1,\dots,n. \tag{5.23}$$

By (5.22) it is clear that $\mu_{i,p} \to 0$ as $p \to +\infty$ and that

$$\liminf_{p \to +\infty} |u_p(x_{i,p})| \geq 1. \tag{5.24}$$

Then we define the concentration set

$$\mathcal{S} = \left\{ \lim_{p \to +\infty} x_{i,p}, \, i = 1, \ldots, n \right\} \subset \bar{\Omega} \tag{5.25}$$

and the function

$$R_{n,p}(x) = \min_{i=1,\ldots,n} |x - x_{i,p}|, \, \forall x \in \Omega. \tag{5.26}$$

Finally, we introduce the following properties.

$(\mathcal{P}_1^n)$ For any $i, j \in \{1, \ldots, n\}$, $i \neq j$,

$$\lim_{p \to +\infty} \frac{|x_{i,p} - x_{j,p}|}{\mu_{i,p}} = +\infty.$$

$(\mathcal{P}_2^n)$ For any $i = 1, \ldots, n$, for $x \in \tilde{\Omega}_{i,p} := \{x \in \mathbb{R}^2 : x_{i,p} + \mu_{i,p} x \in \Omega\}$,

$$v_{i,p}(x) := \frac{p}{u_p(x_{i,p})} (u_p(x_{i,p} + \mu_{i,p} x) - u_p(x_{i,p})) \longrightarrow U(x) \tag{5.27}$$

in $C^2_{\mathrm{loc}}(\mathbb{R}^2)$ as $p \to +\infty$, where U is the same function as in (5.12).

$(\mathcal{P}_3^n)$ There exists $C > 0$ such that

$$p R_{n,p}(x)^2 |u_p(x)|^{p-1} \leq C$$

for all $p > 1$ and all $x \in \Omega$.

$(\mathcal{P}_4^n)$ There exists $C > 0$ such that

$$p R_{n,p}(x) |\nabla u_p(x)| \leq C$$

for all $p > 1$ and all $x \in \Omega$.

Lemma 5.7 *If there exists $n \in \mathbb{N} \setminus \{0\}$ such that the properties $(\mathcal{P}_1^n)$ and $(\mathcal{P}_2^n)$ hold for families $(x_{i,p})_{i=1,\ldots,n}$ of points satisfying (5.22), then*

$$p \int_\Omega |\nabla u_p|^2 \, dx \geq 8\pi \sum_{i=1}^n \alpha_i^2 + o_p(1) \quad as \, p \to +\infty,$$

$$\tag{5.24}$$

where $\alpha_i := \liminf_{p \to +\infty} |u_p(x_{i,p})| \, (\geq 1)$.

Proof Let us write, for any $R > 0$

$$p \int_{B_{R\mu_{i,p}}(x_{i,p})} |u_p|^{p+1} \, dx = \int_{B_R(0)} \frac{|u_p(x_{i,p} + \mu_{i,p} y)|^{p+1}}{|u_p(x_{i,p})|^{p-1}} \, dy$$

$$= u_p^2(x_{i,p}) \int_{B_R(0)} \left| 1 + \frac{v_{i,p}(y)}{p} \right|^{p+1} \, dy. \tag{5.28}$$

Thanks to $(\mathcal{P}_2^n)$, we have

$$\|v_{i,p} - U\|_{L^\infty(B_R(0))} = o_p(1) \quad \text{as } p \to +\infty. \tag{5.29}$$

Thus by (5.28), (5.29) and Fatou's lemma

$$\liminf_{p \to +\infty} \left(p \int_{B_{R\mu_{i,p}}(x_{i,p})} |u_p|^{p+1} \, dx \right) \geq \alpha_i^2 \int_{B_R(0)} e^U \, dx. \tag{5.30}$$

Moreover, by virtue of $(\mathcal{P}_1^n)$ it is not hard to see that $B_{R\mu_{i,p}}(x_{i,p}) \cap B_{R\mu_{j,p}}(x_{j,p}) = \emptyset$ for all $i \neq j$. Hence, in particular, thanks to (5.30),

$$\liminf_{p \to +\infty} \left(p \int_\Omega |u_p|^{p+1} \, dx \right) \geq \sum_{i=1}^n \left(\alpha_i^2 \int_{B_R(0)} e^U \, dx \right).$$

At last, since this holds for any $R > 0$, we get

$$p \int_\Omega |\nabla u_p|^2 \, dx = p \int_\Omega |u_p|^{p+1} \, dx \geq \sum_{i=1}^n \alpha_i^2 \int_{\mathbb{R}^2} e^U \, dx + o(1)$$

$$= 8\pi \sum_{i=1}^n \alpha_i^2 + o(1) \quad \text{as } p \to +\infty.$$

$\square$

The next result shows that the solutions concentrate at a finite number of points and also establishes the existence of a maximal number of "bubbles".

Theorem 5.8 *Let (u_p) be a family of solutions to (5.2) and assume that (5.6) holds. Then there exist $k \in \mathbb{N} \setminus \{0\}$ and k families of points $(x_{i,p})$ in Ω $i = 1, \ldots, k$ such that, after passing to a sequence, $(\mathcal{P}_1^k)$, $(\mathcal{P}_2^k)$ and $(\mathcal{P}_3^k)$ hold. Moreover, $x_{1,p} = x_p^+$ and, given any family of points $x_{k+1,p}$, it is impossible to extract a new sequence from the previous one such that $(\mathcal{P}_1^{k+1})$, $(\mathcal{P}_2^{k+1})$ and $(\mathcal{P}_3^{k+1})$ hold with the sequences $(x_{i,p})$, $i = 1, \ldots, k+1$. Lastly, we have*

$$\sqrt{p} u_p \to 0 \quad \text{in } C^2_{\text{loc}}(\bar{\Omega} \setminus \mathcal{S}) \text{ as } p \to +\infty. \tag{5.31}$$

Moreover, there exists $v \in C^2(\bar{\Omega} \setminus \mathcal{S})$ such that

$$p u_p \to v \quad \text{in } C^2_{\text{loc}}(\bar{\Omega} \setminus \mathcal{S}) \text{ as } p \to +\infty, \tag{5.32}$$

and $(\mathcal{P}_4^k)$ holds.

Proof This result is mainly contained in [20]. The proof is inspired by the one in [25, Proposition 1] (see also [43]), but we have to deal with an extra difficulty because we allow the solutions u_p to be sign-changing. We divide the proof into several steps and all the claims are up to a subsequence.

STEP 1. *We show that there exists a family* $(x_{1,p})$ *of points in* Ω *such that, after passing to a sequence* $(\mathcal{P}_2^1)$ *holds.*

We let x_p^+ be a point in Ω where $|u_p|$ achieves its maximum. The proof then follows by taking $x_{1,p} := x_p^+$ and using the results in Theorem 5.6-(ii).

STEP 2. *We assume that* $(\mathcal{P}_1^n)$ *and* $(\mathcal{P}_2^n)$ *hold for some* $n \in \mathbb{N} \setminus \{0\}$. *Then we show that either* $(\mathcal{P}_1^{n+1})$ *and* $(\mathcal{P}_2^{n+1})$ *hold or* $(\mathcal{P}_3^n)$ *holds, specifically there exists* $C > 0$ *such that*

$$pR_{n,p}(x)^2 |u_p(x)|^{p-1} \leq C$$

for all $x \in \Omega$, *with* $R_{n,p}$ *defined as in (5.26).*

Let $n \in \mathbb{N} \setminus \{0\}$ and assume that $(\mathcal{P}_1^n)$ and $(\mathcal{P}_2^n)$ hold while

$$\sup_{x\in\Omega} \left(pR_{n,p}(x)^2 |u_p(x)|^{p-1}\right) \to +\infty \quad \text{as } p \to +\infty. \tag{5.33}$$

We will prove that $(\mathcal{P}_1^{n+1})$ and $(\mathcal{P}_2^{n+1})$ hold.

We let $x_{n+1,p} \in \bar{\Omega}$ be such that

$$pR_{n,p}(x_{n+1,p})^2 |u_p(x_{n+1,p})|^{p-1} = \sup_{x\in\Omega} \left(pR_{n,p}(x)^2 |u_p(x)|^{p-1}\right). \tag{5.34}$$

Clearly $x_{n+1,p} \in \Omega$ because $u_p = 0$ on $\partial\Omega$. By (5.34) and since Ω is bounded it is clear that

$$p|u_p(x_{n+1,p})|^{p-1} \to +\infty \quad \text{as } p \to +\infty.$$

We claim that

$$\frac{|x_{i,p} - x_{n+1,p}|}{\mu_{i,p}} \to +\infty \quad \text{as } p \to +\infty \tag{5.35}$$

for all $i = 1,\ldots,n$ and $\mu_{i,p}$ as in (5.23). In fact, assuming by contradiction that there exists $i \in \{1,\ldots,n\}$ such that $|x_{i,p} - x_{n+1,p}|/\mu_{i,p} \to R$ as $p \to +\infty$ for some $R \geq 0$, thanks to $(\mathcal{P}_2^n)$, we get

$$\lim_{p\to+\infty} p|x_{i,p} - x_{n+1,p}|^2 |u_p(x_{n+1,p})|^{p-1} = R^2 \left(\frac{1}{1+\frac{1}{8}R^2}\right)^2 < +\infty,$$

against (5.34). Setting

$$(\mu_{n+1,p})^{-2} := p|u_p(x_{n+1,p})|^{p-1}, \tag{5.36}$$

by (5.33) and (5.34) we deduce that

$$\frac{R_{n,p}(x_{n+1,p})}{\mu_{n+1,p}} \to +\infty \quad \text{as } p \to +\infty. \tag{5.37}$$

Then (5.36), (5.37) and $(\mathcal{P}_1^n)$ imply that $(\mathcal{P}_1^{n+1})$ holds with the added sequence $(x_{n+1,p})$.

Next we show that also $(\mathcal{P}_2^{n+1})$ holds with the added sequence $(x_{n+1,p})$. Let us define the scaled domain

$$\widetilde{\Omega}_{n+1,p} = \{x \in \mathbb{R}^2 : x_{n+1,p} + \mu_{n+1,p}x \in \Omega\},$$

and, for $x \in \widetilde{\Omega}_{n+1,p}$, the rescaled functions

$$v_{n+1,p}(x) = \frac{p}{u_p(x_{n+1,p})}(u_p(x_{n+1,p} + \mu_{n+1,p}x) - u_p(x_{n+1,p})), \qquad (5.38)$$

which, by (5.2), satisfy

$$-\Delta v_{n+1,p}(x) = \frac{|u_p(x_{n+1,p} + \mu_{n+1,p}x)|^{p-1}u_p(x_{n+1,p} + \mu_{n+1,p}x)}{|u_p(x_{n+1,p})|^{p-1}u_p(x_{n+1,p})} \qquad \text{in } \widetilde{\Omega}_{n+1,p}, \qquad (5.39)$$

or equivalently

$$-\Delta v_{n+1,p}(x) = \left|1 + \frac{v_{n+1,p}(x)}{p}\right|^{p-1}\left(1 + \frac{v_{n+1,p}(x)}{p}\right) \qquad \text{in } \widetilde{\Omega}_{n+1,p}. \quad (5.40)$$

Fix $R > 0$ and let (z_p) be any point in $\widetilde{\Omega}_{n+1,p} \cap B_R(0)$, whose corresponding points in Ω are

$$x_p = x_{n+1,p} + \mu_{n+1,p}z_p.$$

Thanks to the definition of $x_{n+1,p}$ we have

$$pR_{n,p}(x_p)^2|u_p(x_p)|^{p-1} \leq pR_{n,p}(x_{n+1,p})^2|u_p(x_{n+1,p})|^{p-1}. \qquad (5.41)$$

Since $|x_p - x_{n+1,p}| \leq R\mu_{n+1,p}$ we have

$$R_{n,p}(x_p) \geq \min_{i=1,\dots,n}|x_{n+1,p} - x_{i,p}| - |x_p - x_{n+1,p}|$$
$$\geq R_{n,p}(x_{n+1,p}) - R\mu_{n+1,p}$$

and, analogously,

$$R_{n,p}(x_p) \leq R_{n,p}(x_{n+1,p}) + R\mu_{n+1,p}.$$

Thus, by (5.37) we get

$$R_{n,p}(x_p) = (1 + o(1))R_{n,p}(x_{n+1,p})$$

and in turn from (5.41)

$$|u_p(x_p)|^{p-1} \leq (1 + o(1))|u_p(x_{n+1,p})|^{p-1}. \qquad (5.42)$$

In the following we show that for any compact subset K of $\mathbb{R}^2$

$$-1 + o(1) \leq -\Delta v_{n+1,p} \leq 1 + o(1) \qquad \text{in } \widetilde{\Omega}_{n+1,p} \cap K \qquad (5.43)$$

and

$$\limsup_{p \to +\infty} \sup_{\tilde{\Omega}_{n+1,p} \cap K} v_{n+1,p} \le 0. \tag{5.44}$$

In order to prove (5.43) and (5.44) we will distinguish two cases.

(*i*) *Assume that*

$$\frac{u_p(x_p)}{u_p(x_{n+1,p})} = \frac{|u_p(x_p)|}{|u_p(x_{n+1,p})|}. \tag{5.45}$$

Then by (5.42) we get $|u_p(x_p)|^p \le (1+o(1))|u_p(x_{n+1,p})|^p$ and so by (5.39)

$$(0 \le) - \Delta v_{n+1,p}(z_p) \overset{(5.45)}{=} \frac{|u_p(x_p)|^p}{|u_p(x_{n+1,p})|^p} \le 1 + o(1). \tag{5.46}$$

Moreover, since (5.40) implies $-\Delta v_{n+1,p}(z_p) = e^{v_{n+1,p}(z_p)} + o(1)$, we get

$$\limsup_{p \to +\infty} v_{n+1,p}(z_p) \le 0. \tag{5.47}$$

(*ii*) *Assume that*

$$\frac{u_p(x_p)}{u_p(x_{n+1,p})} = -\frac{|u_p(x_p)|}{|u_p(x_{n+1,p})|}. \tag{5.48}$$

Then, by the expression of $v_{n+1,p}$ necessarily

$$v_{n+1,p}(z_p) \le 0, \tag{5.49}$$

and moreover by (5.42)

$$0 \ge -\Delta v_{n+1,p}(z_p) = -\frac{|u_p(x_p)|^p}{|u_p(x_{n+1,p})|^p} \ge -1 + o(1). \tag{5.50}$$

(5.46) and (5.50) imply (5.43), while (5.49), (5.47) and the arbitrariness of z_p give (5.44).

Using (5.43) and (5.44) we can prove, similarly to the proof of Theorem 5.6-(ii), that

$$\tilde{\Omega}_{n+1,p} \to \mathbb{R}^2 \text{ as } p \to +\infty. \tag{5.51}$$

Then for any $R > 0$, $B_R(0) \subset \tilde{\Omega}_{n+1,p}$ for p large enough and $v_{n+1,p}$ are functions with uniformly bounded Laplacian in $B_R(0)$ and with $v_{n+1,p}(0) = 0$. So, by the Harnack inequality, $v_{n+1,p}$ is uniformly bounded in $B_R(0)$ for all $R > 0$ and then by standard elliptic regularity $v_{n+1,p} \to U$ in $C^2_{loc}(\mathbb{R}^2)$ as $p \to +\infty$ with $U \in C^2(\mathbb{R}^2)$, $U(0) = 0$ and, by (5.44), $U \le 0$. Passing to the limit in (5.40) we get that U is a solution of $-\Delta U = e^U$ in $\mathbb{R}^2$. Then by Fatou's lemma, as in the proof of Theorem 5.6-(ii), we get that $e^U \in L^1(\mathbb{R}^2)$ and so by the classification result in [8] we have the explicit expression of U.

This proves that $(\mathcal{P}_2^{n+1})$ holds with the added points $(x_{n+1,p})$, thus *STEP 2* is proved.

STEP 3. We complete the proof of Theorem 5.8.

From *STEP 1* we have that $(\mathcal{P}_1^1)$ and $(\mathcal{P}_2^1)$ hold. Then, by *STEP 2*, either $(\mathcal{P}_1^2)$ and $(\mathcal{P}_2^2)$ hold or $(\mathcal{P}_3^1)$ holds. In the last case the assertion is proved with $k = 1$. In the first case we go on and proceed with the same alternative until we reach a number $k \in \mathbb{N} \setminus \{0\}$ for which $(\mathcal{P}_1^k)$, $(\mathcal{P}_2^k)$ and $(\mathcal{P}_3^k)$ hold up to a sequence. Note that this is possible because the solutions u_p satisfy (5.6) and Lemma 5.7 holds and hence the maximal number k of families of points for which $(\mathcal{P}_1^k)$, $(\mathcal{P}_2^k)$ hold must be finite.

Moreover, given any other family of points $x_{k+1,p}$, it is impossible to extract a new sequence from it such that $(\mathcal{P}_1^{k+1})$, $(\mathcal{P}_2^{k+1})$ and $(\mathcal{P}_3^{k+1})$ hold together with the points $(x_{i,p})_{i=1,\ldots,k+1}$. Indeed, if $(\mathcal{P}_1^{k+1})$ was verified then

$$\frac{|x_{k+1,p} - x_{i,p}|}{\mu_{k+1,p}} \to +\infty \quad \text{as } p \to +\infty, \text{ for any } i \in \{1,\ldots,k\},$$

but this would contradict $(\mathcal{P}_3^k)$.

Finally, the proofs of (5.31) and (5.32) are a direct consequence of $(\mathcal{P}_3^k)$. Indeed, if K is a compact subset of $\bar{\Omega} \setminus S$ by $(\mathcal{P}_3^k)$ we have that there exists $C_K > 0$ such that

$$p|u_p(x)|^{p-1} \le C_K \quad \text{for all } x \in K.$$

Then by (5.2) $\|\Delta(\sqrt{p}u_p)\|_{L^\infty(K)} \le C_K \frac{\|u_p\|_{L^\infty(K)}}{\sqrt{p}} \to 0$, as $p \to +\infty$. Hence standard elliptic theory shows that $\sqrt{p}u_p \to w$ in $C^2(K)$, for some w. But by Theorem 5.6 we know that $\sqrt{p}u_p \rightharpoonup 0$, so $w = 0$ and (5.31) is proved. Iterating, we then have $\|\Delta(pu_p)\|_{L^\infty(K)} \le C_K\|u_p\|_{L^\infty(K)} \to 0$, as $p \to +\infty$ by (5.31). And so by standard elliptic theory we have that $pu_p \to v$ in $C^2(K)$, for some v. The arbitrariness of the compact set K ends the proof of (5.32).

It remains to prove $(\mathcal{P}_4^k)$. Green's representation gives

$$p|\nabla u_p(x)| = p\left|\int_\Omega \nabla G(x,y)u_p(y)^p dy\right| \le p\int_\Omega |\nabla G(x,y)||u_p(y)|^p dy$$
$$\le Cp\int_\Omega \frac{|u_p(y)|^p}{|x-y|} dy, \qquad (5.52)$$

where G is the Green's function of $-\Delta$ in Ω with Dirichlet boundary conditions, and in the last estimate we have used that $|\nabla_x G(x,y)| \le \frac{C}{|x-y|}$ $\forall x, y \in \Omega$, $x \ne y$ (see for instance [12]). Let $R_{k,p}(x) = \min_{i=1,\ldots,k} |x - x_{i,p}|$

and $\Omega_{i,p} = \{x \in \Omega : |x - x_{i,p}| = R_{i,p}(x)\}$, $i = 1, \ldots, k$. We then have

$$p \int_{\Omega_{i,p}} |x-y|^{-1} |u_p(y)|^p \, dy = p \int_{\Omega_{i,p} \cap B_{\frac{|x-x_{i,p}|}{2}}(x_{i,p})} |x-y|^{-1} |u_p(y)|^p \, dy$$

$$+ p \int_{\Omega_{i,p} \setminus B_{\frac{|x-x_{i,p}|}{2}}(x_{i,p})} |x-y|^{-1} |u_p(y)|^p \, dy.$$

Note that by (5.14) and $(\mathcal{P}_3^k)$ for $y \in \Omega_{i,p} \setminus B_{\frac{|x-x_{i,p}|}{2}}(x_{i,p})$ we have

$$\frac{p |u_p(y)|^p}{|x-y|} \le C \frac{p |u_p(y)|^{p-1}}{|x-y|} \le \frac{C}{|x-y||y-x_{i,p}|^2} \le \frac{C}{|x-y||x-x_{i,p}|^2},$$

and hence

$$\int_{\Omega_{i,p} \setminus B_{\frac{|x-x_{i,p}|}{2}}(x_{i,p})} \frac{p |u_p(y)|^p}{|x-y|} \, dy \le \frac{1}{|x-x_{i,p}|^2} \int_{|x-y| \le |x-x_{i,p}|} \frac{C}{|x-y|} \, dy$$

$$+ \frac{1}{|x-x_{i,p}|} \int_{\Omega_{i,p}} p |u_p(y)|^p \, dy$$

$$\overset{(5.15)}{\le} \frac{C}{|x-x_{i,p}|}.$$

For $y \in \Omega_{i,p} \cap B_{\frac{|x-x_{i,p}|}{2}}(x_{i,p})$, $|x-y| \ge |x-x_{i,p}| - |y-x_{i,p}| \ge |x-x_{i,p}|/2$, and hence by (5.14) and (5.15) we get

$$p \int_{\Omega_{i,p} \cap B_{\frac{|x-x_{i,p}|}{2}}(x_{i,p})} \frac{|u_p(y)|^p}{|x-y|} \, dy \le \frac{C}{|x-x_{i,p}|}, \qquad \text{for } i = 1, \ldots, k$$

so that $(\mathcal{P}_4^k)$ is proved. $\qquad \square$

In the rest of this section we derive some consequences of Theorem 5.8.

Remark 5.9 *Under the assumptions of Theorem 5.8 we have*

$$\frac{dist(x_{i,p}, \partial\Omega)}{\mu_{i,p}} \xrightarrow[p \to +\infty]{} +\infty \qquad \text{for all } i \in \{1, \ldots, k\}.$$

Corollary 5.10 *Under the assumptions of Theorem 5.8, if u_p is sign-changing it follows that*

$$\frac{dist(x_{i,p}, NL_p)}{\mu_{i,p}} \xrightarrow[p \to +\infty]{} +\infty \qquad \text{for all } i \in \{1, \ldots, k\},$$

where NL_p denotes the nodal line of u_p.
As a consequence, for any $i \in \{1, \ldots, k\}$, letting $\mathcal{N}_{i,p} \subset \Omega$ be the nodal domain

of u_p containing $x_{i,p}$ and setting $u_p^i := u_p \chi_{\mathcal{N}_{i,p}}$ (χ_A is the characteristic function of the set A), then the scaling of u_p^i around $x_{i,p}$:

$$z_{i,p}(x) := \frac{p}{u_p(x_{i,p})}(u_p^i(x_{i,p} + \mu_{i,p}x) - u_p(x_{i,p})),$$

defined on $\tilde{\mathcal{N}}_{i,p} := \frac{\mathcal{N}_{i,p} - x_{i,p}}{\mu_{i,p}}$, converges to U in $C^2_{loc}(\mathbb{R}^2)$, where U is the same function defined in $(\mathcal{P}_2^k)$.

Proof Let us suppose by contradiction that

$$\frac{dist(x_{i,p}, NL_p)}{\mu_{i,p}} \xrightarrow[p \to +\infty]{} \ell \geq 0,$$

then there exist $y_p \in NL_p$ such that $\frac{|x_{i,p} - y_p|}{\mu_{i,p}} \to \ell$ as $p \to +\infty$. Setting

$$v_{i,p}(x) := \frac{p}{u_p(x_{i,p})}(u_p(x_{i,p} + \mu_{i,p}x) - u_p(x_{i,p})),$$

on the one hand

$$v_{i,p}(\frac{y_p - x_{i,p}}{\mu_{i,p}}) = -p \xrightarrow[p \to +\infty]{} -\infty,$$

on the other hand by $(\mathcal{P}_2^k)$ and up to subsequences

$$v_{i,p}\left(\frac{y_p - x_{i,p}}{\mu_{i,p}}\right) \xrightarrow[p \to +\infty]{} U(x_\infty) > -\infty,$$

where $x_\infty = \lim_{p \to +\infty} \frac{y_p - x_{i,p}}{\mu_{i,p}} \in \mathbb{R}^2$ and so $|x_\infty| = \ell$. Thus we have obtained a contradiction which proves the assertion. $\square$

For a family of points $(x_p)_p \subset \Omega$ we denote by $\mu(x_p)$ the numbers defined by

$$[\mu(x_p)]^{-2} := p|u_p(x_p)|^{p-1}. \tag{5.53}$$

Proposition 5.11 *Let $(x_p)_p \subset \Omega$ be a family of points such that $p|u_p(x_p)|^{p-1} \to +\infty$ and let $\mu(x_p)$ be as in (5.53). By $(\mathcal{P}_3^k)$, up to a sequence, $R_{k,p}(x_p) = |x_{i,p} - x_p|$, for a certain $i \in \{1, \ldots, k\}$. Then*

$$\limsup_{p \to +\infty} \frac{\mu_{i,p}}{\mu(x_p)} \leq 1.$$

Proof To shorten the notation let us denote $\mu(x_p)$ simply by μ_p. Let us start by proving that $\frac{\mu_{i,p}}{\mu_p}$ is bounded. So we assume by contradiction that there exists a sequence $p_n \to +\infty$, as $n \to +\infty$, such that

$$\frac{\mu_{i,p_n}}{\mu_{p_n}} \to +\infty \quad \text{as } n \to +\infty. \tag{5.54}$$

By $(\mathcal{P}_3^k)$ and (5.54) we then have

$$\frac{|x_{p_n} - x_{i,p_n}|}{\mu_{i,p_n}} = \frac{|x_{p_n} - x_{i,p_n}|}{\mu_{p_n}} \frac{\mu_{p_n}}{\mu_{i,p_n}} \to 0 \quad \text{as } n \to +\infty,$$

so that by $(\mathcal{P}_2^k)$

$$v_{i,p_n}\left(\frac{x_{p_n} - x_{i,p_n}}{\mu_{i,p_n}}\right) \to U(0) = 0 \quad \text{as } n \to +\infty.$$

As a consequence

$$\frac{\mu_{i,p_n}}{\mu_{p_n}} = \left(\frac{u_{p_n}(x_{p_n})}{u_{p_n}(x_{i,p_n})}\right)^{p_n-1} = \left(1 + \frac{v_{i,p_n}\left(\frac{x_{p_n}-x_{i,p_n}}{\mu_{i,p_n}}\right)}{p_n}\right)^{p_n-1} \to e^{U(0)} = 1$$

$$\text{as } n \to +\infty,$$

which contradicts with (5.54). Hence we have proved that $\frac{\mu_{i,p}}{\mu_p}$ is bounded.

Next we show that $\frac{\mu_{i,p}}{\mu_p} \le 1$. Assume by contradiction that there exist $\ell > 1$ and a sequence $p_n \to +\infty$ as $n \to +\infty$ such that

$$\frac{\mu_{i,p_n}}{\mu_{p_n}} \to \ell \quad \text{as } n \to +\infty. \tag{5.55}$$

By $(\mathcal{P}_3^k)$ and (5.55) we then have

$$\frac{|x_{p_n} - x_{i,p_n}|}{\mu_{i,p_n}} = \frac{|x_{p_n} - x_{i,p_n}|}{\mu_{p_n}} \frac{\mu_{p_n}}{\mu_{i,p_n}} \le \frac{2\sqrt{C}}{\ell}$$

for n large, so that by $(\mathcal{P}_2^k)$ there exists $x_\infty \in \mathbb{R}^2$, $|x_\infty| \le \frac{2\sqrt{C}}{\ell}$ such that, up to a subsequence,

$$v_{i,p_n}\left(\frac{x_{p_n} - x_{i,p_n}}{\mu_{i,p_n}}\right) \to U(x_\infty) \le 0 \quad \text{as } n \to +\infty.$$

As a consequence

$$\frac{\mu_{i,p_n}}{\mu_{p_n}} = \left(\frac{u_{p_n}(x_{p_n})}{u_{p_n}(x_{i,p_n})}\right)^{p_n-1} = \left(1 + \frac{v_{i,p_n}\left(\frac{x_{p_n}-x_{i,p_n}}{\mu_{i,p_n}}\right)}{p_n}\right)^{p_n-1} \to e^{U(x_\infty)}$$

$$\text{as } n \to +\infty.$$

By (5.55) and the assumption $\ell > 1$ we deduce

$$U(x_\infty) = \log \ell + o_n(1) > 0,$$

reaching a contradiction. $\qquad\square$

Proposition 5.12 *Let x_p and $x_{i,p}$ be as in the statement of Proposition 5.11. If*

$$\frac{|x_p - x_{i,p}|}{\mu_{i,p}} \to +\infty \quad \text{as } p \to +\infty, \tag{5.56}$$

then

$$\frac{\mu_{i,p}}{\mu(x_p)} \to 0 \quad \text{as } p \to +\infty,$$

where $\mu(x_p)$ is defined in (5.53).

Proof By Proposition 5.11 we know that

$$\frac{\mu_{i,p}}{\mu(x_p)} \leq 1 + o(1).$$

Assume by contradiction that (5.56) holds but there exist $0 < \ell \leq 1$ and a sequence $p_n \to +\infty$ such that

$$\frac{\mu_{i,p_n}}{\mu(x_{p_n})} \to \ell, \quad \text{as } n \to +\infty. \tag{5.57}$$

Then (5.57) and $(\mathcal{P}_3^k)$ imply

$$\frac{|x_{p_n} - x_{i,p_n}|}{\mu_{i,p_n}} = \frac{|x_{p_n} - x_{i,p_n}|}{\ell \, \mu(x_{p_n})} + o_n(1) \leq \frac{C}{\ell} + o_n(1) \quad \text{as } n \to +\infty,$$

which contradicts (5.56). $\qquad\square$

Remark 5.13 *If $u_p(x_p)$ and $u_p(x_{i,p})$ have opposite sign, i.e.*

$$u_p(x_p)u_p(x_{i,p}) < 0,$$

then, by Corollary 5.10, necessarily (5.56) holds. Hence in this case

$$\frac{\mu_{i,p}}{\mu(x_p)} \to 0 \ \text{ as } p \to +\infty.$$

The next result characterizes in different ways the concentration set $\mathcal{S}$.

Proposition 5.14 (Characterizations of $\mathcal{S}$) *Let (u_p) be a family of solutions to (5.2) satisfying (5.6). Then the following hold:*

$$\mathcal{S} = \left\{ x \in \overline{\Omega} : \forall r_0 > 0, \ \forall p_0 > 1, \ \exists p > p_0 \text{ s.t. } p \int_{B_{r_0}(x) \cap \Omega} |u_p(x)|^{p+1} \, dx \geq 1 \right\};$$

$$\tag{5.58}$$

$$\mathcal{S} = \left\{ x \in \overline{\Omega} : \exists \text{ a subsequence of } (u_p) \text{ and a sequence } x_p \to_p x \right.$$
$$\left. \text{s.t. } p|u_p(x_p)| \to_p +\infty \right\}. \tag{5.59}$$

Proof Proof of (5.58): by $(\mathcal{P}_3^k)$ it is immediate to see that if $x \notin \mathcal{S}$ then $p \int_{B_r(x) \cap \Omega} |u_p(x)|^{p+1} \, dx \to 0$ as $r \to 0^+$, uniformly in p. On the other hand if $x \in S$, i.e. $x = \lim_{p \to +\infty} x_{i,p}$ for some $i = 1, \ldots, k$, we fix $R > 0$ such that $\int_{B_R(0)} e^U \, dx > 1$ (where U is defined in (5.12)) and then for p large, reasoning as in the proof of Lemma 5.7, we get:

$$p \int_{B_{r_0}(x) \cap \Omega} |u_p(x)|^{p+1} dx \geq p \int_{B_{R\mu_{i,p}}(x_{i,p})} |u_p(x)|^{p+1} dx$$

$$= |u_p(x_{i,p})|^2 \int_{B_R(0)} \left(1 + \frac{v_{j,p}(y)}{p} \right)^{p+1} dy,$$

where by Fatou's lemma

$$\liminf_{p \to +\infty} |u_p(x_{i,p})|^2 \int_{B_R(0)} \left(1 + \frac{v_{j,p}(y)}{p} \right)^{p+1} dy \;\; \geq \;\; \liminf_{p \to +\infty} |u_p(x_{i,p})|^2 \int_{B_R(0)} e^{U(y)} dy$$

$$\overset{(5.24)}{\geq} \int_{B_R(0)} e^{U(y)} dy > 1.$$

Proof of (5.59): if $x \notin S$, then by $(\mathcal{P}_4^k)$, which holds by Theorem 5.8, $p|u_p|$ is uniformly bounded in $L^\infty(K)$ for some compact set K containing x and then there can not exist a sequence $x_p \to x$ such that $p|u_p(x_p)| \to +\infty$. Conversely, if $x \in S$, i.e. $x = \lim_{p \to +\infty} x_{i,p}$ for some $i = 1, \ldots, k$, and by (5.22) we know that $|u_p(x_{i,p})| \geq \frac{1}{2}$ for p large, therefore $p|u_p(x_{i,p})| \to +\infty$. This proves (5.59). $\qquad \square$

We conclude this section with a result for positive solutions that we have recently obtained (see [22] and the latest improvement in [17]) carrying on the asymptotic analysis started in Theorem 5.8.

Theorem 5.15 *Let (u_p) be a family of positive solutions to (5.2) and assume that (5.6) holds. Let $k \in \mathbb{N} \setminus \{0\}$ and $(x_{i,p})$, $i = 1, \ldots, k$, be the integer and the families of points of Ω introduced in Theorem 5.8. One has:*

$$\lim_{p \to +\infty} \|u_p\|_\infty = \sqrt{e}.$$

Moreover, there exist k distinct points $x_i \in \Omega$, $i = 1, \ldots, k$ such that:

(i) *up to a subsequence $\lim_{p \to +\infty} x_{i,p} = x_i$, for any $i = 1, \ldots, k$; therefore the concentration set S, introduced in (5.25), consists of the k points*

$$S = \{x_1, \ldots, x_k\} \subset \Omega;$$

(ii)

$$pu_p(x) \to 8\pi \sqrt{e} \sum_{i=1}^{k} G(x, x_i) \text{ as } p \to +\infty, \text{ in } C^2_{loc}(\bar{\Omega} \setminus S),$$

where G is the Green's function of $-\Delta$ in Ω under Dirichlet boundary conditions;

(iii)

$$p \int_\Omega |\nabla u_p(x)|^2 \, dx \to 8\pi e \cdot k, \quad as \ p \to +\infty;$$

(iv) *the concentration points* x_i, $i = 1, \ldots, k$ *satisfy*

$$\nabla_x H(x_i, x_i) + \sum_{i \neq \ell} \nabla_x G(x_i, x_\ell) = 0, \tag{5.60}$$

where

$$H(x, y) = G(x, y) + \frac{\log(|x - y|)}{2\pi} \tag{5.61}$$

is the regular part of the Green's function G;

(v)

$$\lim_{p \to +\infty} \|u_p\|_{L^\infty(\overline{B_\delta(x_i)})} = \sqrt{e}, \quad i = 1, \ldots, k$$

for any $\delta > 0$ *such that* $B_\delta(x_i)$ *does not contain any other* x_j, $j \neq i$.

Remark 5.16 *Observe that* $k = 1$ *for the least energy solutions and that in this case the results in the previous theorem were already known (see [41], [42] and [1]).*

5.3 The *G*-symmetric Case

In this section we focus on sign-changing solutions. Of course all the results in Section 5.2 hold true if assumption (5.6) is satisfied, in particular Theorems 5.6 and 5.8.

Hence letting $x_p^\pm$ be the family of points where $|u_p(x_p^\pm)| = \|u_p^\pm\|_\infty$, then from Theorem 5.6-$(i)$ we know that for $x_p^\pm$ the analogs of (5.22) and (5.24) hold and so we have

$$\mu_p^\pm := \left(p|u_p(x_p^\pm)|^{p-1} \right)^{-\frac{1}{2}} \to 0 \ as \ p \to +\infty. \tag{5.62}$$

From now on w.l.g. we assume that the L^∞-norm of u_p is assumed at a maximum point, namely that $u_p(x_p^+) = \|u_p^+\|_\infty = \|u_p\|_\infty$ and that $-u_p(x_p^-) = \|u_p^-\|_\infty$.

So by Theorem 5.6-(ii) we already know that scaling u_p about the maximum point x_p^+ as in (5.11) gives a first "bubble" converging to the function U defined in (5.12).

In general, for sign-changing solutions, one would like to investigate the behavior of u_p when scaling about the minima x_p^- and understand whether x_p^- coincides with one of the k sequences in Theorem 5.8 or not. Moreover, one would like to describe the set of concentration $\mathcal{S}$.

Recall that if x_p^- is one of the sequences of Theorem 5.8 then by Corollary 5.10 one has

$$\frac{dist(x_p^-, NL_p)}{\mu_p^-} \to +\infty, \quad \text{as } p \to +\infty, \tag{5.63}$$

where NL_p denotes the nodal line of u_p. On the contrary, it is easy to see that if (5.63) is satisfied, then one can use the same ideas as in the proof of Theorem 5.6-(*ii*) also for the scaling about the minimum, which we define in the natural way as

$$v_p^-(x) := p\frac{u_p(x_p^- + \mu_p^- x) - u_p(x_p^-)}{u_p(x_p^-)}, \quad x \in \widetilde{\Omega}_p^- := \{x \in \mathbb{R}^2 : x_p^- + \mu_p^- x \in \Omega\}, \tag{5.64}$$

and obtain that $v_p^- \to U$ in $C^2_{loc}(\mathbb{R}^2)$ (this has been done for instance in [31] for the case of low-energy sign-changing solutions under some additional assumptions).

Here we analyze the case when u_p belongs to a family of G-symmetric sign-changing solutions satisfying the same properties as the ones in Theorem 5.3 recalled in the Introduction and show that a different phenomenon appears.

All the results of this section are mainly based on the work [20], the existence result (Theorem 5.3) is instead contained in [19].

We prove the following.

Theorem 5.17 *Let $\Omega \subset \mathbb{R}^2$ be a connected bounded smooth domain, invariant under the action of a cyclic group G of rotations about the origin, with $|G| \geq 4e$ ($|G|$ is the order of G) and such that the origin $O \in \Omega$. Let (u_p) be a family of sign-changing G-symmetric solutions of (5.2) with two nodal regions, $NL_p \cap \partial\Omega = \emptyset$, $O \notin NL_p$ and*

$$p\int_\Omega |\nabla u_p|^2 dx \leq \alpha \, 8\pi e \tag{5.65}$$

for some $\alpha < 5$ and p large. Then, assuming w.l.g. that $\|u_p\|_\infty = \|u_p^+\|_\infty$, we have:

(i) *$S = \{O\}$ and $k = 1$ where S and k are the ones in Theorem 5.8;*

(ii) *$|x_p^+| \to O$ as $p \to +\infty$;*

(iii) *$|x_p^-| \to O$ as $p \to +\infty$;*

(iv) *NL_p shrinks to the origin as $p \to +\infty$;*

(v) *There exists $x_\infty \in \mathbb{R}^2 \setminus \{0\}$ such that, up to a subsequence, $\frac{x_p^-}{\mu_p^-} \to -x_\infty$ and*

$$v_p^-(x) \longrightarrow V(x - x_\infty) \quad in \ C^2_{loc}(\mathbb{R}^2 \setminus \{x_\infty\}) \ as \ p \to +\infty,$$

where

$$V(x) := \log\left(\frac{2\alpha^2 \beta^\alpha |x|^{\alpha-2}}{(\beta^\alpha + |x|^\alpha)^2}\right), \tag{5.66}$$

with $\alpha = \alpha(|x_\infty|) = \sqrt{2|x_\infty|^2 + 4}$, $\beta = \beta(|x_\infty|) = |x_\infty|\left(\frac{\alpha+2}{\alpha-2}\right)^{\frac{1}{\alpha}}$, *is a singular radial solution of*

$$\begin{cases} -\Delta V = e^V + H\delta_0 & \text{in } \mathbb{R}^2, \\ \int_{\mathbb{R}^2} e^V dx < \infty, \end{cases} \tag{5.67}$$

where $H = H(|x_\infty|) < 0$ *is a suitable constant and* δ_0 *is the Dirac measure centered at* 0.

Observe that the existence of families of solutions u_p having all the properties as in the assumptions of Theorem 5.17 has been proved in [19] when Ω is simply connected and when $|G| > 4$ (see Theorem 5.3 in Section 5.1).

We also recall that in [32] the case of least energy sign-changing radial solutions in a ball has been studied, proving a result similar to that in Theorem 5.17 with precise estimates of α, β and H.

5.3.1 Proofs of (i) − (ii) of Theorem 5.17

Let us introduce the following notation:

- $\mathcal{N}_p^\pm \subset \Omega$ denotes the positive/negative nodal domain of u_p,
- $\tilde{\mathcal{N}}_p^\pm$ are the rescaled nodal domains about the points $x_p^\pm$ by the parameters $\mu_p^\pm$ defined in the introduction, i.e.

$$\tilde{\mathcal{N}}_p^\pm := \frac{\mathcal{N}_p^\pm - x_p^\pm}{\mu_p^\pm} = \{x \in \mathbb{R}^2 : x_p^\pm + \mu_p^\pm x \in \mathcal{N}_p^\pm\}.$$

Let k, $(x_{i,p})$, $i = 1,\ldots,k$ and S be as in Theorem 5.8 then, defining $\mu_{i,p}$ as in (5.23), we get the following.

Proposition 5.18 *Under the assumptions of Theorem 5.17,*

$$\frac{|x_{i,p}|}{\mu_{i,p}} \text{ is bounded}, \quad \forall i = 1,\ldots,k.$$

In particular $|x_{i,p}| \to 0$, $\forall i = 1,\ldots,k$, *as* $p \to +\infty$, *so that* $S = \{O\}$.

Proof W.l.g. we can assume that for each $i = 1,\ldots,k$ either $(x_{i,p})_p \subset \mathcal{N}_p^+$ or $(x_{i,p})_p \subset \mathcal{N}_p^-$. We prove the result in the case $(x_{i,p})_p \subset \mathcal{N}_p^+$, the other case being similar. Moreover, in order to simplify the notation we drop the dependence on i, that is we set $x_p := x_{i,p}$ and $\mu_p := \mu_{i,p}$.

Let $h := |G|$ and let us denote by g^j, $j = 0, \ldots, h-1$, the elements of G. We consider the rescaled nodal domains

$$\widetilde{\mathcal{N}}_p^{j,+} := \{x \in \mathbb{R}^2 \ : \ \mu_p x + g^j x_p \in \mathcal{N}_p^+\}, \quad j = 0, \ldots, h-1,$$

and the rescaled functions $z_p^{j,+}(x) : \widetilde{\mathcal{N}}_p^{j,+} \to \mathbb{R}$ defined by

$$z_p^{j,+}(x) := \frac{p}{u_p^+(x_p)} \left(u_p^+(\mu_p x + g^j x_p) - u_p^+(x_p) \right), \quad j = 0, \ldots, h-1. \quad (5.68)$$

Hence it's not difficult to see (as in Corollary 5.10) that each $z_p^{j,+}$ converges to U in $C_{loc}^2(\mathbb{R}^2)$ as $p \to \infty$, where U is the function in (5.12). Assume by contradiction that there exists a sequence $p_n \to +\infty$ as $n \to +\infty$ such that $\frac{|x_{p_n}|}{\mu_{p_n}} \to +\infty$. Then, since the h distinct points $g^j x_{p_n}, j = 0, \ldots, h-1$, are the vertex of a regular polygon centered in O, $d_n := |g^j x_{p_n} - g^{j+1} x_{p_n}| = 2\widetilde{d}_n \sin \frac{\pi}{h}$, where $\widetilde{d}_n := |g^j x_{p_n}|$, $j = 0, \ldots, h-1$, and so we also have that $\frac{d_n}{\mu_{p_n}} \to +\infty$ as $n \to +\infty$. Let

$$R_n := \min \left\{ \frac{d_n}{3}, \frac{d(x_{p_n}, \partial\Omega)}{2}, \frac{d(x_{p_n}, NL_{p_n})}{2} \right\}, \quad (5.69)$$

then by construction $B_{R_n}(g^j x_{p_n}) \subseteq \mathcal{N}_{p_n}^+$ for $j = 0, \ldots, h-1$,

$$B_{R_n}(g^j x_{p_n}) \cap B_{R_n}(g^l x_{p_n}) = \emptyset, \quad \text{for } j \neq l \quad (5.70)$$

and

$$\frac{R_n}{\mu_{p_n}} \to +\infty \quad \text{as } n \to +\infty. \quad (5.71)$$

Using (5.71), the convergence of $z_{p_n}^{j,+}$ to U, (5.24) and Fatou's lemma, we have

$$8\pi = \int_{\mathbb{R}^2} e^U \, dx \leq \lim_n \int_{B_{\frac{R_n}{\mu_{p_n}}}(0)} \left| 1 + \frac{z_{p_n}^{j,+}}{p_n} \right|^{(p_n+1)} dx$$

$$= \lim_n \frac{p_n}{|u_{p_n}^+(x_{p_n})|^2} \int_{B_{R_n}(g^j x_{p_n})} |u_{p_n}^+|^{(p_n+1)} \, dx$$

$$\overset{(5.24)}{\leq} \lim_n p_n \int_{B_{R_n}(g^j x_{p_n})} |u_{p_n}^+|^{(p_n+1)} \, dx. \quad (5.72)$$

Summing on $j = 0, \ldots, h-1$, using (5.70), (5.65), (5.8) and (5.9) we get:

$$h \cdot 8\pi \leq \lim_n p_n \sum_{j=0}^{h-1} \int_{B_{R_n}(g^j x_{p_n})} |u_{p_n}^+|^{(p_n+1)} \, dx \overset{(5.70)}{\leq} \lim_n p_n \int_{\mathcal{N}_{p_n}^+} |u_{p_n}^+|^{(p_n+1)} \, dx$$

$$= \lim_n \left(p_n \int_\Omega |u_{p_n}|^{(p_n+1)} \, dx - p_n \int_{\mathcal{N}_{p_n}^-} |u_{p_n}^-|^{(p_n+1)} \, dx \right)$$

$$\overset{(5.65)+(5.8)}{\leq} (\alpha - 1) \cdot 8\pi e \overset{\alpha < 5}{<} 4 \cdot 8\pi e,$$

which contradicts the assumption $|G| \geq 4e$.   □

Remark 5.19 *If we knew that $\|u_p\|_\infty \geq \sqrt{e}$, then we would obtain a better estimate in (5.72), and so Proposition 5.18 would hold under the weaker symmetry assumption $|G| \geq 4$ (recall that $|G| \geq 4$ is the assumption under which one can prove Theorem 5.3).*

It is also possible to prove the following (see [20, Corollary 3.5] for more details).

Corollary 5.20 *Under the assumptions of Theorem 5.17*

(i) $O \in \mathcal{N}_p^+$ for p large;
(ii) let $i \in \{1,\ldots,k\}$, then $x_{i,p} \in \mathcal{N}_p^+$ for p large.

Proposition 5.21 *Under the assumptions of Theorem 5.17, the maximal number k of families of points $(x_{i,p})$, $i = 1,\ldots,k$, for which (P_1^k), (P_2^k) and (P_3^k) hold is 1.*

Proof Let us assume by contradiction that $k > 1$ and set $x_p^+ = x_{1,p}$. For a family $(x_{j,p})\, j \in \{2,\ldots,k\}$ by Proposition 5.18, there exists $C > 0$ such that

$$\frac{|x_{1,p}|}{\mu_{1,p}} \leq C \quad \text{and} \quad \frac{|x_{j,p}|}{\mu_{j,p}} \leq C.$$

Thus, since by definition $\mu_p^+ = \mu_{1,p} \leq \mu_{j,p}$, also

$$\frac{|x_{1,p}|}{\mu_{j,p}} \leq C.$$

Hence

$$\frac{|x_{1,p} - x_{j,p}|}{\mu_{j,p}} \leq \frac{|x_{1,p}| + |x_{j,p}|}{\mu_{j,p}} \leq C,$$

which contradicts (P_1^k) when $p \to +\infty$.   □

Then we easily get the following.

Proposition 5.22 *Under the assumptions of Theorem 5.17 there exists $C > 0$ such that*

$$\frac{|x_p|}{\mu(x_p)} \leq C \tag{5.73}$$

for any family $(x_p)_p \subset \Omega$, where $\mu(x_p)$ is defined as in (5.53). In particular, since by (5.62) $\mu_p^- \to 0$, then $|x_p^-| \to 0$.

Proof (5.73) holds for x_p^+ by Proposition 5.18. Moreover, $k = 1$ by Proposition 5.21, so applying $(\mathcal{P}_3^1)$ to the points (x_p), for $x_p \neq x_p^+$, we have

$$\frac{|x_p - x_p^+|}{\mu(x_p)} \leq C.$$

By definition, $\mu_p^+ \leq \mu(x_p)$, hence we get

$$\frac{|x_p|}{\mu(x_p)} \leq \frac{|x_p - x_p^+|}{\mu(x_p)} + \frac{|x_p^+|}{\mu(x_p)} \leq \frac{|x_p - x_p^+|}{\mu(x_p)} + \frac{|x_p^+|}{\mu_p^+} \leq C.$$

□

Lemma 5.23 *Let the assumptions of Theorem 5.17 be satisfied and let $(x_p) \subset \Omega$ be such that $p|u_p(x_p)|^{p-1} \to +\infty$ and $\mu(x_p)$ be as in (5.53). Assume also that the rescaled functions $v_p(x) := \frac{p}{u_p(x_p)}(u_p(x_p + \mu(x_p)x) - u_p(x_p))$ converge to U in $C_{loc}^2(\mathbb{R}^2 \setminus \{-\lim_p \frac{x_p}{\mu(x_p)}\})$ as $p \to +\infty$ (U as in (5.12)). Then*

$$\frac{|x_p|}{\mu(x_p)} \to 0 \quad \text{as } p \to +\infty. \tag{5.74}$$

As a byproduct we deduce that $v_p \to U$ in $C_{loc}^2(\mathbb{R}^2 \setminus \{0\})$, as $p \to +\infty$.

Proof By Proposition 5.22 we know that $\frac{|x_p|}{\mu(x_p)} \leq C$. Assume by contradiction that $\frac{|x_p|}{\mu(x_p)} \to \ell > 0$. Let $g \in G$ such that $|x_p - gx_p| = C_g|x_p|$ with constant $C_g > 1$ (such a g exists because G is a group of rotation about the origin). Hence

$$\frac{|x_p - gx_p|}{\mu(x_p)} = C_g \frac{|x_p|}{\mu(x_p)} \to C_g\ell > \ell.$$

Then $x_0 := \lim_{p \to +\infty} \frac{gx_p - x_p}{\mu(x_p)} \in \mathbb{R}^2 \setminus \{-\lim_p \frac{x_p}{\mu(x_p)}\}$ and so by the C_{loc}^2 convergence we get

$$v_p\left(\frac{gx_p - x_p}{\mu(x_p)}\right) \to U(x_0) < 0 \quad \text{as } p \to +\infty.$$

On the other hand, for any $g \in G$, one also has

$$v_p\left(\frac{gx_p - x_p}{\mu(x_p)}\right) = \frac{p}{u_p(x_p)}(u_p(gx_p) - u_p(x_p)) = 0,$$

by the symmetry of u_p and this gives a contradiction. □

5.3.2 Asymptotic Analysis about the Minimum Points x_p^- and Study of NL_p

Proposition 5.21 implies that $(\mathcal{P}_3^1)$ holds, from which

$$\frac{|x_p^+ - x_p^-|}{\mu_p^-} \leq C, \tag{5.75}$$

with μ_p^- as in (5.62). Moreover, since we already know that $\frac{d(x_p^+,NL_p)}{\mu_p^+} \to +\infty$ as $p \to +\infty$, we deduce that $\frac{|x_p^+ - x_p^-|}{\mu_p^+} \to +\infty$ as $p \to +\infty$, and in turn by (5.75) we get

$$\frac{\mu_p^+}{\mu_p^-} \to 0 \quad \text{as } p \to +\infty. \tag{5.76}$$

Note that (5.75) and (5.76) hold more generally for any family of points (x_p) such that $u_p(x_p) < 0$ and $p|u_p(x_p)|^{p-1} \to +\infty$.

By Proposition 5.22 we have

$$\frac{|x_p^-|}{\mu_p^-} \leq C, \tag{5.77}$$

so there are two possibilities: either $\frac{|x_p^-|}{\mu_p^-} \to \ell > 0$ or $\frac{|x_p^-|}{\mu_p^-} \to 0$ as $p \to +\infty$, up to subsequences. We will exclude the latter case. We start with a preliminary result.

Lemma 5.24 *For $x \in \frac{\Omega}{|x_p^-|} := \{y \in \mathbb{R}^2 \ : \ y|x_p^-| \in \Omega\}$ let us define the rescaled function*

$$w_p^-(x) := \frac{p}{u_p(x_p^-)} \left(u_p(|x_p^-|x) - u_p(x_p^-) \right).$$

Then

$$w_p^- \to \gamma \quad \text{in } C^2_{loc}(\mathbb{R}^2 \setminus \{0\}) \text{ as } p \to +\infty, \tag{5.78}$$

where $\gamma \in C^2(\mathbb{R}^2 \setminus \{0\})$, $\gamma \leq 0$, $\gamma(x_\infty) = 0$ for a point $x_\infty \in \partial B_1(0)$ and it is a solution to

$$-\Delta\gamma = \ell^2 e^\gamma \quad \text{in } R^2 \setminus \{0\}.$$

In particular $\gamma \equiv 0$ when $\ell = 0$.

Proof (5.77) implies that $|x_p^-| \to 0$ as $p \to +\infty$, so it follows that the set $\frac{\Omega}{|x_p^-|} \to \mathbb{R}^2$ as $p \to +\infty$.

By definition we have

$$w_p^- \leq 0, \quad w_p\left(\frac{x_p^-}{|x_p^-|}\right) = 0 \quad \text{and} \quad w_p^- = -p \quad \text{on } \partial\left(\frac{\Omega}{|x_p^-|}\right). \tag{5.79}$$

Moreover, for $x \in \frac{\Omega}{|x_p^-|}$ we define $\xi_p := |x_p^-|x$ and μ_{ξ_p} as $\mu_{\xi_p}^{-2} := p|u_p(\xi_p)|^{p-1}$. Thanks to (5.2) we then have

$$|-\Delta w_p^-(x)| = \frac{p|x_p^-|^2|u_p(\xi_p)|^p}{|u_p(x_p^-)|} = \frac{|u_p(\xi_p)|}{|u_p(x_p^-)|}\frac{|x_p^-|^2}{\mu_{\xi_p}^2} \leq c_\infty \frac{|x_p^-|^2}{\mu_{\xi_p}^2}, \tag{5.80}$$

where $c_\infty := \lim_p \|u_p\|_\infty$. Then, observing that $\frac{|x_p^-|}{\mu_{\xi_p}} \leq \frac{C}{|x|}$ by Proposition 5.22 applied to ξ_p, we have

$$| - \Delta w_p^-(x)| \leq \frac{c_\infty C^2}{|x|^2}.$$

Specifically, for any $R > 0$

$$| - \Delta w_p^-| \leq c_\infty C^2 R^2 \quad \text{in} \quad \frac{\Omega}{|x_p^-|} \setminus B_{\frac{1}{R}}(0). \tag{5.81}$$

So, similarly to the proof of Theorem 5.6(*ii*) (using now that $w_p^-(\frac{x_p^-}{|x_p^-|}) = 0$), it follows that for any $R > 1$ ($\frac{x_p^-}{|x_p^-|} \in \partial B_1(0) \subset B_R(0) \setminus B_{\frac{1}{R}}(0)$ for $R > 1$), w_p^- is uniformly bounded in $B_R(0) \setminus B_{\frac{1}{R}}(0)$.

After passing to a subsequence, standard elliptic theory applied to the following equation

$$- \Delta w_p^-(x) = \frac{|x_p^-|^2}{(\mu_p^-)^2} \left(1 + \frac{w_p^-(x)}{p}\right) \left|1 + \frac{w_p^-(x)}{p}\right|^{p-1} \tag{5.82}$$

gives that w_p^- is bounded in $C_{\text{loc}}^{2,\alpha}(\mathbb{R}^2 \setminus \{0\})$. Hence (5.78) and the properties of γ follow.

In particular when $\ell = 0$ it follows that γ is harmonic in $\mathbb{R}^2 \setminus \{0\}$ and $\gamma(x_\infty) = 0$ for some point $x_\infty \in \partial B_1(0)$, therefore by the maximum principle we obtain $\gamma \equiv 0$. $\qquad\square$

Proposition 5.25 *There exists $\ell > 0$ such that*

$$\frac{|x_p^-|}{\mu_p^-} \to \ell \quad \text{as } p \to +\infty.$$

Proof By Proposition 5.22 we know that $\frac{|x_p^-|}{\mu_p^-} \to \ell \in [0, +\infty)$ as $p \to +\infty$. Let us suppose by contradiction that $\ell = 0$. Then Lemma 5.24 implies that

$$w_p^- \to 0 \quad \text{in} \quad C_{\text{loc}}^2(\mathbb{R}^2 \setminus \{0\}) \quad \text{as } p \to +\infty. \tag{5.83}$$

By (5.2), applying the divergence theorem in $B_{|x_p^-|}(0)$ we get

$$p \int_{\partial B_{|x_p^-|}(0)} \nabla u_p(y) \cdot \frac{y}{|y|} \, d\sigma(y) = p \int_{B_{|x_p^-|}(0) \cap \mathcal{N}_p^-} |u_p(x)|^p \, dx$$

$$- p \int_{B_{|x_p^-|}(0) \cap \mathcal{N}_p^+} |u_p(x)|^p \, dx. \tag{5.84}$$

Scaling u_p with respect to $|x_p^-|$ as in Lemma 5.24, by (5.83) we obtain

$$\left| p \int_{\partial B_{|x_p^-|}(0)} \nabla u_p(y) \cdot \frac{y}{|y|} \, d\sigma(y) \right| = \left| p \int_{\partial B_1(0)} |x_p^-| \nabla u_p(|x_p^-|x) \cdot \frac{x}{|x|} \, d\sigma(x) \right|$$

$$= \left| \int_{\partial B_1(0)} u_p(x_p^-) \nabla w_p^-(x) \cdot \frac{x}{|x|} \, d\sigma(x) \right| \le |u_p(x_p^-)| 2\pi \sup_{|x|=1} |\nabla w_p^-(x)| = o_p(1).$$

$$(5.85)$$

Now we want to estimate the right-hand side in (5.84). We first observe that scaling around $|x_p^-|$ with respect to μ_p^- we get

$$p \int_{B_{|x_p^-|}(0) \cap \mathcal{N}_p^-} |u_p(x)|^p \, dx = p \int_{B_1(0) \cap \frac{\mathcal{N}_p^-}{|x_p^-|}} |u_p(|x_p^-|y)|^p |x_p^-|^2 \, dy$$

$$\le c_\infty \int_{B_1(0) \cap \frac{\mathcal{N}_p^-}{|x_p^-|}} \frac{|u_p(|x_p^-|y)|^{p-1}}{|u_p(x_p^-)|^{p-1}} \frac{|x_p^-|^2}{(\mu_p^-)^2} \, dy = o_p(1),$$

$$(5.86)$$

where in the last equality we have used that $\frac{|u_p(|x_p^-|y)|^{p-1}}{|u_p(x_p^-)|^{p-1}} \le 1$, since $|x_p^-|y \in \mathcal{N}_p^-$ and that by assumption $\frac{|x_p^-|}{\mu_p^-} \to 0$ as $p \to +\infty$.

Next we claim that there exists $\bar{p} > 1$ such that for any $p \ge \bar{p}$

$$B_{\mu_p^+}(x_p^+) \subset B_{|x_p^-|}(0). \qquad (5.87)$$

Indeed, Corollary 5.10 implies that

$$+\infty = \lim_p \frac{d(x_p^+, NL_p)}{\mu_p^+} \le \lim_p \frac{|x_p^+ - x_p^-|}{\mu_p^+} \le \lim_p \frac{|x_p^+|}{\mu_p^+} + \lim_p \frac{|x_p^-|}{\mu_p^+} = \lim_p \frac{|x_p^-|}{\mu_p^+},$$

where the last equality follows from Lemma 5.23 (i.e. $\frac{|x_p^+|}{\mu_p^+} \to 0$). Hence, for any $x \in B_1(0)$ we have

$$\frac{|x_p^+ + \mu_p^+ x|}{|x_p^-|} \le \frac{|x_p^+|}{|x_p^-|} + \frac{\mu_p^+}{|x_p^-|} \le \frac{2\mu_p^+}{|x_p^-|} \to 0 \quad \text{as } p \to +\infty,$$

and so (5.87) is proved.

Hence, by (5.87) and scaling around x_p^+ with respect to μ_p^+ we obtain

$$p \int_{B_{|x_p^-|}(0) \cap \mathcal{N}_p^+} |u_p(x)|^p \, dx \ge p \int_{B_{\mu_p^+}(x_p^+)} |u_p(x)|^p dx = c_\infty \int_{B_1(0)} e^U dx + o_p(1).$$

$$(5.88)$$

Collecting (5.84), (5.85), (5.86) and (5.88) we clearly get a contradiction.

$\square$

Next we show that the nodal line shrinks to the origin faster than μ_p^- as $p \to +\infty$.

Proposition 5.26 *We have*

$$\frac{\max\limits_{y_p \in NL_p} |y_p|}{\mu_p^-} \to 0 \quad as \ p \to +\infty.$$

Proof By Proposition 5.25 it is enough to prove that

$$\frac{\max\limits_{y_p \in NL_p} |y_p|}{|x_p^-|} \to 0 \quad as \ p \to +\infty.$$

First we show that, for any $y_p \in NL_p$, the following alternatives hold:

$$\text{either} \quad \frac{|y_p|}{|x_p^-|} \to 0 \quad \text{or} \quad \frac{|y_p|}{|x_p^-|} \to +\infty \quad as \ p \to +\infty. \tag{5.89}$$

Indeed, assume by contradiction that $\frac{|y_p|}{|x_p^-|} \to m \in (0, +\infty)$ as $p \to +\infty$. Then $w_p^-(\frac{y_p}{|x_p^-|}) = -p \to -\infty$ as $p \to +\infty$. But we have proved in Lemma 5.24 that $w_p^-(\frac{y_p}{|x_p^-|}) \to \gamma(y_m) \in \mathbb{R}$, where y_m is such that $|y_m| = m > 0$, and this gives a contradiction.

To conclude the proof we have then to exclude the second alternative in (5.89). For $y_p \in NL_p$, let us assume by contradiction that $\frac{|y_p|}{|x_p^-|} \to +\infty$ as $p \to +\infty$ and let us observe that

$$\exists \, z_p \in NL_p \text{ such that } \frac{|z_p|}{|x_p^-|} \to 0 \quad as \ p \to +\infty. \tag{5.90}$$

Indeed, in the previous section we have shown that $O \in \mathcal{N}_p^+$, hence there exists $t_p \in (0, 1)$ such that $z_p := t_p x_p^- \in NL_p$. Since $\frac{|z_p|}{|x_p^-|} < 1$, by (5.89) we get (5.90).

Then for any $M > 0$ there exists $\alpha_p^M \in NL_p$ such that $\frac{|\alpha_p^M|}{|x_p^-|} \to M$ as $p \to +\infty$ and this is in contradiction with (5.89). $\qquad\qquad\square$

Finally, we can analyze the local behavior of u_p around the minimum point x_p^-. Note that by Lemma 5.23 and Proposition 5.25 we can already claim that the rescaling v_p^- about x_p^- (see (5.64)) cannot converge to U in $\mathbb{R}^2 \setminus \{0\}$, where U is the function in (5.12), indeed we have the following.

Proposition 5.27 *Passing to a subsequence*

$$v_p^-(x) \longrightarrow V(x - x_\infty) \quad in \ C^2_{\text{loc}}(\mathbb{R}^2 \setminus \{x_\infty\}), \ as \ p \to +\infty, \tag{5.91}$$

where V is the radial singular function in (5.66) which satisfies the Liouville equation (5.67) and $x_\infty := \lim_p \frac{x_p^-}{\mu_p^-}$.

Proof Let us consider the translations of v_p^-:

$$s_p^-(x) := v_p^-\left(x - \frac{x_p^-}{\mu_p^-}\right) = \frac{p}{u_p(x_p^-)}(u_p(\mu_p^- x) - u_p(x_p^-)), \qquad x \in \frac{\Omega}{\mu_p^-},$$

which solve

$$-\Delta s_p^-(x) = \left|1 + \frac{s_p^-(x)}{p}\right|^{p-1}\left(1 + \frac{s_p^-(x)}{p}\right), \qquad s_p^-\left(\frac{x_p^-}{\mu_p^-}\right) = 0, \qquad s_p^- \le 0.$$

Observe that $\frac{\Omega}{\mu_p^-} \to \mathbb{R}^2$ as $p \to +\infty$.

We claim that for any fixed $r > 0$, $|-\Delta s_p^-|$ is bounded in $\frac{\Omega}{\mu_p^-} \setminus B_r(0)$.

Indeed, Proposition 5.26 implies that if $x \in \frac{\mathcal{N}_p^+}{\mu_p^-}$, then $|x| \le \frac{\max\limits_{z_p \in NL_p} |z_p|}{\mu_p^-} < r$, for p large, hence

$$\left(\frac{\Omega}{\mu_p^-} \setminus B_r(0)\right) \subset \frac{\mathcal{N}_p^-}{\mu_p^-} \qquad \text{for } p \text{ large}$$

and so the claim follows by observing that for $x \in \frac{\mathcal{N}_p^-}{\mu_p^-}$, then $|-\Delta s_p^-(x)| \le 1$.

Hence, by the arbitrariness of $r > 0$, $s_p^- \to V$ in $C^2_{\text{loc}}(\mathbb{R}^2 \setminus \{0\})$ as $p \to +\infty$, where V is a solution of

$$-\Delta V = e^V \quad \text{in } \mathbb{R}^2 \setminus \{0\}$$

which satisfies $V \le 0$ and $V(x_\ell) = 0$, where $x_\ell := \lim_p \frac{x_p^-}{\mu_p^-}$ and $|x_\ell| = \ell$ by Proposition 5.25. Moreover, by virtue of Theorem 5.6-(*i*) and by (5.65) it can be seen that $e^V \in L^1(\mathbb{R}^2)$.

Observe that if V was a classical solution of $-\Delta V = e^V$ in the whole $\mathbb{R}^2$ then necessarily $V(x) = U(x - x_\ell)$. As a consequence $v_p^-(x) = s_p^-(x + \frac{x_p^-}{\mu_p^-}) \to V(x + x_\ell) = U(x)$ in $C^2_{\text{loc}}(\mathbb{R}^2 \setminus \{-x_\ell\})$ as $p \to +\infty$. But then Lemma 5.23 would imply that $|x_\ell| = \frac{|x_p^-|}{\mu_p^-} \to 0$ as $p \to +\infty$, which is in contradiction with Proposition 5.25. Thus, by [9, 10, 11] and the classification in [8] we have that V solves, for some $\eta > 0$, the following entire equation:

$$\begin{cases} -\Delta V = e^V - 4\pi \eta \delta_0 & \text{in } \mathbb{R}^2, \\ \int_{\mathbb{R}^2} e^V dx = 8\pi(1+\eta), \end{cases}$$

where δ_0 denotes the Dirac measure centered at the origin.

By the classification given in [40], we have that either V is radial, or $\eta \in \mathbb{N}$ and V is $(\eta + 1)$-symmetric. Actually it turns out that the energy bound (5.65) forces V to be a radial solution, $V(r)$, satisfying

$$\begin{cases} -V'' - \frac{1}{r}V' = e^V & \text{in } (0,+\infty), \\ V \leq 0, \\ V(\ell) = V'(\ell) = 0. \end{cases}$$

The solution of this problem is

$$V(r) = \log\left(\frac{2\alpha^2 \beta^\alpha r^{\alpha-2}}{(\beta^\alpha + r^\alpha)^2}\right),$$

where $\alpha = \sqrt{2\ell^2 + 4}$ and $\beta = \ell\left(\frac{\alpha+2}{\alpha-2}\right)^{1/\alpha}$. The conclusion follows by observing that $v_p^-(x) = s_p^-\left(x + \frac{x_p^-}{\mu_p}\right)$ and setting $x_\infty = -x_\ell$. $\square$

5.3.3 Conclusion of the Proof of Theorem 5.17

It follows by combining all the previous results. More precisely (*i*) and (*ii*) follow from Propositions 5.18 and 5.21. (*iii*) is due to Proposition 5.22. (*iv*) is in Proposition 5.26. Finally, (*v*) comes from Proposition 5.27.

References

[1] Adimurthi, M. Grossi, *Asymptotic estimates for a two-dimensional problem with polynomial nonlinearity*, Proc. Amer. Math. Soc. 132 (2004), no. 4, 1013–1019.

[2] Adimurthi, M. Struwe, *Global compactness properties of semilinear elliptic equations with critical exponential growth*, J. Funct. Anal. 175 (2000), no. 1, 125–167.

[3] Adimurthi, S.L. Yadava, *An elementary proof of the uniqueness of positive radial solutions of a quasilinear Dirichlet problem*, Arch. Rational Mech. Anal. 127 (1994), no. 3, 219–229.

[4] A. Aftalion, F. Pacella, *Qualitative properties of nodal solutions of semilinear elliptic equations in radially symmetric domains*, C. R. Math. Acad. Sci. Paris 339 (2004), no. 5, 339–344.

[5] T. Bartsch, T. Weth, *A note on additional properties of sign changing solutions to superlinear elliptic equations*, Topological Methods in Nonlinear Analysis 22 (2003), 1–14.

[6] A. Castro, J. Cossio, J. Neuberger, *A sign-changing solution for a superlinear Dirichlet problem*, Rocky Mountain Journal of Mathematics 27, no. 4 (1997) 1041–1053.

[7] T. Cazenave, F. Dickstein, F.B. Weissler, *Sign-changing stationary solutions and blowup for the nonlinear heat equation in a ball*, Math. Ann. 344 (2009), no. 2, 431–449.

[8] W.X. Chen, C. Li, *Classification of solutions of some nonlinear elliptic equations*, Duke Math. J. 63 (1991), 615–622.

[9] W.X. Chen, C. Li, *Qualitative properties of solutions of some nonlinear elliptic equations*, Duke Math. J. 71 (1993), 427–439.

[10] K.S. Chou, T. Wan, *Asymptotic radial symmetry for solutions of* $\Delta u + e^u = 0$ *in a punctured disc*, Pacific J. Math. 163 (1994), no. 2, 269–276.

[11] K.S. Chou, T. Wan, *Correction to: "Asymptotic radial symmetry for solutions of* $\Delta u + e^u = 0$ *in a punctured disc" [Pacific J. Math. 163 (1994), no. 2, 269–276]*, Pacific J. Math. 171 (1995), no. 2, 589–590.

[12] A. Dall'Acqua, G. Sweers, *Estimates for Green function and Poisson kernels of higher order Dirichlet boundary value problems*, J. Differential Equations 205 (2004), no. 2, 466–487.

[13] L. Damascelli, *On the nodal set of the second eigenfunction of the Laplacian in symmetric domains in* R^N, Atti Accad. Naz. Lincei Cl. Sci. Fis. Mat. Natur. Rend. Lincei (9) Mat. Appl. 11 (2000), no. 3, 175–181.

[14] L. Damascelli, M. Grossi, F. Pacella, *Qualitative properties of positive solutions of semilinear elliptic equations in symmetric domains via the maximum principle*, Ann. Inst. H. Poincaré Anal. Non Linéaire 16 (1999), no. 5, 631–652.

[15] E.N. Dancer, *The effect of domain shape on the number of positive solutions of certain nonlinear equations*, J. Differential Equations 74 (1988), no. 1, 120–156.

[16] F. De Marchis, M. Grossi, I. Ianni, F. Pacella, *Morse index and uniqueness of positive solutions of the Lane–Emden problem in planar domains*, arXiV:1804.03499.

[17] F. De Marchis, M. Grossi, I. Ianni, F. Pacella, L^∞-*norm and energy quantization for the planar Lane–Emden problem with large exponent*, to appear on Archiv der Mathematik.

[18] F. De Marchis, I. Ianni, *Blow up of solutions of semilinear heat equations in non radial domains of* $\mathbb{R}^2$, Discrete Contin. Dyn. Syst. 35 (2015), no. 3, 891–907.

[19] F. De Marchis, I. Ianni, F. Pacella, *Sign changing solutions of Lane Emden problems with interior nodal line and semilinear heat equations*, J. Differential Equations 254 (2013), 3596–3614.

[20] F. De Marchis, I. Ianni, F. Pacella, *Asymptotic analysis and sign changing bubble towers for Lane–Emden problems*, J. Europ. Math. Soc. (2015) 17, no. 8, 2037–2068.

[21] F. De Marchis, I. Ianni, F. Pacella, *Morse index and sign-changing bubble towers for Lane–Emden problems*, Annali di Matematica Pura ed Applicata 195 (2016), no. 2, 357–369.

[22] F. De Marchis, I. Ianni, F. Pacella, *Asymptotic profile of positive solutions of Lane–Emden problems in dimension two*, Journal of Fixed Point Theory and Applications 19 (2017), no. 1, 889–916.

[23] F. De Marchis, I. Ianni, F. Pacella, *Exact Morse index computation for nodal radial solutions of Lane–Emden problems*, Mathematische Annalen 367 (2017), no. 1, 185–227.

[24] F. Dickstein, F. Pacella, B. Sciunzi, *Sign-changing stationary solutions and blowup for a nonlinear heat equation in dimension two*, J. Evol. Equ. 14 (2014), no. 3, 617–633.

[25] O. Druet, *Multibumps analysis in dimension 2: quantification of blow-up levels*, Duke Math. J. 132 (2006), no. 2, 217–269.

[26] O. Druet, E. Hebey, F. Robert, *Blow-up theory for elliptic PDEs in Riemannian geometry*, Mathematical Notes, 45, Princeton University Press, Princeton, NJ, 2004. ISBN: 0-691-11953-8

[27] P. Esposito, M. Musso, A. Pistoia, *Concentrating solutions for a planar elliptic problem involving nonlinearities with large exponent*, J. Differential Equations 227 (2006), no. 1, 29–68.

[28] P. Esposito, M. Musso, A. Pistoia, *On the existence and profile of nodal solutions for a two-dimensional elliptic problem with large exponent in nonlinearity*, Proc. Lond. Math. Soc. (3) 94 (2007), no. 2, 497–519.

[29] B. Gidas, W.M. Ni, L. Nirenberg, *Symmetry and related properties via the maximum principle*, Comm. Math. Phys. 68 (1979), no. 3, 209–243.

[30] F. Gladiali, F. Pacella, T. Weth, *Symmetry and nonexistence of low Morse index solutions in unbounded domains*, J. Math. Pures Appl. (9) 93 (2010), no. 5, 536–558.

[31] M. Grossi, C. Grumiau, F. Pacella, *Lane Emden problems: asymptotic behavior of low energy nodal solutions*, Ann. Inst. H. Poincaré Anal. Non Linéaire 30 (2013), no. 1, 121–140.

[32] M. Grossi, C. Grumiau, F. Pacella, *Lane Emden problems with large exponents and singular Liouville equations*, Journal Math. Pure and Appl. 101 (2014), no. 6, 735–754.

[33] C.S. Lin, *On the second eigenfunctions of the Laplacian in R^2*, Comm. Math. Phys. 111 (1987), no. 2, 161–166.

[34] C.S. Lin, *Uniqueness of least energy solutions to a semilinear elliptic equation in R^2*, Manuscripta Math. 84 (1994), no. 1, 13–19.

[35] V. Marino, F. Pacella, B. Sciunzi, *Blow up of solutions of semilinear heat equations in general domains*, Commun. Contemp. Math. 17 (2015), no. 2, 1350042, 17 pp.

[36] A.D. Melas, *On the nodal line of the second eigenfunction of the Laplacian in R^2*, J. Differential Geom. 35 (1992), no. 1, 255–263.

[37] F. Pacella, *Symmetry results for solutions of semilinear elliptic equations with convex nonlinearities*, J. Funct. Anal. 192 (2002), no. 1, 271–282.

[38] F. Pacella, *Uniqueness of positive solutions of semilinear elliptic equations and related eigenvalue problems*, Milan J. Math. 73 (2005), 221–236.

[39] F. Pacella, T. Weth, *Symmetry of solutions to semilinear elliptic equations via Morse index*, Proc. Amer. Math. Soc. 135 (2007), no. 6, 1753–1762.

[40] J. Prajapat, G. Tarantello, *On a class of elliptic problems in $\mathbb{R}^2$: Symmetry and uniqueness results*, Proc. Roy. Soc. Edinburgh 131A (2001), 967–985.

[41] X. Ren, Xiaofeng, J. Wei, *On a two-dimensional elliptic problem with large exponent in nonlinearity*, Trans. Amer. Math. Soc. 343 (1994), no. 2, 749–763.

[42] X. Ren, Xiaofeng, J. Wei, *Single-point condensation and least-energy solutions*, Proc. Amer. Math. Soc. 124 (1996), no. 1, 111–120.

[43] S. Santra, J. Wei, *Asymptotic behavior of solutions of a biharmonic Dirichlet problem with large exponents*, J. Anal. Math. 115 (2011), 1–31.

[44] S. Solimini, *Morse index estimates in min-max theorems*, Manuscripta Math. 63 (1989), no. 4, 421–453.

[45] P.N. Srikanth, *Uniqueness of solutions of nonlinear Dirichlet problems*, Differential Integral Equations 6 (1993), no. 3, 663–670.

[46] M. Willem, *Minmax theorems*, Progress in Nonlinear Differential Equations and Their Applications, Vol. 24, Birkäuser, Boston MA, 1996.

* P.le Aldo Moro 5, 00185 Roma, Italy
† V.le Lincoln 5, 81100 Caserta, Italy
‡ P.le Aldo Moro 5, 00185 Roma, Italy

6

A Data Assimilation Algorithm: the Paradigm of the 3D Leray-α Model of Turbulence

Aseel Farhat[*], Evelyn Lunasin[†] and Edriss S. Titi[‡]

In this paper we survey the various implementations of a new data assimilation (downscaling) algorithm based on spatial coarse mesh measurements. As a paradigm, we demonstrate the application of this algorithm to the 3D Leray-α subgrid scale turbulence model. Most importantly, we use this paradigm to show that it is not always necessary to collect coarse mesh measurements of all the state variables that are involved in the underlying evolutionary system, in order to recover the corresponding exact reference solution. Specifically, we show that in the case of the 3D Leray-α model of turbulence, the solutions of the algorithm, constructed using only coarse mesh observations of *any two components of the three-dimensional velocity field*, and without any information on the third component, converge, at an exponential rate in time, to the corresponding exact reference solution of the 3D Leray-α model. This study serves as an addendum to our recent work on abridged continuous data assimilation for the 2D Navier–Stokes equations. Notably, similar results have also been recently established for the 3D viscous Planetary Geostrophic circulation model in which we show that coarse mesh measurements of the temperature alone are sufficient for recovering, through our data assimilation algorithm, the full solution; i.e. the three components of velocity vector field and the temperature. Consequently, this proves the Charney conjecture for the 3D Planetary Geostrophic model; namely, that the history of the large spatial scales of temperature is sufficient for determining all the other quantities (state variables) of the model.

This paper is dedicated to the memory of Professor Abbas Bahri

MSC Subject Classifications: 35Q30, 93C20, 37C50, 76B75, 34D06.
Keywords: 3D Leray-α-model, subgrid scale turbulence models, continuous data assimilation, downscaling, Charney conjecture, coarse measurements of velocity.

[*] Department of Mathematics, University of Virginia.
[†] Department of Mathematics, United States Naval Academy.
[‡] Department of Mathematics, Texas A&M University and The Science Program, Texas A&M University at Qatar.

6.1 Introduction

Data assimilation is a methodology to estimate weather or ocean variables by combining (synchronizing) information from observational data with a numerical dynamical (forecast) model. In the context of control engineering, tracing back since the early 1970s, data assimilation was also applied to simpler models in [59, 69, 77]. One of the classical methods of continuous data assimilation, see e.g., [20, 37], is to insert observational measurements directly into a model as the latter is being integrated in time. For example, one can insert Fourier low mode observables into the evolution equation for the high modes, then the values for the low modes and high modes are combined to form a complete approximation of the state of the system. This resulting state value is then used as an initial condition to evolve the forecast model using high resolution simulation. For the 2D Navier–Stokes, this approach was considered in [9, 10, 39, 41, 52, 58, 70, 72]. The problem when some state variable observations are not available as an input to the numerical forecast model was studied in [13, 21, 37, 38, 42, 61] for simplified numerical forecast models.

Recent studies in [23, 24, 25, 26, 27] have established rigorous analytical support pertaining to a continuous data assimilation algorithm similar to that introduced in [4], which is a feedback control algorithm applied to data assimilation (see also [3, 64]), previously known as nudging or Newtonian relaxation. In these cases, the authors have analyzed a data assimilation algorithm for different systems when some of the state variable observations are not supplied in the algorithm. In other words, a rigorous analytical support to a data assimilation algorithm that can identify the full state of the system knowing only coarse spatial mesh observational measurements not of full state variables of the model, but only of the selected state variables in the system, were analytically justified. In this article we summarize our recent results to provide support to the applicability of the scheme in several model equations. We then demonstrate the application of this algorithm to the Leray-α subgrid scale turbulence model which serves as an addendum to our recent work on abridged continuous data assimilation for the 2D Navier–Stokes equations.

Starting from the work of [4], the recent works [23, 24, 25, 26, 27], mentioned above provided some stepping stones to rigorous justification to some of the earlier conjectures of meteorologists in numerical weather prediction. For example, the systematic theoretical framework of the proposed global control scheme allowed the authors in [4] and [23] to provide sufficient conditions on the spatial resolution of the collected spatial coarse mesh data and the relaxation parameter that guarantees that the approximating solution obtained from this algorithm converges to the corresponding unknown reference solution over time (with the assumption that the observational data measurements are free of noise). Without access to concrete theoretical analysis, earlier implementation of the "nudging" algorithm left geophysicists searching for the optimal or suitable nudging coefficient (or relaxation parameter) through expensive numerical experiments. Naturally, one wishes for the availability of a sharper estimate for the operational characteristic parameters than what the theoretical results give. However, these recent analytical results, although not sharp, may allow one to track the correct parameter ranges without the expensive numerical experiments. Computational studies implementing these algorithms under drastically more relaxed conditions were demonstrated in [2, 36] for example.

To understand the value of the development of these analytical results stemming from the series of studies, we should mention its valuable impact in meteorology. In weather prediction, we've mentioned earlier the need to analyze the success of a data assimilation algorithm when some state variable observations are not available. Charney's question in [13, 37, 38] of whether temperature observations are enough to determine all the dynamical state of the system is an important problem in meteorology and engineering. In [38], an analytical argument suggested that the Charney's conjecture is correct, in particular, for the shallow water model. The authors of [38] derived a diagnostic system for the velocity field that gives the velocity in terms of first- and second-time derivatives of the mass field (the geopotential of the free surface of the fluid). A mathematical argument was then presented to justify that the mass field and its first- and second-time derivatives determine the velocity field fully. Similar diagnostic systems can be derived for other simple primitive equation models. The work in [38] gave a precise theoretical formulation of the Charney conjecture for certain simple atmospheric models.

Numerical tests in [38] suggested that in practice, it can be hard to implement this method to solve for the velocity field using only measurements on the mass field. Further numerical testing in [37] affirmed that it is not certain whether assimilation with temperature data alone will yield initial states of arbitrary accuracy. The authors of [37] considered the primitive equations (the main weather forecast model) and tested and compared different time-continuous data assimilation methods using temperature data alone. In their numerical experiments, they concluded that the accuracy of the assimilation is highly dependent on the assimilation method used and on the integrity of the measured observational temperature data. A relevant recent numerical study on a data assimilation algorithm for the 2D Bénard system [2], inspired by the work in [23], showed that it is sufficient to use coarse velocity measurements in the algorithm to recover the full true state of the system. On the other hand, it was concluded in [2] that data assimilation using coarse temperature measurements only will not always recover the true state of the system. It was observed that the convergence to the true state using temperature measurements only is actually sensitive to the amount of noise in the measured data as well as to the spacing (the sparsity of the collected data) and the time-frequency of such measured temperature data. These results in [2] are consistent with the results of the earlier numerical experiments in [38] and [37].

These results may indicate that the answer to Charney's question is negative for the practical issues we have with our measuring equipments or our numerical solving techniques. More recently, in [27], we proposed an improved continuous data assimilation algorithm for the 3D Planetary Geostrophic model that requires observations of the temperature only. We provided a rigorous mathematical argument that temperature history of the atmosphere will determine other state variables for this specific planetary scale model, thus justifying theoretically Charney's conjecture for this model. Numerical implementation of our algorithm for the 3D Planetary Geostrophic model is subject to new work to compare with the numerical results obtained in [2], [37] and [38].

We will review relevant results starting from the algorithm introduced in [4]. The algorithm in [4] can be formally described as follows: suppose that $u(t)$ represents a solution of some dynamical system governed by an evolution equation of the type

$$\frac{\mathrm{d}u}{\mathrm{d}t} = F(u), \tag{6.1}$$

where the initial data $u(0) = u_{\mathrm{in}}$ is missing. Let $I_h(u(t))$ represent an interpolant operator based on the observations of the system at a coarse spatial resolution of size h, for $t \in [0, T]$. The algorithm proposed in [4] is to construct a solution $v(t)$ from the observations that satisfies the equations

$$\frac{\mathrm{d}v}{\mathrm{d}t} = F(v) - \mu(I_h(v) - I_h(u)), \tag{6.2a}$$

$$v(0) = v_{\mathrm{in}}, \tag{6.2b}$$

where $\mu > 0$ is a relaxation (nudging) parameter and v_{in} is taken to be arbitrary initial data. As mentioned in the previous literature, this particular algorithm was designed to work for general dissipative dynamical systems of the form (6.1) that are known to have global, in time, solutions, a finite-dimensional global attractor and a finite set of determining parameters (see, e.g., [18, 30, 32, 33, 34, 35, 44, 47, 48] and references therein). Lower bounds on $\mu > 0$

and upper bounds on $h > 0$ can be derived such that the approximate solution $v(t)$ converges in time to the reference solution $u(t)$. These estimates are not sharp (see the numerical results in [2] and [36]) since their derivation uses the existing estimates for the global solution in the global attractor of these dissipative systems.

In the context of the incompressible 2D Navier–Stokes equations (NSE), the authors of [4] studied the conditions under which the approximate solution $v(t)$ obtained by the algorithm in (6.2) converges to the reference solution $u(t)$ over time (see also [36]). In [1], it was shown that approximate solutions constructed using observations on all three components of the unfiltered velocity field converge in time to the reference solution of the 3D Navier–Stokes-α model. Another application of this algorithm for the three-dimensional Brinkman–Forchheimer–Extended Darcy model was introduced in [65]. The authors of [46] studied the convergence of the algorithm to the reference solution in the case of the two-dimensional subcritical surface quasi-geostrophic (SQG) equation. The convergence of this synchronization algorithm for the 2D NSE, in higher-order (Gevery class) norm and in L^∞ norm, was later studied in [8]. An extension of the approach in [4] to the case when the observational data contains stochastic noise was analyzed in [7]. A study of the the algorithm for the 2D NSE when the measurements are obtained discretely in time and are contaminated by systematic error is presented in [31]. More recently in [68], the authors obtain uniform in time estimates for the error between the numerical approximation given by the Post-Processing Galerkin method of the downscaling algorithm and the reference solution, for the 2D NSE. Notably, this uniform in time error estimates provide strong evidence for the practical reliable numerical implementation of this algorithm.

In [23], a continuous data assimilation scheme for the two-dimensional incompressible Bénard convection problem was introduced. The data assimilation algorithm in [23] constructed the approximate solutions for the velocity u and temperature fluctuations using only the observational data of the velocity field and without any measurements for the temperature fluctuations. In [24], we introduced an abridged dynamic continuous data assimilation for the 2D NSE inspired by the recent algorithms introduced in [4, 23]. There we establish convergence results for the improved algorithm where the observational data needing to be measured and inserted into the model equation are reduced or subsampled. Our algorithm required observational measurements of only one component of the velocity vector field. The underlying analysis was made feasible by taking advantage of the divergence-free condition on the velocity field. Our work in [24] was then applied and extended for the convergence analysis for a 2D Bénard convection problem, where the approximate solutions constructed using observations in only the horizontal component of the two-dimensional velocity field and without any measurements on the temperature converge in time to the reference solution of the 2D Bénard system. This was a progression from the recent result in [23] where convergence results were established, given that observations are known on all of the components of the velocity field and without any measurements of the temperature. In [25] we propose that a data assimilation algorithm based on temperature measurements alone can be designed for the Bénard convection in a porous medium. In this work it was established that requiring a sufficient amount of coarse spatial observational measurements of only the temperature measurements as input is able to recover the full state of the system. Subsequently, in [27] we proposed an improved data assimilation algorithm for recovering the exact full reference solution (velocity and temperature field) of the 3D Planetary Geostrophic model, at an exponential rate in time, by employing coarse spatial mesh observations of the temperature alone. In particular, we presented a rigorous justification of an earlier conjecture of Charney which states that the temperature history of the atmosphere, for certain simple atmospheric models, determines all the other state variables.

6.2 Application to Turbulence Models

All the analysis of the proposed data assimilation algorithm assumes the global existence of the underlying model and uses previously-known estimates. It is for this reason that we are not able to prove similar results for the 3D NSE case, even though numerical testing may be applicable and feasible. Note, however, that we are able to formulate the analytical setting for a family of globally well-posed subgrid scale turbulence models belonging to a family called α-models of turbulence. These are simplified models through an averaging process that is designed to capture the large scale dynamics of the flow and at the same time provide a reliable closure model for the averaged equations. The first member of the family was introduced in the late 1990s and called the Navier–Stokes-α (NS-α) model (also known as the Lagrangian averaged Navier–Stokes-α (LANS-α) or viscous Camassa–Holm equations [14, 15, 16, 28, 29]). It is written as follows:

$$\partial_t v + u \cdot \nabla v + \nabla u \cdot v + \nabla p = \nu \Delta v + f, \tag{6.3a}$$

$$\nabla \cdot u = 0, \text{ and } v = u - \alpha^2 \Delta u. \tag{6.3b}$$

Unlike other subgrid closure models which normally add some additional dissipative process, this new modeling approach regularizes the NSE by

Table 6.1. *Some special cases of the model (6.4) with $\alpha > 0$, and with $S = (I - \alpha^2\Delta)^{-1}$ and $S_{\theta_2} = [I + (-\alpha^2\Delta)^{\theta_2}]^{-1}$.*

Model	NSE	Leray-α	ML-α	SBM	NSV	NS-α	NS-α-like
A	$-\nu\Delta$	$-\nu\Delta$	$-\nu\Delta$	$-\nu\Delta$	$-\nu\Delta S$	$-\nu\Delta$	$\nu(-\Delta)^\theta$
M	I	S	I	S	S	S	S_{θ_2}
N	I	I	S	S	S	I	I
χ	0	0	0	0	0	1	1

restructuring the distribution of the energy in the wave number $k > 1/\alpha$ of the inertial range [28]. In other words, NS-α smooths the nonlinearity of the NSE, instead of enhancing dissipation. Many other α-models, such as the Leray-α [17], the Clark-α [11], the Navier–Stokes–Voigt (NSV) equation [49, 50, 73, 74], and the models we have introduced, namely, the modified Leray-α (ML-α) [45], the simplified Bardina model (SBM) [12, 55], and the NS-α-like models [71], were inspired by this regularization technique. These models can be represented by a generalized model of the form

$$\partial_t u + Au + (Mu \cdot \nabla)(Nu) + \chi \nabla(Mu)^{\mathrm{T}} \cdot (Nu) + \nabla p = f(x), \tag{6.4a}$$

$$\nabla \cdot u = 0, \tag{6.4b}$$

$$u(0,x) = u_{\mathrm{in}}(x), \tag{6.4c}$$

where A, M, and N are bounded linear operators having certain mapping properties, χ is either 1 or 0, θ controls the strength of the dissipation operator A, and the two parameters which control the degree of smoothing in the operators M and N, respectively, are θ_1 and θ_2. Table 6.1 summarizes certain α-models of turbulence.

All of the models just mentioned have global regular solutions and possess fewer degrees of freedom than the NSE. Moreover, mathematical analysis also proved that the solutions to these models converge to the solution of NSE in the limit as the filter width parameter α tends to zero. In addition, several of the α-models of turbulence have been tested against averaged empirical data collected from turbulent channels and pipes, for a wide range of Reynolds numbers (up to 17×10^6) [14, 15, 16]. The successful analytical, empirical and computational aspects (see for example [28, 40, 62, 63] and references therein) of the alpha turbulence models have attracted numerous applications, see for example [5] for application to the quasi-geostrophic equations, [51] for application to the Birkhoff–Rott approximation dynamics of vortex sheets of the 2D Euler equations, and [60, 66, 67] for applications to incompressible magnetohydrodynamic equations. See also [53, 54] for the α-regularization of the inviscid 3D Boussinesq equation. A unified analysis of an additional family of α-type regularized models, also called a general family of regularized Navier–Stokes and MHD models on n-dimensional smooth compact Riemannian manifolds with or without boundary, with $n \geq 2$, is studied in [43]. For approximate deconvolution models of turbulence see [56, 57]. For other closure models see [6] and references therein.

The proposed algorithms in [23, 24, 26, 27, 25] sparked an idea that perhaps for this α-model one can construct approximate solutions, using only observations in the horizontal components and without any measurements on the vertical component of the velocity field, which converge in time to the reference solution. This is indeed the case, made possible by taking advantage of the divergence-free condition for the velocity field. Similar results can be claimed for certain other α-models. In this paper, we apply the data assimilation algorithm for the case of the 3D Leray-α model which we recall below [17]:

$$\partial_t v - v\Delta v + (u \cdot \nabla)v = -\nabla p + f, \quad (6.5a)$$

$$\nabla \cdot u = \nabla \cdot v = 0, \quad (6.5b)$$

$$v = u - \alpha^2 \Delta u, \quad (6.5c)$$

$$v(0, x, y, z) = v_{\text{in}}(x, y, z), \quad (6.5d)$$

where u, v and p are periodic, with basic periodic box $\Omega = [0, L]^3 = \mathbb{T}^3$.

$$(6.5e)$$

The nonlinearity is advected by the smoother velocity field, and notice that, consistent with all the other alpha models, the above system is the Navier–Stokes system of equations when $\alpha = 0$, i.e. $u = v$.

Recall that $I_h(\varphi)$ represents an interpolant operator based on the observational measurements of the scalar function φ at a coarse spatial resolution of size h. Given the viscosity v, the proposed algorithm for reconstructing $u(t)$ and $v(t)$ from only the horizontal observational measurements, which are represented by means of the interpolant operators $I_h(v_1(t))$ and $I_h(v_2(t))$ for $t \in [0, T]$, is given by the system

$$\partial_t v_1^* - v\Delta v_1^* + (u^* \cdot \nabla)v_1^* = -\partial_x p^* + \mu\left(I_h(v_1) - I_h(v_1^*)\right) + f_1, \quad (6.6a)$$

$$\partial_t v_2^* - v\Delta v_2^* + (u^* \cdot \nabla)v_2^* = -\partial_y p^* + \mu\left(I_h(v_2) - I_h(v_2^*)\right) + f_2, \quad (6.6b)$$

$$\partial_t v_3^* - v\Delta v_3^* + (u^* \cdot \nabla)v_3^* = -\partial_x p^* + f_3, \quad (6.6c)$$

$$\nabla \cdot u^* = \nabla \cdot v^* = 0, \quad (6.6d)$$

$$v^* = u^* - \alpha^2 \Delta u^*, \quad (6.6e)$$

$$v^*(0, x, y, z) = v_{\text{in}}^*(x, y, z), \quad (6.6f)$$

supplemented with periodic boundary conditions, where μ is again a positive parameter which relaxes (nudges) the coarse spatial scales of v^* toward those of the observed data.

Consequently, $v^*(t,x,y,z)$ is the approximating velocity field, with $v^*(0,x,y,z) = v_{in}^*(x,y,z)$ taken to be arbitrary. We note that any data assimilation algorithm using two out of three components of the velocity field also works. Here, observational data of the horizontal components $I_h(v_1(t))$ and $I_h(v_2(t))$ were chosen as an example.

We will consider interpolant observables given by linear interpolant operators $I_h : H^1(\Omega) \to L^2(\Omega)$, which approximate the identity and satisfy the approximation property

$$\|\varphi - I_h(\varphi)\|_{L^2(\Omega)} \leq \gamma_0 h \|\varphi\|_{H^1(\Omega)}, \tag{6.7}$$

for every φ in the Sobolev space $H^1(\Omega)$. One example of an interpolant observable of this type is the orthogonal projection onto the low Fourier modes with wave numbers k such that $|k| \leq 1/h$. A more physical example is the volume elements that were studied in [47]. A second type of linear interpolant operators $I_h : H^2(\Omega) \to L^2(\Omega)$, which satisfy the approximation property

$$\|\varphi - I_h(\varphi)\|_{L^2(\Omega)} \leq \gamma_1 h \|\varphi\|_{H^1(\Omega)} + \gamma_2 h^2 \|\varphi\|_{H^2(\Omega)} \tag{6.8}$$

for every φ in the Sobolev space $H^2(\Omega)$, can be considered with this algorithm. An example of this type of interpolant observables is given by the measurements at a discrete set of nodal points in Ω (see Appendix A in [4]). The treatment for the second type of interpolant is slightly more technical (see, e.g. [23, 24]), and thus we won't consider it here for the sake of keeping this note concise.

We prove an analytic upper bound on the spatial resolution h of the observational measurements and an analytic lower bound on the relaxation parameter μ that is needed in order for the proposed algorithm (6.6) to recover the reference solution of the 3D Leray-α system (6.5) that corresponds to the coarse measurements. These bounds depend on physical parameters of the system, the Grashof number as an example. We remark that extensions of algorithm (6.6), for the cases of measurements with stochastic noise and of discrete spatio-temporal measurements with systematic error, can be established by combining the ideas we present here with the techniques reported in [7] and [31], respectively.

6.3 Preliminaries

We define $\mathcal{F}$ to be the set of divergence-free L-periodic trigonometric polynomial vector fields from $\mathbb{R}^3 \to \mathbb{R}^3$, with spatial average zero over Ω. We denote by $L^2(\Omega)$, $W^{s,p}(\Omega)$, and $H^s(\Omega) \equiv W^{s,2}(\Omega)$ the usual Sobolev spaces in three dimensions, and we denote by H and V the closure of $\mathcal{F}$ in $L^2(\Omega)$ and $H^1(\Omega)$, respectively.

We denote the dual of V by V' and the Helmholz–Leray projector from $L^2(\Omega)$ onto H by P_σ. The Stokes operator $A : V \to V'$ can now be expressed as

$$Au = -P_\sigma \Delta u,$$

for each $u, v \in V$. We observe that in the periodic boundary condition case $A = -\Delta$. The linear operator A is self-adjoint and positive definite with compact inverse $A^{-1} : H \to H$. Thus, there exists a complete orthonormal set of eigenfunctions w_i in H such that $Aw_i = \lambda_i w_i$, where $0 < \lambda_i \le \lambda_{i+1}$ for $i \in \mathbb{N}$. The domain of A will be written as $\mathcal{D}(A) = \{u \in V : Au \in H\}$.

We define the inner products on H and V respectively by

$$(\mathbf{u}, \mathbf{v}) = \sum_{i=1}^{3} \int_{\mathbb{T}^3} u_i v_i \, d\mathbf{x} \quad \text{and} \quad ((\mathbf{u}, \mathbf{v})) = \sum_{i,j=1}^{3} \int_{\mathbb{T}^3} \partial_j u_i \partial_j v_i \, d\mathbf{x},$$

and their associated norms $(\mathbf{u}, \mathbf{u})^{1/2} = \|u\|_{L^2(\Omega)}$, $((\mathbf{u}, \mathbf{u}))^{1/2} = \left\|A^{1/2}\mathbf{u}\right\|_{L^2(\Omega)}$. Note that $((\cdot, \cdot))$ is a norm due to the Poincaré inequality

$$\|\phi\|_{L^2(\Omega)}^2 \le \lambda_1^{-1} \|\nabla\phi\|_{L^2(\Omega)}^2, \quad \text{for all } \phi \in V, \tag{6.9}$$

where λ_1 is the smallest eigenvalue of the operator A in three dimensions, subject to periodic boundary conditions.

We use the following inner products in $H^1(\Omega)$ and $H^2(\Omega)$, respectively

$$((u, v))_{H^1(\Omega)} = \lambda_1 \left[(u, v) + \alpha^2 (A^{1/2}u, A^{1/2}v) \right]$$

and

$$((u, v))_{H^2(\Omega)} = \lambda_1^2 \left[(u, v) + 2\alpha^2 (A^{1/2}u, A^{1/2}v) + \alpha^4 (Au, Av) \right].$$

The above inner products were used in [17] so that the norms in $H^1(\Omega)$ and $H^2(\Omega)$ are dimensionally homogeneous. Using these definitions, one can observe that

$$\lambda_1 \|v\|_{L^2(\Omega)} \le \|u\|_{H^2(\Omega)} \le 2\lambda_1 \|v\|_{L^2(\Omega)}, \tag{6.10}$$

where $v = u - \alpha^2 \Delta u$.

Remark 6.1 We will use these notations indiscriminately for both scalars and vectors, which should not be a source of confusion.

Let Y be a Banach space. We denote by $L^p([0,T];Y)$ the space of (Bochner) measurable functions $t \mapsto w(t)$, where $w(t) \in Y$, for a.e. $t \in [0,T]$, such that the integral $\int_0^T \|w(t)\|_Y^p \, dt$ is finite.

Hereafter, c denotes a universal dimensionless positive constant. Our estimates for the nonlinear terms will involve the Sobolev inequality in three dimensions:

$$\|u\|_{L^\infty(\Omega)} \le c\lambda_1^{-1/4} \|u\|_{H^2(\Omega)}. \tag{6.11}$$

Furthermore, inequality (6.7) implies that

$$\|w - I_h(w)\|_{L^2(\Omega)}^2 \le c_0^2 h^2 \left\|A^{1/2}w\right\|_{L^2(\Omega)}^2, \tag{6.12}$$

for every $w \in V$, where $c_0 = \gamma_0$.

Here, G denotes the the Grashof number in three dimensions

$$G = \frac{\|f\|_{L^2(\Omega)}}{\nu^2 \lambda_1^{3/4}}. \tag{6.13}$$

We recall that the 3D Leray-α model (6.5), subject to periodic boundary conditions, is well-posed and possesses a finite-dimensional compact global attractor.

Theorem 6.2 (Existence and uniqueness) *[17] If $v_{\text{in}} \in V$ and $f \in H$, then, for any $T > 0$, the 3D Leray-α model (6.5) has a unique global strong solution $v(t,x,y,z) = (v_1(t,x,y,z), v_2(t,x,y,z), v_3(t,x,y,z))$ that satisfies*

$$v \in C([0,T];V) \cap L^2([0,T];\mathcal{D}(A)), \qquad \text{and} \qquad \frac{dv}{dt} \in L^2([0,T];H).$$

Moreover, the system admits a finite-dimensional global attractor $\mathcal{A}$ that is compact in H.

The following bounds on solutions v of (6.5) can be proved using the estimates obtained in [17].

Proposition 6.3 *[17] Let $\tau > 0$ be arbitrary, and let G be the Grashof number given in (6.13). Suppose that v is a solution of (6.5), then there exists a time $t_0 > 0$ such that for all $t \ge t_0$ we have*

$$\|v(t)\|_{L^2(\Omega)}^2 \le 2\nu^2 \lambda_1^{-1/2} G^2 \tag{6.14a}$$

and

$$\int_t^{t+\tau} \|\nabla v(s)\|_{L^2(\Omega)}^2 \, ds \leq 2(1 + \tau \nu \lambda_1^{1/2}) \nu G^2. \tag{6.14b}$$

We also recall the following bound on the solutions v in the global attractor of (6.5) that was proved in [22]. This estimate improves the estimate in [17] on the enstrophy $\left\|A^{1/2}v\right\|_{L^2(\Omega)}^2$.

Proposition 6.4 *[22] Suppose that v is a solution in the global attractor of (6.5), then*

$$\left\|A^{1/2}v(t)\right\|_{L^2(\Omega)}^2 \leq \tilde{c} \frac{\nu^2 G^4}{\alpha^4 \lambda_1^{3/2}}, \tag{6.15}$$

for large $t > 0$, for some dimensionless constant $\tilde{c} > 0$.

6.4 Analysis of the Data Assimilation Algorithm

We will prove that under certain conditions on μ, the approximate solution (v_1^*, v_2^*, v_3^*) of the data assimilation system (6.6) converges to the solution (v_1, v_2, v_3) of the 3D Leray-α (6.5), as $t \to \infty$, when the observable operators satisfy (6.7).

Theorem 6.5 *Suppose I_h satisfy (6.7) and $\mu > 0$ and $h > 0$ are chosen such that $\mu c_0^2 h^2 \leq \nu$, where c_0 is the constant in (6.7). Let $v(t,x,y,z)$ be a strong solution of the Leray-α model (6.5) and choose $\mu > 0$ large enough such that*

$$\mu \geq 2 \frac{c\tilde{c}\nu G^4}{\alpha^4 \lambda_1}, \tag{6.16}$$

and $h > 0$ small enough such that $\mu c_0^2 h^2 \leq \nu$, where the constants $c, \tilde{c}$, and c_0 appear in (6.30), (6.15) and (6.12), respectively.

If the initial data $v_{in}^* \in V$ and $f \in H$, then the continuous data assimilation system (6.6) possesses a unique global strong solution $v^*(t,x,y,z) = (v_1^*(t,x,y,z), v_2^*(t,x,y,z), v_3^*(t,x,y,z))$ that satisfies

$$v^* \in C([0,T]; V) \cap L^2([0,T]; \mathcal{D}(A)), \qquad \text{and} \qquad \frac{dv^*}{dt} \in L^2([0,T]; H).$$

Moreover, the solution $v^*(t,x,y,z)$ depends continuously on the initial data v_{in}^* and it satisfies

$$\|v(t) - v^*(t)\|_{L^2(\Omega)} \to 0,$$

at exponential rate, as $t \to \infty$, where $v = (v_1, v_2, v_3)$ is the solution of the 3D Leray-α (6.5), with observed data on v_1 and v_2 only.

Proof Define $\tilde{p} = p - p^*$, $\tilde{u} = u - u^*$, and $\tilde{v} = v - v^*$, thus $\tilde{v} = \tilde{u} - \alpha^2 \Delta \tilde{u}$. Then $\tilde{v}_1$, $\tilde{v}_2$ and $\tilde{v}_3$ satisfy the equations

$$\frac{\partial \tilde{v}_1}{\partial t} - \Delta \tilde{v}_1 + \tilde{u}_1 \partial_x v_1 + \tilde{u}_2 \partial_y v_1 + \tilde{u}_3 \partial_z v_1 + (u^* \cdot \nabla) \tilde{v}_1 + \partial_x \tilde{p} = -\mu I_h(\tilde{v}_1),$$

$$\text{(6.17a)}$$

$$\frac{\partial \tilde{v}_2}{\partial t} - \Delta \tilde{v}_2 + \tilde{u}_1 \partial_x v_2 + \tilde{u}_2 \partial_y v_2 + \tilde{u}_3 \partial_z v_2 + (u^* \cdot \nabla) \tilde{v}_2 + \partial_y \tilde{p} = -\mu I_h(\tilde{v}_2),$$

$$\text{(6.17b)}$$

$$\frac{\partial \tilde{v}_3}{\partial t} - \Delta \tilde{v}_3 + \tilde{u}_1 \partial_x v_3 + \tilde{u}_2 \partial_y v_3 + \tilde{u}_3 \partial_z v_3 + (u^* \cdot \nabla) \tilde{v}_3 + \partial_z \tilde{p} = 0, \quad \text{(6.17c)}$$

$$\partial_x \tilde{v}_1 + \partial_y \tilde{v}_2 + \partial_z \tilde{v}_3 = \partial_x \tilde{u}_1 + \partial_y \tilde{u}_2 + \partial_z \tilde{u}_3 = 0. \quad \text{(6.17d)}$$

Since we assume that v is a reference solution of system (6.5), then it is enough to show the existence and uniqueness of the difference $\tilde{v}$. In the proof below, we will derive formal a priori bounds on $\tilde{v}$, under appropriate conditions on μ and h. These a priori estimates, together with the global existence and uniqueness of the solution v, form the key elements for showing the global existence of the solution v^* of system (6.6). The convergence of the approximate solution v^* to the exact reference solution v will also be established under the tighter condition on the nudging parameter μ as stated in (6.16). Uniqueness can then be obtained using similar energy estimates.

The estimates we provide in this proof are formal, but can be justified by the Galerkin approximation procedure and then passing to the limit while using the relevant compactness theorems. We will omit the rigorous details of this standard procedure (see, e.g., [19, 75, 76]) and provide only the formal a priori estimates.

Taking the $L^2(\Omega)$-inner product of (6.17a), (6.17b) and (6.17c) with $\tilde{v}_1$, $\tilde{v}_2$ and $\tilde{v}_3$, respectively, we obtain

$$\frac{1}{2} \frac{d}{dt} \|\tilde{v}_1\|_{L^2(\Omega)}^2 + \nu \|A^{1/2} \tilde{v}_1\|_{L^2(\Omega)}^2 \le |J_1| - (\partial_x \tilde{p}, \tilde{v}_1) - \mu(I_h(\tilde{v}_1), \tilde{v}_1),$$

$$\frac{1}{2} \frac{d}{dt} \|\tilde{v}_2\|_{L^2(\Omega)}^2 + \nu \|A^{1/2} \tilde{v}_2\|_{L^2(\Omega)}^2 \le |J_2| - (\partial_y \tilde{p}, \tilde{v}_2) - \mu(I_h(\tilde{v}_2), \tilde{v}_2),$$

$$\frac{1}{2} \frac{d}{dt} \|\tilde{v}_3\|_{L^2(\Omega)}^2 + \nu \|A^{1/2} \tilde{v}_1\|_{L^2(\Omega)}^2 \le |J_3| - (\partial_z \tilde{p}, \tilde{v}_3),$$

where

$$J_1 := J_{1a} + J_{1b} + J_{1c} := (\tilde{u}_1 \partial_x v_1, \tilde{v}_1) + (\tilde{u}_2 \partial_y v_1, \tilde{v}_1) + (\tilde{u}_3 \partial_z v_1, \tilde{v}_1),$$

$$J_2 := J_{2a} + J_{2b} + J_{2c} := (\tilde{u}_1 \partial_x v_2, \tilde{v}_2) + (\tilde{u}_2 \partial_y v_2, \tilde{v}_2) + (\tilde{u}_3 \partial_z v_2, \tilde{v}_2),$$

$$J_3 := J_{3a} + J_{3b} + J_{3c} := (\tilde{u}_1 \partial_x v_3, \tilde{v}_3) + (\tilde{u}_2 \partial_y v_3, \tilde{v}_3) + (\tilde{u}_3 \partial_z v_3, \tilde{v}_3).$$

By the Hölder inequality, inequality (6.10) and the Sobolev inequality (6.11), we can show that

$$
\begin{aligned}
|J_{1a}| = |(\tilde{u}_1 \partial_x v_1, \tilde{v}_1)| &\leq \|\partial_x v_1\|_{L^2(\Omega)} \|\tilde{u}_1\|_{L^\infty(\Omega)} \|\tilde{v}_1\|_{L^2(\Omega)} \\
&\leq c\lambda_1^{-1/4} \|\partial_x v_1\|_{L^2(\Omega)} \|\tilde{u}_1\|_{H^2(\Omega)} \|\tilde{v}_1\|_{L^2(\Omega)} \\
&\leq c\lambda_1^{3/4} \|\partial_x v_1\|_{L^2(\Omega)} \|\tilde{v}_1\|_{L^2(\Omega)}^2 \\
&\leq c\lambda_1^{1/4} \|\partial_x v_1\|_{L^2(\Omega)} \|\tilde{v}_1\|_{L^2(\Omega)} \|A^{1/2}\tilde{v}_1\|_{L^2(\Omega)} \\
&\leq \frac{\nu}{8} \|A^{1/2}\tilde{v}_1\|_{L^2(\Omega)}^2 + \frac{c\lambda_1^{1/2}}{\nu} \|\partial_x v_1\|_{L^2(\Omega)}^2 \|\tilde{v}_1\|_{L^2(\Omega)}^2 .
\end{aligned}
\tag{6.18}
$$

Using similar analysis to that used above, we obtain the following estimates:

$$
|J_{1b}| = \left|(\tilde{u}_2 \partial_y v_1, \tilde{v}_1)\right| \leq \frac{\nu}{8} \|A^{1/2}\tilde{v}_2\|_{L^2(\Omega)}^2 + \frac{c\lambda_1^{1/2}}{\nu} \|\partial_y v_1\|_{L^2(\Omega)}^2 \|\tilde{v}_1\|_{L^2(\Omega)}^2 ,
\tag{6.19}
$$

$$
|J_{1c}| = |(\tilde{u}_3 \partial_z v_1, \tilde{v}_1)| \leq \frac{\nu}{20} \|A^{1/2}\tilde{v}_3\|_{L^2(\Omega)}^2 + \frac{c\lambda_1^{1/2}}{\nu} \|\partial_z v_1\|_{L^2(\Omega)}^2 \|\tilde{v}_1\|_{L^2(\Omega)}^2 ,
\tag{6.20}
$$

$$
|J_{2a}| = |(\tilde{u}_1 \partial_x v_2, \tilde{v}_2)| \leq \frac{\nu}{8} \|A^{1/2}\tilde{v}_1\|_{L^2(\Omega)}^2 + \frac{c\lambda_1^{1/2}}{\nu} \|\partial_x v_2\|_{L^2(\Omega)}^2 \|\tilde{v}_2\|_{L^2(\Omega)}^2 ,
\tag{6.21}
$$

$$
|J_{2b}| = \left|(\tilde{u}_2 \partial_y v_2, \tilde{v}_2)\right| \leq \frac{\nu}{8} \|A^{1/2}\tilde{v}_2\|_{L^2(\Omega)}^2 + \frac{c\lambda_1^{1/2}}{\nu} \|\partial_y v_2\|_{L^2(\Omega)}^2 \|\tilde{v}_2\|_{L^2(\Omega)}^2 ,
\tag{6.22}
$$

$$
|J_{2c}| = |(\tilde{u}_3 \partial_z v_2, \tilde{v}_2)| \leq \frac{\nu}{20} \|A^{1/2}\tilde{v}_3\|_{L^2(\Omega)}^2 + \frac{c\lambda_1^{1/2}}{\nu} \|\partial_z v_2\|_{L^2(\Omega)}^2 \|\tilde{v}_2\|_{L^2(\Omega)}^2 ,
\tag{6.23}
$$

$$
|J_{3a}| = |(\tilde{u}_1 \partial_x v_3, \tilde{v}_3)| \leq \frac{\nu}{20} \|A^{1/2}\tilde{v}_3\|_{L^2(\Omega)}^2 + \frac{c\lambda_1^{1/2}}{\nu} \|\partial_x v_3\|_{L^2(\Omega)}^2 \|\tilde{v}_1\|_{L^2(\Omega)}^2 ,
\tag{6.24}
$$

$$
|J_{3b}| = \left|(\tilde{u}_2 \partial_y v_3, \tilde{v}_3)\right| \leq \frac{\nu}{20} \|A^{1/2}\tilde{v}_3\|_{L^2(\Omega)}^2 + \frac{c\lambda_1^{1/2}}{\nu} \|\partial_y v_3\|_{L^2(\Omega)}^2 \|\tilde{v}_2\|_{L^2(\Omega)}^2 .
\tag{6.25}
$$

Next, using integration by parts and the divergence-free condition (6.17d), we obtain

$$
J_{3c} = (\tilde{u}_3 \partial_z v_3, \tilde{v}_3) = -(v_3, \partial_z(\tilde{u}_3 \tilde{v}_3))
$$

$$= -(v_3, \partial_z \tilde{u}_3 \tilde{v}_3) - (v_3, \tilde{u}_3 \partial_z \tilde{v}_3)$$
$$= (v_3, (\partial_x \tilde{u}_1 + \partial_y \tilde{u}_2)\tilde{v}_3) + (v_3, \tilde{u}_3(\partial_x v_1 + \partial_y v_2))$$
$$=: J_{3d} + J_{3e}.$$

Integration by parts once again implies

$$J_{3d} = (v_3, (\partial_x \tilde{u}_1 + \partial_y \tilde{u}_2)\tilde{v}_3)$$
$$= -(v_3, \tilde{u}_1 \partial_x \tilde{v}_3) - (v_3, \tilde{u}_2 \partial_y \tilde{v}_3) - (\partial_x v_3, \tilde{u}_1 \tilde{v}_3) - (\partial_y v_3, \tilde{u}_2 \tilde{v}_3)$$
$$=: J_{3d1} + J_{3d2} + J_{3d3} + J_{3d4}$$

and

$$J_{3e} = (v_3, \tilde{u}_3(\partial_x v_1 + \partial_y v_2))$$
$$= -(v_3, \partial_x \tilde{u}_3 \tilde{v}_1) - (v_3, \partial_y \tilde{u}_3 \tilde{v}_2) - (\partial_x v_3, \tilde{u}_3 \tilde{v}_1) - (\partial_y v_3, \tilde{u}_3 \tilde{v}_2)$$
$$=: J_{3e1} + J_{3e2} + J_{3e3} + J_{3e4}.$$

Using the Hölder inequality, inequality (6.10) and the Sobolev inequality (6.11), we have

$$
\begin{aligned}
|J_{3d1}| = |(v_3, \tilde{u}_1 \partial_x \tilde{v}_3)| &\leq \|v_3\|_{L^2(\Omega)} \|\tilde{u}_1\|_{L^\infty(\Omega)} \|\partial_x \tilde{v}_3\|_{L^2(\Omega)} \\
&\leq c\lambda_1^{-1/4} \|v_3\|_{L^2(\Omega)} \|\tilde{u}_1\|_{H^2(\Omega)} \|\partial_x \tilde{v}_3\|_{L^2(\Omega)} \\
&\leq c\lambda_1^{3/4} \|v_3\|_{L^2(\Omega)} \|\tilde{v}_1\|_{L^2(\Omega)} \|\partial_x \tilde{v}_3\|_{L^2(\Omega)} \\
&\leq c\lambda_1^{1/4} \|A^{1/2} v_3\|_{L^2(\Omega)} \|\tilde{v}_1\|_{L^2(\Omega)} \|\partial_x \tilde{v}_3\|_{L^2(\Omega)} \\
&\leq \frac{\nu}{20} \|\partial_x \tilde{v}_3\|_{L^2(\Omega)}^2 + \frac{c\lambda_1^{1/2}}{\nu} \|A^{1/2} v_3\|_{L^2(\Omega)}^2 \|\tilde{v}_1\|_{L^2(\Omega)}^2
\end{aligned}
$$

and similarly,

$$|J_{3d2}| = |(v_3, \tilde{u}_2 \partial_y \tilde{v}_3)| \leq \frac{\nu}{20} \|\partial_y \tilde{v}_3\|_{L^2(\Omega)}^2 + \frac{c\lambda_1^{1/2}}{\nu} \|A^{1/2} v_3\|_{L^2(\Omega)}^2 \|\tilde{v}_2\|_{L^2(\Omega)}^2.$$

By a similar argument as in (6.18), we can show that

$$|J_{3d3}| = |(\partial_x v_3, \tilde{u}_1 \tilde{v}_3)| \leq \frac{\nu}{20} \|A^{1/2} \tilde{v}_3\|_{L^2(\Omega)}^2 + \frac{c\lambda_1^{1/2}}{\nu} \|\partial_x v_3\|_{L^2(\Omega)}^2 \|\tilde{v}_1\|_{L^2(\Omega)}^2$$

and

$$|J_{3d4}| = |(\partial_y v_3, \tilde{u}_3 \tilde{v}_2)| \leq \frac{\nu}{20} \|A^{1/2} \tilde{v}_3\|_{L^2(\Omega)}^2 + \frac{c\lambda_1^{1/2}}{\nu} \|\partial_y v_3\|_{L^2(\Omega)}^2 \|\tilde{v}_2\|_{L^2(\Omega)}^2.$$

Thus,

$$|J_{3d}| \leq \frac{\nu}{5} \|A^{1/2} \tilde{v}_3\|_{L^2(\Omega)}^2 + \frac{c\lambda_1^{1/2}}{\nu} \|A^{1/2} v_3\|_{L^2(\Omega)}^2 \left(\|\tilde{v}_1\|_{L^2(\Omega)}^2 + \|\tilde{v}_2\|_{L^2(\Omega)}^2 \right).$$

We apply similar calculations to J_{3e} and obtain

$$|J_{3e}| \leq \frac{\nu}{5} \left\|A^{1/2}\tilde{v}_3\right\|^2_{L^2(\Omega)} + \frac{c\lambda_1^{1/2}}{\nu} \left\|A^{1/2}v_3\right\|^2_{L^2(\Omega)} \left(\|\tilde{v}_1\|^2_{L^2(\Omega)} + \|\tilde{v}_2\|^2_{L^2(\Omega)}\right).$$

This yields

$$|J_{3c}| = |(\tilde{u}_3 \partial_z v_3, \tilde{v}_3)|$$

$$\leq \frac{2\nu}{5} \left\|A^{1/2}\tilde{v}_3\right\|^2_{L^2(\Omega)} + \frac{c\lambda_1^{1/2}}{\nu} \left\|A^{1/2}v_3\right\|^2_{L^2(\Omega)} \left(\|\tilde{v}_1\|^2_{L^2(\Omega)} + \|\tilde{v}_2\|^2_{L^2(\Omega)}\right).$$
(6.26)

Young's inequality and the assumption $\mu c_0^2 h^2 \leq \nu$ imply that

$$-\mu(I_h(\tilde{v}_i), \tilde{v}_i) = -\mu(I_h(\tilde{v}_i) - \tilde{v}_i, \tilde{v}_i) - \mu \|\tilde{v}_i\|^2_{L^2(\Omega)}$$

$$\leq \mu c_0 h \|\tilde{v}_i\|_{L^2(\Omega)} \left\|A^{1/2}\tilde{v}_i\right\|_{L^2(\Omega)} - \mu \|\tilde{v}_i\|^2_{L^2(\Omega)}$$

$$\leq \frac{\nu}{2} \left\|A^{1/2}\tilde{v}_i\right\|^2_{L^2(\Omega)} - \frac{\mu}{2} \|\tilde{v}_i\|^2_{L^2(\Omega)}, \qquad i = 1, 2.$$
(6.27)

Also we note that

$$(\partial_x \tilde{p}, \tilde{v}_1) + (\partial_y \tilde{p}, \tilde{v}_2) + (\partial_z \tilde{p}, \tilde{v}_3) = 0,$$
(6.28)

due to integration by parts, the boundary conditions, and the divergence-free condition (6.17d).

Combining all the bounds (6.18)–(6.28) and denoting $\|\tilde{v}_H\|^2_{L^2(\Omega)} := \|\tilde{v}_1\|^2_{L^2(\Omega)} + \|\tilde{v}_2\|^2_{L^2(\Omega)}$, we obtain

$$\frac{d}{dt} \|\tilde{v}\|^2_{L^2(\Omega)} + \frac{\nu}{2} \left\|A^{1/2}\tilde{v}\right\|^2_{L^2(\Omega)} \leq \left(\frac{c\lambda_1^{1/2}}{\nu} \left\|A^{1/2}v\right\|^2_{L^2(\Omega)} - \mu\right) \|\tilde{v}_H\|^2_{L^2(\Omega)},$$

or, using the Poincaré inequality (6.9), we have

$$\frac{d}{dt} \|\tilde{v}\|^2_{L^2(\Omega)} + \frac{\nu\lambda_1}{2} \|\tilde{v}\|^2_{L^2(\Omega)} + \beta(t) \|\tilde{v}_H\|^2_{L^2(\Omega)} \leq 0,$$
(6.29)

where

$$\beta(t) := \mu - \frac{c\lambda_1^{1/2}}{\nu} \left\|A^{1/2}v\right\|^2_{L^2(\Omega)}.$$
(6.30)

Since $v(t)$ is a solution in the global attractor of (6.5), then $\left\|A^{1/2}v\right\|^2_{L^2(\Omega)}$ satisfies the bound (6.15) for $t > t_0$, for some large enough $t_0 > 0$. Now, the assumption (6.16) yields

$$\frac{d}{dt} \|\tilde{v}\|^2_{L^2(\Omega)} + \min\left\{\frac{\nu\lambda_1}{2}, \frac{\mu}{2}\right\} \|\tilde{v}\|^2_{L^2(\Omega)} \leq 0,$$

for $t > t_0$. By Gronwall's lemma, we obtain

$$\left\| v(t) - v^*(t) \right\|_{L^2(\Omega)}^2 = \left\| \tilde{v}(t) \right\|_{L^2(\Omega)} \to 0,$$

at an exponential rate, as $t \to \infty$. □

Acknowledgements

We would like to thank Professor M. Ghil for the stimulating exchange concerning the Charney conjecture and for pointing out to us some of the relevant references. E.S.T. is thankful for the kind hospitality of ICERM, Brown University, where part of this work was completed. The work of A.F. is supported in part by the NSF grant DMS-1418911. The work of E.L. is supported in part by the ONR grant N0001416WX01475, N0001416WX00796 and HPC grant W81EF61205768. The work of E.S.T. is supported in part by the ONR grant N00014-15-1-2333 and the NSF grants DMS-1109640 and DMS-1109645.

References

[1] D. Albanez, H. Nussenzveig-Lopes and E. S. Titi, *Continuous data assimilation for the three-dimensional Navier–Stokes-α model*, Asymptotic Analysis **97**, issues 1–2, (2016), 139–164.

[2] M. U. Altaf, E. S. Titi, T. Gebrael, O. Knio, L. Zhao, M. F. McCabe and I. Hoteit, *Downscaling the 2D Bénard convection equations using continuous data assimilation*, Computational Geosciences (COMG), **21**, issue 3, (2017), 393–410. DOI 10.1007/s10596-017-9619-2.

[3] A. Azouani and E. S. Titi, *Feedback control of nonlinear dissipative systems by finite determining parameters - a reaction-diffusion paradigm*, Evolution Equations and Control Theory (EECT) **3(4)**, (2014), 579–594.

[4] A. Azouani, E. Olson and E. S. Titi, *Continuous data assimilation using general interpolant observables*, J. Nonlinear Sci. **24(2)**, (2014), 277–304.

[5] C. Bardos, J. Linshiz, and E. S. Titi. *Global regularity for a Birkhoff–Rott-α approximation of the dynamics of vortex sheets of the 2D Euler equations*, Phys. D: Nonlinear Phenomena **237**, issues 14–17, (2008), 1905–1911.

[6] L. C. Berselli, T. Iliescu and W. J. Layton, *Mathematics of Large Eddy Simulation of Turbulent Flows*. Springer, Berlin, Heidelberg, New York, 2006.

[7] H. Bessaih, E. Olson and E. S. Titi, *Continuous assimilation of data with stochastic noise*, Nonlinearity **28** (2015), 729–753.

[8] A. Biswas and V. Martinez, *Higher-order synchronization for a data assimilation algorithm for the 2D Navier–Stokes equations*, Nonlinear Analysis: Real World Applications, **35** (2017), 132–157.

[9] D. Blömker, K. J. H. Law, A. M. Stuart and K. C. Zygalakis, *Accuracy and stability of the continuous-times 3DVAR filter for the Navier–Stokes equations*, Nonlinearity **26**, (2013), 2193–2219.

[10] G. L. Browning, W. D. Henshaw and H. O. Kreiss, *A numerical investigation of the interaction between the large scales and small scales of the two-dimensional incompressible Navier–Stokes equations*, Research Report LA-UR-98-1712, Los Alamos National Laboratory. (1998).

[11] C. Cao, D. Holm and E. S. Titi, *On the Clark-α model of turbulence: global regularity and long-time dynamics*, Journal of Turbulence **6(20)**, (2005), 1–11.

[12] Y. Cao, E. Lunasin and E. S. Titi, *Global well-posedness of the three-dimensional viscous and inviscid simplified Bardina turbulence models*, Communications in Mathematical Sciences **4(4)**, (2006), 823–848.

[13] J. Charney, J. Halem and M. Jastrow, *Use of incomplete historical data to infer the present state of the atmosphere*, Journal of Atmospheric Science **26**, (1969), 1160–1163.

[14] S. Chen, C. Foias, D. D. Holm, E. Olson, E. S. Titi and S. Wynne, *The Camassa–Holm equations and turbulence.* Phys. D **133(1–4)**, (1999), 49–65.

[15] S. Chen, C. Foias, D. D. Holm, E. Olson, E. S. Titi and S. Wynne, *A connection between the Camassa-Holm equations and turbulent flows in channels and pipes.* Phys. Fluids **11(8)**, (1999), 2343–2353.

[16] S. Chen, C. Foias, D. D. Holm, E. Olson, E. S. Titi and S. Wynne, *Camassa–Holm equations as a closure model for turbulent channel and pipe flow.* Phys. Rev. Lett. **81(24)**, (1998), 5338–5341.

[17] A. Cheskidov, D. D. Holm, E. Olson and E. S. Titi, *On a Leray-α model of turbulence,* Royal Soc. A, Mathematical, Physical and Engineering Sciences **461**, (2005), 629–649.

[18] B. Cockburn, D. A. Jones and E. S. Titi, *Estimating the number of asymptotic degrees of freedom for nonlinear dissipative systems*, Mathematics of Computation, **66**, (1997), 1073–1087.

[19] P. Constantin and C. Foias, *Navier–Stokes Equations*, Chicago Lectures in Mathematics, University of Chicago Press, Chicago, IL, 1988. MR 972259 (90b:35190)

[20] R. Daley, *Atmospheric Data Analysis*, Cambridge Atmospheric and Space Science Series, Cambridge University Press (1991).

[21] R. Errico, D. Baumhefner, *Predictability experiments using a high-resolution limited-area model*, Monthly Weather Review **115**, (1986), 488–505.

[22] A. Farhat, M. S. Jolly and E. Lunasin, *Bounds on energy and enstrophy for the 3D Navier–Stokes-α and Leray-α models*, Communications on Pure and Applied Analysis **13**, (2014), 2127–2140.

[23] A. Farhat, M. S. Jolly and E. S. Titi, *Continuous data assimilation for the 2D Bénard convection through velocity measurements alone*, Phys. D: Nonlinear Phenomena **303** (2015), 59–66.

[24] A. Farhat, E. Lunasin and E. S. Titi, *Abridged continuous data assimilation for the 2D Navier–Stokes equations utilizing measurements of only one component of the velocity field*, J. Math. Fluid Mech. **18(1)**, (2015), 1–23.

[25] A. Farhat, E. Lunasin and E. S. Titi, *Data assimilation algorithm for 3D Bénard convection in porous media employing only temperature measurements*, Jour. Math. Anal. Appl. **438(1)**, (2016), 492–506.

[26] A. Farhat, E. Lunasin and E. S. Titi, *Continuous data assimilation for a 2D Bénard convection system through horizontal velocity measurements alone*, J. Nonlinear Sci., **27**, (2017), 1065–1087.

[27] A. Farhat, E. Lunasin and E. S. Titi, *On the Charney conjecture of data assimilation employing temperature measurements alone: the paradigm of 3D Planetary Geostrophic model*, Math. Clim. Weather Forecast. **2(1)**, (2016), 61–74.

[28] C. Foias, D. D. Holm and E. S. Titi, *The Navier–Stokes-alpha model of fluid turbulence. Advances in nonlinear mathematics and science*, Phys. D **152/153**, (2001), 505–519.

[29] C. Foias, D. D. Holm and E. S. Titi, *The three-dimensional viscous Camassa–Holm equations, and their relation to the Navier–Stokes equations and turbulence theory*, J. Dynam. Differential Equations **14**, (2002), 1–35.

[30] C. Foias, O. Manley, R. Rosa and R. Temam, *Navier–Stokes Equations and Turbulence*, Encyclopedia of Mathematics and Its Applications **83**, Cambridge University Press (2001).

[31] C. Foias, C. Mondaini and E. S. Titi, *A discrete data assimilation scheme for the solutions of the 2D Navier–Stokes equations and their statistics*, SIAM Journal on Applied Dynamical Systems **15(4)**, (2016), 2109–2142.

[32] C. Foias and G. Prodi, *Sur le comportement global des solutions non-stationnaires des équations de Navier–Stokes en dimension 2*, Rend. Sem. Mat. Univ. Padova **39**, (1967), 1–34.

[33] C. Foias and R. Temam, *Asymptotic numerical analysis for the Navier–Stokes equations*, Nonlinear Dynamics and Turbulence, Eds G. I. Barenblatt, G. Iooss and D. D. Joseph, Boston: Pitman Advanced Pub. Prog., 1983.

[34] C. Foias and R. Temam, *Determination of the solutions of the Navier–Stokes equations by a set of nodal values*, Math. Comp. **43**, (1984), 117–133.

[35] C. Foias and E. S. Titi, *Determining nodes, finite difference schemes and inertial manifolds*, Nonlinearity **4(1)**, (1991), 135–153.

[36] M. Gesho, E. Olson and E. S. Titi, *A computational study of a data assimilation algorithm for the two-dimensional Navier–Stokes equations*, Communications in Computational Physics **19**, (2016), 1094–1110.

[37] M. Ghil, M. Halem and R. Atlas, *Time-continuous assimilation of remote-sounding data and its effect on weather forecasting*, Mon. Wea. Rev **107**, (1978), 140–171.

[38] M. Ghil, B. Shkoller and V. Yangarber, *A balanced diagnostic system compatible with a barotropic prognostic model.*, Mon. Wea. Rev **105**, (1977), 1223–1238.

[39] K. Hayden, E. Olson and E. S. Titi, *Discrete data assimilation in the Lorenz and 2D Navier–Stokes equations*, Physica D **240**, (2011), 1416–1425.

[40] M. Hecht, D. D. Holm, M. Petersen and B. Wingate, *Implementation of the LANS-α turbulence model in a primitive equation ocean model*, J. Comp. Physics, **227**, (2008), 5691–5716.

[41] W. D. Henshaw, H. O. Kreiss and J. Yström, *Numerical experiments on the interaction between the large- and small-scale motion of the Navier–Stokes equations*, SIAM J. Multiscale Modeling & Simulation **1**, (2003), 119–149.

[42] J. Hoke and R. Anthes, *The initialization of numerical models by a dynamic relaxation technique*, Mon. Wea. Rev **104**, (1976), 1551–1556.

[43] M. Holst, E. Lunasin and G. Tsogtgerel, *Analytical study of generalized α-models of turbulence*, Journal of Nonlinear Sci. **20(5)**, (2010), 523–567.

[44] M. J. Holst and E. S. Titi, *Determining projections and functionals for weak solutions of the Navier–Stokes equations*, Contemporary Mathematics **204**, (1997), 125–138.

[45] A. Ilyin, E. Lunasin and E. S. Titi, *A modified-Leray-α subgrid scale model of turbulence*, Nonlinearity **19**, (2006), 879–897.

[46] M. Jolly, V. Martinez and E. S. Titi, *A data assimilation algorithm for the subcritical surface quasi-geostrophic equation*, Advanced Nonlinear Studies, **17(1)** (2017), 167–192.

[47] D. A. Jones and E. S. Titi, *Determining finite volume elements for the 2D Navier–Stokes equations*, Physica D **60**, (1992), 165–174.

[48] D. A. Jones and E. S. Titi, *Upper bounds on the number of determining modes, nodes and volume elements for the Navier–Stokes equations*, Indiana Univ. Math. J. **42(3)**, (1993), 875–887.

[49] V. K. Kalantarov, B. Levant and E. S. Titi, *Gevrey regularity of the global attractor of the 3D Navier–Stokes–Voight equations*, J. Nonlinear Sci. **19**, (2009), 133–152.

[50] V. K. Kalantarov and E. S. Titi. *Global attractors and estimates of the number of degrees of freedom of determining modes for the 3D Navier–Stokes–Voight equations*, Chin. Ann. Math. **30**, B(6), (2009), 697–714.

[51] B. Khouider and E. S. Titi, *An inviscid regularization for the surface quasi-geostrophic equation*, Comm. Pure Appl. Math **61(10)**, (2008), 1331–1346.

[52] P. Korn, *Data assimilation for the Navier–Stokes-α equations*, Physica D **238**, (2009), 1957–1974.

[53] A. Larios, E. Lunasin and E. S. Titi, *Global well-posedness for the 2d Boussinesq system without heat diffusion and with either anisotropic viscosity or inviscid Voigt-regularization*, J. Differ. Equations **255**, (2013), 2636–2654.

[54] A. Larios, E. Lunasin and E. S. Titi, *Global well-posedness for the 2D Boussinesq system without heat diffusion and with either anisotropic viscosity or inviscid Voigt-α regularization*, (2010), arXiv:1010.5024 [math.AP].

[55] W. Layton and R. Lewandowski, *On a well-posed turbulence model*, Discrete and Continuous Dyn. System B, **6**, (2006), 111–128.

[56] W. Layton and M. Neda, *A similarity theory of approximate deconvolution models of turbulence*, J. Math. Anal. and Appl., **333(1)**, (2007), 416–429.

[57] W. Layton and L. Rebholz, *Approximate Deconvolution Models of Turbulence: Analysis, Phenomenology and Numerical Analysis*, Lecture Notes in Mathematics, Springer, 2010.

[58] K. J. H Law, A. Shukla and A. M. Stuart, *Analysis of the 3DVAR filter for the partially observed Lorenz'63 model*, Disc and Cont. Dyn. Sys. **34**, (2014), 1061–1078.

[59] D. Leunberger, *An introduction to observers*, IEEE. T. A. Control **16**, (1971), 596–602.

[60] J. Linshiz and E. S. Titi, *Analytical study of certain magnetohydrodynamic-alpha models*, J. Math. Phys. **48(6)**, (2007).

[61] A. Lorenc, W. Adams and J. Eyre, *The treatment of humidity in ECMWF's data assimilation scheme*, Atmospheric Water Vapor, Academic Press New York, (1980), 497–512.

[62] E. Lunasin, S. Kurien, M. Taylor and E. S. Titi, *A study of the Navier–Stokes-α model for two-dimensional turbulence*, Journal of Turbulence **8**, (2007), 751–778.

[63] E. Lunasin, S. Kurien and E. S. Titi, *Spectral scaling of the Leray-α model for two-dimensional turbulence*, Journal of Physics A: Math. Theor. **41**, (2008), 344014.

[64] E. Lunasin and E. S. Titi, *Finite determining parameters feedback control for distributed nonlinear dissipative systems – a computational study*, Evolution Equations and Control Theory **6(4)**, (2017), 535–557.

[65] P. Markowich, E. S. Titi, and S. Trabelsi, *Continuous data assimilation for the three-dimensional Brinkman–Forchheimer–Extended Darcy model*, Nonlinearity **29(4)**, (2016), 1292–1328.

[66] P. D. Mininni, D. C. Montgomery and A. Pouquet, *Numerical solutions of the three-dimensional magnetohydrodynamic alpha-model*, Phys. Rev. E **71**, (2005), 046304.

[67] P. D. Mininni, D. C. Montgomery and A. Pouquet, *A numerical study of the alpha model for two dimensional magnetohydrodynamic turbulent flows*, Phys. Fluids **17**, (2005), 035112.

[68] C. Mondaini and E. S. Titi, *Uniform in time error estimates for the postprocessing Galerkin method applied to a data assimilation algorithm*, SIAM Journal on Numerical Analysis **56(1)**, (2018), 78–110.

[69] H. Nijmeijer, *A dynamic control view of synchronization*, Physica D **154**, (2001), 219–228.

[70] E. Olson and E. S. Titi, *Determining modes for continuous data assimilation in 2D turbulence*, Journal of Statistical Physics **113(5–6)**, (2003), 799–840.

[71] E. Olson and E. S. Titi, *Viscosity versus vorticity stretching: global well-posedness for a family of Navier–Stokes-alpha-like models*, Nonlinear Anal. **66(11)**, (2007) 2427–2458.

[72] E. Olson and E. S. Titi, *Determining modes and Grashoff number in 2D turbulence*, Theoretical and Computation Fluid Dynamics **22(5)**, (2008), 327–339.

[73] A. P. Oskolkov, *The uniqueness and solvability in the large of the boundary value problems for the equations of motion of aqueous solutions of polymers*, Zap. Naucn. Sem. Leningrad. Otdel. Mat. Inst. Steklov (LOMI) **38**, (1973), 98–136.

[74] A. P. Oskolkov, *On the theory of Voight fluids*, Zap. Naucn. Sem. Leningrad. Otdel. Mat. Inst. Steklov (LOMI) **98**, (1980), 233–236.

[75] J. C. Robinson, *Infinite-dimensional Dynamical Systems. An Introduction to Dissipative Parabolic PDEs and the Theory of Global Attractors*. Cambridge Texts in Applied Mathematics. Cambridge University Press, Cambridge, 2001.

[76] R. Temam, *Navier–Stokes Equations: Theory and Numerical Analysis*, AMS Chelsea Publishing, Providence, RI, 2001, Reprint of the 1984 edition.

[77] F.E. Thau, *Observing the state of non-linear dynamic systems*, Int. J. Control **17** (1973), 471–479.

* Aseel Farhat, Department of Mathematics, University of Virginia, Charlottesville, VA 22904, USA. af7py@virginia.edu.

† Evelyn Lunasin, Department of Mathematics, United States Naval Academy, Annapolis, MD, 21402 USA. lunasin@usna.edu.

‡ Edriss S. Titi, Department of Mathematics, Texas A&M University, 3368 TAMU, College Station, TX 77843-3368, USA. Also, The Science Program, Texas A&M University at Qatar, Doha, Qatar. titi@math.tamu.edu

7

Critical Points at Infinity Methods in CR Geometry

Najoua Gamara*

Sub-Riemannian spaces are spaces whose metric structure may be viewed as a constrained geometry, where motion is only possible along a given set of directions, changing from point to point. The simplest example of such spaces is given by the so-called Heisenberg group. The characteristic constrained motion of sub-Riemannian spaces has numerous applications ranging from robotic control in engineering to neurobiology where it arises naturally in functional magnetic resonance imaging (FMRI). It also arises naturally in other branches of pure mathematics as Cauchy–Riemann geometry, complex hyperbolic spaces, and jet spaces. In this paper, we review the use of the relationship between Heisenberg geometry and Cauchy–Riemann (CR) geometry. More precisely, we focus on the resolution of the Yamabe Conjecture which was definitely solved by techniques related to the theory of critical points at infinity. These techniques were first introduced by A. Bahri and H. Brezis for the Yamabe conjecture in the Riemannian settings. We also review the problem of the prescription of the scalar curvature using the same techniques which were studied first by A. Bahri and J. M. Coron as well as the multiplicity of solutions. Finally, we announce in this direction new existence results for Cauchy–Riemann spheres.

Key words: Yamabe Problem, CR manifold, Webster scalar curvature, Critical points at infinity, Euler–Poincaré Characteristic, Flatness condition.
MSC[2000]: 58E05, 57R70, 53C21, 53C15.

7.1 Introduction

In 1995, Professor A. Bahri proposed to R. Yacoub and N. Gamara to solve the remaining cases left open by D. Jerison and J.M. Lee of the Cauchy–Riemann–Yamabe Conjecture [36, 37, 38]. In 1987, D. Jerison and J.M. Lee formulated in [38] the CR Yamabe conjecture and developed

*College of Science, Taibah University, Saudi Arabia and Faculty of Science, University Tunis El Manar, Tunisia.

the analogy between it and the Yamabe problem in conformal Riemannian geometry, which had already been solved by T. Aubin [1] and R. Schoen [44]. Besides the proof of T. Aubin and R. Schoen, another proof by A. Bahri [4], A. Bahri and H. Brézis [5] was available using methods related to the theory of critical points at infinity. Based on his experience and his numerous works in that direction [2, 3, 4, 5, 6, 7], A. Bahri was convinced that these topological methods are well adapted to solve this conjecture in the Cauchy–Riemann settings. This theory involves variational methods, algebraic topology and Morse theory. In 2001, R. Yacoub and N. Gamara [33] solved the spherical case of the CR Yamabe Conjecture and N. Gamara finalized the resolution of the CR Yamabe conjecture [24].

This paper has three purposes, first, we give a review of the Yamabe problem [40, 49] on compact Cauchy–Riemannian manifolds and present its extension to the problem of the prescription of the scalar curvature on compact Cauchy–Riemannian manifolds without boundary. Multiplicity results for conformal contact forms admitting a given prescribed scalar curvature are also reviewed. Second, we announce recent improvements of the scalar curvature prescription problem in the case of CR spheres.

To fix the notation in our context, let M be a real orientable $2n + 1$-dimensional manifold. A CR structure on M is given by a distinguished n-dimensional complex subbundle of the complexified tangent bundle of M, $T_{1,0} \subset \mathbb{C}TM$, called the holomorphic tangent bundle satisfying $T_{1,0} \cap \overline{T_{1,0}} = 0$, where overbars denote complex conjugation and $[T_{1,0}, T_{1,0}] \subset T_{1,0}$; this axiom is often referred to as the Frobenius integrability property of the CR structure. Standard examples of CR manifolds are those of real hypersurface type, in this case M is a hypersurface of $\mathbb{C}^{n+1}$ and the CR structure defined on M is the one induced by the CR structure of the ambient space

$$T_{1,0}(M)_x = \mathbb{C}TM_x \cap T_{1,0}(\mathbb{C}^{n+1})_x, \ x \in M.$$

$T_{1,0}(\mathbb{C}^{n+1})$ denotes the holomorphic tangent bundle over $\mathbb{C}^{n+1}$: the span of $\{\frac{\partial}{\partial z^j}, 1 \leq j \leq n+1\}$, where $(z^1, \ldots, z^{n+1})$ are the cartesian complex coordinates on $\mathbb{C}^{n+1}$.

The Lévy distribution of the CR manifold M is $H = \text{Re}(T_{1,0} + \overline{T_{1,0}})$. Since M is orientable H is oriented by its complex structure($J : H \to H$, $J(v + \bar{v}) = i(v - \bar{v})$) and $H^\perp$ has a global non-vanishing section θ. The Lévy form of θ is the non-degenerate hermitian form defined by

$$L_\theta(X, \bar{Y}) = -2id\theta(X \wedge \bar{Y})X, \ Y \in T_{1,0}.$$

If L_θ is positive definite, M is said to be strictly pseudoconvex and θ defines a contact structure on M; $\theta \wedge d\theta^n$ is a volume form on M.

A pseudohermitian structure on M is a CR structure together with a contact form θ. There is a unique globally defined tangent vector field T transverse to the Lévy distribution, determined by

$$\theta(T) = 1, \ T \lrcorner d\theta = 0$$

(where $\lrcorner$ means interior product, so this equation means $d\theta(T,X) = 0$ for all vector fields X on M). T is referred to as the Reeb vector field of (M,θ). For a strictly pseudoconvex CR manifold the Reeb vector field extends the Lévy form L_θ to a Riemannian metric g_θ on M : the Webster metric on (M,θ), given by

$$g_\theta(X,Y) = L_\theta(X,Y), \ g_\theta(X,T) = 0, \ g_\theta(T,T) = 1, \ X,Y \in H.$$

The horizontal gradient is given by $\nabla^H u = \prod_H \nabla u$, where $\prod_H \nabla : T(M) \rightarrow H$ is the projection associated with the natural sum decomposition $T(M) = H \oplus \mathbb{R}T$ and the gradient of u, ∇u, is given by $g_\theta(\nabla u, Y) = Y(u)$ for any vector field Y on M. If div denote the divergence operator associated with the volume form $\theta \wedge d\theta^n$, we define a natural self adjoint, positive, second-order differential operator

$$\Delta_b u = -\text{div}(\nabla^H u), \ u \in C^2(M),$$

called the sub-Laplacian of (M,θ). If $\{W_\alpha\}$, $\alpha = 1,\ldots,n$ is any local frame for $T_{1,0}$, the admissible coframe dual to W_α is the collection of $(1,0)$ forms

$$\{\theta^\alpha\}, \ \theta^\alpha(W_\beta) = \delta^\alpha_\beta, \theta^\alpha(T) = \theta^\beta(W_{\bar\alpha}) = 0,$$

where $\bar\alpha = \alpha + n$, and $W_{\bar\alpha} = \overline{W}_\alpha$. $\{T, W_\alpha, W_{\bar\beta}\}$ is a frame for $\mathbb{C}TM$ and its admissible dual frame is $\{\theta, \theta^\alpha, \theta^{\bar\beta}\}$. $d\theta = ih_{\alpha\bar\beta}\theta^\alpha \wedge \theta^{\bar\beta}$, $L_\theta(X^\alpha W_\alpha, Y^{\bar\beta} W_{\bar\beta}) = h_{\alpha\bar\beta}X^\alpha Y^{\bar\beta}$.

If we denote again $\{W_\alpha\}$, $\alpha = 1,\ldots,2n$ a local $L_\theta-$ orthonormal frame in H, then the sub-Laplacian operator is locally given by

$$\Delta_b u = -\sum_{\alpha=1}^{2n}(W_\alpha(W_\alpha u) - \nabla_{W_\alpha} W_\alpha(u)).$$

More precisely, in a local coordinate (U, w^i) system on M, for $\alpha = 1,\ldots,2n$, we have

$$W_\alpha = a^i_\alpha \frac{\partial}{\partial w^i}, \ a^i_\alpha \in C^\infty(U,\mathbb{R}), \ 1 \leq i \leq 2n+1$$

and

$$\Delta_b u = -\sum_{ij=1}^{2n+1} \frac{\partial}{\partial w^i}\left(\alpha^{ij}\frac{\partial u}{\partial w^j}\right) + \sum_{j=1}^{2n+1}\alpha^j\frac{\partial u}{\partial w^j},$$

with $\alpha^{ij} = \sum_{\alpha=1}^{2n} a_\alpha^i a_\alpha^j$ and $\alpha^j = \frac{\partial \alpha^{ij}}{\partial w^i} + \alpha^{ik}\Gamma_{ik}^j$, where Γ_{ik}^j are the Christoffel symbols associated with the CR connection. Now, if we denote by W_α^* the formal adjoint of the vector field W_α given by

$$W_\alpha^* u = -\frac{\partial}{\partial w^i}(a_\alpha^i u) - a_\alpha^i \Gamma_{ik}^j u, \; u \in C_0^1(U,\mathbb{R})$$

and use the *Hörmander* operator associated with the system of vector fields $W = \{W_\alpha, \; \alpha = 1,\ldots,2n\}$ given by

$$H_W u = \sum_{\alpha=1}^{2n} W_\alpha W_\alpha^* u,$$

it is straightforward that locally, $\Delta_b = H_W$, see [22].

As an example, we consider the Heisenberg group H^n, it is the Lie group whose underlying manifold is $\mathbb{C}^n \times \mathbb{R}$ with the following group law:

$$(z,t) * (z',t') = (z+z', t+t'+2\text{Im}(zz')) \; \forall \, z,z' \in \mathbb{C}^n \text{ and } t,t' \in \mathbb{R},$$

where $zz' = \sum_{1 \leq j \leq n} z^j \bar{z}'^j$. We define a norm in H^n by

$$|(z,t)|_{H^n} = (|z|^4 + t^2)^{\frac{1}{4}}.$$

The complex vectors fields on H^n,

$$Z_j = \frac{\partial}{\partial z^j} + i\bar{z}^j \frac{\partial}{\partial t},$$

$$\frac{\partial}{\partial z^j} = \frac{1}{2}\left(\frac{\partial}{\partial x^j} - \frac{\partial}{\partial y^j}\right), \; z^j = x^j + iy^j, \quad 1 \leq j \leq n,$$

are left invariant with respect to the group law and homogeneous with degree -1 with respect to the dilations

$$(z,t) \rightarrow (\lambda z, \lambda^2 t), (\lambda \in \mathbb{R}).$$

The space $(T_{1,0})_{(z,t)}$ is spanned by $Z_{j,(z,t)}$, $1 \leq j \leq n$ and gives a left invariant CR structure on H^n. The form

$$\theta_0 = dt + 2 \sum_{1 \leq j \leq n} (x_j dy_j - y_j dx_j)$$

annihilates $T_{1,0}$; we take it to be the contact form of the CR structure on H^n. The sub-laplacian operator on H^n associated with the contact form θ_0 is

$$\Delta_b = \mathcal{L}_{\theta_0} = -\frac{1}{2}\sum_{j=1}^{j=n}(Z_j \bar{Z}_j + \bar{Z}_j Z_j).$$

7.2 The Yamabe Problem and Related Topics

The Yamabe problem goes back to Yamabe himself [49] who claimed in 1960 to have a solution, but in 1968, N. Trudinger [46] discover an error in his proof and corrected Yamabe's proof. T. Aubin [1] improved Trudinger's result, using variational methods and Weyl's tensor characteristics. In 1984, R. Schoen [44] solved the remaining cases using variational methods and the positive mass theorem. We have also to point out the work of J.M. Lee and T.H. Parker in [40], which is a detailed discussion on the Yamabe problem unifying the work of T. Aubin [1] with that of R. Schoen [44]. Besides the proof by Aubin and Schoen for the Riemannian Yamabe conjecture, another proof by A. Bahri [4], A. Bahri and H. Brézis [5] was available by techniques related to the theory of critical points at infinity.

7.2.1 The Riemannian Yamabe Problem: Overview

Given a compact Riemannian manifold (M, g) without boundary, of dimension $n \geq 3$, the Yamabe conjecture states that there is a metric $\widetilde{g}$ conformal to g which has a constant scalar curvature $R_{\widetilde{g}} = \lambda$. We write $\widetilde{g} = u^{\frac{4}{n-2}} g$, $u > 0$. We obtain the following transformation law for the scalar curvature of the metrics g and $\widetilde{g}$:

$$R_{\widetilde{g}} = u^{-\frac{n+2}{n-2}} (c_n \Delta u + Ru).$$

Hence the Yamabe problem is equivalent to solving

$$c_n \Delta u + Ru = \lambda u^{\frac{n+2}{n-2}}, \ u > 0.$$

Let $p = \frac{2n}{n-2}$ and $L = c_n \Delta + R$ be the conformal Laplacian of (M, g).
The last equation can be rewritten as

$$Lu = \lambda u^{p-1}, \ u > 0.$$

The Yamabe problem has a variational formulation with Euler functional

$$J(u) = \frac{\int_M uLu \, dv_g}{\left(\int_M u^p \, dv_g \right)^{\frac{2}{p}}}.$$

Let u be a positive function in $C^\infty(M)$ and a critical point of J, then $\lambda = J(u)$.
The infimum of the functional J,

$$\lambda(M) = \inf \left\{ J(u) \, / \, u \in C^\infty(M), u > 0 \right\},$$

is a conformal invariant called the Yamabe invariant of (M, g).

We have the following results.

Theorem 7.1 (Yamabe, Trudinger, Aubin) *For any compact Riemannian manifold (M,g) without boundary, we have*

$$\lambda(M) \leq \lambda(S^n) = n(n-1)\omega_n^{\frac{2}{n}}.$$

Theorem 7.2 *(Yamabe, Trudinger, Aubin) The Yamabe problem can be solved on any compact manifold M if $\lambda(M) < \lambda(S^n)$.*

Hence, Theorem 7.2 reduces the resolution of the Yamabe problem to the estimate of the Yamabe invariant $\lambda(M)$. In this way, T. Aubin proved the conjecture in the cases where (M,g) is not a conformally flat compact Riemannian manifold of dimension $n \geq 6$. All the remaining cases of the Yamabe problem were solved by R. Schoen using the positive mass theorem.

7.2.2 CR Yamabe Problem

Let (M,θ) be a real compact orientable and integrable pseudohermitian manifold of dimension $2n + 1$. We denote by $L = L_\theta = (2 + \frac{2}{n})\Delta_b + R_\theta$ the conformal CR Laplacian on M, where Δ_b is the sub-Laplacian operator and R_θ the Webster scalar curvature associated with θ.

The CR Yamabe conjecture states that there is a contact form $\tilde{\theta}$ on M, CR conformal to θ, which has a constant Webster scalar curvature $R_{\tilde{\theta}}$. We have then to find $\tilde{\theta}$ in the form $\tilde{\theta} = u^{\frac{2}{n}}\theta$, where u is a positive function defined on M. This problem is equivalent to solving the following partial differential equation:

$$(P): \begin{cases} Lu & = & u^{1+\frac{2}{n}} \quad \text{on } M, \\ u & > & 0. \end{cases}$$

This equation is a particular case of more general equations of the type:

$$Lu + Qu = u^{1+\frac{2}{n}}, \quad u > 0 \text{ on } M, \ Q \in L^\infty(M).$$

In [38], D. Jerison and J.M. Lee have extensively studied the CR Yamabe problem and showed that there is a deep analogy between the CR Yamabe problem and the Riemannian one. Their results can be formally compared to the partial completion of the proof of the Riemannian Yamabe conjecture by

T. Aubin. In 1986, D. Jerison and J.M. Lee proved some properties of the CR Yamabe invariant:

$$\lambda(M) = \inf_{u \in S_1^2(M)} \{A_\theta(u)/B_\theta(u) = 1\},$$

where $S_1^2(M)$ is a Folland–Stein space, $A_\theta(u) = \int_M Lu \, u \, \theta \wedge d\theta^n$; $B_\theta(u) = \int_M |u|^{2+\frac{2}{n}} \theta \wedge d\theta^n$, and gave a necessary condition on it to have the existence of solutions for the CR Yamabe problem.

Theorem 7.3 1. $\lambda(M)$ *depends on the CR structure on M, not on the choice of θ.*

2. $\lambda(M) \leq \lambda(S^{2n+1})$, *where $S^{2n+1} \subset \mathbb{C}^{n+1}$ is the unit sphere with its standard CR structure.*

3. *If $\lambda(M) < \lambda(S^{2n+1})$, then equation (P) has a solution.*

In 1987, D. Jerison and J.M. Lee proved the following result.

Theorem 7.4 *Let M be a compact strictly pseudoconvex $2n + 1$-dimensional CR manifold, $n \geq 2$, not locally CR equivalent to S^{2n+1}, then $\lambda(M) < \lambda(S^{2n+1})$. Hence, the CR Yamabe problem can be solved on M.*

The remaining cases left open by D. Jerison and J.M. Lee should by analogy be solved by using some CR positive mass theorem. Unfortunately, such a CR version of the positive mass theorem did not exist at that time. Besides the proof of T. Aubin and R. Schoen of the Riemannian Yamabe conjecture, another proof by A. Bahri [4] and A. Bahri and H. Brézis [5] of the same conjecture was available by techniques related to the theory of critical points at infinity. This proof is completely different in spirit as well in techniques and details from the proof of T. Aubin and R. Schoen. It does not require the use of any theory of minimal surfaces, nor the use of a CR positive mass theorem. It turns out that this proof can be carried to the CR framework.

The cases left open by D. Jerison and J.M. Lee have been the subject of two papers.

1. In 2001, R. Yacoub and N. Gamara in [33] solved the CR Yamabe problem for spherical CR manifolds.

2. In the same year, N. Gamara in [24] completed the resolution of the CR Yamabe conjecture for all dimensions by solving the three-dimensional subcase of the non-conformally flat case.

The proofs of the results of [33] and [24] are based on a contradiction argument.

7.2.3 Critical Points at Infinity Method

We consider the subspace of $S_1^2(M)$, defined by

$$H = \left\{ u \in S_1^2(M) / \int_M |du|_\theta^2 \, \theta \wedge d\theta^n < \infty, \ \int_M |u|^{2+\frac{2}{n}} \theta \wedge d\theta^n < \infty \right\}.$$

Let $\Sigma = \{u \in H, s.t. \|u\|_H = 1\}$, $\|u\|_H = \left(\int_M ((2 + \frac{2}{n}) |du|_\theta^2 + R_\theta u^2) \theta \wedge d\theta^n \right)^{\frac{1}{2}}$, and let

$$\Sigma_+ = \{u \in \Sigma, /u > 0\}.$$

For $u \in H$, we define the CR Yamabe functional:

$$J(u) = \frac{\int_M ((2 + \frac{2}{n}) |du|_\theta^2 + R_\theta u^2) \theta \wedge d\theta^n}{(\int_M |u|^{2+\frac{2}{n}} \theta \wedge d\theta^n)^{\frac{n}{n+1}}}.$$

If u is a critical point of J on Σ_+, then $J(u)^{\frac{n}{2}} u$ is a solution of the Yamabe equation (P).

Let us recall that the standard solutions of the CR Yamabe equation on the Heisenberg group $\mathbb{H}^n$ are obtained by left translations and dilations

$$(z, t) \rightarrow (\lambda z, \lambda^2 t), (\lambda \in \mathbb{R})$$

of the functions $u(z, t) = K |w + i|^{-n}$, $w = t + i |z|^2$ $(z, t) \in H^n$, $K \in \mathbb{C}$.

Since the injection $S_1^2(H^n) \rightarrow L^{2+\frac{2}{n}}(H^n)$ is continuous but not compact, the functional J does not satisfy the Palais–Smale condition denoted by (PS). More precisely, one can see that the standard solutions on H^n after superposition are the good candidate sequences which violate (PS). Therefore, the classical variational theory based on compactness arguments does not apply in this case.

7.2.4 The Case of a Spherical CR Manifold

General Settings

Let (M, θ) be a compact spherical CR manifold, we show the existence of a conformal factor $\tilde{u}_a^{\frac{2}{n}}$ depending differentiably on $a \in M$, such that if θ is replaced by $\tilde{u}_a^{\frac{2}{n}} \theta$ in a ball $B(a, \rho)$, then $(M, \tilde{u}_a^{\frac{2}{n}} \theta)$ is locally (H^n, θ_0). We may use in $B(a, \rho)$ the usual multiplication of H^n and the standard solutions of the CR Yamabe problem, which we denote by $\delta(b, \lambda)$, where $\lambda \in \mathbb{R}$. The function $\delta(b, \lambda)$ satisfies

$$\begin{cases} L_{\theta_0} \delta(b, \lambda) = \delta(b, \lambda)^{1+\frac{2}{n}} & \text{on } B(a, \rho), \\ b \in B(a, \rho). \end{cases}$$

We then define on M a family of "almost solutions", which we denote by $\hat{\delta}(a,\lambda)$. These functions are the solutions of

$$L\hat{\delta}(a,\lambda) = \delta'(a,\lambda)^{1+\frac{2}{n}},$$

where

$$\begin{cases} \delta'(a,\lambda) = \omega_a \tilde{u}_a \delta(a,\lambda) \text{ on } & B(a,\rho), \\ \delta'(a,\lambda) = 0 & \text{on } B^c(a,\rho). \end{cases}$$

Here ω_a is a cut-off function used to localize our function near the base point a as λ goes to infinity, we show that these "almost solutions" closely approximate at infinity the Yamabe solutions of the Heisenberg group

$$\left| \hat{\delta}(a,\lambda) - \delta'(a,\lambda) \right| = O\left(\frac{1}{\lambda^n}\right),$$

$$\left| \hat{\delta}(a,\lambda) - \delta'(a,\lambda) \right|_{C^2} = O\left(\frac{1}{\lambda^n}\right), \text{ when } \lambda \to \infty.$$

Neighborhoods of Critical Points at Infinity

Following A. Bahri (see [2, 4]), we set the following definitions and notations.

Definition 7.5 *A critical point at infinity of J on Σ_+ is a limit of a flow line $u(s)$ of the equation:*

$$\begin{cases} \dfrac{\partial u}{\partial s} = -\partial J(u), \\ u(0) = u_0. \end{cases}$$

We define neighborhoods of critical points at infinity of J as follows:

$$V(p,\varepsilon) = \left\{ \begin{array}{l} u \in \Sigma_+ \ s.t. \text{ there exist } p \text{ concentration} \\ \text{points } a_1,\ldots,a_p \text{ in } M \text{ and} \\ p \text{ concentrations } \lambda_1,\ldots,\lambda_p \ s.t. \\ \left\| u - \frac{1}{p^{\frac{1}{2}} S^{\frac{n}{2}}} \sum_{i=1}^{i=p} \hat{\delta}(a_i,\lambda_i) \right\|_H < \varepsilon \\ \text{with } \lambda_i > \frac{1}{\varepsilon}, \text{and for } i \neq j \\ \varepsilon_{ij} = (\frac{\lambda_i}{\lambda_j} + \frac{\lambda_j}{\lambda_i} + \lambda_i \lambda_j \tilde{d}^2(a_i,a_j))^{-n} \geq \frac{1}{\varepsilon}, \end{array} \right\}$$

where $\tilde{d}(x,y)$, if x and y are in a small ball of M of radius ρ is $\left\| \exp_x^{-1}(y) \right\|_{H^n}$ $\exp_x$ is the CR exponential map for the point x and $\tilde{d}(x,y)$ is equal to $\frac{\rho}{2}$ otherwise. (S is the Sobolev constant for the inclusion $S_1^2(H^n) \to L^{2+\frac{2}{n}}(H^n)$).

Because of the estimates of the "almost solutions" given above, we can replace, in the analysis of the (PS) condition, the functions δ' or δ by the functions $\widehat{\delta}$. Hence, we will be able to characterize the sequences of functions which violate the Palais–Smale condition.

In fact, we prove that, if (u_k) is a sequence of H satisfying $\partial J(u_k) \to 0$ and $J(u_k)$ is bounded, then (u_k) has a weak limit $\bar{u}$ in H. Hence, if $\bar{u}$ is non-zero, we prove that $\bar{u}$ is a critical point of J. Since we will prove the CR Yamabe problem using a contradiction argument, we suppose that equation (P) has no solutions. Then we have the following characterization of the sequences failing the (PS) condition.

Proposition 7.6 *Let $\{u_k\}$ be a sequence such that $\partial J(u_k) \to 0$ and $J(u_k)$ is bounded. Then there exist an integer $p \in \mathbb{N}^+$, a sequence $\varepsilon_k \to 0$ $(\varepsilon_k > 0)$ and an extracted subsequence of (u_k) such that $\frac{u_k}{\|u_k\|} \in V(p, \varepsilon_k)$.*

This proposition was first introduced in the Riemannian settings in [5, 2]; the proof follows from iterated blow-up around the concentration points. For the CR settings, a complete proof is given in [33].

The CR Yamabe Problem: Ideas about the Proof

Considering for $p \in \mathbb{N}$ the formal barycentric sets:

$$B_0(M) = \emptyset,$$

$$B_p(M) = \left\{ \sum_1^p \alpha_i \delta_{x_i}, \sum_{i=1}^p \alpha_i = 1, \alpha_i > 0, x_i \in M \right\},$$

where δ_{x_i} is the Dirac mass at the point x_i, and the following level sets of the functional J:

$$W_p = \left\{ u \in \Sigma_+ \, / \quad J(u) < (p+1)^{\frac{1}{n}} S \right\}.$$

We define a map $f_p(\lambda)$ from $B_p(M)$ to Σ_+ by

$$f_p(\lambda) \left(\sum_{i=1}^{i=p} \alpha_i \delta_{x_i} \right) = \frac{\sum_{i=1}^{i=p} \alpha_i \widehat{\delta}(x_i, \lambda_i)}{\| \sum_{i=1}^{i=p} \alpha_i \widehat{\delta}(x_i, \lambda_i) \|}.$$

The following theorem is proved in [33].

Theorem 7.7 1. *For any integer $p \geq 1$, there exists a real $\lambda_p > 0$, such that $f_p(\lambda)$ sends $B_p(M)$ in W_p, for any $\lambda > \lambda_p$.*

2. *There exists an integer $p_0 \geq 1$, such that for any integer $p \geq p_0$ and for any $\lambda > \lambda_{p_0}$, the map of pairs $f_p(\lambda) : (B_p(M), B_{p-1}(M)) \to (W_p, W_{p-1})$ is homologically trivial, i.e.*

$$f_{p*}(\lambda) = 0,$$

where

$$f_{p*}(\lambda) : H_*(B_p(M), B_{p-1}(M)) \to H_*(W_p, W_{p-1})$$

and $H_*(\bullet)$ is the homology group with $\mathbb{Z}/2\mathbb{Z}$ coefficients of $\bullet$.

On the other hand, arguing by contradiction, we will assume that the weak limit $\bar{u}$ of any (PS) sequences (u_k) of H satisfying $\partial J(u_k) \to 0$ with $J(u_k)$ bounded is zero; otherwise, our problem would be solved, since we would have found a solution. Then, assuming that (u_k) is non-negative, we prove that we can extract from (u_k) a subsequence denoted again by (u_k), such that $\frac{u_k}{\|u_k\|_H} \in V(p, \varepsilon_k)$ with $\varepsilon_k > 0$ and $\lim_{k\to\infty} \varepsilon_k = 0$.

In this case, we proved that the pair (W_p, W_{p-1}) retracts by deformation on the pair $(W_{p-1} \cup A_p, W_{p-1})$, where $A_p \subset V(p, \varepsilon)$. More precisely, we prove that the elements of A_p are of the form $\sum_{i=1}^{i=p} \alpha_i \widehat{\delta}_{x_i, \lambda_i} + v$, with v small in the norm $\|\ \|_H$. Therefore, the model $\ldots \subset B_{p-1}(M) \subset B_p(M) \subset \ldots$ can be compared via f_p to $\ldots \subset W_{p-1} \subset W_p$, and we proved that $f_{p*}(\lambda) \neq 0$, for every $p \in \mathbb{N}^*$, which is a contradiction with the result of Theorem 7.7 and therefore achieves the proof of the CR Yamabe problem in this case.

7.2.5 The CR 3-dimensional Non-spherical Case

In the paper [24], it is shown how the techniques of critical points at infinity can settle the case of a strictly pseudoconvex CR manifold (M, θ) of dimension $2n + 1$, without assuming that M is locally conformally flat. In fact, in [24] we focus on the case $n = 1$, but the techniques apply to higher CR dimensions with no more assumptions. We just follow the sketch of the proof given for the case $n = 1$, introducing where required some modifications due to the dimension of the CR manifold.

Here, we will give some ideas about the proof for the CR Yamabe problem in the case of a generic 3-dimensional non-spherical CR manifold.

The proof of the result in this case is similar to the one given for the CR spherical case. It is obtained by using a contradiction argument.

We use the same techniques given by A. Bahri and H. Brézis in [5]. However, in this case the study of "the almost solutions" $\widehat{\delta}_a$ is not straightforward as in the CR spherical case, where we have locally a relation between the conformal Laplacians of M and H^n. Here, we have to use the Green's function associated with L to derive a good asymptotic expansion of the Yamabe funtional J near the sets of its critical points at infinity. Finally, to compute the numerator and denominator of J, we used the approach of D. Jerison and J.M. Lee who refined in [36] the notion of normal coordinates by constructing new intrinsic CR normal coordinates for an abstract CR manifold. These coordinates are called pseudohermitian normal coordinates. The notions and results introduced and proved by D. Jerison and J.M. Lee are parallel, with drastically different techniques to the ones introduced by J.M. Lee and T. Parker [40] for the Riemannian Yamabe problem. In pseudohermitian normal coordinates, D. Jerison and J.M. Lee gave the Taylor series of θ and $\{\theta^\alpha\}$ to high order at a point $q \in M$, in terms of the pseudohermitian curvature and torsion. Since the problem is CR invariant they had to choose θ so as to simplify the curvature and the torsion at a base point q as much as possible. Using the results of [36], we proved the following estimates in pseudohermitian normal coordinates near a base point $q \in M$:

$$R = O(2), \quad W = Z + O(3), \quad \overline{W} = \overline{Z} + O(3), \quad L = -2(Z\overline{Z} + \overline{Z}Z) + O(2),$$
$$G_q(z,t) = C(\rho^{-2}(z,t)) + A + O(\rho(z,t)), \quad G_q > 0,$$

where $O(m)$ is a homogenous polynomial in ρ of degree a least m. Using these estimates and the topological method based on the theory of critical points at infinity explained earlier, we derive the result in this case.

7.2.6 Kazdan–Warner-like Problems

For the Yamabe problem, we have to look for a new contact form of the Cauchy–Riemannian manifold, conformal to the basic one which has a constant scalar curvature. In this section, we prescribe on a compact Cauchy–Riemannian manifold without boundary positive functions with different behavior and we explore the relationship between the critical points of such functions and the existence of solutions for the problem.

For the Riemannian settings, the problem of prescribing the scalar curvature is known as the Kazdan–Warner problem, and it has been studied by various authors for dimensions 2, 3 and 4 as well as in high dimensions. There are many papers devoted to this problem as well as to the multiplicity of solutions for the related differential equation, we can mention [12, 13, 14, 15, 16, 34, 35, 39, 41]. For the CR settings, see [42] and [23]. Here, we will merely refer to the most directly related literature, using the method based on the theory of critical points at infinity. For the Riemannian settings, we refer to [3, 6, 8, 10, 11] and more recently [17]. Concerning the Cauchy–Riemann setting the pioneer article is due to N. Gamara [25] and for the recent ones, see [20, 21, 26, 27, 30, 31, 32].

In this section, we will review the first work in this direction in CR manifolds: the case of a compact spherical CR manifold (M, θ) of dimension 3, without boundary, as displayed in [25]. Let $K : M \to R_+^*$ be a positive C^2 function; our objective is to find suitable conditions on K for which there exists a contact form $\widetilde{\theta}$ conformal to θ such that K is its Webster scalar curvature: $K = R_{\widetilde{\theta}}$, $\widetilde{\theta} = u^2\theta$, where $u : M \to R$ is positive. The problem of the scalar curvature is equivalent to the resolution of the partial differential equation

$$(P_K): \begin{cases} Lu & = & Ku^3 \quad \text{on } M, \\ u & > & 0, \end{cases}$$

where L is the conformal Laplacian of the manifold M.

Problem (P_K) has a variational structure, with associated Euler functional: $J(u), u \in S_1^2(M)$. A solution u of (P_K) is a critical point of J. As for the Yamabe problem, the functional J fails to satisfy the Palais–Smale condition, that is, there exist non-compact sequences along which the functional is bounded and its gradient goes to zero. The failure of the (PS) condition has been analyzed for the Riemannian case throughout the works of [2, 3, 6, 8, 10, 39, 41, 43, 45]. For the CR case, a complete description of sequences failing to satisfy (PS) is given in [33].

Since this problem has been formulated, obstructions have to be pointed out. The main difficulty encountered when one tries to solve equations of the type (P_K) consists of the failure of the Palais–Smale condition, which leads to the failure of classical existence mechanisms. We will use a gradient flow to overcome the non-compactness. Thinking of the sequences failing to satisfy the Palais–Smale condition as "critical points", our objective was to try to find suitable parameters, in order to complete a Morse lemma at infinity analogous to the one given for the Riemannian case. The Morse lemma is crucial to prove the existence of solution for equation (P_K); more precisely, the method we used to prove the existence of solutions for problem (P_K) is based on the work of A. Bahri [3, 6, 7]. This method involves a Morse lemma at infinity, which establishes, near the set of critical points at infinity of the functional J, a change of variables in the space $(a_i, \alpha_i, \lambda_i, v)$, $1 \le i \le p$ to $(\widetilde{a}_i, \widetilde{\alpha}_i, \widetilde{\lambda}_i, V)$, $(\widetilde{\alpha}_i = \alpha_i)$, where V is a variable completely independent of $\widetilde{a}_i$ and $\widetilde{\lambda}_i$ such that $J(\sum \alpha_i \delta_{a_i, \lambda_i})$ behaves like $J(\sum \alpha_i \delta_{\widetilde{a}_i, \widetilde{\lambda}_i}) + \|V\|^2$. The Morse lemma relies on the construction of a suitable pseudogradient for the associated variational problem, which is based on the expansion of J and its gradient ∂J near infinity. We define also a pseudogradient for the V variable with the aim of making this variable disappear by setting $\frac{\partial V}{\partial s} = -\nu V$, where ν is taken to be a very large constant. Then at $s = 1$, $V(s) = \exp(-\nu s)V(0)$ will be as small as we wish. This shows that, in order to define our deformation, we can work as if V was zero. The deformation will be extended immediately with the same properties to a neighborhood of zero in the V variable.

We prove that the Palais–Smale condition is satisfied along the decreasing flow lines of this pseudogradient, as long as these flow lines do not enter the neighborhood of a finite number of critical points of K. This method allows us to study the critical points at infinity of the variational problem, by computing their total index and comparing this total index to the Euler–Poincaré characteristic of the space of variations. This procedure was extensively used in earlier Riemannian work and has displayed the role of the Green's function in solving equations of the type (P_K).

It is important to recall that for the case we review, we have a balance phenomenon between the self interactions and interactions between the functions failing to satisfy the Palais–Smale condition.

To state our results, we set up the following conditions and notation.

Let $G(a,)$ be a Green's function for L at $a \in M$ and A_a the value of the regular part of G evaluated at a.

We assume that K has only non-degenerate critical points $\xi_1, \xi_2, \ldots, \xi_r$ such that

$$-\frac{\Delta_\theta K(\xi_i)}{3K(\xi_i)} - 2A_{\xi_i} \neq 0, \quad i = 1, \ldots, r.$$

Assume that ξ_i, $i = 1, \ldots, r_1$ are all the critical points of K with $-\frac{\Delta_\theta K(\xi_i)}{3K(\xi_i)} - 2A_{\xi_i} > 0$. Let $\tau_l = (i_1, \ldots, i_l)$ denote any l-tuple of $(1, \ldots, r_1)$, $1 \leq l \leq r_1$. We define the following matrix $M(\tau_l) = (M_{st})$ with

$$M_{ss} = -\frac{\Delta_\theta K(\xi_s)}{3K^2(\xi_s)} - 2\frac{A_{\xi_s}}{K(\xi_s)},$$

$$M_{st} = -2\frac{G(\xi_s, \xi_t)}{\sqrt{K(\xi_s)K(\xi_t)}}, \quad \text{for } 1 \leq s \neq t \leq l.$$

Assume that for any τ_l, $1 \leq l \leq r_1$, $M(\tau_l)$ is non-degenerate. If we denote by k_{i_j} the index of the critical point ξ_{i_j} with respect to K, $i(\tau_l) = 4l - 1 - \sum_{j=1}^{l} k_{i_j}$ is the index of the critical point at infinity τ_l. We obtain the following result.

Theorem 7.8

$$\text{If } \sum_{l=1}^{r_1} \sum_{\tau_l, M(\tau_l) > 0} (-1)^{i(\tau_l)} \neq 1,$$

then (P_K) has a solution.

This result means that if the total contribution of the critical points at infinity to the topology of the level sets of the associated functional J is not trivial, then we have a solution for (P_K).

To close this section, let us recall some results concerning multiplicity results for problem (P_K). The first paper in this direction is due to H. Chtioui, M. Ould Ahmedou and R. Yacoub. In [20] the authors generalized the result of N. Gamara [25]; they addressed the case where the total sum given above is equal to 1 but a partial sum is not equal to 1, and proved existence and multiplicity results for (P_K). More precisely, we have the following result.

Theorem 7.9 *Let $\rho(\tau_l)$ denotes the least eigenvalue of the matrix $M(\tau_l)$ if there exists a positive integer $k \in \mathbb{N}$ such that:*

1. $\displaystyle\sum_{\substack{\tau_l, \rho(\tau_l) > 0 \\ i(\tau_l) \le k-1}} (-1)^{i(\tau_l)} \neq 1,$

2. $\forall \tau_l$ *such that* $\rho(\tau_l) > 0, \ i(\tau_l) \neq k.$

Then, there exists a solution ω to the problem (P_K) such that

$$m(\omega) \le k,$$

where $m(\omega)$ denotes the Morse index of ω, defined as the dimension of the space of negativity of the linearized operator $\mathcal{L}(\hat{\delta}) := L_\theta(\hat{\delta}) - 3\omega^2\hat{\delta}$.

Moreover, for generic K, if

$$\sum_{\tau_l, \rho(\tau_l) > 0} (-1)^{i(\tau_l)} \neq 1$$

and if we denote by $\mathcal{S}$ the set of all the solutions of (P_K), a lower bound of $\mathcal{S}$ is given by

$$\#\mathcal{S} \ge \left| 1 - \sum_{\tau_l, \rho(\tau_l) > 0} (-1)^{i(\tau_l)} \right|.$$

Recall that many mathematicians have worked on these problems in the Riemannian settings as well in the CR settings, we can mention for example [6, 8, 9, 10, 11, 18, 19] and [21, 26, 27, 29, 47, 48, 30, 31, 32].

7.3 Curvature "Flatness Condition" on CR Spheres

In this section, we announce new existence results on Cauchy–Riemann spheres concerning the prescription of the scalar curvature [28]. Our aim is to prescribe on $\mathbb{S}^{2n+1}$, the unit sphere of $\mathbb{C}^{n+1}$ endowed with its standard contact form θ_1 a C^2 positive function K satisfying the so called β-flatness condition. As we have seen in Section 7.2, this problem is equivalent to solving the following partial differential equation:

$$(P_\beta): \begin{cases} L_{\theta_1} u = K u^{1+\frac{2}{n}} & \text{on } \mathbb{S}^{2n+1}, \\ u > 0, \end{cases} \tag{7.1}$$

where L_{θ_1} is the conformal Laplacian of $\mathbb{S}^{2n+1}$, $L_{\theta_1} = (2 + \frac{2}{n})\Delta_{\theta_1} + R_{\theta_1}$, where $\Delta_{\theta_1} = \Delta_{\mathbb{S}^{2n+1}}$ and $R_{\theta_1} = \frac{n(n+1)}{2}$ are respectively the sub-Laplacian operator and the Webster scalar curvature of $(\mathbb{S}^{2n+1}, \theta_1)$.

We will focus here on the case $n = 1$: the 3-dimensional CR sphere. To state our results, we set up the following conditions and notation.

Let $G(a,)$ be a Green's function for L at $a \in \mathbb{S}^3$.

We denote by

$$\mathcal{K} = \left\{ (\xi_i)_{(1 \le i \le r)}, \text{ such that } \nabla K(\xi_i) = 0 \right\}$$

the set of all critical points of K. We say that K satisfies the β-**flatness condition** if for all $\xi_i \in \mathcal{K}$, there exist

$$2 \le \beta = \beta(\xi_i) < 4 \quad \text{and} \quad b_1 = b_1(\xi_i), \ b_2 = b_2(\xi_i), \ b_0 = b_0(\xi_i) \in \mathbb{R}^*$$

such that in some pseudohermitian normal coordinates system centered at ξ_i, we have

$$K(x) = K(\xi_i) + b_1 |x_1|^\beta + b_2 |x_2|^\beta + b_0 |t|^{\frac{\beta}{2}} + \mathcal{R}(x), \tag{7.2}$$

where $\sum_{k=1}^{2} b_k + \kappa b_0 \ne 0$, $\sum_{k=1}^{2} b_k + \kappa' b_0 \ne 0$ with

$$\kappa = \frac{\displaystyle\int_{\mathbb{H}^1} |t|^{\frac{\beta}{2}} \frac{1 - ||z|^2 - it|^2}{\left| 1 + |z|^2 - it \right|^6} \theta_0 \wedge d\theta_0}{\displaystyle\int_{\mathbb{H}^1} |x_1|^\beta \frac{1 - ||z|^2 - it|^2}{\left| 1 + |z|^2 - it \right|^6} \theta_0 \wedge d\theta_0}, \qquad \kappa' = \frac{\displaystyle\int_{\mathbb{H}^1} \frac{|t|^{\frac{\beta}{2}}}{\left| 1 + |z|^2 - it \right|^4} \theta_0 \wedge d\theta_0}{\displaystyle\int_{\mathbb{H}^1} \frac{|x_1|^\beta}{\left| 1 + |z|^2 - it \right|^4} \theta_0 \wedge d\theta_0}.$$

The function $\sum_{p=0}^{[\beta]} |\nabla^p \mathcal{R}(x)| \, \|x\|_{\mathbb{H}^1}^{-\beta - r} = o(1)$ as x approaches ξ_i, ∇^r denotes all possible partial derivatives of order r and $[\beta]$ is the integer part of β.

Let

$$\mathcal{K}_1 := \left\{ \xi_i \in \mathcal{K} \text{ such that } \beta = \beta(\xi_i) = 2 \text{ and } b_1 + b_2 + \kappa' b_0 < 0 \right\},$$

$$\mathcal{K}_2 := \left\{ \xi_i \in \mathcal{K} \text{ such that } \beta = \beta(\xi_i) > 2 \text{ and } b_1 + b_2 + \kappa' b_0 < 0 \right\}.$$

The index of the function K at $\xi_i \in \mathcal{K}$, denoted by $m(\xi_i)$, is the number of strictly negative coefficients $b_k(\xi_i)$:

$$m(\xi_i) = \#\left\{ b_k(\xi_i); b_k(\xi_i) < 0 \right\}.$$

For each p-tuple $(\xi_{i_1}, \ldots, \xi_{i_p}) \in (\mathcal{K}_1)^p$ ($\xi_{i_l} \neq \xi_{i_j}$ if $l \neq j$), we associate the matrix $M(\xi_{i_1}, \ldots, \xi_{i_p}) = (M_{st})_{1 \leq s, t \leq p}$:

$$
\begin{aligned}
M_{ss} &= -\frac{\sum_{k=1}^{2} b_k + \kappa' b_0}{3 K^2(\xi_s)}, \\
M_{st} &= -2 \frac{G(\xi_s, \xi_t)}{[K(\xi_s) K(\xi_t)]^1/2}, \quad \text{for } s \neq t.
\end{aligned}
\tag{7.3}
$$

We say that K satisfies condition **(C)** if

for each p-tuple $(\xi_{i_1}, \ldots, \xi_{i_p}) \in (\mathcal{K}_1)^p$ the corresponding matrix (M_{st})

is non degenerate. $\tag{7.4}$

We define the sets

$$
\bullet \; \mathcal{K}_1^+ := \bigcup_p \left\{ (\xi_{i_1}, \ldots, \xi_{i_p}) \in (\mathcal{K}_1)^p, \; \varrho(\xi_{i_1}, \ldots, \xi_{i_p}) > 0 \right\}
$$

and

$$
\bullet \; l^+ := \max \left\{ p \in \mathbb{N} \; \text{s.t} \; \exists \; (\xi_{i_1}, \ldots, \xi_{i_p}) \in \mathcal{K}_1^+ \right\}.
$$

For $(\xi_{i_1}, \ldots, \xi_{i_p}) \in \mathcal{K}_1^+$, let $i(\xi_{i_1}, \ldots, \xi_{i_p}) := 4p - 1 - \sum_{j=1}^{p} m(\xi_{i_j})$.

The main result of this paper is the following.

Theorem 7.10 *Let K be a C^2 positive function on $\mathbb{S}^3$ satisfying the β-flatness condition and condition* **(C)**, *if*

$$
\sum_{\xi \in \mathcal{K}_2} (-1)^{3-m(\xi)} + \sum_{p=1}^{l^+} \sum_{(\xi_{i_1} \ldots \xi_{i_p}) \in \mathcal{K}_1^+} (-1)^{i(\xi_{i_1} \ldots \xi_{i_p})} \neq 1.
$$

Then, there exists at least one solution of (P_β).

In the present case, we have the presence of multiple blow-up points. In fact, looking at the possible formations of blow-up points, it comes out that the interaction of two different bubbles given by $< \delta_{a_i, \lambda_i}, \delta_{a_j, \lambda_j} >_L, i \neq j$ dominates the self interaction $< \delta_{a_i, \lambda_i}, \delta_{a_i, \lambda_i} >_L$, in the case where $2 < \beta < 4$, while in the case where $\beta = 2$, we have a balance phenomenon, that is any interaction of two bubbles is of the same order with respect to the self interaction.

Problem (P_β) has a nice variational structure, with associated Euler functional

$$
J(u) = \frac{\int_{\mathbb{S}^{2n+1}} Lu \, u \, \theta \wedge d\theta}{\left(\int_{\mathbb{S}^{2n+1}} K \, u^{2+\frac{2}{n}} \, \theta \wedge d\theta \right)^{\frac{1}{2}}}, \quad u \in S_1^2(\mathbb{S}^{2n+1}).
$$

As in the case reviewed in Section 7.2, the functional J fails to satisfy the Palais–Smale condition on the set $\Sigma^+ = \left\{ u \in S_1^2(\mathbb{S}^{2n+1}) / \|u\| = 1 \; u \geq 0 \right\}$. Using the CR equivalence F induced by the Cayley transform (see definition below) between $\mathbb{S}^{2n+1}$ minus a point and the Heisenberg group $\mathbb{H}^n$, equation (P_β) is equivalent up to an influent constant to

$$(P_{\mathbb{H}^n}) : \begin{cases} (2 + \frac{2}{n}) \Delta_{\mathbb{H}^n} u = \tilde{K} \, u^{1 + \frac{2}{n}} & \text{on } \mathbb{H}^n, \\ u > 0, \end{cases}$$

where $\Delta_{\mathbb{H}^n}$ is the sub-Laplacian of $\mathbb{H}^n$ and $\tilde{K} = K \circ F^{-1}$.

Next, we will introduce the Cayley transform.

Let $B^{n+1} = \left\{ z \in \mathbb{C}^{n+1} / |z| < 1 \right\}$ be the unit ball in $\mathbb{C}^{n+1}$ and $\mathcal{D}_{n+1} = \left\{ (z, w) \in \mathbb{C}^n \times \mathbb{C} / \text{Im}(w) > |z|^2 \right\}$ be the Siegel domain, where $\partial \mathcal{D}_{n+1} = \left\{ (z, w) \in \mathbb{C}^n \times \mathbb{C} / \text{Im}(w) = |z|^2 \right\}$.

Definition 7.11 *[22] The Cayley transform is the correspondence between the unit ball B^{n+1} in $\mathbb{C}^{n+1}$ and the Siegel domain $\mathcal{D}_{n+1}$, given by*

$$\mathcal{C}(\zeta) = \left(\frac{\zeta'}{1 + \zeta_{n+1}} , \; i \frac{1 - \zeta_{n+1}}{1 + \zeta_{n+1}} \right) ; \quad \zeta = (\zeta', \zeta_{n+1}), \quad 1 + \zeta_{n+1} \neq 0.$$

The Cayley transform gives a biholomorphism of B^{n+1} onto the Siegel domain $\mathcal{D}_{n+1}$. Moreover, when restricted to the sphere minus a point, $\mathcal{C}$ gives a CR diffeomorphism:

$$\mathcal{C} : \mathbb{S}^{2n+1} \backslash (0, \ldots, 0, -1) \longrightarrow \partial \mathcal{D}_{n+1}.$$

Let us recall the CR diffeomorphism

$$\begin{array}{rcl} f : & \mathbb{H}^n & \longrightarrow \quad \partial \mathcal{D}_{n+1}, \\ & (z, t) & \longmapsto \quad f(z, t) = (z, t + i|z|^2), \end{array}$$

with the obvious inverse $f^{-1}(z, w) = (z, \text{Re}(w))$, $z \in \mathbb{C}^n$, $w \in \mathbb{C}$. We obtain the CR equivalence with this mapping:

$$\begin{array}{rcl} F : & \mathbb{S}^{2n+1} \backslash (0, \ldots, 0, -1) & \longrightarrow \quad \mathbb{H}^n, \\ & \zeta = (\zeta_1, \ldots, \zeta_{n+1}) & \longmapsto \quad (z, t) = \left(\frac{\zeta_1}{1 + \zeta_{n+1}}, \ldots, \frac{\zeta_n}{1 + \zeta_{n+1}}, \; i \frac{2 \text{Im} \zeta_{n+1}}{|1 + \zeta_{n+1}|^2} \right) \end{array}$$

with inverse

$$\begin{array}{rcl} F^{-1} : & \mathbb{H}^n & \longrightarrow \quad \mathbb{S}^{2n+1} \backslash (0, \ldots, 0, -1) \\ & (z, t) & \longmapsto \quad \zeta = \left(\frac{2z_1}{1 + |z|^2 - it}, \ldots, \frac{2z_n}{1 + |z|^2 - it}, \; i \frac{1 - |z|^2 + it}{1 + |z|^2 - it} \right), \end{array}$$

and choose the standard contact form of $\mathbb{S}^{2n+1}$ as

$$\theta_1 = i \sum_{j=1}^{n+1} (\zeta_j d\bar{\zeta}_j - \bar{\zeta}_j d\zeta_j).$$

Then, we have $F^*(4(c_0^{-1}\delta_{(0,1)})^{\frac{2}{n}}\theta_0) = \theta_1$.

Let us differentiate and take into account that $\delta_{(0,1)}(F(\zeta)) = c_0|1 + \zeta_{n+1}|^2$; we obtain

$$d\theta_1 = \left(\frac{d\zeta_{n+1}}{1 + \zeta_{n+1}} + \frac{d\bar{\zeta}_{n+1}}{1 + \bar{\zeta}_{n+1}}\right) \wedge \theta_1 + |1 + \zeta_{n+1}|^2 F^*(d\theta_0)$$

and

$$\theta_1 \wedge d\theta_1^n = |1 + \zeta_{n+1}|^{2(n+1)} F^*(\theta_0 \wedge d\theta_0^n).$$

We introduce the following function for each (ζ_0, λ) on $\mathbb{S}^{2n+1} \times]0, +\infty[$:

$$\omega_{(\zeta_0,\lambda)}(\zeta) = |1 + \zeta_{n+1}|^{-n} \delta_{(F(\zeta_0),\lambda)} \circ F(\zeta).$$

We have $L_{\theta_1}\omega_{(\zeta_0,\lambda)} = \omega_{(\zeta_0,\lambda)}^{1+\frac{2}{n}}$, i.e. $\omega_{(\zeta_0,\lambda)}$ is a solution of the Yamabe problem on $\mathbb{S}^{2n+1}$.

We also have

$$\int_{\mathbb{S}^{2n+1}} L_{\theta_1}\omega_{(\zeta_0,\lambda)} \, \omega_{(\zeta_0,\lambda)} \, \theta_1 \wedge d\theta_1^n = \int_{\mathbb{H}^n} L_{\theta_0}\delta_{(g_0,\lambda)} \, \delta_{(g_0,\lambda)} \, \theta_0 \wedge d\theta_0^n,$$

and

$$\int_{\mathbb{S}^{2n+1}} |\omega_{(\zeta_0,\lambda)}|^{2+\frac{2}{n}} \theta_1 \wedge d\theta_1^n = \int_{\mathbb{H}^n} |\delta_{(g_0,\lambda)}|^{2+\frac{2}{n}} \theta_0 \wedge d\theta_0^n,$$

where $g_0 = F(\zeta_0)$, and $g = F(\zeta)$.

As a consequence, the variational formulation for (P_β) is equivalent to the variational formulation for $(P_{\mathbb{H}^n})$. So, in this case, we don't need to construct "almost solutions" to solve (P_β), we go through the Heisenberg group using directly the solutions of the Yamabe problem in $\mathbb{H}^n$.

Acknowledgements

The author would like to express her deep gratitude to Professor Abbas Bahri for his valuable research supervision during the planning and development of the resolution of the CR Yamabe conjecture. His willingness to give his time so generously has been very much appreciated. She also would like to thank the Laboratory of Non-Linear Analysis of the Mathematics Department, Rutgers University, New Jersey, USA, where most of this work was accomplished.

References

[1] T. Aubin: Equations differentielles non linéaires et probleme de Yamabe concernant la courbure scalaire, J. Math. Pures Appl. 55, (1976), 269–296.

[2] A. Bahri: Critical points at infinity in some variational problems, Pitman Research Notes in Mathematics, Series 182 (Longman), 1989, MR 91h:58022, Zbl 676.58021.

[3] A. Bahri: An invariant for Yamabe-type flows with application to scalar curvature problems in high dimensions, Duke. Math. J 281, (1996), 323–466.

[4] A. Bahri: Proof of the Yamabe conjecture for locally conformally flat manifolds. Non Lin. Anal, T. M. A, 20 (10), (1993), 1261–1278, MR 94e:53033, Zbl 782.53027.

[5] A. Bahri and H. Brézis: Nonlinear elliptic equations, in "Topics in Geometry in memory of Joseph d'Atri", Simon Gindiken, ed., Birkhauser. (1996), 1–100, Zbl 863.35037.

[6] A. Bahri and J.M. Coron: The scalar curvature problem on the standard three-dimensional sphere, J. Funct. Anal. 95 (1991), 106–172.

[7] A. Bahri and P.H. Rabinowitz: Periodic solutions of 3-body problems, Ann. Inst. H. Poincaré Anal. Non linéaire. 8 (1991), 561–649.

[8] M. Ben Ayed, Y. Chen, H. Chtioui and M. Hammami: On the prescribed scalar curvature problem on 4-manifolds, Duke Math. J. vol 84, no.3, (1996), 633–677.

[9] M. Ben Ayed and H. Chtioui: Topological tools in prescribing the scalar curvature on the half sphere. Adv. Nonlinear Studies 4 (2004), 121–148.

[10] M. Ben Ayed, H. Chtioui and M. Hammami: The scalar curvature problem on higher dimensional spheres, Duke Math. J. 93 (1998), 379–424.

[11] M. Ben Ayed and M. Ould Ahmedou: Multiplicity results for the prescribed scalar curvature on low spheres, Anna. Scuola. Sup. di. Piza.(5), vol.vii (2008), 1–26.

[12] K.C. Chang and J.Q. Liu: On Nirenberg's problems, Internat. J. Math. 4, (1993), 35–58.

[13] S.Y. Chang and P. Yang: A perturbation result in prescribing scalar curvature on S^n, Duke. Math. J. 64 (1991), 27–69.

[14] S.Y. Chang and P. Yang: Conformal deformation of metrics on S2, J. Diff. Geom. 27 (1988), 259–296.

[15] S.Y. Chang and P. Yang: Prescribing Gaussian curvature on $S2$, Acta Math. 159 (1987), 215–259.

[16] W. Chen and W. Ding: Scalar curvatures on S2, Trans. Amer. Math. Soc. 303, (1987), 365–382.

[17] H. Chtioui, W. Abdelhedi and H. Hajaiej: A complete study of the lack of compactness and existence results of a Fractional Nirenberg Equation via a flatness hypothesis: Part I, arXiv:1409.5884v1 [math.AP] 20 September 2014.

[18] H. Chtioui and M. Ould Ahmedou: Conformal metrics of prescribed scalar curvature on 4 manifolds: the degree zero case. Arabian Journal of Mathematics, 6(3), 127–136.

[19] H. Chtioui, M. Ould Ahmedou and W. Abdelhedi: A Morse theoretical approach for the boundary mean curvature problem on B4, Journal of Functional Analysis, 254 (2008), 1307–1341.

[20] H. Chtioui, M. Ould Ahmedou and R. Yacoub: Existence and multiplicity results for the prescribed Webster scalar curvature on 3 CR manifolds, J. Geo. Anal., 23(2), (2013), 878–894.

[21] H. Chtioui, M. Ould Ahmedou and R. Yacoub: Topological methods for the prescribed Webster scalar curvature problem on CR manifolds, Dif. Geo. and Appl. 28 (2010), 264–281.

[22] S. Dragomir and G. Tomassini: Differential Geometry and Analysis on CR Manifolds, Progress in Mathematics, Volume 246.

[23] V. Felli and F. Uguzzoni: Some existence results for the Webster scalar curvature problem in presence of symmetry, Ann. Math. 183 (2004), 469–493.

[24] N. Gamara: The CR Yamabe conjecture in the case $n = 1$, J. Eur. Math. Soc. 3, (2001), 105–137.

[25] N. Gamara: The prescribed scalar curvature on a 3-dimensional CR manifold, Advanced Nonlinear Studies 2 (2002), 193–235.

[26] N. Gamara, H. Chtioui and K. Elmehdi: The Webster scalar curvature problem on the three dimentional CR manifolds, Bull. Sci. Math. 131, (2007), 361–374.

[27] N. Gamara and S. El Jazi: The Webster scalar curvature revisited – the case of the three dimensional CR sphere, Calculus of Variations and Partial Differential equations Vol. 42, no. 1–2 (2011), 107–136.

[28] N. Gamara and B. Hafassa: The β-Flatness Condition in CR Spheres, Adv. Nonlinear Stud. v.17, issue 1, (2017), 193–213.

[29] N. Gamara, H. Guemri and A. Amri: Optimal control in prescribing Webster scalar curvature on 3-dimensional pseudoHermitian manifold, Nonlin. Anal. TMA, 127 (2015), 235–262.

[30] N. Gamara and M. Riahi: Multiplicity results for the prescribed Webster scalar curvature on the three CR sphere under "flatness condition", Bull. Sci. Math., 136, (2012), 72–95.

[31] N. Gamara and M. Riahi: The impact of the flatness condition on the prescribed Webster scalar curvature, Arab. J. Math., King Fahd University, (2013) 2, 381–392.

[32] N. Gamara and M. Riahi: The interplay between the CR flatness condition and existence results for the prescribed Webster scalar curvature, Adv. Pure. App. Math., APAM, v. 3, issue 3, (2012), 281–291.

[33] N. Gamara and R. Yacoub: CR Yamabe conjecture: the conformally flat case, Pac. J. Math., vol. 201, no. 1 (2001), 121–175.

[34] Z. Han: Prescribing Gaussian curvature on S2, Duke. Math. J. 61 (1990), 679–703.

[35] E. Hebey: Changements de métriques conformes sur la sphère, le problème de Nirenberg, Bull. Sci. Math. 114 (1990), 215–242.

[36] D. Jerison and J.M. Lee: Extremals for the Sobolev inequality on the Heisenberg group and the CR Yamabe problem, J. Amer. Math. Soc., 1, (1988), 1–13.

[37] D. Jerison and J.M. Lee: Intrinsic CR normal coordinates and the CR Yamabe problem, J. Diff. Geo., 29 (1989), 303–343.

[38] D. Jerison and J.M. Lee: The Yamabe problem on CR manifolds, J. Diff. Geom., 25 (1987), 167–197. MR 88i:58162, Zbl 661 32026.

[39] J. Kazdan and F. Warner: Existence and conformal deformation of metrics with prescribed Gaussian and scalar curvature, Ann. Math. (2) 101 (1975), 317–331.

[40] J.M. Lee and T.H. Parker: The Yamabe problem, Bull. Amer. Math. Soc. (NS) 17 (1987),
no.1, 37–91.

[41] Y.Y. Li: Prescribing scalar curvature on S^n and related problems, Part I, J. Diff. Equa. 120 (1995), 319–410.

[42] A. Malchiodi and F. Uguzzoni: A perturbation result for the Webster scalar curvature problem on the CR sphere, J. Math. Pures. Appl., 81 (2002), 983–997.

[43] J. Sacks and K. Uhlenbeck: The existence of minimal immersions of 2-spheres, Ann. Math. 113 (1981), 1–24.

[44] R. Schoen: Conformal deformation of a metric to constant scalar curvature, J. Diff. Geo. 20 (1984), 479–495.

[45] M. Struwe: A global compactness result for elliptic boundary value problems involving limiting nonlinearities, Math. Z. 187 (1984), 511–517.

[46] N.S. Trudinger: Remarks concerning the conformal deformation of Riemannian structures on comapact manifolds, Ann. Scuola Norm. Sup. Pisa (3) 22 (1968), 265–274.

[47] R. Yacoub: On the Scalar Curvature Equations in high dimension, Adv. Non-lin. Studies 2(4) (2002), 373–393.

[48] R. Yacoub: Existence results for the Prescribed Webster Scalar Curvature on higher dimensional CR manifolds, Adv. Non-lin. Studies 13 (3) (2013), 625–661.

[49] H. Yamabe: On a deformation of Riemannian structures on compact manifolds, Osaka Math. J. 12 (1960), 21–37.

*ngamara7@gmail.com, ngamara@taibahu.edu.sa

8

Some Simple Problems for the Next Generations

Alain Haraux*

A list of open problems on global behavior in time of some evolution systems, mainly governed by partial differential equations, is given together with some background information explaining the context in which these problems appeared. The common characteristic of these problems is that they appeared a long time ago in the personal research of the author and received almost no answer till now, with the exception of very partial results which are listed to help the readers' understanding of the difficulties involved.

AMS classification numbers: 35B15, 35B40, 35L10, 37L05, 37L15
Keywords: Evolution equations, bounded solutions, compactness, oscillation theory, almost periodicity, weak convergence, rate of decay.

8.1 Introduction

Will future generations go on studying mathematical problems? This in itself is an open question, but the growing importance of computer applications in everyday life together with the fundamental intricacies of computer science, abstract mathematical logic and the developments of new mathematical methods makes the positive answer rather probable.

This text does not comply with the usual standards of mathematical papers for two reasons: it is a survey paper in which no new result will be presented and the results which we recall to motivate the open questions will be given without proof.

It is not so easy to introduce an open question in a few lines. Giving the statement of the question is not enough, we must also justify why we consider the question important and explain why it could not be solved until now. Both points are delicate because the importance of a problem is always questionable and the difficulty somehow disappears when the problem is solved.

*Sorbonne Universités and Laboratoire Jacques-Louis Lions, Paris

The questions presented here concern the theory of differential equations and mostly the case of PDEs. They were encountered by the author during his research and some of them are already 40 years old. They might be considered purely academic by some of our colleagues more concerned with real world applications, but they are selected, among a much wider range of open questions, since their solution probably requires completely new approaches and will likely open the door towards a new mathematical landscape.

8.2 Compactness and Almost Periodicity

Throughout this section, the terms "maximal monotone operator" and "almost periodic function" will be used without having been defined. Although both terms are by now rather well known, the definitions and main properties of these objects will be found respectively in the reference texts [8, 2, 6, 29]. To help the understanding of readers who are not experts in the field, we just point out that the concept of *maximal monotone operator* is a generalization in the Hilbert space framework of the idea of a nondecreasing continuous function, sufficiently large to encompass all positive or skew-adjoint (possibly unbounded) linear operators. On the other hand, the concept of an *almost periodic function* is a topological extension for the finite sum of periodic functions with arbitrary periods. An almost periodic function does not enjoy any specific local property and can be "recognized" or identified only when examined on an infinite interval, a half-line or the whole real line.

One of my first fields of investigation was, in connection with the abstract oscillation theory, the relationship between (pre-)compactness and asymptotic almost periodicity for the trajectories of an almost periodic contractive process. The case of autonomous processes (contraction semigroups on a metric space) had been studied earlier in the Hilbert space framework by Dafermos and Slemrod [15], the underlying idea being that on the omega-limit set of a precompact trajectory, the semi-group becomes an isometry group. Then the situation resembles the simpler case of the isometry group generated on a Hilbert space H by the equation

$$u' + Au(t) = 0$$

with

$$A^* = -A,$$

for which almost periodicity of precompact trajectories was known already from L. Amerio quite a while ago (the case of vibrating membranes and vibrating plates with fixed bounded edge are special cases of this general result).

Remark 8.1 The problem solved by Dafermos and Slemrod [15] gives a general answer to asymptotic behavior of mechanical systems subject to only interior elastic forces and (exterior or interior) damping forces. When an exterior force is added, the problem under study becomes the adaptation of the response to the exterior source and can become very complicated; resonance phenomena may occur if the damping is incomplete or too weak.

As expected, the case of a nonautonomous process, associated with a time-dependent evolution equation of the form

$$u' + A(t)u(t) \ni 0$$

is not so good in general. In [17] (see also [18] for a related new almost periodicity criterion) I established an almost periodicity result for precompact trajectories of a *periodic* contraction process on a complete metric space, and in the same paper I exhibited a simple almost periodic (linear) isometry process on $\mathbb{R}^2$, generated by an equation of the form

$$u' + c(t)Ju(t) = 0$$

with J a $\frac{\pi}{2}$-rotation around 0, for which no trajectory except 0 is almost periodic.

Actually, while writing my thesis dissertation, I was specifically interested in the so-called "quasi-autonomous" problem, and I met the following general question.

Problem 8.2 (1977) Let A be a maximal monotone operator on a real Hilbert space H, let $f : \mathbb{R} \longrightarrow H$ be almost periodic and let u be a solution of

$$u' + Au(t) \ni f(t)$$

on $[0, +\infty)$ with a **precompact range**. Can we conclude that u is asymptotically almost periodic?

After studying a lot of particular cases in which the answer is positive ($A = L$ linear, A a subdifferential $\partial\Phi$ and some operators of the form $L + \partial\Phi$), I proved in [19] that the answer is positive if $H = \mathbb{R}^N$ with $N \leq 2$. But the answer is unknown for general maximal monotone operators even if $H = \mathbb{R}^3$.

Remark 8.3 In [28] it is stated that the answer is positive for all N, but there is a mistake in the proof, relying on a geometrical property which is not valid in higher dimensions, more specifically in 3D the intersection of the (relative) interiors of two arbitrarily close isometric proper triangles can be empty. Therefore the argument from [19] cannot be used in the same way for $N \geq 3$.

Remark 8.4 The problem is also open even when $A \in C^1(H,H)$, in which case the monotonicity just means

$$\forall u \in H, \quad \forall v \in H, \quad (A'(u),v) \geq 0.$$

Remark 8.5 The answer is positive if f is periodic, as a particular case of the main result of [17].

Since an almost periodic function has precompact range, studying the existence of almost periodic solutions requires some criteria for precompactness of bounded orbits. In the case of an evolution PDE, precompactness is classically derived from higher regularity theory. For parabolic equations the smoothing effect provides some higher-order regularity for $t > 0$ for bounded semi-orbits defined on $\mathbb{R}^+$. In the hyperbolic case, although there is no smoothing effect in finite time, precompactness of orbits was derived by Amerio and Prouse [1] from higher regularity of the source and strong coercivity of the damping operator g in the case of the semilinear hyperbolic problem

$$u_{tt} - \Delta u + g(u_t) = f(t,x) \text{ in } \mathbb{R}^+ \times \Omega, \quad u = 0 \text{ on } \mathbb{R}^+ \times \partial\Omega,$$

where Ω is a bounded domain of $\mathbb{R}^N$. But this method does not apply even in the simple case $g(v) = cv^3$ for $c > 0, N \leq 3$, a case where boundedness of all trajectories is known. The following question makes sense even when the source term is periodic in t and g is globally Lipschitz continuous.

Problem 8.6 (1978) Let Ω be a bounded domain of $\mathbb{R}^N$ and g a nonincreasing Lipschitz function. We consider the semilinear hyperbolic problem

$$u_{tt} - \Delta u + g(u_t) = f(t,x) \text{ in } \mathbb{R}^+ \times \Omega, \quad u = 0 \text{ on } \mathbb{R}^+ \times \partial\Omega.$$

We assume that $f : \mathbb{R} \longrightarrow L^2(\Omega)$ is continuous and periodic in t. Assuming

$$u \in C_b(\mathbb{R}^+, H_0^1(\Omega)) \cap C_b^1(\mathbb{R}^+, L^2(\Omega)),$$

can we conclude that

$$\bigcup_{t \geq 0} \{(u(t,.),u_t(t,.))\} \text{ is precompact in } H_0^1(\Omega) \times L^2(\Omega)?$$

Remark 8.7 The answer is positive in the following extreme cases.

(1) If $g = 0$ (by Browder–Petryshyn's theorem, there is a periodic solution, hence compact, and all the others are precompact by addition).

(2) If g^{-1} is uniformly continuous, see [22], the result does not require Lipschitz continuity of g and applies for instance to $g(v) = cv^3$ for $c > 0, N \leq 3$.

It would be tempting to "interpolate", but even the case $g(v) = v^+$ and $N = 1$ already seems to be nontrivial.

Remark 8.8 The same question is of course also relevant when f is almost periodic, and the result of [22] is true in this more general context. Moreover, precompactness of bounded trajectories when $g = 0$ is also true when f is almost periodic. This is related to a fundamental result of Amerio stating that if the primitive of an almost periodic function: $\mathbb{R} \longrightarrow H$ is bounded, it is also almost periodic. More precisely, if H is a Hilbert space and A is a (possibly unbounded) skew-adjoint linear operator with compact resolvent, let us consider a bounded solution (on $\mathbb{R}$ with values in H) of the equation

$$U' + AU = F,$$

where $F : \mathbb{R} \longrightarrow H$ is almost periodic. Then $\exp(tA)U := V$ is a bounded solution of

$$V' = \exp(tA)F$$

and, since $\exp(tA)\psi$ is almost periodic as well as $\exp(-tA)\psi$ for any $\psi \in H$, by a density argument on generalized trigonometric polynomials, it is immediate to check that a function $W : \mathbb{R} \longrightarrow H$ is almost periodic if and only if $\exp(tA)W : \mathbb{R} \longrightarrow H$ is almost periodic. Then Amerio's Theorem applied to V gives the result, and this property applies in particular to the wave equation written as a system in the usual energy space. Then, starting from a solution bounded on $\mathbb{R}^+$, a classical translation-(weak)compactness argument of Amerio gives a solution bounded on $\mathbb{R}$ of the same equation. We skip the details since this remark is mainly intended for experts in the field.

Remark 8.9 Historically, the work of Amerio and Prouse [1] was motivated by the study of vibrations of a membrane under the action of elastic forces, an exterior oscillating force and a damping term of the form

$$g(u_t) = c_1 u_t + c_2 |u_t| u_t.$$

For small velocities the viscous damping $g_1(u_t) = c_1 u_t$ is prevalent while for large velocities the quadratic hydrodynamical response $g_2(u_t) = c_2 |u_t| u_t$ prevails. When studying the autonomous damped oscillations, generally only $g_1(u_t) = c_1 u_t$ is significant and the hydrodynamical term is important for the so-called *transient* oscillations leading, after a time depending on the initial state, to a situation where the total energy is small enough for g_1 to become prevalent.

8.3 Oscillation Theory

Apart from the almost periodicity of solutions, which provides a starting point to describe precisely the global time behavior of vibrating strings and membranes with fixed edge, it is natural to try a description of sign changes of the solutions on some subset of the domain. Let us first consider the basic equation

$$u'' + Au(t) = 0, \tag{8.1}$$

where V is a real Hilbert space, $A \in L(V, V')$ is a symmetric, positive, coercive operator and there is a second real Hilbert space H for which $V \hookrightarrow H = H' \hookrightarrow V'$ where the imbedding on the left is compact. In this case it is well known that all solutions $u \in C(\mathbb{R}, V) \cap C^1(\mathbb{R}, H)$ of (8.1) are almost periodic : $\mathbb{R} \longrightarrow V$ with mean-value 0. Then for any form $\zeta \in V'$, the function $g(t) := \langle \zeta, u(t) \rangle$ is a real-valued continuous almost periodic function with mean-value 0. It is then easy to show that either $g \equiv 0$, or there exists $M > 0$ such that on each interval J with $|J| \geq M$, g takes both positive and negative values. We shall say that a number $M > 0$ is a strong oscillation length for a numerical function $g \in L^1_{loc}(\mathbb{R})$ if the following alternative holds: either $g(t) = 0$ almost everywhere, or for any interval J with $|J| \geq M$, we have

$$\text{meas}\{t \in J, f(t) > 0\} > 0 \quad \text{and} \quad \text{meas}\{t \in J, f(t) < 0\} > 0.$$

As a consequence of the previous argument, under the above conditions on H, V and A, for any solution $u \in C(\mathbb{R}, V) \cap C^1(\mathbb{R}, H)$ of (8.1) and for any $\zeta \in V'$, the function $g(t) := \langle \zeta, u(t) \rangle$ has some finite strong oscillation length $M = M(u, \zeta)$.

In the papers [9, 10, 21, 26] the main objective was to obtain a strong oscillation length independent of the solution and the observation in various cases, including nonlinear perturbations of equation (8.1). A basic example is the vibrating string equation

$$u_{tt} - u_{xx} + g(t, u) = 0 \quad \text{in } \mathbb{R} \times (0, l), \quad u = 0 \text{ on } \mathbb{R} \times \{0, l\}, \tag{8.2}$$

where $l > 0$ and $g(t, .)$ is an odd nondecreasing function of u for all t. Here the function spaces are $H = L^2(0, l)$ and $V = H^1_0(0, l)$. Since any function of V is continuous, a natural form $\zeta \in V'$ is the Dirac mass δ_{x_0} for some $x_0 \in (0, l)$. It turns out that $2l$ is a strong oscillation length independent of the solution and the observation point x_0, exactly as in the special case $g = 0$, the ordinary vibrating string. Since in this case all solutions are $2l$-periodic with mean-value 0 functions with values in V, it is clear that $2l$ is a strong oscillation length independent of the solution and the observation point x_0. The slightly more complicated $g(t, u) = au$ with $a > 0$ is immediately more difficult since the general solution is no longer time-periodic, it is only almost periodic in t. The time-periodicity is too unstable and for an almost periodic function, the determination of strong oscillation lengths is not easy in general, as was exemplified in [26]. The oscillation result of [9, 10] is consequently not so immediate even in the linear case. In the nonlinear case, it becomes even more interesting because the solutions are no longer known to be almost periodic.

In dimensions $N \geq 2$, even the linear case becomes difficult. It has been established in [26] that even for analytic solutions of the usual wave equation in a rectangle, there is no uniform pointwise oscillation length common to all solutions at some points of the domain. One would imagine that it becomes true if the point is replaced by an open subset of the domain, but apparently nobody knows the answer to the following exceedingly simpler question.

Problem 8.10 (1985) Let $\Omega = (0, 2l) \times (0, 2l) \subset \mathbb{R}^2$. We consider the linear wave equation

$$u_{tt} - \Delta u = 0 \text{ in } \mathbb{R} \times \Omega, \quad u = 0 \text{ on } \mathbb{R} \times \partial \Omega.$$

Given $T > 0$, can we find a solution u for which

$$\forall (t, x) \in [0, T] \times (0, l) \times (0, l), \quad u(t, x) > 0?$$

Or does this become impossible for T large enough?

Another simple-looking intriguing question concerns the pointwise oscillation of solutions to semilinear beam equations, since the solutions of the corresponding linear problem oscillate at least as fast as those of the string equation.

Problem 8.11 (1985) We consider the semilinear beam equation

$$u_{tt} + u_{xxxx} + g(u) = 0 \text{ in } \mathbb{R} \times (0, 1), \quad u = u_{xx} = 0 \text{ on } \mathbb{R} \times \{0, 1\}$$

with g odd and nonincreasing with respect to u. Is it possible for a solution $u(t, .)$ to remain positive at some point x_0 on an arbitrarily long (possibly unbounded) time interval?

Finally, let us mention a question on spatial oscillation of solutions to parabolic problems. Since the heat equation has a very strong smoothing effect on the data, and all solutions are analytic inside the domain for $t > 0$, it seems natural to think that they do no accumulate oscillations and, for instance in 1D, the zeroes of $u(t,.)$ will be isolated for $t > 0$. A very general result of this type, valid for semilinear problems as well, has been proved by Angenent [3]. But as soon as $N \geq 2$, even the linear case is not quite understood. The answer to the following question seems to be unknown.

Problem 8.12 (1997) Let $\Omega \subset \mathbb{R}^N$ be a bounded open domain and $f \in C^1(\mathbb{R})$. We consider the heat equation

$$u_t - \Delta u + f(u) = 0 \text{ in } \mathbb{R} \times \Omega, \quad u = 0 \text{ on } \mathbb{R} \times \partial\Omega.$$

For $t > 0$, we consider

$$\mathcal{E} = \{x \in \Omega, \, u(t,x) \neq 0\}.$$

Is it true that $\mathcal{E}$ has a finite number of connected components?

Remark 8.13 The solutions u of the *elliptic* problem

$$-\Delta u + f(u) = 0 \text{ in } \Omega, \quad u = 0 \text{ on } \partial\Omega$$

are such that $\{x \in \Omega, \, u(x) \neq 0\}$ has a finite number of connected components for a large class of functions f, see e.g. [14]. Hence stationary solutions cannot provide a counterexample.

Remark 8.14 Actually Problem 8.12 is open even for $f = 0$ and $N = 2$, the most difficult aspect being the behavior of the solution near the boundary.

8.4 A Semilinear String Equation

There are in the literature a lot of results on global behavior of solutions to Hamiltonian equations in finite and infinite dimensions. Apart from Poincaré's recurrence theorem and the classical results of Liouville on quasi-periodicilty for most solutions of completely integrable finite-dimensional Hamiltonians, none of the recent results is easy and there is essentially nothing on PDEs except in 1D. Even the case of semilinear string equations is not at all well understood. While looking for almost-periodic solutions (trying to generalize the Rabinowitz theorem on nontrivial periodic solution) I realized that even precompactness of general solutions is unknown for the simplest semilinear string equation in the usual energy space.

Problem 8.15 (1976) For the simple equation

$$u_{tt} - u_{xx} + u^3 = 0 \text{ in } \mathbb{R} \times (0,1), \, u = 0 \text{ on } \mathbb{R} \times \{0,1\}$$

the following simple-looking questions seem to be still open.

Question 1. Are there solutions which converge weakly to 0 as time goes to infinity?

Question 2. If $(u(0,.), u_t(0,.)) \in H^2((0,1)) \cap H_0^1((0,1)) \times H_0^1((0,1)) := \mathcal{V}$, does $(u(t,.), u_t(t,.))$ remain bounded in $\mathcal{V}$ for all times?

Remark 8.16 To understand the difficulty of the problem, let us just mention that the equation

$$iu_t + |u|^2 u = 0 \text{ in } \mathbb{R} \times (0,1), \quad u = 0 \text{ on } \mathbb{R} \times \{0,1\}$$

has many solutions tending weakly to 0 and, although the calculations are less obvious, the same thing probably happens to

$$u_{tt} + u^3 = 0 \text{ in } \mathbb{R} \times (0,1), \; u = 0 \text{ on } \mathbb{R} \times \{0,1\}.$$

Hence the problem appears as a competition between the "good" behavior of the linear string equation and the bad behavior of the distributed ODE associated with the cubic term.

Remark 8.17 If the answer to question 2 is negative, it means that, following the terminology of Bourgain [7], the cubic wave equation on an interval is a weakly turbulent system. Besides, weak convergence to 0 might correspond to an accumulation of steep spatial oscillations of weak amplitude, not contradictory with the energy conservation of solutions.

Remark 8.18 In [11, 12, 13], the authors investigated the problem

$$u_{tt} - u_{xx} + u \int_0^l u^2(t,x)dx = 0 \quad \text{in } \mathbb{R} \times (0,l), \quad u = 0 \text{ on } \mathbb{R} \times \{0,l\}, \quad (8.3)$$

which can be viewed as a simplified model to aid understanding of the above equation. In this case, there is no solution tending weakly to 0, and the answer to question 2 is positive. Interestingly enough, in this case the distributed ODE takes the form $u_{tt} + c^2(t)u = 0$, so that the solution has the form $a(x)u_1(t) + b(x)u_2(t)$ and remains in a two-dimensional vector space! This precludes both weak convergence to 0 and weak turbulence.

8.5 Rate of Decay for Damped Wave Equations

Let us consider the semilinear hyperbolic problem

$$u_{tt} - \Delta u + g(u_t) = 0 \text{ in } \mathbb{R}^+ \times \Omega, \quad u = 0 \text{ on } \mathbb{R}^+ \times \partial\Omega,$$

where Ω is a bounded domain of $\mathbb{R}^N$ and g is a nondecreasing function with $g(0) = 0$. Under some natural growth conditions on g, the initial value problem is well-posed and can be put in the framework of evolution equations generated by a maximal monotone operator in the energy space

$$H_0^1(\Omega) \times L^2(\Omega).$$

An immediate observation is the formal identity

$$\frac{d}{dt}[\int_\Omega (u_t^2 + |\nabla u|^2)dx] = -2\int_\Omega g(u_t)u_t dx \leq 0,$$

showing that the energy of the solution is nonincreasing. When $g(s) = cs$ with $c > 0$, one can prove the exponential decay of the energy by a simple calculation involving a modified energy function

$$E_\varepsilon(t) = \int_\Omega (u_t^2 + |\nabla u|^2)dx + \varepsilon \int_\Omega uu_t dx.$$

The exponential decay is of course optimal since

$$\frac{d}{dt}[\int_\Omega (u_t^2 + |\nabla u|^2)dx] = -2\int_\Omega cu_t^2 dx \geq -2c\int_\Omega (u_t^2 + |\nabla u|^2)dx.$$

A similar calculation can be performed if $0 < c \leq g'(s) \leq C$, and the result is even still valid for $g(s) = cs + a|s|^\alpha s$ under a restriction on $\alpha > 0$ depending on the dimension.

More difficult, and somehow more interesting, is the case

$$g(s) = a|s|^\alpha s, \quad a > 0, \quad \alpha > 0,$$

in which under a restriction relating α and N, various authors (see e.g. [30], [27] and the references therein) obtained the energy estimate

$$\int_\Omega (u_t^2 + |\nabla u|^2)dx \leq C(1+t)^{-\frac{2}{\alpha}}.$$

But now the energy identity only gives

$$\frac{d}{dt}[\int_\Omega (u_t^2 + |\nabla u|^2)dx] = -2\int_\Omega a|u_t|^{\alpha+2}dx,$$

while to prove the optimality of the decay we would need something like

$$\frac{d}{dt}[\int_\Omega (u_t^2 + |\nabla u|^2)dx] \geq -C(\int_\Omega u_t^2 dx)^{1+\frac{\alpha}{2}}.$$

Unfortunately the norm of u_t in $L^{\alpha+2}$ cannot be controlled in terms of the L^2 norm, even if strong restrictions on u_t are known. If u_t is known to be bounded in a strong norm, let us say an L^p norm with p large, we can derive a lower estimate of the type

$$[\int_\Omega (u_t^2 + |\nabla u|^2)dx] \geq \delta(1+t)^{-\beta}$$

for some $\beta > \frac{2}{\alpha}$. But even $p = \infty$ does not allow us to reach the right exponent.

In 1994, using special Lyapunov functions only valid for $N = 1$, the author (see [20]) showed that for all sufficiently regular nontrivial initial data, we have the estimate

$$\int_\Omega (u_t^2 + |\nabla u|^2)dx \geq C(1+t)^{-\frac{3}{\alpha}}.$$

In general, for $N > 2$, some estimate of the form

$$\int_\Omega (u_t^2 + |\nabla u|^2)dx \geq C(1+t)^{-K}$$

will be obtained if the initial data belong to $D(-\Delta) \times H_0^1(\Omega)$ and $\alpha < \frac{4}{N-2}$. But we shall have in all cases $K > \frac{4}{\alpha}$ and K tends to infinity when α approaches the value $\frac{4}{N-2}$.

Remark 8.19 It is perfectly clear that none of the above partial results is satisfactory, since for analogous systems in finite dimensions, of the type

$$u'' + Au + g(u')$$

with A symmetric, coercive, $(g(v), v) \geq c|v|^{\alpha+2}$ and $|g(v)| \leq C|v|^{\alpha+1}$, the exact asymptotic of any nontrivial solution is

$$|u'|^2 + |u|^2 \sim (1+t)^{-\frac{2}{\alpha}}.$$

Moreover, an optimality result of the decay estimate has been obtained in 1D by J. Vancostenoble and P. Martinez [33] in the case of a boundary damping for which the same upper estimate holds. The difference is that inside the domain, an explicit formula gives a lot of information on the solution.

Problem 8.20 For the equation

$$u_{tt} - \Delta u + g(u_t) = 0 \text{ in } \mathbb{R}^+ \times \Omega, \quad u = 0 \text{ on } \mathbb{R}^+ \times \partial\Omega$$

with

$$g(s) = a|s|^\alpha s, \quad a > 0, \alpha > 0,$$

Question 1. can we find a solution u for which

$$|\int_\Omega (u_t^2 + |\nabla u|^2)dx \sim (1+t)^{-\frac{2}{\alpha}}?$$

Question 2. can we find a solution u for which the above property is *not* satisfied?

Remark 8.21 Both questions seem to be still open for any domain and any $\alpha > 0$.

Remark 8.22 One might ask why we are interested in arbitrary values of $\alpha > 0$, since the basic mechanical model corresponds to the damping induced by air and given by $g(u_t) = c_1 u_t + c_2 |u_t| u_t$, in which case we may simplify by considering only one of the two terms. However, for "average" values of the velocity a more accurate approximation would be $c|u_t|^\alpha u_t$ for some $\alpha \in (0,1)$.

8.6 The Resonance Problem for Damped Wave Equations with Source Term

To close this short list, we consider the semilinear hyperbolic problem with source term

$$u_{tt} - \Delta u + g(u_t) = f(t,x) \text{ in } \mathbb{R}^+ \times \Omega, \quad u = 0 \text{ on } \mathbb{R}^+ \times \partial\Omega,$$

where Ω is a bounded domain of $\mathbb{R}^N$. We assume that the exterior force $f(t,x)$ is bounded with values in $L^2(\Omega)$, In this case, all solutions $U = (u, u_t)$ are locally bounded on $(0,T)$ with values in the energy space $H_0^1(\Omega) \times L^2(\Omega)$. The question is what happens as t tends to infinity.

When $g(s)$ behaves like a super-linear power $|s|^\alpha s$ for large values of the velocity, it follows from a method introduced by G. Prouse [32] and extended successively by many authors, among which are M. Biroli [4], [5] and the author of this survey in [25], that the energy of any weak solution remains bounded for t large, under the restriction $\alpha(N-2) \le 4$. Then many attempts were tried to avoid this growth assumption. Many partial results were obtained under additional conditions (f bounded in stronger norms, f anti-periodic, higher growths for $N \le 2$, see e.g. [23], [24], [16]). But the following basic question remains open.

Problem 8.23 Assume $N \ge 3$,

$$g(s) = a|s|^\alpha s, \quad a > 0, \ \alpha > \frac{4}{N-2}.$$

Is it still true that the energy of all solutions remains bounded for any exterior force $f(t,x)$ bounded with values in $L^2(\Omega)$?

Remark 8.24 One might ask why we are interested in arbitrary values of $\alpha > 0$. The same observation as in Remark 8.22 applies here. In addition, the existence of 3D loudspeakers (which did not exist at the time of Amerio and Prouse's seminal paper) gives us another motivation to extend the study to higher dimensions and large values of α, even if the problem with $\alpha > 2$ looks for the moment purely "academic". We consider it important to understand what happens and a counterexample proving that overdamping may occur here would change substantially our understanding of mechanical vibrations and energy transfer phenomena.

Remark 8.25 The positive boundedness results require a weaker boundedness condition on f; it is sufficient that it belongs to a Stepanov space $S^p(\mathbb{R}, L^2(\Omega)$ with $p > 1$. The first results in this direction were actually published by G. Prodi in 1956 (see [31]), so that the problem is about 60 years old.

References

[1] L. Amerio and G. Prouse, *Uniqueness and almost-periodicity theorems for a non linear wave equation*, Atti Accad. Naz. Lincei Rend. Cl. Sci. Fis. Mat. Natur. **8**, 46 (1969) 1–8.

[2] L. Amerio and G. Prouse, *Almost-periodic functions and functional equations*, Van Nostrand Reinhold Co., New York–Toronto, Ont.–Melbourne 1971 viii+184 pp.

[3] S. Angenent, *The zero set of a solution of a parabolic equation*, J. Reine Angew. Math. **390** (1988), 79–96.

[4] M. Biroli, *Bounded or almost periodic solution of the non linear vibrating membrane equation*, Ricerche Mat. **22** (1973), 190–202.

[5] M. Biroli and A. Haraux, *Asymptotic behavior for an almost periodic, strongly dissipative wave equation*, J. Differential Equations **38** (1980), no. 3, 422–440.

[6] S. Bochner and J. Von Neumann, *On compact solutions of operational-differential equations*, Ann. of Math. **2**, 36 (1935), 255–291.

[7] Jean Bourgain, *On the growth in time of higher Sobolev norms of smooth solutions of Hamiltonian*, Internat. Math. Res. Notices **6** (1996), 277–304.

[8] H. Brézis, *Opérateurs maximaux monotones et semi-groupes de contractions dans les espaces de Hilbert* (in French), North-Holland Mathematics Studies, No. 5., Amsterdam–London; American Elsevier Publishing Co., Inc., New York, 1973. vi+183 pp.

[9] T. Cazenave and A. Haraux, *Propriétés oscillatoires des solutions de certaines équations des ondes semi-linéaires*, C. R. Acad. Sci. Paris Ser. I Math. **298** (1984), no. 18, 449–452.

[10] T. Cazenave and A. Haraux, *Oscillatory phenomena associated to semilinear wave equations in one spatial dimension*, Trans. Amer. Math. Soc. **300** (1987), no. 1, 207–233.

[11] T. Cazenave, A. Haraux and F.B. Weissler, *Une équation des ondes complètement intégrable avec non-linéarité homogène de degré trois*, C. R. Acad. Sci. Paris Ser. I Math. **313** (1991), no. 5, 237–241.

[12] T. Cazenave, A. Haraux and F.B. Weissler, *A class of nonlinear, completely integrable abstract wave equations*, J. Dynam. Differential Equations **5** (1993), no. 1, 129–154.

[13] T. Cazenave, A. Haraux and F.B. Weissler, *Detailed asymptotics for a convex Hamiltonian system with two degrees of freedom.*, J. Dynam. Differential Equations **5** (1993), no. 1, 155–187.

[14] M. Comte, A. Haraux and P. Mironescu, *Multiplicity and stability topics in semilinear parabolic equations*, Differential Integral Equations **13** (2000), no. 7–9, 801–811.

[15] C.M. Dafermos and M. Slemrod, *Asymptotic behavior of nonlinear contraction semigroups*, J. Functional Analysis **13** (1973), 97–106.

[16] T. Gallouet, *Sur les injections entre espaces de Sobolev et espaces d'Orlicz et application au comportement à l'infini pour des équations des ondes semi-linéaires* (in French), Portugal. Math. **42** (1983/84), no. 1, 97–112 (1985).

[17] A. Haraux, *Asymptotic behavior of trajectories for some nonautonomous, almost periodic processes*, J. Diff. Eq. **49**(1983), no. 3, 473–483.

[18] A. Haraux, *A simple almost-periodicity criterion and applications*, J. Diff. Eq. **66** (1987), no. 1, 51–61.

[19] A. Haraux, *Asymptotic behavior for two-dimensional, quasi-autonomous, almost periodic evolution equations*, J. Diff. Eq. **66** (1987), no. 1, 62–70.

[20] A. Haraux, *L^p estimates of solutions to some non-linear wave equations in one space dimension*, Int. J. Math. Modelling and Numerical Optimization **1** (2009), Nos 1–2, p. 146–154.

[21] A. Haraux, *On the strong oscillatory behavior of all solutions to some second order evolution equations*, Port. Math. **72** (2015), no. 2, 193–206.

[22] A. Haraux, *Almost-periodic forcing for a wave equation with a nonlinear, local damping term*, Proc. Roy. Soc. Edinburgh Sect. A **94** (1983), no. 3–4, 195–212.

[23] A. Haraux, *Nonresonance for a strongly dissipative wave equation in higher dimensions*, Manuscripta Math. **53** (1985), no. 1–2, 145–166.

[24] A. Haraux, *Anti-periodic solutions of some nonlinear evolution equations*, Manuscripta Math. 63 (1989), no. 4, 479–505.

[25] A. Haraux, *Semi-linear hyperbolic problems in bounded domains*, Math. Rep. 3 (1987), no. 1, i–xxiv and 1–281.

[26] A. Haraux and V. Komornik, *Oscillations of anharmonic Fourier series and the wave equation*, Rev. Mat. Iberoamericana **1** (1985), no. 4, 57–77.

[27] A. Haraux and E. Zuazua, *Decay estimates for some semilinear damped hyperbolic problems*, Arch. Rational Mech. Anal. **100** (1988), no. 2, 191–206.

[28] Z. Hu and A.B. Mingarelli, *Almost periodicity of solutions for almost periodic evolution equations*, Differential Integral Equations **18** (2005), no. 4, 469–480.

[29] C. F. Muckenhoupt, *Almost periodic functions and vibrating systems*, Journal of Mathematical Physics **8** (1928–1929), 163–199.

[30] M. Nakao, *Asymptotic stability of the bounded or almost periodic solution of the wave equation with nonlinear dissipative term*, J. Math. Anal. Appl. **58** (1977), no. 2, 336–343.

[31] G. Prodi, *Soluzioni periodiche di equazioni a derivate parziali di tipo iperbolico non lineari* (in Italian), Ann. Mat. Pura Appl. (4) **42** (1956), 25–49.

[32] G. Prouse, *Soluzioni limitate dell'equazione delle onde non omogenea con termine dissipativo quadratico* (in Italian), Ricerche Mat. **14** (1965), 41–48.

[33] J. Vancostenoble and P. Martinez, *Optimality of energy estimates for the wave equation with nonlinear boundary velocity feedbacks*, SIAM J. Control Optim. 39 (2000), no. 3, 776–797 (electronic).

*UPMC Université Paris 06, CNRS, UMR 7598. 4, place Jussieu 75005, Paris, France.
haraux@ann.jussieu.fr

9

Clustering Phenomena for Linear Perturbation of the Yamabe Equation

Angela Pistoia[*] and Giusi Vaira[†]

This paper is warmly dedicated to Professor Abbas Bahri
on the occasion of his 60th birthday

Let (M,g) be a non-locally conformally flat compact Riemannian manifold with dimension $N \geq 7$. We are interested in finding positive solutions to the linear perturbation of the Yamabe problem

$$-\mathcal{L}_g u + \epsilon u = u^{\frac{N+2}{N-2}} \text{ in } (M,g),$$

where the first eigenvalue of the conformal Laplacian $-\mathcal{L}_g$ is positive and ϵ is a small positive parameter. We prove that for any point $\xi_0 \in M$ which is non-degenerate and non-vanishing minimum point of the Weyl's tensor and for any integer k there exists a family of solutions developing k peaks collapsing at ξ_0 as ϵ goes to zero. In particular, ξ_0 is a non-isolated blow-up point.

Keywords: Yamabe problem, linear perturbation, blow-up points

AMS subject classification: 35J35, 35J60

9.1 Introduction

Let (M,g) be a smooth, compact Riemannian manifold of dimension $N \geq 3$. The Yamabe problem consists of finding metrics of constant scalar curvature in the conformal class of g. It is equivalent to finding a positive solution to the problem

$$\mathcal{L}_g u + \kappa u^{\frac{N+2}{N-2}} = 0 \text{ in } M, \tag{9.1}$$

[*] Angela Pistoia, Dipartimento di Scienze di Base e Applicate per l'Ingegneria, Sapienza Università di Roma
[†] Giusi Vaira, Dipartimento di Scienze di Base e Applicate per l'Ingegneria, Sapienza Università di Roma

for some constant κ. Here $\mathcal{L}_g u := \Delta_g u - \frac{N-2}{4(N-1)} R_g u$ is the conformal Laplacian, Δ_g is the Laplace–Beltrami operator and R_g is the scalar curvature of the manifold .

In particular, if u solves (9.1), then the scalar curvature of the metric $\tilde{g} = u^{\frac{4}{N-2}} g$ is nothing but $\frac{4(N-1)}{N-2} \kappa$. The Yamabe problem was completely solved by Yamabe [25], Aubin [1], Trudinger [24] and Schoen [19] (see also the proof given by Bahri [2]). The solution is unique in the case of negative scalar curvature and it is unique (up to a constant factor) in the case of zero scalar curvature. The uniqueness is not true any more in the case of positive scalar curvature. Indeed, Schoen [20] and Pollack in [15] exhibit examples where a large number of high energy solutions with high Morse index exist. Thus it is natural to ask if the set of solutions is compact or not, as was raised by Schoen in [21]. It is also useful to point out that in the case of the round sphere $(\mathbb{S}^N, g_0)$ the compactness does not hold (see Obata in [14]). Indeed, the scalar curvature $R_{g_0} = N(N-1)$ and the Yamabe problem (9.1) reads as

$$-\Delta_{g_0} u + \frac{N(N-2)}{4} u = u^{\frac{N+2}{N-2}} \text{ in } (\mathbb{S}^N, g_0),$$

which is equivalent (via the stereographic projection) to the equation in Euclidean space

$$-\Delta U = U^{\frac{N+2}{N-2}} \text{ in } \mathbb{R}^N. \tag{9.2}$$

It is known that (9.2) has infinitely many solutions, the so-called *standard bubbles*,

$$U_{\mu,y}(x) = \mu^{-\frac{N-2}{2}} U\left(\frac{x-y}{\mu}\right), x, y \in \mathbb{R}^N, \mu > 0, \text{ where } U(x) := \alpha_N \frac{1}{\left(1 + |x|^2\right)^{\frac{N-2}{2}}}. \tag{9.3}$$

Here $\alpha_N := [N(N-2)]^{\frac{N-2}{4}}$.

The compactness turns out to be true when the dimension of the manifold satisfies $3 \leq N \leq 24$, as was shown by Khuri, Marques and Schoen [9]) (previous results were obtained by Schoen [22], Schoen and Zhang [23], Li and Zhu [11], Li and Zhang [10], Marques [12] and Druet [6]), while it is false when $N \geq 25$ thanks to the examples built by Brendle [4] and Brendle and Marques [5]. The proof of compactness strongly relies on proving sharp pointwise estimates at a blow-up point of the solution. In particular, when compactness holds every sequence of unbounded solutions to (9.1) must blow-up at some points of the manifold which are necessarily isolated and simple, i.e. around each blow-up point ξ_0 the solution can be approximated by a standard bubble (see (9.3))

$$u_n(x) \sim \alpha_N \frac{\mu_n^{\frac{N-2}{2}}}{\left(\mu_n^2 + (d_g(x,\xi_n))^2\right)^{\frac{N-2}{2}}} \quad \text{for some } \xi_n \to \xi_0 \text{ and } \mu_n \to 0.$$

More precisely, let u_n be a sequence of solutions to problem (9.1). We say that u_n blows-up at a point $\xi_0 \in M$ if there exists $\xi_n \in M$ such that $\xi_n \to \xi_0$ and $u_n(\xi_n) \to +\infty$. ξ_0 is said to be a *blow-up point* for u_n. Blow-up points can be classified according to the definitions introduced by Schoen in [21]. $\xi_0 \in M$ is an *isolated blow-up point* for u_n if there exists $\xi_n \in M$ such that ξ_n is a local maximum of u_n, $\xi_n \to \xi_0$, $u_n(\xi_n) \to +\infty$ and there exist $c > 0$ and $R > 0$ such that

$$0 < u_n(x) \le c \frac{1}{d_g(x,\xi_n)^{\frac{N-2}{2}}} \quad \text{for any } x \in B(\xi_0, R).$$

Moreover, $\xi_0 \in M$ is an *isolated simple blow-up point* for u_n if the function

$$\hat{u}_n(r) := r^{\frac{N-2}{2}} \frac{1}{|\partial B(\xi_n, r)|_g} \int_{\partial B(\xi_n, r)} u_n d\sigma_g$$

has exactly one critical point in $(0, R)$.

Motivated by the previous consideration, we are led to study the linear perturbation of the Yamabe problem

$$-\mathcal{L}_g u + \epsilon u = u^{\frac{N+2}{N-2}}, \ u > 0, \ \text{in } (M, g), \tag{9.4}$$

where the first eigenvalue of $-\mathcal{L}_g$ is positive and ϵ is a small parameter. In particular, we address the following questions.

(i) Do there exist solutions to (9.4) which blow-up as $\epsilon \to 0$?
(ii) Do there exist solutions to (9.4) with non-isolated blow-up points, namely with *clustering* blow-up points?
(iii) Do there exist solutions to (9.4) with non-isolated simple blow-up points, specifically with *towering* blow-up points?

Concerning question (i), Druet in [6] proved that equation (9.4) does not have any blowing-up solution when $\epsilon < 0$ and $N = 3, 4, 5$ (except when the manifold is conformally equivalent to the round sphere). It is completely open in the case when the dimension is $N \ge 6$. The situation is completely different when $\epsilon > 0$. Indeed, if $N = 3$ no blowing-up solutions exist, as proved by Li and Zhu [11], while if $N \ge 4$ blowing-up solutions do exist, as shown by Esposito, Pistoia and Vetois in [8]. In particular, if the dimension $N \ge 6$ and the manifold is not locally conformally flat, Esposito, Pistoia and Vetois built solutions which blow-up at non-vanishing stable critical points ξ_0 of the Weyl's tensor, i.e. $|\text{Weyl}_g(\xi_0)|_g \neq 0$. In this paper, we show that the blowing-up point ξ_0 is not isolated as soon as it is a non-degenerate minimum point of the Weyl's tensor. This result gives a positive answer to question (ii). Finally, a positive answer to question (iii) has been given by Morabito, Pistoia and Vaira in [13].

Now, let us state the main result obtained in this paper.

Theorem 9.1 *Let (M, g) be not locally conformally flat and $N \geq 7$. Let $\xi_0 \in M$ be a non-degenerate minimum point of $\xi \to |\text{Weyl}_g(\xi)|_g^2$. Then, for any $k \in \mathbb{N}$, there exist $\xi_\varepsilon^j \in M$ for $j = 1, \ldots, k$ and $\varepsilon_k > 0$ such that for all $\varepsilon \in (0, \varepsilon_k)$ the problem (9.4) has a solution $(u_\varepsilon)_\varepsilon$ with k positive peaks at ξ_ε^j and $\xi_\varepsilon^j \to \xi_0$ as $\varepsilon \to 0$.*

Let us point out that Robert and Vétois in [17] built solutions having clustering blow-up points for a special class of perturbed Yamabe-type equations which look like

$$-\mathcal{L}_g u + \epsilon H u = u^{\frac{N+2}{N-2}}, \ u > 0, \ \text{in } (M, g), \tag{9.5}$$

where the potential H is chosen with k distinct strict local maxima concentrating at a point ξ_0 with $|\text{Weyl}_g(\xi_0)|_g \neq 0$. Indeed, these maxima points generate solutions with k positive peaks collapsing to ξ_0 as ϵ goes to zero. Their result is related to a suitable choice of the potential H, but actually our result shows that the clustering phenomena are intrinsic in the geometry of the manifold.

Let us give an example. The warped product $\left(S^n \times S^m, g_{S^n} \otimes f^2 g_{S^m}\right)$ is the Riemannian manifold $S^n \times S^m$ equipped with the metric $g = g_{S^n} \otimes f^2 g_{S^m}$. Here $f : S^n \to \mathbb{R}$ is a positive function called the warping function. It is easy to see that if the warping function $f \equiv 1$ then the product manifold $(S^n \times S^m, g_{S^n} \otimes g_{S^m})$ has Weyl tensor different from zero at any point. Using similar arguments to the ones used in [16], we can prove that for generic warping functions f close to the constant 1, the Weyl tensor has a non-degenerate and non-vanishing minimum point.

The proof of our result relies on a finite-dimensional Lyapunov–Schmidt reduction, whose main steps are described in Section 9.3 and their proofs are postponed until Section 9.4. Section 9.2 is devoted to recalling some known results.

9.2 Preliminaries

We provide the Sobolev space $H_g^1(M)$ with the scalar product

$$\langle u, v \rangle = \int_M \langle \nabla u, \nabla v \rangle_g \, dv_g + \beta_N \int_M R_g \, uv \, dv_g, \tag{9.6}$$

where dv_g is the volume element of the manifold. Here $\beta_N := \frac{N-2}{4(N-1)}$. We let $\|\cdot\|$ be the norm induced by $\langle\cdot,\cdot\rangle$. Moreover, for any function u in $L^q(M)$, we denote the L^q-norm of u by $\|u\|_q = \left(\int_M |u|^q dv_g\right)^{1/q}$.

We let $\imath^* : L^{\frac{2N}{N+2}}(M) \to H^1_g(M)$ be the adjoint operator of the embedding $\imath : H^1_g(M) \hookrightarrow L^{2^*}(M)$, where $2^* = \frac{2N}{N-2}$, i.e. for any w in $L^{\frac{2N}{N+2}}(M)$, the function $u = \imath^*(w)$ in $H^1_g(M)$ is the unique solution of the equation $-\Delta_g u + \beta_N R_g u = w$ in M. By the continuity of the embedding of $H^1_g(M)$ into $L^{2^*}(M)$, we get

$$\left\|\imath^*(w)\right\| \le C \|w\|_{\frac{2N}{N+2}} \tag{9.7}$$

for some positive constant C independent of w. We rewrite problem (9.4) as

$$u = \imath^*[f(u) - \varepsilon u], \qquad u \in H^1_g(M), \tag{9.8}$$

where we set $f(u) := (u^+)^p$ with $p = \frac{N+2}{N-2}$.

We also define the energy $J_\epsilon : H^1_g(M) \to \mathbb{R}$ by

$$J_\epsilon(u) := \frac{1}{2} \int_M \left(|\nabla_g u|^2 + \beta_N R_g u^2 + \epsilon u^2\right) dv_g - \frac{1}{p+1} \int_M (u^+)^{p+1} dv_g, \tag{9.9}$$

whose critical points are solutions to the problem (9.4).

We are going to use the Euclidean bubble defined in (9.3) on the manifold via a geodesic normal coordinate system around a point $\xi \in M$, i.e.

$$\mathcal{U}_{\mu,\xi}(z) = U_{\mu,0}\left(\exp_\xi^{-1}(z)\right) = \mu^{-\frac{N-2}{2}} U\left(\frac{\exp_\xi^{-1}(z)}{\mu}\right), \ z \in B_g(\xi, r).$$

It is necessary to write the conformal Laplacian in geodesic normal coordinates around the point ξ. In particular, if $x \in B(0, r)$, using standard properties of the exponential map we can write

$$-\Delta_g u = -\Delta u - (g^{ij} - \delta^{ij})\partial^2_{ij} u + g^{ij}\Gamma^k_{ij}\partial_k u, \tag{9.10}$$

with

$$g^{ij}(x) = \delta^{ij}(x) - \frac{1}{3}R_{iabj}(\xi)x_a x_b + O(|x|^3) \text{ and } g^{ij}(x)\Gamma^k_{ij}(x) = \partial_l \Gamma^k_{ii}(\xi)x_l + O(|x|^2). \tag{9.11}$$

Here R_{iabj} denotes the Riemann curvature tensor and Γ^k_{ii} the Christoffel symbols. Therefore, if we compare the conformal Laplacian with the Euclidean

Laplacian of the bubble the error at main order looks like

$$\mathcal{L}_g \mathcal{U}_{\mu,\xi} - \Delta \mathcal{U}_{\mu,\xi} \sim -\frac{1}{3} \sum_{a,b,i,j=1}^{N} R_{iabj}(\xi) x_a x_b \partial_{ij}^2 \mathcal{U}_{\mu,\xi}$$

$$- \sum_{i,l,k=1}^{N} \partial_l \Gamma_{ii}^k(\xi) x_l \partial_k \mathcal{U}_{\mu,\xi} - \beta_N R_g(\xi) \mathcal{U}_{\mu,\xi}.$$

For later purposes, it is necessary to kill this main term by adding to the bubble a higher-order term V which is defined as follows. First, we recall that any solution of the linear equation (see [3])

$$- \Delta v = p U^{p-1} v \quad \text{in } \mathbb{R}^N \tag{9.12}$$

is a linear combination of the functions

$$\psi^0(x) = x \cdot \nabla U(x) + \frac{N-2}{2} U(x) \text{ and } \psi^i(x) = \partial_i U(x), \ i = 1,\dots,N. \tag{9.13}$$

Next, we introduce the higher-order term V, which has been defined in Section 2 in [7].

Proposition 9.2 *For any point* $\xi \in M$, *there exist* $v(\xi) \in \mathbb{R}$ *and a function* $V \in \mathcal{D}^{1,2}(\mathbb{R}^N)$, *the solution to*

$$-\Delta V - f'(U)V = - \sum_{a,b,i,j=1}^{N} \frac{1}{3} R_{iabj}(\xi) x_a x_b \partial_{ij}^2 U - \sum_{i,l,k=1}^{N} \partial_l \Gamma_{ii}^k(\xi) x_l \partial_k U$$

$$- \beta_N R_g(\xi) U + v(\xi) \psi^0 \text{ in } \mathbb{R}^N, \tag{9.14}$$

with

$$\int_{\mathbb{R}^N} V(x) \psi^i(x) dx = 0, \ i = 0,1,\dots,N.$$

Moreover, there exists $C \in \mathbb{R}$ *such that*

$$|V(x)| + |x| |\partial_k V(x)| + |x|^2 |\partial_{ij}^2 V(x)| \leq C \frac{1}{\left(1+|x|^2\right)^{\frac{N-4}{2}}}, \quad x \in \mathbb{R}^N. \tag{9.15}$$

9.3 Clustering

9.3.1 The Ansatz: the Cluster

Let r_0 be a positive real number less than the injectivity radius of M and χ be a smooth cut-off function such that $0 \leq \chi \leq 1$ in $\mathbb{R}$, $\chi \equiv 1$ in $[-r_0/2, r_0/2]$, and $\chi \equiv 0$ out of $[-r_0, r_0]$. Let also η be a smooth cut-off function such that $0 \leq \eta \leq 1$ in $\mathbb{R}$, $\eta \equiv 1$ in $[-1,1]$, and $\eta \equiv 0$ out of $[-2,2]$.

Let $k \geq 1$ be a fixed integer. Assume that $\xi_0 \in M$ is a non-degenerate minimum point of $\xi \to |\text{Weyl}_g(\xi)|_g^2$ with $|\text{Weyl}_g(\xi_0)| \neq 0$, i.e.

$$\nabla_g |\text{Weyl}_g(\xi_0)|_g^2 = 0 \text{ and the quadratic form } \mathcal{Q}(\xi_0)$$

$$:= D_g^2 |\text{Weyl}_g(\xi_0)|_g^2 \text{ is positive definite.} \tag{9.16}$$

Set

$$d_0 := \left(\frac{B_N}{2A_N |\text{Weyl}_g(\xi_0)|_g^2} \right)^{1/2}$$

$$(A_N \text{ and } B_N \text{ are positive constants defined in (9.36))} \tag{9.17}$$

and let us choose

$$\tau_1, \dots, \tau_k \in \mathbb{R}^N \text{ with } \tau_i \neq \tau_j \text{ if } i \neq j \tag{9.18}$$

and for any $i = 1, \dots, k$,

$$\mu_i = \varepsilon^\alpha \left(d_0 + d_i \varepsilon^\beta \right), \text{ where } d_1, \dots, d_k \in [0, +\infty), \ \alpha := \frac{1}{2}, \ \beta := \frac{N-6}{2N}. \tag{9.19}$$

Then, let us define

$$\mathcal{W}_i(z) := \chi(d_g(z, \xi_0)) \mu_i^{-\frac{N-2}{2}} U\left(\frac{\exp_{\xi_0}^{-1}(z) - \varepsilon^\beta \tau_i}{\mu_i} \right) + \mu_i^2 \eta \left(\frac{\left| \exp_{\xi_0}^{-1}(z) - \varepsilon^\beta \tau_i \right|}{\mu_i} \right)$$

$$\times \chi(d_g(z, \xi_0)) \mu_i^{-\frac{N-2}{2}} V\left(\frac{\exp_{\xi_0}^{-1}(z) - \varepsilon^\beta \tau_i}{\mu_i} \right), \ z \in M, \tag{9.20}$$

where the functions U and V are defined, respectively, in (9.3) and (9.14). Set

$$\mathcal{C} := \{(\tau_1, \dots, \tau_k) \in \mathbb{R}^{kN} : \tau_i \neq \tau_j \text{ if } i \neq j\}.$$

We look for solutions of equation (9.4) or (9.8) of the form

$$u_\varepsilon(z) = \sum_{i=1}^k \mathcal{W}_i(z) + \phi_\varepsilon(z), \tag{9.21}$$

where the remainder term ϕ_ε belongs to the space $\mathcal{K}^\perp$ defined as follows. For any $i = 1, \dots, k$ we introduce the functions

$$Z_{j,i}(z) = \chi\left(d_g(z, \xi_0) \right) \mu_i^{-\frac{N-2}{2}} \psi^j \left(\frac{\exp_{\xi_0}^{-1}(z) - \varepsilon^\beta \tau_i}{\mu_i} \right), \ j = 0, 1, \dots, N, \tag{9.22}$$

where the functions ψ^j are defined in (9.13). We define the subspaces

$$\mathcal{K} := \mathrm{Span}\left\{\imath^*\left(Z_{j,i}\right), j = 0, 1, \ldots, N, \ i = 1, \ldots, k\right\}$$

and

$$\mathcal{K}^\perp := \left\{\phi \in H^1_g(M) : \left\langle\phi, \imath^*\left(Z_{j,i}\right)\right\rangle = 0, \quad j = 0, \ldots, N, \ i = 1, \ldots, k\right\}$$

and we also define the projections Π and $\Pi^\perp$ of $H^1_g(M)$ onto $\mathcal{K}$ and $\mathcal{K}^\perp$, respectively.

Therefore, equation (9.8) turns out to be equivalent to the system

$$\Pi^\perp\{u_\varepsilon - \imath^*\left[f(u_\varepsilon) - \varepsilon u_\varepsilon\right]\} = 0, \tag{9.23}$$

$$\Pi\{u_\varepsilon - \imath^*\left[f(u_\varepsilon) - \varepsilon u_\varepsilon\right]\} = 0. \tag{9.24}$$

where u_ε is given in (9.21).

9.3.2 The Remainder Term: Solving Equation (9.23)

In order to find the remainder term ϕ_ε we rewrite (9.23) as

$$\mathcal{E} + \mathcal{L}(\phi_\varepsilon) + \mathcal{N}(\phi_\varepsilon) = 0,$$

where the error term $\mathcal{E}$ is defined by

$$\mathcal{E} := \Pi^\perp\left\{\sum_{i=1}^k W_i - \imath^*\left[f\left(\sum_{i=1}^k W_i\right) - \varepsilon\sum_{i=1}^k W_i\right]\right\}, \tag{9.25}$$

the linear operator $\mathcal{L}$ is defined by

$$\mathcal{L}(\phi_\varepsilon) := \Pi^\perp\left\{\phi_\varepsilon - \imath^*\left[f'\left(\sum_{i=1}^k W_i\right)\phi_\varepsilon - \varepsilon\phi_\varepsilon\right]\right\} \tag{9.26}$$

and the higher-order term $\mathcal{N}$ is defined by

$$\mathcal{N} := \Pi^\perp\left\{-\imath^*\left[f\left(\sum_{i=1}^k W_i + \phi_\varepsilon\right) - f\left(\sum_{i=1}^k W_i\right) - f'\left(\sum_{i=1}^k W_i\right)\phi_\varepsilon\right]\right\}. \tag{9.27}$$

In order to solve equation (9.23), first of all we need to evaluate the $H^1_g(M)-$ norm of the error term $\mathcal{E}$. This is done in the following lemma whose proof is postponed until Section 9.4.

Lemma 9.3 *For any compact subset $A \subset [0, +\infty)^k \times \mathcal{C}$ there exist a positive constant C and $\varepsilon_0 > 0$ such that for any $(d_1, \ldots, d_k, \tau_1, \ldots, \tau_k) \in A$ and for any*

$\varepsilon \in (0, \varepsilon_0)$ *it holds that*

$$\|\mathcal{E}\| \leq C \begin{cases} \varepsilon^{\frac{5}{4}} & \text{if } N = 7, \\ \varepsilon^{\frac{3}{2}} |\ln \varepsilon|^{\frac{5}{8}} & \text{if } N = 8, \\ \varepsilon^{\frac{3}{2}} & \text{if } N \geq 9. \end{cases} \tag{9.28}$$

Next, we need to understand the invertibility of the linear operators $\mathcal{L}$. This is done in the following lemma whose proof can be carried out as in [18].

Lemma 9.4 *For any compact subset $A \subset [0, +\infty)^k \times \mathcal{C}$ there exist a positive constant C and $\varepsilon_0 > 0$ such that for any $(d_1, \ldots, d_k, \tau_1, \ldots, \tau_k) \in A$ and for any $\varepsilon \in (0, \varepsilon_0)$ it holds that*

$$\|\mathcal{L}(\phi)\| \geq C\|\phi\| \text{ for any } \phi \in \mathcal{K}^\perp. \tag{9.29}$$

Finally, we are able to solve equation (9.23). This is done in the following proposition, whose proof is omitted since it relies on a standard contraction mapping argument (see, for instance, [7]).

Proposition 9.5 *For any compact subset $A \subset (0, +\infty)^k \times \mathcal{C}$ there exist a positive constant C and ε_0 such that for $\varepsilon \in (0, \varepsilon_0)$ and for any $(d_1, \ldots, d_k, \tau_1, \ldots, \tau_k) \in A$ there exists a unique function $\phi_\varepsilon \in \mathcal{K}^\perp$ which solves equation (9.23) such that*

$$\|\phi_\varepsilon\| \leq C\varepsilon^{\frac{3(N-2)}{2N} + \zeta} \tag{9.30}$$

for some $\zeta > 0$. Moreover, the map $(d_1, \ldots, d_\ell, \tau_1, \ldots, \tau_k) \to \phi_{\ell, \varepsilon}(d_1, \ldots, d_\ell, \tau_1, \ldots, \tau_k)$ is of class C^1 and

$$\|\nabla_{(d_1, \ldots, d_\ell, \tau_1, \ldots, \tau_k)} \phi_\varepsilon\| \leq C\varepsilon^{\frac{3(N-2)}{2N} + \zeta}$$

for some positive constants C and ζ.

9.3.3 The Reduced Problem: Proof of Theorem 9.1

Let us introduce the reduced energy, defined by

$$\tilde{J}_\varepsilon(d_1, \ldots, d_k, \tau_1, \ldots, \tau_k) := J_\varepsilon\left(\sum_{i=1}^k W_i + \phi_\varepsilon\right),$$
$$(d_1, \ldots, d_k, \tau_1, \ldots, \tau_k) \in [0, +\infty)^k \times (\mathbb{R}^N)^k, \tag{9.31}$$

where the remainder term ϕ_ε is defined in Proposition 9.5.

The following result allows us as usual to reduce our problem to a finite-dimensional one. The proof is standard and it is postponed until Section 9.4.

Proposition 9.6 (i) $\sum_{i=1}^{k} W_i + \phi_\varepsilon$ *is a solution to (9.4) if and only if* $(d_1, \ldots, d_k, \tau_1, \ldots, \tau_k) \in [0, +\infty)^k \times (\mathbb{R}^N)^k$ *is a critical point of the reduced energy (9.31).*

(ii) *The following expansion holds true:*

$$\widetilde{J}_\varepsilon(d_1, \ldots, d_k, \tau_1, \ldots, \tau_k) := kD_N + c(\xi_0)\varepsilon^2 + \varepsilon^{3\frac{N-2}{N}} \mathfrak{J}(d_1, \ldots, d_k, \tau_1, \ldots, \tau_k)$$
$$+ o\left(\varepsilon^{3\frac{N-2}{N}}\right)$$

(9.32)

as $\varepsilon \to 0$, C^0-uniformly with respect to $(d_1, \ldots, d_k, \tau_1, \ldots, \tau_k)$ in compact subsets of $[0, +\infty)^k \times C$. Here $c(\xi_0) := k\left[-A_N |\mathrm{Weyl}_g(\xi_0)|_g^2 d_0^4 + B_N d_0^2\right]$, A_N, B_N, D_N and E_N are positive constants defined in (9.36) and

$$\mathfrak{J}(d_1, \ldots, d_k, \tau_1, \ldots, \tau_k) := -\frac{1}{2} A_N d_0^4 \sum_{i=1}^{k} \mathcal{Q}(\xi_0)(\tau_i, \tau_i)$$

$$- E_N d_0^{N-2} \sum_{\substack{i,j=1 \\ i \neq j}}^{k} \frac{1}{|\tau_i - \tau_j|^{N-2}} - B_N \sum_{i=1}^{k} d_i^2. \quad (9.33)$$

Proof of Theorem 9.1 By (i) of Proposition 9.6, it suffices to find a critical point of the reduced energy $\widetilde{J}_\varepsilon$. Now, the function $\mathfrak{J}$ defined in (9.33) has a maximum point which is stable under C^0-perturbations. Therefore, by (ii) of Proposition 9.6, we deduce that if ε is small enough there exists $(d_{1\varepsilon}, \ldots, d_{k\varepsilon}, \tau_{1\varepsilon}, \ldots, \tau_{k\varepsilon})$, a critical point of $\widetilde{J}_\varepsilon$. That concludes the proof. $\square$

9.4 Appendix

For any $i = 1, \ldots, k$, we set

$$W_i(x) := \mu_i^{-\frac{N-2}{2}} U\left(\frac{x - \varepsilon^\beta \tau_i}{\mu_i}\right) + \eta\left(\frac{|x - \varepsilon^\beta \tau_i|}{\mu_i}\right)$$

$$\times \chi(d_g(z, \xi_0)) \mu_i^{-\frac{N-6}{2}} V\left(\frac{x - \varepsilon^\beta \tau_i}{\mu_i}\right), \quad x \in \mathbb{R}^N.$$

It is important to point out that there exists $c > 0$ such that

$$|W_i(x)| \leq c \frac{\mu_i^{\frac{N-2}{2}}}{|x - \varepsilon^\beta \tau_i|^{N-2}} \quad \forall \, x \in \mathbb{R}^N. \tag{9.34}$$

9.4.1 Proof of Lemma 9.3

It is easy to see that ($\nu(\xi)$ is defined in (9.14))

$$\|\mathcal{E}\| \leq c \sum_{i=1}^k \left| -\Delta_g W_i + (\beta_N R_g + \varepsilon) W_i - \nu(\xi) Z_{0,i} - f(W_i) \right|_{\frac{2N}{N+2}}$$

$$+ c \left| f \left(\sum_{i=1}^k W_i \right) - \sum_{i=1}^k f(W_i) \right|_{\frac{2N}{N+2}}.$$

Arguing exactly as in Lemma 3.1 of [7], we can estimate each term:

$$\left| -\Delta_g W_i + (\beta_N R_g + \varepsilon) W_i - \nu(\xi) Z_{0,i} - f(W_i) \right|_{\frac{2N}{N+2}}$$

$$= \begin{cases} O\left(\varepsilon^{\frac{5}{4}} \right) & \text{if } N = 7, \\ O\left(\varepsilon^{\frac{3}{2}} |\ln \varepsilon|^{\frac{5}{8}} \right) & \text{if } N = 8, \\ O\left(\varepsilon^{\frac{3}{2}} \right) & \text{if } N \geq 9. \end{cases}$$

Next, we show that

$$\left| f \left(\sum_{i=1}^k W_i \right) - \sum_{i=1}^k f(W_i) \right|_{\frac{2N}{N+2}} = O\left(\varepsilon^{3 \frac{N+2}{2N}} \right).$$

Set for any $h = 1, \ldots, k$ $B_h := B(\varepsilon^\beta \tau_h, \varepsilon^\beta \sigma / 2)$, where $\sigma > 0$ and small enough. For (9.18) $B_h \subset B(0, r_0)$ and they are disjoint. We write

$$\left| f \left(\sum_{i=1}^k W_i \right) - \sum_{i=1}^k f(W_i) \right|_{\frac{2N}{N+2}} \leq c \left[\int_{B(0,r_0)} (1 - \chi^{p+1}(|x|)) | \cdots |^{\frac{2N}{N+2}} |g(x)|^{\frac{1}{2}} \, dx \right]^{\frac{N+2}{2N}}$$

$$+ c \left[\int_{B(0,r_0) \backslash \cup_h B_h} | \cdots |^{\frac{2N}{N+2}} |g(x)|^{\frac{1}{2}} \, dx \right]^{\frac{N+2}{2N}} + c \sum_{h=1}^k \left[\int_{B_h} | \cdots |^{\frac{2N}{N+2}} |g(x)|^{\frac{1}{2}} \, dx \right]^{\frac{N+2}{2N}}$$

$$\leq c \sum_{i=1}^k \left[\int_{B(0,r_0)} (1 - \chi^{p+1}(|x|)) |W_i|^{\frac{2N}{N-2}} |g(x)|^{\frac{1}{2}} \, dx \right]^{\frac{N+2}{2N}}$$

$$+ c \left[\int_{B(0,r_0) \backslash \cup_h B_h} |W_i|^{\frac{2N}{N-2}} |g(x)|^{\frac{1}{2}} \, dx \right]^{\frac{N+2}{2N}}$$

$$
+ c \sum_{h=1}^{k} \left[\int_{B_h} \left| W_h^{p-1} \sum_{i \neq h} W_i \right|^{\frac{2N}{N+2}} |g(x)|^{\frac{1}{2}} \, dx \right]^{\frac{N+2}{2N}}
$$

$$
+ c \sum_{h=1}^{k} \left[\int_{B_h} \left| \sum_{i \neq h} W_i \right|^{\frac{2N}{N-2}} |g(x)|^{\frac{1}{2}} \, dx \right]^{\frac{N+2}{2N}},
$$

where $\cdots$ stands for

$$
f\left(\sum_i W_i \right) - \sum_i f(W_i).
$$

Let us estimate each term in the previous expression. We use (9.34).

$$
\sum_{i=1}^{k} \left[\int_{B(0,r_0)} (1 - \chi^{p+1}(|x|)) |W_i|^{\frac{2N}{N-2}} |g(x)|^{\frac{1}{2}} \, dx \right]^{\frac{N+2}{2N}}
$$

$$
\leq c \sum_{i=1}^{k} \left[\int_{\mathbb{R}^N \setminus B(0,r_0)} \frac{\mu_i^N}{|x - \varepsilon^\beta \tau_i|^{2N}} \, dx \right]^{\frac{N+2}{2N}}
$$

$$
\leq c \sum_{i=1}^{k} \frac{\mu_i^{\frac{N+2}{2}}}{\varepsilon^{\beta \frac{N+2}{2}}} \left[\int_{\mathbb{R}^N \setminus B(0,r_0/\varepsilon^\beta)} \frac{1}{|y - \tau_i|^{2N}} \, dy \right]^{\frac{N+2}{2N}}
$$

$$
\leq c \varepsilon^{(\alpha - \beta)\frac{N+2}{2} + \alpha \frac{N+2}{2}} \leq c \varepsilon^{3 \frac{N+2}{2N}},
$$

$$
\left[\int_{B(0,r_0) \setminus \cup_h B_h} |W_i|^{\frac{2N}{N+2}} |g(x)|^{\frac{1}{2}} \, dx \right]^{\frac{N+2}{2N}}
$$

$$
\leq c \frac{\mu_i^{\frac{N+2}{2}}}{\varepsilon^{\beta \frac{N+2}{2}}} \left[\int_{B(0,r_0/\varepsilon^\beta) \setminus \cup_h B(\tau_h, \sigma/2)} \frac{1}{|y - \tau_i|^{2N}} \, dy \right]^{\frac{N+2}{2N}}
$$

$$
\leq C \varepsilon^{(\alpha - \beta)\frac{N+2}{2}} \leq C \varepsilon^{3 \frac{N+2}{2N}},
$$

$$
\sum_{h=1}^{k} \left[\int_{B_h} \left| W_h^{p-1} \sum_{i \neq h} W_i \right|^{\frac{2N}{N+2}} |g(x)|^{\frac{1}{2}} \, dx \right]^{\frac{N+2}{2N}}
$$

$$
\leq c \sum_{h=1}^{k} \sum_{i \neq h} \left[\int_{B_h} \frac{\mu_h^{\frac{4N}{N+2}}}{|x - \varepsilon^\beta \tau_h|^{\frac{8N}{N+2}}} \frac{\mu_i^{\frac{N(N-2)}{N+2}}}{|x - \varepsilon^\beta \tau_i|^{\frac{2N(N-2)}{N+2}}} \, dx \right]^{\frac{N+2}{2N}}
$$

$$\le c \sum_{h=1}^{k} \sum_{i \ne h} \mu_h^2 \mu_i^{\frac{N-2}{2}} \left[\int_{B_h} \frac{1}{|x - \varepsilon^\beta \tau_h|^{\frac{8N}{N+2}}} \, dx \right]^{\frac{N+2}{2N}}$$

$$\le c \sum_{h=1}^{k} \sum_{i \ne h} \mu_h^2 \mu_i^{\frac{N-2}{2}} \left[\int_{B(0, \varepsilon^\beta \sigma/2)} \frac{1}{|y|^{\frac{8N}{N+2}}} \, dy \right]^{\frac{N+2}{2N}}$$

$$\le c \sum_{h=1}^{k} \sum_{i \ne h} \mu_h^2 \mu_i^{\frac{N-2}{2}} \varepsilon^{\beta \frac{N-6}{2}} \le c \varepsilon^{3 \frac{N+2}{2N}}$$

and

$$\sum_{h=1}^{k} \left[\int_{B_h} \left| \sum_{i \ne h} W_i \right|^{\frac{2N}{N-2}} |g(x)|^{\frac{1}{2}} \, dx \right]^{\frac{N+2}{2N}} \le c \sum_{h=1}^{k} \sum_{i \ne h} \left[\int_{B_h} \frac{\mu_i^N}{|x - \varepsilon^\beta \tau_i|^{2N}} \, dx \right]^{\frac{N+2}{2N}}$$

$$\le C \frac{\mu_i^{\frac{N+2}{2}}}{\varepsilon^{\beta \frac{N+2}{2}}} \left[\int_{B(\tau_h, \sigma/2)} \frac{1}{|y - \tau_i|^{2N}} \, dy \right]^{\frac{N+2}{2N}}$$

$$\le c \varepsilon^{3 \frac{N+2}{2N}}.$$

9.4.2 Proof of Proposition 9.6

It is quite standard to prove that

$$J_\varepsilon \left(\sum_{i=1}^{k} \mathcal{W}_i + \phi_\varepsilon \right) = J_\varepsilon \left(\sum_{i=1}^{k} \mathcal{W}_i \right) + \Theta$$

C^0-uniformly with respect to $(d_1, \ldots, d_k, \tau_1, \ldots, \tau_k)$ in a compact subset of $(0, +\infty)^k \times \mathcal{C}$, where Θ is a smooth function such that $|\Theta|, |\nabla \Theta| = O\left(\varepsilon^{3 \frac{N-2}{N} + \zeta} \right)$ for some small $\zeta > 0$. We shall prove that

$$J_\varepsilon \left(\sum_{i=1}^{k} \mathcal{W}_i \right) = k D_N + k \varepsilon^2 \left[-A_N \left| \mathrm{Weyl}_g(\xi_0) \right|_g^2 + B_N d_0^2 \right]$$

$$+ \varepsilon^{3 \frac{N-2}{N}} \left[-\frac{1}{2} A_N d_0^4 \sum_{i=1}^{k} \mathcal{Q}(\xi_0)(\tau_i, \tau_i) - E_N d_0^{N-2} \sum_{\substack{i,j=1 \\ i \ne j}}^{k} \frac{1}{|\tau_i - \tau_j|^{N-2}} - B_N \sum_{i=1}^{k} d_i^2 \right]$$

$$+ \Theta, \tag{9.35}$$

where

$$A_N := \frac{K_N^{-N}}{24N\,(N-4)\,(N-6)}, \quad B_N := \frac{2\,(N-1)\,K_N^{-N}}{N\,(N-2)\,(N-4)}, \quad D_N := \frac{K_N^{-N}}{N},$$

$$E_N := \alpha_N \int_{\mathbb{R}^N} U^p(y)dy \qquad (9.36)$$

and K_N is the best constant for the embedding of $D^{1,2}\left(\mathbb{R}^N\right)$ into $L^{2^*}\left(\mathbb{R}^N\right)$. Here Θ is a smooth function such that $|\Theta|, |\nabla\Theta| = O(\varepsilon^{3\frac{N-2}{N}+\zeta})$ for some small $\zeta > 0$. Let us prove (9.35).

$$J_\varepsilon\left(\sum_{i=1}^k \mathcal{W}_i\right) = \underbrace{\sum_{i=1}^k J_\varepsilon\left(\mathcal{W}_i\right)}_{I} - \underbrace{\sum_{j<i} \int_M f\left(\mathcal{W}_i\right)\mathcal{W}_j dv_g}_{II}$$

$$+ \sum_{i<j} \int_M \left[\nabla_g\mathcal{W}_i\nabla_g\mathcal{W}_j + \beta_N\,R_g\,\mathcal{W}_i\mathcal{W}_j - f\left(\mathcal{W}_i\right)\mathcal{W}_j\right]dv_g$$

$$- \int_M \left[F\left(\sum_{i=1}^k \mathcal{W}_i\right) - \sum_{i=1}^k F\left(\mathcal{W}_i\right) - \sum_{i\neq j} f\left(\mathcal{W}_i\right)\mathcal{W}_j\right]dv_g$$

$$+ \varepsilon\sum_{i<j} \int_M \mathcal{W}_i\mathcal{W}_j dv_g, \qquad (9.37)$$

where $F(t) = \int_0^t f(s)\,ds$.

First of all, we estimate the two leading terms I and II in (9.37).

The term I is given by the contribution of each bubble. Indeed, in Section 4 of [7] it was proved that for any $i = 1, \ldots, k$,

$$J_\varepsilon\left(\mathcal{W}_i\right) = D_N - A_N\left|\text{Weyl}_g\left(\xi_i\right)\right|_g^2 \mu_i^4 + \varepsilon B_N \mu_i^2$$

$$+ \left\{O(\varepsilon^{\frac{5}{2}}) \text{ if } N = 7, \ O(\varepsilon^3|\ln\varepsilon|^3) \text{ if } N = 8, \ O(\varepsilon^3) \text{ if } N \geq 9\right\}. \tag{9.38}$$

Now, by the choice of d_0 in (9.17) and the choice of μ_i, α and β in (9.19), we get

$$\left|\text{Weyl}_g\left(\xi_i\right)\right|^2 = \left|\text{Weyl}_g\left(\xi_0\right)\right|^2 + \frac{1}{2}Q(\xi_0)[\tau_i,\tau_i]\varepsilon^{2\beta} + O\left(\varepsilon^{3\beta}\right),$$

$$\mu_i^4 = \varepsilon^{4\alpha}\left[d_0^4 + 4d_0^3 d_i\varepsilon^\beta + 6d_0^2 d_i^2\varepsilon^{2\beta} + O\left(\varepsilon^{3\beta}\right)\right],$$

$$\mu_i^2 = \varepsilon^{2\alpha}\left[d_0^2 + 2d_0 d_i\varepsilon^\beta + d_i^2\varepsilon^{2\beta}\right].$$

Therefore, a straightforward computation shows that

$$-A_N \left|\text{Weyl}_g\left(\xi_i\right)\right|_g^2 \mu_i^4 + \varepsilon B_N \mu_i^2 = \varepsilon^2 \left[-A_N \left|\text{Weyl}_g\left(\xi_0\right)\right|_g^2 + B_N d_0^2\right]$$

$$+ \varepsilon^{3\frac{N-2}{N}} \left[-\frac{1}{2} A_N d_0^4 Q(\xi_0)(\tau_i, \tau_i) - B_N \sum_{i=1}^{k} d_i^2\right] + O\left(\varepsilon^{\frac{7N-18}{2N}}\right). \quad (9.39)$$

By (9.38) and (9.39) we deduce the estimate of I.

The term II is given by the interaction of different bubbles. For any $h = 1, \dots, k$ let $B_h := B(\varepsilon^\beta \tau_h, \varepsilon^\beta \sigma/2)$. By (9.18) we deduce that $B_h \subset B(0, r_0)$ provided σ is small enough and they are disjoint. Therefore, if $i \neq j$,

$$\int_M f(W_i) W_j \, dv_g = \int_{B_i} f(W_i(x)) W_j(x) |g(x)|^{1/2} \, dx$$

$$+ \int_{B(0,r_0) \setminus B_i} f(W_i(x)) W_j(x) |g(x)|^{1/2} \, dx$$

$$+ \int_{B(0,r_0)} \left[1 - \chi^{p+1}(|x|)\right] f(W_i(x)) W_j(x) |g(x)|^{1/2} \, dx$$

$$= E_N d_0^{N-2} \frac{1}{|\tau_i - \tau_j|^{N-2}} \varepsilon^{3\frac{N-2}{N}} + O\left(\varepsilon^3\right). \quad (9.40)$$

Indeed, the main term of (9.40) is given by

$$\int_{B_i} f(W_i(x)) W_j(x) |g(x)|^{1/2} \, dx$$

$$= \int_{B_i} f\left(\mu_i^{-\frac{N-2}{2}} U\left(\frac{x - \varepsilon^\beta \tau_i}{\mu_i}\right) + \mu_i^{-\frac{N-6}{2}} \eta\left(\frac{x - \varepsilon^\beta \tau_i}{\mu_i}\right) V\left(\frac{x - \varepsilon^\beta \tau_i}{\mu_i}\right)\right)$$

$$\times \left(\mu_j^{-\frac{N-2}{2}} U\left(\frac{x - \varepsilon^\beta \tau_j}{\mu_j}\right) + \mu_i^{-\frac{N-6}{2}} \eta\left(\frac{x - \varepsilon^\beta \tau_j}{\mu_i}\right) V\left(\frac{x - \varepsilon^\beta \tau_j}{\mu_j}\right)\right) |g(x)|^{1/2} \, dx$$

$$\left(\eta\left(\frac{x - \varepsilon^\beta \tau_j}{\mu_j}\right) = 0 \text{ if } x \in B_i \text{ and } \varepsilon \text{ is small enough}\right)$$

$$= \int_{B_i} f\left(\mu_i^{-\frac{N-2}{2}} U\left(\frac{x - \varepsilon^\beta \tau_i}{\mu_i}\right) + \mu_i^{-\frac{N-6}{2}} \eta\left(\frac{x - \varepsilon^\beta \tau_i}{\mu_i}\right) V\left(\frac{x - \varepsilon^\beta \tau_i}{\mu_i}\right)\right)$$

$$\times \left(\mu_j^{-\frac{N-2}{2}} U\left(\frac{x - \varepsilon^\beta \tau_j}{\mu_j}\right)\right) |g(x)|^{1/2} \, dx$$

(setting $x - \varepsilon^\beta \tau_i = \mu_i y$)

$$= \mu_i^{\frac{N-2}{2}} \int_{B(0,\varepsilon^\beta \sigma/2\mu_i)} f\left(U(y) + \mu_i^2 \eta(|y|) V(y)\right) \frac{\alpha_N \mu_j^{\frac{N-2}{2}}}{\left(\mu_j^2 + |\mu_j y + \varepsilon^\beta(\tau_i - \tau_j)|^2\right)^{\frac{N-2}{2}}}$$

$$\times \left|g(\mu_i y + \varepsilon^\beta \tau_i)\right|^{1/2} dy$$

$$= \alpha_N \frac{\mu_i^{\frac{N-2}{2}} \mu_j^{\frac{N-2}{2}}}{\varepsilon^{\beta(N-2)} |\tau_i - \tau_j|^{N-2}} \left[\int_{\mathbb{R}^N} U^p(y) dy + O(\mu_i^2) + O\left(\left(\frac{\mu_i}{\varepsilon^\beta}\right)^N\right)\right]$$

$$= \varepsilon^{3\frac{N-2}{N}} \frac{d_0^{N-2}}{|\tau_i - \tau_j|^{N-2}} \alpha_N \int_{\mathbb{R}^N} U^p(y) dy + O(\varepsilon^3), \tag{9.41}$$

because of the choice of μ_i in (9.19). Moreover, by (9.34)

$$\left|\int_{B(0,r_0)\setminus B_i} f(W_i(x)) W_j(x) |g(x)|^{1/2} dx\right|$$

$$\leq c \int_{\mathbb{R}^N \setminus B_i} \frac{\mu_i^{\frac{N+2}{2}}}{|x - \varepsilon^\beta \tau_i|^{N+2}} \frac{\mu_j^{\frac{N-2}{2}}}{|x - \varepsilon^\beta \tau_j|^{N-2}} dx \ (\text{setting } x = \varepsilon^\beta y)$$

$$\leq c \frac{\mu_i^{\frac{N+2}{2}} \mu_j^{\frac{N-2}{2}}}{\beta^N} \int_{\mathbb{R}^N \setminus B(\tau_i, \sigma/2)} \frac{1}{|y - \tau_i|^{N+2}} \frac{\mu_j^{\frac{N-2}{2}}}{|y - \tau_j|^{N-2}} dy$$

$$= O\left(\frac{\mu_i^{\frac{N+2}{2}} \mu_j^{\frac{N-2}{2}}}{\varepsilon^{\beta N}}\right) = O(\varepsilon^3)$$

and

$$\left|\int_{B(0,r_0)} \left[1 - \chi^{p+1}(|x|)\right] f(W_i(x)) W_j(x) |g(x)|^{1/2} dx\right| = O\left(\varepsilon^{\frac{N}{2}}\right).$$

Finally, let us prove that all the other terms in (9.37) are of higher order. By (9.10) and (9.11), we deduce that

$$\left|\Delta_g W_i + \beta_N R_g W_i - f(W_i)\right| (\exp_{\xi_0}(x)) \leq c \frac{\mu_i^{\frac{N-2}{2}}}{\left(\mu_i^2 + |x - \varepsilon^\beta \tau_i|^2\right)^{\frac{N-2}{2}}}$$

and so by (9.34) if $i \neq j$ we have

$$\left| \int\limits_M \left[\nabla_g \mathcal{W}_i \nabla_g \mathcal{W}_j + \beta_N \operatorname{R}_g \mathcal{W}_i \mathcal{W}_j - f(\mathcal{W}_i) \mathcal{W}_j \right] dv_g \right|$$

$$= \left| \int\limits_M \left[\Delta_g \mathcal{W}_i + \beta_N \operatorname{R}_g \mathcal{W}_i - f(\mathcal{W}_i) \right] \mathcal{W}_j dv_g \right|$$

$$\leq c \int\limits_{B(0,r_0)} \frac{\mu_i^{\frac{N-2}{2}}}{|x - \varepsilon^\beta \tau_i|^{N-2}} \frac{\mu_j^{\frac{N-2}{2}}}{|x - \varepsilon^\beta \tau_j|^{N-2}} dx \text{ (setting } x = \varepsilon^\beta y)$$

$$\leq c \frac{\mu_i^{\frac{N-2}{2}} \mu_j^{\frac{N-2}{2}}}{\varepsilon^{\beta N - 4}} \int\limits_{\mathbb{R}^N} \frac{1}{|y - \tau_i|^{N-2}} \frac{1}{|y - \tau_j|^{N-2}} dx = O(\varepsilon^3).$$

Moreover, if $i \neq j$,

$$\left| \int\limits_M \mathcal{W}_i \mathcal{W}_j dv_g \right| = \left| \int\limits_{B(0,r_0)} \mathcal{W}_i(x) \mathcal{W}_j(x) |g(x)|^{1/2} dx \right|$$

$$\leq c \int\limits_{B(0,r_0)} \frac{\mu_i^{\frac{N-2}{2}}}{|x - \varepsilon^\beta \tau_i|^{N-2}} \frac{\mu_j^{\frac{N-2}{2}}}{|x - \varepsilon^\beta \tau_j|^{N-2}} dx \text{ (setting } x = \varepsilon^\beta y)$$

$$\leq c \frac{\mu_i^{\frac{N-2}{2}} \mu_j^{\frac{N-2}{2}}}{\varepsilon^{\beta N - 4}} \int\limits_{\mathbb{R}^N} \frac{1}{|y - \tau_i|^{N-2}} \frac{1}{|y - \tau_j|^{N-2}} dx = O(\varepsilon^3).$$

Finally, we have

$$\int\limits_M \left[F\left(\sum_{i=1}^k \mathcal{W}_i \right) - \sum_{i=1}^k F(\mathcal{W}_i) - \sum_{i \neq j} f(\mathcal{W}_i) \mathcal{W}_j \right] dv_g$$

$$= \sum_{h=1}^k \int\limits_{B_h} \left[F\left(\sum_{i=1}^k \mathcal{W}_i \right) - \sum_{i=1}^k F(\mathcal{W}_i) - \sum_{i \neq j} f(\mathcal{W}_i) \mathcal{W}_j \right] |g(x)|^{1/2} dx$$

$$+ \int\limits_{B(0,r_0) \setminus \cup_h B_h} \left[F\left(\sum_{i=1}^k \mathcal{W}_i \right) - \sum_{i=1}^k F(\mathcal{W}_i) - \sum_{i \neq j} f(\mathcal{W}_i) \mathcal{W}_j \right] |g(x)|^{1/2} dx$$

$$+ \int\limits_{B(0,r_0)} [1 - \chi^{p+1}(|x|)] \left[F\left(\sum_{i=1}^k \mathcal{W}_i \right) - \sum_{i=1}^k F(\mathcal{W}_i) - \sum_{i \neq j} f(\mathcal{W}_i) \mathcal{W}_j \right] |g(x)|^{1/2} dx.$$

It is immediate that

$$
\int\limits_{B(0,r_0)} \left[1 - \chi^{p+1}\left(|x|\right)\right] \left[F\left(\sum_{i=1}^{k} W_i\right) - \sum_{i=1}^{k} F\left(W_i\right) - \sum_{i \neq j} f\left(W_i\right) W_j\right] |g(x)|^{1/2} \, dx
$$
$$
= O\left(\varepsilon^{\frac{N}{2}}\right).
$$

Moreover, outside the k balls we get

$$
\int\limits_{B(0,r_0)\setminus\cup_h B_h} \left| F\left(\sum_{i=1}^{k} W_i\right) - \sum_{i=1}^{k} F\left(W_i\right) - \sum_{i \neq j} f\left(W_i\right) W_j \right| |g(x)|^{1/2} \, dx
$$
$$
\leq c \sum_{i \neq j} \int\limits_{B(0,r_0)\setminus\cup_h B_h} \left(|W_i|^{p-1} W_j^2 + |W_j|^{p-1} W_i^2\right) dx = O\left(\varepsilon^3\right),
$$

because if $2 < q = p+1 < 3$,

$$
\left|(a+b)^q - a^q - b^q - q a^{q-1} b - q a b^{q-1}\right| \leq c\left(a^2 b^{q-2} + a^{q-2} b^2\right) \text{ for any } a, b > 0
$$

and if $j \neq i$,

$$
\int\limits_{B(0,r_0)\setminus\cup_h B_h} |W_i|^{p-1} W_j^2 \, dx \leq c \int\limits_{B(0,r_0)\setminus\cup_h B_h} \frac{\mu_i^2}{|x - \varepsilon^\beta \tau_i|^4} \frac{\mu_j^{N-2}}{|x - \varepsilon^\beta \tau_j|^{2(N-2)}} \, dx
$$
$$
\text{(setting } x = \varepsilon^\beta y)
$$
$$
\leq c \frac{\mu_i^2 \mu_j^{N-2}}{\varepsilon^{\beta N}} \int\limits_{\mathbb{R}^N \setminus \cup_h B(\tau_h, \sigma/2)} \frac{1}{|y - \tau_i|^4} \frac{1}{|y - \tau_j|^{2(N-2)}} \, dx.
$$

On each ball B_h we also have

$$
\int\limits_{B_h} \left| F\left(\sum_{i=1}^{k} W_i\right) - \sum_{i=1}^{k} F\left(W_i\right) - \sum_{i \neq j} f\left(W_i\right) W_j \right| |g(x)\|^{1/2} \, dx
$$
$$
\leq \int\limits_{B_h} \left| F\left(W_h + \sum_{i \neq h} W_i\right) - F\left(W_h\right) - \sum_{j \neq h} f\left(W_h\right) W_j \right| dx
$$
$$
+ \sum_{i \neq h} \int\limits_{B_h} |F\left(W_i\right)| \, dx + \sum_{\substack{i \neq h \\ j \neq i}} \int\limits_{B_h} |f\left(W_i\right) W_j| \, dx
$$
$$
\leq c \sum_{i \neq h} \int\limits_{B_h} W_h^{p-1} W_i^2 \, dx + c \sum_{i \neq h} \int\limits_{B_h} W_i^{p+1} \, dx + c \sum_{\substack{i \neq h \\ j \neq i}} \int\limits_{B_h} W_i^p W_j \, dx, \quad (9.42)
$$

because if $q = p + 1 \geq 1$,

$$\left| (a+b)^q - a^q - qa^{q-1}b \right| \leq c \left(b^q + a^{q-2}b^2 \right) \text{ for any } a, b > 0.$$

Now we use (9.34) and we get if $i \neq h$,

$$\int_{B_h} W_h^{p-1} W_i^2 dx \leq c \int_{B_h} \frac{\mu_h^2}{|x - \varepsilon^\beta \tau_h|^4} \frac{\mu_i^{N-2}}{|x - \varepsilon^\beta \tau_i|^{2(N-2)}} dx \text{ (setting } x = \varepsilon^\beta y)$$

$$\leq c \frac{\mu_h^2 \mu_i^{N-2}}{\varepsilon^{\beta N}} \int_{B(\tau_h, \sigma/2)} \frac{1}{|y - \tau_h|^4} \frac{1}{|y - \tau_i|^{2(N-2)}} dy = O\left(\varepsilon^3\right),$$

if $j, i \neq h$,

$$\int_{B_h} |W_i|^p W_j dx \leq c \int_{B_h} \frac{\mu_i^{\frac{N+2}{2}}}{|x - \varepsilon^\beta \tau_i|^{N+2}} \frac{\mu_j^{\frac{N-2}{2}}}{|x - \varepsilon^\beta \tau_j|^{N-2}} dx$$

$$\leq c \frac{\mu_i^{\frac{N+2}{2}} \mu_j^{\frac{N-2}{2}}}{\varepsilon^{2\beta N}} |B_h| \leq c \frac{\mu_i^{\frac{N+2}{2}} \mu_j^{\frac{N-2}{2}}}{\varepsilon^{\beta N}} = O\left(\varepsilon^3\right),$$

if $i \neq h$,

$$\int_{B_h} |W_i|^p W_h dx \leq c \int_{B_h} \frac{\mu_i^{\frac{N+2}{2}}}{|x - \varepsilon^\beta \tau_i|^{N+2}} \frac{\mu_h^{\frac{N-2}{2}}}{|x - \varepsilon^\beta \tau_h|^{N-2}} dx$$

$$\leq c \frac{\mu_i^{\frac{N-2}{2}}}{\varepsilon^{\beta(N+2)}} \int_{B_h} \frac{1}{|x - \varepsilon^\beta \tau_h|^{N-2}} dx \leq c \frac{\mu_i^{\frac{N-2}{2}} \mu_h^{\frac{N-2}{2}}}{\varepsilon^{\beta N}} = O\left(\varepsilon^3\right)$$

and if $i \neq h$,

$$\int_{B_h} W_i^{p+1} dx \leq c \int_{B_h} \frac{\mu_i^N}{|x - \varepsilon^\beta \tau_i|^{2N}} dx \leq c \frac{\mu_i^N}{\varepsilon^{2\beta N}} |B_h| \leq c \frac{\mu_i^N}{\varepsilon^{\beta N}} = O\left(\varepsilon^3\right).$$

That concludes the proof.

References

[1] T. Aubin, *Equations différentielles non linéaires et problème de Yamabe concernant la courbure scalaire*, J. Math. Pures Appl. (9) (1976) 269–296.

[2] A. Bahri, *Proof of the Yamabe conjecture, without the positive mass theorem, for locally conformally flat manifolds. Einstein metrics and Yang-Mills connections* (Sanda, 1990), 1–26, Lecture Notes in Pure and Appl. Math., 145, Dekker, New York, 1993.

[3] G. Bianchi and H. Egnell, *A note on the Sobolev inequality*, J. Funct. Anal. 100 (1991) 1, 18–24.

[4] S. Brendle, *Blow-up phenomena for the Yamabe equation*, J. Amer. Math. Soc. 21 (2008), no. 4, 951–979.

[5] S. Brendle and F.C. Marques, *Blow-up phenomena for the Yamabe equation. II*, J. Differential Geom. 81 (2009), no. 2, 225–250.

[6] O. Druet, *Compactness for Yamabe metrics in low dimensions*, Int. Math. Res. Not. 23 (2004), 1143–1191.

[7] P. Esposito and A. Pistoia, *Blowing-up solutions for the Yamabe equation*, Port. Math. 71 (2014), no. 3–4, 249–276.

[8] P. Esposito, A. Pistoia, and J. Vétois, *The effect of linear perturbations on the Yamabe problem*, Math. Ann. 358 (2014), no. 1–2, 511–560.

[9] M.A. Khuri, F.C. Marques and R.M. Schoen, *A compactness theorem for the Yamabe problem*, J. Differential Geom. 81 (2009), no. 1, 143–196.

[10] Y.Y. Li and L. Zhang, *Compactness of solutions to the Yamabe problem. III*, J. Funct. Anal. 245 (2007), no. 2, 438–474.

[11] Y.Y. Li and M.J. Zhu, *Yamabe type equations on three-dimensional Riemannian manifolds*, Commun. Contemp. Math. 1 (1999), no. 1, 1–50.

[12] F.C. Marques, *A priori estimates for the Yamabe problem in the non-locally conformally flat case*, J. Differential Geom. 71 (2005), no. 2, 315–346.

[13] F. Morabito, A. Pistoia and G. Vaira, *Towering phenomena for the Yamabe equation on symmetric manifolds*, Potential Analysis 47 (2017), no. 1, 53–102.

[14] M. Obata, *The conjectures on conformal transformations of Riemannian manifolds*, J. Differential Geom. 6 (1972) 247–258.

[15] D. Pollack, *Nonuniqueness and high energy solutions for a conformally invariant scalar curvature equation*, Comm. Anal. and Geom. 1 (1993) 347–414.

[16] A. Pistoia and G. Vaira, *From periodic ODE's to supercritical PDE's*. Nonlinear Anal. 119 (2015), 330–340.

[17] F. Robert and J. Vétois, *Examples of non-isolated blow-up for perturbations of the scalar curvature equation on non-locally conformally flat manifolds*, J. Differential Geom. 98 (2014), no. 2, 349–356.

[18] F. Robert and J. Vétois, *A general theorem for the construction of blowing-up solutions to some elliptic nonlinear equations via Lyapunov–Schmidt's reduction*, Concentration Compactness and Profile Decomposition (Bangalore, 2011), Trends in Mathematics, Springer, Basel, 2014, pp. 85–116.

[19] R. Schoen, *Conformal deformation of a Riemannian metric to constant scalar curvature*, J. Differential Geometry 20 (1984) 479–495,

[20] R. Schoen, *Variational theory for the total scalar curvature functional for Riemannian metrics and related topics*, in Topics in Calculus of Variations, Berlin Lecture Notes in Mathematics, Springer-Verlag, (Montecatini Terme, 1987), 1365, 1989.

[21] R.M. Schoen, Notes from graduate lecture in Stanford University (1988). http://www.math.washington.edu/pollack/research/Schoen-1988-notes.html.

[22] R.M. Schoen, *On the number of constant scalar curvature metrics in a conformal class*, Differential geometry, Pitman Monogr. Surveys Pure Appl. Math., 52, Longman Sci. Tech., Harlow, 1991, 311–320.

[23] R.M. Schoen and D. Zhang, *Prescribed scalar curvature on the n-sphere*, Calc. Var. and PDEs 4 (1996) 1–25.

[24] N.S. Trudinger, *Remarks concerning the conformal deformation of Riemannian structures on compact manifolds*, Ann. Scuola Norm. Sup. Pisa (3) (1968) 265–274.

[25] H. Yamabe *On a deformation of Riemannian structures on compact manifolds* Osaka Math. J. 12 (1960) 21–37.

*via Antonio Scarpa 16, 00161 Roma, Italy, angela.pistoia@uniroma1.it
†via Antonio Scarpa 16, 00161 Roma, Italy, vaira.giusi@gmail.com

10

Towards Better Mathematical Models for Physics

Luc Tartar*

10.1 Introduction

There are two physical effects which have been noticed by men since time immemorial: *light* and *gravitation*. These two questions of physics took a long time to be put into a mathematical framework, because one had to invent mathematics first, and then define what physics is: it seems that the term physics meant philosophy for Aristotle (384 BCE–322 BCE)[1] and at the time of Newton (1643–1727) physics was called "natural philosophy". However, the questions *what is light?* and *what is gravitation?* are still *not so well understood*, even though mathematicians play with models proposed by physicists, which often involve geometry. Some find it useful to work on questions studied by physicists, who in some cases are somewhat misled, so that *it has become a mathematician's duty to analyze how to improve some models in physics.*

This text was prepared for the conference *Nonlinear Partial Differential Equations arising from Geometry and Physics*, held on March 20–29, 2015, in Hammamet, Tunisia, with the support of MIMS (The Mediterranean Institute for the Mathematical Sciences) and CIMPA (Centre International de Mathématiques Pures et Appliquées), for the occasion of the 60th birthday of Abbas Bahri, whom I warmly congratulate for all his contributions. It was my first visit to Tunisia and it was a pleasure to see Abbas again and to meet a few of his friends who had organized the conference, where I gave six lectures.

* University Professor of Mathematics emeritus, Carnegie Mellon University, Pittsburgh, PA.
[1] BCE = before common era.

10.2 What is a mirror?

Light was first considered a question of *geometry*, maybe because of the *law of reflection* on a mirror, so that leads to the first question: what is a mirror?

I read some archaeological/historical information about mirrors on the internet, with early information coming from an article [2] by J. M. Enoch: the earliest mirrors were pieces of *polished* stone such as obsidian, a naturally occurring volcanic glass, and some obsidian mirrors found in Anatolia (now part of Turkey) were dated to around 6000 BCE, but polished stone mirrors from Central and South America only appear around 2000 BCE.

Mirrors of *polished* copper appear in Mesopotamia around 4000 BCE and in Egypt around 3000 BCE. Mirrors of polished bronze (an alloy of copper and tin) appear in China around 2000 BCE, and mirrors made of other metal *alloys* like speculum metal (two-thirds copper and one-third tin, *white* and highly *reflective*) may have also been produced in China and India.

Metal-coated glass mirrors are said to have been invented in Sidon (now in Lebanon) in the first century, and glass mirrors backed with gold leaf are mentioned by Pliny (23–79) in his *Natural History*, written around 77 AD. The Romans made crude mirrors by coating blown glass with molten lead. The Chinese began making mirrors using silver/mercury amalgams around 500 AD, and the Venetians found a better method of coating glass with a tin/mercury amalgam in the sixteenth century. The invention of the silvered-glass mirror is due to Liebig (1803–1873), but nowadays mirrors are usually produced by the vacuum deposition of aluminum onto the glass.

A thin flat glass is coated on one side with a much thinner metallic substance, *it is the metallic coating which is the mirror*, and it is then held by the transparent glass in front of it, and I wonder why the necessary presence of metal in a mirror had not given a hint earlier that *light must have something to do with electricity*, since metals are known for their high conductivity (of heat or electricity).

Why do I emphasize the words alloy, polished, reflective, white? Because they are linked to physical properties of light, which actually show that *reflection of light is not a question of geometry*!

In 1666, using two[2] prisms, Newton observed that (white) light is composed of the colours of the rainbow, but in 1740 Castel (1688–1757) observed that the sequence of colours depends upon the distance from the prism, which

---

[2] Adding lead to the melted glass gives a glass with a higher index of refraction, and a greater dependence of the index of refraction with respect to colour, so that one prism of such a glass may suffice.

in 1810 led Goethe (1749–1832) to write that Newton had not understood about colours, failing to perceive that Newton was trying to understand a mathematical/physical concept, while he was interested in the perception of colours, which is a matter of physiology: there is a question of how light falling on the retina is transformed into a signal sent to the brain (three shared the 1967 Nobel Prize in Physiology or Medicine for that), and how the brain processes the signal it receives[3] (two shared the 1981 Nobel Prize in Physiology or Medicine for that).

Newton assigned seven colours to the rainbow, using an analogy with the musical scale: red (C = do), orange (D = ré), yellow (E = mi), green (F = fa), blue (G = sol), indigo (A = la), and violet (B = si), and he also bridged the gap between red and violet by adding purple,[4] hence creating a colour wheel (which he was not the first to imagine), but it took two centuries to discover that this idea is not physical, and that UV (ultra-violet) light and IR (infra-red) light are not the same. He understood that the index of refraction (of water or glass) depends upon colour, and he imagined that colours are the way the human eye perceives "differences in energies". His intuition was that light is made of particles, but in a different way than the *light quanta* imagined by Planck (1858–1947, 1918 Nobel Prize in Physics), which Einstein (1879–1955, 1921 Nobel Prize in Physics) used in 1905 for explaining the photoelectric effect.

Newton invented differential calculus,[5] which Leibniz (1646–1716) improved, and he understood the importance of ODEs (ordinary differential equations), but one needs to go further and understand some PDEs (partial differential equations) in order to define what *wavelength* is, which for light is *perceived* as colours, and for sound is *perceived* as the pitch of a musical note.

In order to avoid light being scattered in too many directions, one must polish a mirror to make the defects of the surface much smaller than the wavelength of light, and for visible light the wavelength is approximately between 0.4 micron and 0.8 micron: a white wall reflects most of the visible colours of sunlight, but since the wall is rough at the scale of a micron, it scatters light in many directions, hence the wall does not act as a mirror! At this point, one may still think that reflection of light could be a question of geometry if the surface is smooth enough.

A coloured wall mostly reflects the colour that one sees and absorbs other colours, while a black wall absorbs all visible colours and reflects none

---

[3] I learned this information from Yves Meyer.

[4] The various nuances of purple are obtained by mixing red and violet with different proportions.

[5] Others in different mathematical schools (Chinese, Indian, Persian/Arab) had found similar results before. Europeans may have been more efficient because they considered problems in mechanics/physics.

(since *black is not a colour* but an absence of colour). The proportions of constituents in an alloy may change its colour (as for bronze and speculum metal), but it is not easy to understand why it is so, or at least, it is not an obvious question of geometry.

Even before inventing a scale for temperature, blacksmiths knew that the colour of a piece of iron changes as one heats it, but it was only at the end of the nineteenth century that one could measure a density of radiation depending upon the wavelength, the so-called *blackbody radiation*, which depends upon the "temperature" of the body (which no longer looks black at higher temperatures): it is well approximated by a formula proposed by Planck, of which Einstein gave a curious explanation in terms of statistical mechanics.

Reflection (and absorption) of light depending upon frequency is therefore a complicated question of *interaction of light and matter*; for the moment, I want to concentrate on what light is, and is not.

10.3 Refraction Effects

Ptolemy (85–165) studied refraction around 140, but his formula is only accurate for small incidence angles. It was only in 1990 that Rashed published [5] about two fragments of a 984 manuscript by Ibn Sahl (940–1000), in which the explanations of a drawing clearly show that Ibn Sahl knew an equivalent form of the *law of refraction*. Before this discovery, it was believed that the law had first been found (but not published) in 1621 by Snell (1580–1626), before Descartes (1596–1650) published it in 1637, although I read that Mydorge (1585–1647) explained in 1626 a way to compute the angle of refraction which implies the *sine-law*, and Hariot (1560–1621) is said to have discovered the law of refraction in 1602, but he never published anything (of his 7000 manuscript pages).

Hariot is also said to have built a telescope and pointed it towards the moon before Galileo (1564–1642, whose last name was Galilei), who is credited with having done that after hearing about a Dutch invention of a spy glass: he had to have a good practical knowledge of refraction of light and lenses to build a telescope magnifying about 20 times.

After publishing his sine-law, Descartes spent almost four years writing letters answering objections made by mathematicians, among them Fermat (1601–1665),[6] who apparently did not focus on the physical or metaphysical considerations of Descartes, although I was told something which is about

---

[6] After 1628, Descartes lived in the Dutch province of Holland (with short visits to France in 1644, 1647, and 1648), while Fermat lived in the south of France, in Toulouse.

physics: Descartes must have mentioned a similarity with the propagation of sound in solids, so that Fermat claimed that it is the opposite, that in denser solids sound goes faster while light goes slower.

Mathematicians must be careful when talking about things that are not defined, as is often the case in physics, because *we only know approximations of the real physical laws that govern the Universe*, despite the optimism of physicists, who usually trust their approximations to be the real laws.

Fermat was talking about the speed of sound in a solid, which is not a matter of its density, but of its elastic coefficients (see Appendix C); however, the theory of *linearized elasticity* would not be defined until the first part of the nineteenth century, by Cauchy (1789–1857) and Lamé (1795–1870) in France, and by Piola (1794–1850) in Italy, a little after an estimation of the speed of sound in solids was made by Chladni (1756–1827) in Germany, using his knowledge of music.[7]

Fermat had guessed that light propagates at a finite speed c in a vacuum (or in air) and at speed $\frac{c}{n}$ in a material of index of refraction n, but the first estimation of c was made at the end of the seventeenth century, just after the opening of the observatory in Paris, directed by Cassini (1625–1712): Rømer (1644–1710) used an idea of Cassini concerning discrepancies about the moons of Jupiter, and a crucial estimation of the size of the solar system, following a joint measurement by Cassini in Paris and another assistant in Cayenne, Guyana.

Fermat knew that the index of refraction of water is about 1.33 (it changes with colour, as shown by rainbows), and that the index of refraction of glass is about 1.5 (addition of lead in the melted glass increases the index of refraction, up to 1.7), but wood is usually lighter than water and not transparent, so its index of refraction is infinite.

Elastic properties are consequences of arrangements of atoms and forces at a *microscopic* level, and light is a question of *electromagnetism*, whose theory was first developed by Maxwell (1831–1879) in the second part of the nineteenth century, and which Heaviside (1850–1925) made more practical afterward (so that I call the Maxwell–Heaviside equation what others call the Maxwell equation), and an index of refraction reflects a question of *interaction of light and matter*, a twentieth-century question which is *not so well understood*.

---

[7] Chladni noticed that hitting a bar made of metal gives a note whose frequency permits one (knowing the length of the bar) to compute the speed of sound in that metal: hitting a bar of copper gives a higher pitch than when hitting a bar of lead, showing that the sound velocity in lead is slower than the speed of sound in copper, hence Fermat was wrong to think that the velocity increases with the density of the metal.

Around 60 AD, Heron (10–75) expressed that, in reflection of light by a mirror, light travels along the path of least length. Of course, *Heron's principle is not correct*, since when there is more than one mirror (or only one adequately curved mirror) one may see various copies of a reflected image, corresponding to light traveling different lengths.

Why then did Fermat restart the controversy in 1657 with the followers of Descartes (who had died in 1650), by proposing a principle of least time (now named after him) for refraction (or reflection) of light? Like Heron's principle, *Fermat's principle is not correct*. As differential calculus was not yet officially invented, although Fermat had a good intuition of it,[8] it was not so easy to show that the principle implies the law of sines, and it was only in 1662 that Fermat clearly showed it.[9]

On a Riemannian manifold, when one looks for the shortest path from one point to another, or just a stationary path for the length under smooth perturbations (with fixed end-points), one finds a second-order differential equation: given a point and a tangent direction chosen at that point, it defines a unique curve called a geodesic (which minimizes length between two nearby points). A *beam of light* is a similar concept: given a point and a direction, the beam of light is defined by solving an ordinary differential equation (using the index of refraction, which may vary in space), and there may be a few directions at a point a for the geodesic to go through a point b, corresponding to different times for going from a to b.

In 1669, Bartholin (1625–1698) discovered an effect of *double refraction*, or birefringence, through a crystal from Iceland (Iceland spar, which is calcite, i.e. crystallized $CaCO_3$): for one incident ray of light, it shows two refracted rays (for most directions). In 1690, Huygens (1629–1695) presented his remarks on the wave nature of light and his derivation of the law of sines, but although he presented his own measurements on a birefringent crystal, he was not able to apply his argument and derive a rule for the two refracted rays.

Huygens's argument (see Appendix A) was to consider a plane wave-front propagating with velocity c in the medium of index 1 (air) with incidence angle i, above[10] a medium of index n (water) where light travels at velocity $\frac{c}{n}$, the front touching the interface at $A(t)$.[11] He imagined that $A(t)$ generates a

---

[8] Although Fermat died before the official invention of differential calculus, one attributes to him the fact that at a minimum of a function (of one real variable) its derivative is 0 and its second derivative is ≥ 0.

[9] Nowadays, functional analysis (developed in the early twentieth century) makes such computations easy.

[10] The drawing in the 984 manuscript of Ibn Sahl has a vertical interface: left is air, right is glass!

[11] $A(t)$ moves at "apparent speed" $\frac{c}{\sin i}$, which is $> c$.

centered wave in the medium below, and then the envelope of the corresponding spherical fronts (generated at times $s < t$) is a plane wave-front making a refraction angle r such that $\sin i = n \sin r$.

It is curious that Huygens forgot that $A(t)$ generates a centered wave in the medium above, and that the envelope of the corresponding spherical fronts above is a reflected wave (with reflection angle i), since if one looks at a pool of (transparent) water one more easily sees the reflection of the sky than what is in the bottom of the pool, although the proportion of reflected and refracted light depends upon the incidence angle, a question which Huygens could not address with his geometrical model. However, this was not the reason why Huygens's contribution was not noticed: he advocated a *wave nature for light*, in opposition to Newton, who advocated a particle nature for light.

I have gathered in Appendix A (*Refraction*) a few technical computations concerning the question of reflection and refraction at a plane interface.

10.4 Diffraction Effects

If one tries to create a beam of light, i.e. make light travel in an extremely narrow (possibly curved) "cylinder", by making light go through a small hole, it does not work. Leonardo (1452–1519) seems to have observed this, but it was Grimaldi (1618–1663) who first studied this effect and called it *diffraction*, his observations being published after his death, in 1665. Newton was aware of Grimaldi's work, but differed in his interpretation, since he considered light to be made of "particles", and because of the cult of personality around him which developed in England, one had to wait until 1802 for (Thomas) Young (1773–1829) to revive the work of Huygens on the *wave nature of light*, in a famous experiment with two nearby tiny holes: he observed *interferences*, considered a decisive "proof" of the wave nature of light.

Académie des Sciences in Paris asked for contributions on the subject of diffraction for an 1819 prize,[12] the jury being headed by Arago (1786–1853), who favoured the wave nature of light; the other members of the jury, Biot (1774–1862), Gay-Lussac (1778–1850), Laplace (1749–1827), and Poisson (1781–1840) were of Newton's point of view.

I am not sure which equations Fresnel (1788–1827) used to study diffraction, but he thought that light propagates by "transverse waves"; Poisson worked out some other consequences besides those Fresnel derived in his memoir, and he deduced an "absurd" consequence of Fresnel's ideas: there would be a bright

---

[12] I learned about this question from Michel Gondran, who had kindly sent me a few chapters of the book [4] that he was writing with his son (Alexandre).

spot in the center of the shadow of a circular (opaque) disk. However, Arago asked for a precise experiment, which showed that indeed the bright spot is there, and Fresnel won the prize. This bright spot is more often named after Poisson or Arago than Fresnel, but it had actually been observed a century earlier by an astronomer, Maraldi (1665–1729), and one also credits his student Delisle (1688–1768).

Since I had read in the mid-1980s about GTD (Geometric Theory of Diffraction), which Joe Keller (1996/97 Wolf Prize) developed in the 1950s, I was not surprised by this effect, although the theory is formal and has not yet been put into a clear mathematical setting. Actually, Joe Keller told me that his theory wrongly predicts an infinite amplitude on caustics, but he also mentioned that the effect of grazing rays (or light creeping into the shadow) in GTD bears some resemblance to the tunneling effect in quantum mechanics, except that *light does not go through the obstacle, but around it*: the Poisson/Arago spot is a good example of that effect, and there are some convincing computations about it in the book [4] that Michel Gondran wrote with his son.

10.5 The Scalar Wave Equation

D'Alembert (1717–1783) wrote the one-dimensional wave equation in 1747, motivated by music, perhaps because he was a friend of Rameau (1683–1764): he only considered periodic initial data; the one-dimensional medium in which the wave propagated was a string in a musical instrument (harpsichord or violin), corresponding to a *transverse wave*: the wave propagates along the string, but the material points move in a direction almost perpendicular to it. Euler (1707–1783) generalized his work in 1749 for arbitrary initial data, which seems to have created controversy, and (Daniel) Bernoulli (1700–1782) joined the argument in 1755. The pitch of the note, hence the velocity of the wave, depends upon the tension in the string, and I heard that D. Bernoulli studied this effect with a model of many small masses connected with springs (see Appendix B), but I do not know when he did it; actually, his model seems more adapted to studying a *longitudinal wave*: the material points move in the direction of propagation of the wave.

I have included in Appendix B (*The vibrating string*) some computations about this idea of D. Bernoulli, and how the one-dimensional wave equation results in the limit, but this uses ideas of functional analysis, which I learned as a student in courses by Jacques-Louis Lions (1928–2001).

I have included in Appendix C (*Linearized elasticity*) similar computations for linearized elasticity in N (≥ 2) dimensions, alongside ideas of Cauchy and Lamé, and I discuss there longitudinal P-waves and transverse S-waves.

In 1766 Euler wrote the three-dimensional wave equation in an isotropic medium (for which the propagation speed is the same in all directions), and in 1818 Poisson found an explicit formula for its solution. I do not know who first considered a wave equation in an anisotropic medium.

10.6 The Polarization of Light

In 1809, Malus (1775–1812) observed polarized light. It is not so easy to make a similar observation nowadays, due to an improvement in glass-making technology: the silica melt must not have been hot enough, so that when one rolled it in one direction, one obtained an anisotropic medium, hence window panes cut out after the glass cooled down had different properties according to their orientation; Malus observed the reflected light of the sun on a distant window through his own window, and there was an angle for which the distant window became black, as if no light was reflected (or the reflected light did not go through his window).

One year later he published a theory for the double refraction of light in some crystals. Maybe it was useful for polarization too, but finding an efficient practical formula does not mean that one understands what is going on: it is acceptable for an engineer, who must become efficient quickly, but it should not be the method of a scientist whose goal is to discover the laws of nature, and who should recognize the approximative character of all the laws one writes due to the limitation of the mathematical knowledge of the time!

10.7 Electromagnetism

In the *MacTutor History of Mathematics archive* created by O'Connor and Robertson,[13] they mention in their article about Heaviside that he greatly simplified 20 equations in 20 variables written by Maxwell, and replaced them by four equations in two variables, and they quote the opinion of Fitzgerald (1851–1901): "Maxwell's treatise is cumbered with the debris of his brilliant lines of assault, of his entrenched camps, of his battles", followed by "Oliver Heaviside has cleared these away, has opened up a direct route, has made a broad road, and has explored a considerable trace of country".

Since what one usually calls the "Maxwell equation" is actually the simpli-fied system obtained by Heaviside, I prefer to call it the *Maxwell–Heaviside* equation instead.

[13] www-history.mcs.st-and.ac.uk

There was a reason for Maxwell to imagine a complicated mechanistic model for unifying electricity and magnetism: like many physicists, he believed in a curious medium called *aether*. I was taught that aether was "needed" because "electromagnetic waves are transverse waves", but the only definition which I was taught afterward about transverse or longitudinal waves was about plane-waves in elasticity.

Maybe the concept of aether followed the approach of Fresnel, who had imagined (before the work of Cauchy and Lamé in linearized elasticity) that light is a transverse wave, and obtained good results in this way for his analysis of diffraction, but in the beginning of his work Maxwell did not know that light is an electromagnetic wave. O'Connor and Robertson write in their article about Maxwell that his 1855 article *On Faraday's lines of force* contained an extension and mathematical formulation of the theories of electricity and magnetic lines of force of Faraday (1791–1867), showing that a few relatively simple mathematical equations could express the behaviour of electric and magnetic fields and their interrelation. It was only in 1862 that Maxwell calculated the speed of propagation of an electromagnetic field and found it similar to the speed of light c, so he proposed that light is an electromagnetic phenomenon, writing "We can scarcely avoid the conclusion that light consists in the transverse undulations of the same medium which is the cause of electric and magnetic phenomena".

As a mathematician I am surprised that physicists often confuse 'A implies B' with 'B implies A', and Maxwell's use of the term "conclusion" is a typical error of this kind (which I call pseudo-logic). Using a hypothetic medium called aether, he had proposed a mechanistic model for explaining electricity and magnetism, so that he used terms pertaining to displacements in solids (as in the theory of elasticity), but the fact that his model predicted something observed does not mean that nature uses this model!

In their article about Maxwell, O'Connor and Robertson also write that the four partial differential equations now known as Maxwell's equations first appeared in fully developed form in *Electricity and Magnetism* (1873). This contradicts what they wrote about Heaviside, who seems to have studied this particular work of Maxwell in order to simplify it, and part of his work was to develop a calculus for vectors, contrary to the fashion at the time, which was to work with the quaternions[14] defined by Hamilton (1805–1865).

---

[14] Quaternions have the form $U = r + x\mathbf{i} + y\mathbf{j} + z\mathbf{k}$ with $r,x,y,z \in \mathbb{R}$, so that it is $\mathbb{R}^4$, but with a bilinear multiplication ($\mathbf{i}^2 = \mathbf{j}^2 = \mathbf{k}^2 = \mathbf{ijk} = -1$, so that $\mathbf{ij} = \mathbf{k} = -\mathbf{ji}$, $\mathbf{jk} = \mathbf{i} = -\mathbf{kj}$, $\mathbf{ki} = \mathbf{j} = -\mathbf{ik}$) which makes it a non-commutative division ring (non-zero elements have an inverse for multiplication).

Although the Maxwell–Heaviside equation permits one to explain the birefringence of some crystals, as well as polarized light, a few other results about light can be described in the simplified framework using a scalar wave equation, which has sometimes lured mathematicians and physicists into using wrong equations.

An example of such a mistake is Einstein's "explanation" of the bending of light rays near the sun, since he used the same approach as Hilbert (1862–1943) who was studying gravitation in a framework of Riemannian manifolds. Light being an electromagnetic phenomenon, he should have used the Maxwell–Heaviside equation and realized that it is natural for grazing rays to behave as if they were bent by refraction effects because there are intense electromagnetic effects near the surface of the sun, but it is a difficult matter to derive a correct list of boundary conditions to use at the surface of the sun, since one does not really know in detail the electromagnetic effects which occur there; however, he should have expected that, as for an index of refraction which depends upon colour, this question is highly frequency dependent.

Besides Fresnel's idea about diffraction, which involved Poisson in a theoretical way, and Arago in an experimental way, I have seen the name of Airy (1801–1892) attached to the question of *light creeping into the shadow*, and he may have introduced the Airy function for that question.[15]

For diffraction, Fresnel defined the *Fresnel number* as $F = \frac{a^2}{L\lambda}$, with a a characteristic size of the aperture, L the distance from the aperture to the screen, and λ the wavelength of the light: the *Fresnel diffraction* (or near-field diffraction) corresponds to $F \gg 1$ (i.e. large), while the *Fraunhofer diffraction* (or far-field diffraction) corresponds to a Fresnel number $F \ll 1$ (i.e. small), but Fraunhofer (1787–1826) is also known for inventing the spectroscope (in 1814) and identifying hundreds of dark "lines" in the solar spectrum, a subject that went on to play an important role in quantum mechanics, which developed in the twentieth century in a quite illogical way.

I am not sure who (in the nineteenth century) introduced the approach (called geometrical optics) of considering solutions of a scalar wave equation (usually in an isotropic medium, with varying index of refraction) having the form $A(x,t) e^{i \psi(x,t)}$ for an amplitude A and a phase ψ, i.e. solutions looking like *distorted plane-waves*, and considering an asymptotic expansion with respect to a (frequency) parameter ν tending to ∞, so that the phase is $\nu \psi_0 + \psi_1 + \nu^{-1}\psi_2 + \cdots$, and the amplitude is $A_0 + \nu^{-1}A_1 + \cdots$. Identifying the terms of order ν in the equation shows that ψ_0 must solve an equation called

---

[15] It is defined by $\mathrm{Ai}(x) = \lim_{m \to \infty} \frac{1}{2\pi} \int_{-m}^{+m} e^{i(xt + t^3/3)} \, dt$ for $x \in \mathbb{R}$: it solves $\frac{d^2 \mathrm{Ai}}{dx^2} + x\,\mathrm{Ai} = 0$ on $\mathbb{R}$, and it decays exponentially fast as $x \to +\infty$.

the *eikonal equation*, which is a Hamilton–Jacobi equation, and ψ_0 usually presents singularities on so called *caustics*;[16] identifying the next term shows that A_0 satisfies a *transport equation*, and since the gradient of ψ_0 appears in the coefficients of this equation,[17] one may expect the approximation to be valid only away from caustics.

Given a point x and a direction ξ, one may define curves in the (x, ξ) space, called bicharacteristic rays (which solve a differential equation depending upon the symbol of the wave equation), along which the amplitude A_0 is transported. These curves are precisely the intuitive "light rays" that appear in the approach of Huygens, and they are curved if the coefficients of the wave equation vary smoothly. These (formal) computations are only about distorted plane-waves, and the caustics are precisely the points around which the solution does not look like a distorted plane-wave, so that it is not a situation where the energy (among other things) is only transported along one curve, as in an idealized "beam of light"!

A mathematician must be careful not to confuse something which is observed with mathematical properties of the solution of an equation, supposed to describe the physical phenomenon under scrutiny: proving (and not guessing) mathematical results incompatible with observations is the only way to reject a model, although if a model has a few properties which are observed, it does not mean that nature uses this model!

One observed that there is some "light inside the shadow",[18] but it is a different matter to prove that the solution of the wave equation corresponding to a plane-wave arriving on a bounded obstacle is not exactly zero in the (geometrical) shadow. I do not know who first proved this, but I learned about it by reading articles by Peter Lax (1987 Wolf Prize, 2005 Abel Prize), Cathleen Morawetz, and Walter Strauss for a graduate course that I taught at UW (University of Wisconsin, Madison, WI) in 1974–1975.

Actually, one should prove such results for the Maxwell–Heaviside equation, since real light is not described by a scalar wave equation!

Unifying a few explicit examples previously computed by different people, Joe Keller developed in the 1950s an extension of the method of geometrical optics, which he called GTD (geometric theory of diffraction), by considering rays hitting a vertex or an edge, but also grazing rays (tangent to a smooth part of the boundary), and how they can follow the (geodesics along the)

---

[16] Near caustics, the solution does not look like a distorted plane-wave.

[17] Only the direction of the (spatial) gradient of ψ_0 matters.

[18] The shadow is defined in terms of geometry, i.e. the points which cannot be seen in a direct line from one of the sources of light, and for a plane-wave the source must be thought of as being at infinity.

boundary for a while on a convex part (so that specialists talk of *creeping rays*) before they leave by taking the tangent, and this description holds for the scalar wave equation, corresponding to questions in acoustics, or for the Maxwell–Heaviside equation, corresponding to applications to antennas and radar waves. GTD is still only at a formal level, and Joe Keller mentioned that his theory wrongly predicts an infinite amplitude at caustics.

The behaviour of creeping rays is dependent upon which boundary conditions are used, such as the Dirichlet condition or the Neumann condition in the acoustic approximation (i.e. for the scalar wave equation). The intensity of grazing rays decreases exponentially with the arc-length s of contact to a convex part of the boundary, with a factor proportional to the cubic root of the *wave number* $k = \frac{2\pi}{\lambda}$ (where λ is the wavelength). Since the argument of an exponential must be a number, and a wave number has for dimension the inverse of a length, I guess that another length scale must appear, maybe related to the radius of curvature. It then seems that there is a boundary layer around an obstacle illuminated by a plane-wave, but there seems to be also another boundary layer, of a different thickness, at the boundary of the shadow.

Joe Keller told me that his computations have some relation with the *tunneling effect* mentioned in quantum mechanics, except that light does not go "through the obstacle", but *around the obstacle*.

10.8 Boundary Conditions

The one-dimensional wave equation for a string comes with a Dirichlet condition, since the variable is a displacement, and the ends of the string are fixed, but for other musical instruments the variable is the pressure (of air) and the three-dimensional wave equation has different boundary conditions; in some cases, one restricts attention to a one-dimensional approximation with Dirichlet or Neumann conditions, but one rarely explains which approximations are made, since the boundary condition on a surface made of wood or metal (like copper or brass) should take into account its elastic properties, for example.

When Fourier (1768–1830) studied questions of conduction of heat at the beginning of the nineteenth century, he used a Dirichlet condition at places where the temperature is given, and a Neumann condition at an interface with a perfect insulator, but he introduced a third type, corresponding to the temperature being given on the outside of a good conducting container. However, instead of calling it a Fourier condition many people speak of a Robin condition, even though Robin (1855–1897) lived in the second half of the nineteenth century.

10.9 Frequency Dependent Coefficients

The two basic unknowns in the Maxwell–Heaviside equation are the *electric field E* and the *magnetic field H*, but it also uses two other important vectors, the (electric) polarization field *D* and the (magnetic) induction field *B*, which in the *vacuum* are related by $D = \varepsilon_0 E$ and $B = \mu_0 H$, where ε_0 is the dielectric permittivity (of the vacuum) and μ_0 is the magnetic susceptibility (of the vacuum). This implies that the speed of light *c* (of propagation of all electromagnetic waves) satisfies $\varepsilon_0 \mu_0 c^2 = 1$.

Although it may seem natural to pair *E* and *D* on one side and *H* and *B* on the other, there is a way to write the Maxwell–Heaviside equation using differential forms (see Appendix M), which I learned from Joel Robbin, where *E* and *B* are the coefficients of a closed 2-form,[19] and *H* and *D* are the coefficients of another 2-form.[20]

In the presence of *matter*, one uses the *constitutive relation* $D = \varepsilon E$ and $B = \mu H$, where the dielectric permittivity ε and the magnetic susceptibility μ are symmetric positive definite tensors, but one always avoids discussing what matter could be, besides the intuition of it which physicists have taught.

One observes that the index of refraction (of transparent isotropic materials) depends upon colour, so that, at least for the wavelengths corresponding to visible light, ε and μ could depend upon frequency.

Costas Dafermos mentioned to me that Heaviside considered a constitutive relation where *D* depends upon previous values of *E* at the same point (and similarly for *B* in terms of *H*), i.e. there is a convolution in *t*: by applying a Laplace transform it corresponds to ε and μ depending upon the Laplace variable *p*.

I consider that these models are just approximations of the real laws that nature follows, since one made a logical mistake, also present in the development of quantum mechanics: *an interaction* (here of light and matter) *cannot be described by a linear equation*!

Since light moves very fast, and macroscopic bodies move slowly, it was a reasonable first approximation to consider that matter is fixed, and that light is just slowed down by the presence of obstacles, but what kind of obstacles are there?

The measurements were often made using monochromatic waves, implicitly assuming that one dealt with a linear (or linearized) equation, and a

---

[19] It is the exterior derivative of a 1-form: its coefficients are the scalar and vector potentials *V* and *A*.

[20] Its exterior derivative is a 3-form whose coefficients are the density of (electric) charge ρ and the density of current *j*, and that this 3-form is closed expresses the conservation of (electric) charge.

decomposition into monochromatic waves would then show what the result would be for a general wave. The Fourier transform gives such a decomposition in the whole space, but which equation is behind it, and what kind of boundary conditions must one use?

Anyway, was matter thought of as the rigid bodies used in eighteenth-century classical mechanics?

10.10 Conservation Laws

In 1807, Poisson studied rarefaction waves in a gas with equation of state $p = C\rho^\gamma$ (maybe at the suggestion of Laplace) for correcting a discrepancy about the speed of sound when one uses the Boyle–Mariotte law $p = C\rho$.[21] In 1848, Challis (1803–1882) noticed something wrong with the solution (in implicit form) found by Poisson when one uses periodic initial data, and Stokes (1819–1903) explained that the profile of a solution gets steeper and steeper until it approaches a discontinuous solution. As a consequence of conservation of mass and the balance of momentum he derived the correct jump conditions for discontinuous solutions, now called the Rankine–Hugoniot conditions, probably because Stokes did not reproduce his 1848 derivation of the jump conditions when he edited his complete works in 1880, apologizing for his "mistake", since he had been (wrongly) convinced by Lord Rayleigh (1842–1919, 2004 Nobel Prize in Physics) and by Thomson (1824–1907, Lord Kelvin after 1892), that his discontinuous solutions were not physical because they did not conserve energy.

It shows that these great scientists did not understand at the time that the "missing energy" had been transformed into *heat*. Heat (or equivalently *internal energy*) is too loose a notion, since it only means the energy which has "disappeared" from our macroscopic level, and which is stored at various mesoscopic levels, in various forms, each moving around according to its nature, and it is unnatural to expect that a sum of disparate things could satisfy an equation. Thermodynamics, which grew out of this question, is not a well-chosen name, since it is not about the dynamics of heat, but more about what happens near equilibrium.

I have included in Appendix D (*Velocity of sound*) a few technical points concerning hyperbolic systems of conservation laws.

---

[21] The sound propagates too fast for equilibrium in temperature to occur, hence the process is not at constant temperature (so that the Boyle–Mariotte law does not apply): there is no time for heat exchange, so that the process is called adiabatic (i.e. without heat transfer), a term equivalent to isentropic (since the second law of thermodynamics is $\delta Q = \theta\, ds$, and $\theta > 0$, so that $\delta Q = 0$ is equivalent to $ds = 0$).

10.11 Particles and Kinetic Theory

Both Maxwell and Boltzmann (1844–1906) developed a kinetic theory of gases, considering a gas made of particles, studying a density $f(x,v,t)$ of particles at $x \in \mathbb{R}^3$ with velocity $v \in \mathbb{R}^3$ and deducing some fluid quantities by integration in v: the density of *mass* is $\rho = \int f(\cdot,v,\cdot)\,dv$, the density of *linear momentum* has components $P_i = \rho\,u_i = \int f(\cdot,v,\cdot)\,v_i\,dv$,[22] the *internal energy* per unit of mass e is defined by $\frac{\rho\,|u|^2}{2} + \rho\,e = \int f(\cdot,v,\cdot)\,\frac{|v|^2}{2}\,dv$, and the *Cauchy stress tensor* has components $\sigma_{i,j} = -\int f(\cdot,v,\cdot)\,(v_i - u_i)\,(v_j - u_j)dv$.

At equilibrium, one has $\sigma_{i,j} = -p\,\delta_{i,j}$, where p is the *pressure* (and $\delta_{i,j}$ the Kronecker symbol), so that $3p = 2\rho\,e$, hence one cannot expect to describe many real gases with this model!

In the Maxwell–Boltzmann equation, a collision kernel summarizes the probabilities of the output of a "collision" among possible resulting pairs of velocities conserving linear momentum and kinetic energy, and precise kernels follow from a given law of interaction (depending upon distance) between two particles, and one hope was that one could deduce which type of interaction exists at microscopic level from some macroscopic measurement, such as the *viscosity* of a gas. A formal computation by Hilbert predicts the Euler equation (of an inviscid fluid), but Chapman (1888–1970) and Enskog (1884–1947) proposed a different formal procedure, which predicts the Navier–Stokes equation with a small viscosity.

In the 1890s, Boltzmann proved his H-theorem, that $\int f \log(f)\,dx\,dv$ is non-increasing in time, in relation with the concept of *entropy* proposed by Clausius (1822–1888) in 1865, which shows that the Maxwell–Boltzmann equation cannot be derived from a conservative mechanistic model!

Boltzmann generalized his results concerning the equilibria of the Maxwell–Boltzmann equation into the principles of statistical mechanics (which Gibbs (1839–1903) also arrived at independently), whose main defect is to be based on a single parameter (the temperature), so that it cannot really be as relevant to non-equilibrium situations. Curiously, in the middle of the twentieth century, it was proposed to deal with non-equilibrium situations by using the Maxwell–Boltzmann equation, as if no one had taken the time to describe its known defects.

Many cases of irreversible behaviours were known, but no one seemed aware that the Maxwell–Boltzmann equation could not explain the origin of any irreversible behaviour, since a particular type of irreversibility was

---

[22] In electromagnetism, the analog of the linear momentum is the density of current j, but one does not use a "macroscopic velocity", since there is a huge difference in mass between (heavy) ions and (light) electrons.

postulated by introducing probabilities for the output of an "almost collision" of two (identical) particles.

If there are only two particles, which should face a near collision if there were no exterior forces, the minimum distance in their (free) paths seems an important impact parameter, and specialists talk of a *scattering cross section* outside which not much interaction takes place, but the smaller the minimum distance is the larger the deflection angle becomes, and since an important acceleration occurs it seems curious to have assumed that "collisions are instantaneous".

For a force in *distance*$^{-2}$, trajectories are hyperbolas, in contrast with elliptic trajectories observed in the solar system since Kepler (1571–1630) introduced his laws, based on the precise measurements of Brahe (1546–1601), and improving the description by circles rolling on circles of the heliocentric model of Copernicus (1473–1543) and the geocentric model of Ptolemy. The reasoning behind the Maxwell–Boltzmann equation then concerns a *rarefied gas*, but as soon as one considers a less rarefied gas it seems important to consider the formation of groups of particles dancing together for a while: the formal idea for describing a fluid to let a parameter like the Knudsen number tend to infinity in the Maxwell–Boltzmann equation then seems a mistake.[23]

10.12 Real and Fake Brownian Motion

In 1785, Ingenhousz (1730–1799) described the irregular movement of coal dust on the surface of alcohol, while in 1827 (Robert) Brown (1773–1858) observed under his microscope that some grains of pollen suspended in water ejected minute particles which went through a continuous irregular motion. Brown then observed the same motion in particles of inorganic matter, ruling out that the effect could be life-related.

Maybe the (real) *Brownian motion* should be called the Ingenhousz–Brown motion, but the reason why I call it real is to distinguish it from a "fake Brownian motion", not related to what Ingenhousz or R. Brown observed, and the widespread confusion between the two notions seems to be Einstein's fault!

In 1900, in order to model the buying and selling at a stock exchange market, Bachelier (1870–1946) used the analogy of a drunkard's random (one-dimensional) walk, in a formal way, and Thiele (1838–1910) seems to have considered a similar random walk game in 1880. Wiener (1894–1964) first developed the complete mathematical theory of this random walk game, which Feller (1906–1970) called a *Bachelier–Wiener* process.

[23] The inverse of the Knudsen number is similar to a "mean free path between collisions".

Other probabilists chose to call it a Wiener process, but much later people started using the wrong name, "Brownian motion": Einstein made a mistake in using the 1900 formal framework of Bachelier and claiming that it represented the irregular motion observed by R. Brown, a sign that he confused the "unphysical" jumps in position used by Bachelier for the "physical" jumps in velocity observed by R. Brown, and since a jump in velocity implies a change of linear momentum, which is a conserved quantity, he should have said that it implies a "collision with something"!

I have heard people say that the particles that R. Brown looked at collided with particles too small to be seen under his microscope, but since microscopes have gone through many improvements since, and I have not read that such tinier particles were observed, it seems that the "collisions" were with waves!

10.13 Relativity

In 1862, Maxwell introduced what is now called the *Lorentz force*, 30 years before H.A. Lorentz (1853–1928, 1902 Nobel Prize in Physics): a particle with an electric charge q in an electromagnetic field feels the force $q(E + v \times B)$, where v is the velocity of the particle.

It is not clear to me why it was thought that electromagnetic waves are transverse, since the only definition of longitudinal or transverse waves that I have heard of involves displacements, such as in elasticity.

I have included in Appendix E (*Are electromagnetic waves transverse?*) some technical remarks based on the "Lorentz force" (without relativistic effects) in order to show that it is somewhat wrong to call electromagnetic waves transverse, since they may sweep some charged particles in front of them.

It seems that aether was thought to exist, and be a "solid", precisely because people thought that electromagnetic waves are transverse, and in elasticity transverse waves exist in solids but not in liquids, although there is no reason to confuse the Cauchy–Lamé equation of (linearized) elasticity with the Maxwell–Heaviside equation of electromagnetism.

In 1887, Michelson (1852–1931, 1907 Nobel Prize in Physics) and Morley (1838–1923) carried out a carefully planned series of measurements of the speed of light c for testing the theory of a "solid" aether: if one moved in a frame with fixed velocity w with respect to a fixed frame, one expected that light traveling with velocity c in either direction in the moving frame would result in its traveling with speed $c + w$ or $c - w$ in the fixed frame. With the idea that a fixed frame called aether existed, they made a quite precise experiment with light moving in perpendicular arms, and creating interferences, and the pattern of the interferences should have changed by rotating the experiment;

to avoid vibrations, the experiment was set up in the basement of a stone building, on top of a huge block of marble floating in a pool of mercury, for the experiment to be rotated easily. Michelson and Morley arrived at a quite intriguing result, since nothing significant was detected, so that the theory of aether was shattered, and it was admitted that light travels with speed c in the vacuum, in every direction, and in every frame moving at constant velocity.

In 1892, Fitzgerald made a bold suggestion: the results of the experiment could be explained by the contraction of a body along its direction of motion. It was then found by Lorentz, who went further by introducing also a local change of time. (Henri) Poincaré (1854–1912) then looked at transformations leaving the wave equation invariant:[24] he called the group of such transformations the *Lorentz group*.

I have put in Appendix F (*Invariants of the wave equation, and the "Lorentz" group of Poincaré*) some technical remarks on this question of group of invariance.

I only read in 1979 in lecture notes [3] by Feynman (1918–1988, 1965 Nobel Prize in Physics) that the theory of (special) relativity was due to Poincaré; I guess that one often avoids mentioning Poincaré's name for political reasons.

Between 1898 and 1900, Poincaré investigated the question that since the Lorentz force is an action, there must be a reaction, which is another way of saying that since it implies a change of linear momentum, which is a conserved quantity, there must be an opposite change of linear momentum somewhere: he concluded in 1900 that *the energy of the electromagnetic field is equivalent to a density of mass*, according to the formula $e = mc^2$, which Einstein used in 1905 for a different reason.

A consequence of the work of Lorentz and then Poincaré is that, since time changes from one frame to another, *instantaneity does not make sense*: a consequence is that *there is no universal time* (hence the term relativity), so that *there cannot be any instantaneous force acting at a distance* (like gravitation)!

A possible explanation is that "particles" give the information where they are located through a messenger field which transports the information at finite speed (probably the speed of light c): it suggests that the basic systems of partial differential equations for describing nature are hyperbolic.

10.14 What is Matter?

I was taught in the mid-1960s that physicists "knew" in the early twentieth century that "elementary particles" are not points, since a point would

---

[24] I believe that Poincaré checked the transformations leaving the Maxwell–Heaviside equation invariant.

"radiate energy" (whatever that means). I was then taught quantum mechanics, without any criticism of the Schrödinger equation, or of the formalism which uses (linear) operators in Hilbert space,[25] following ideas of von Neumann (1903–1957), or the efficient notation of "bras" and "kets" by Dirac (1902–1984, 1933 Nobel Prize in Physics), or the curious use of probabilities, or of "particles" sometimes being waves.

In 1983, I read that Dirac wrote in 1928 a hyperbolic system of equations for one "relativistic equation". I disagreed with his addition of a mass term (using the "rest mass" of the electron) in the equation, thinking it should appear because of a *homogenization* effect, and the mass of the electron would then just be the electromagnetic energy stored inside the wave which one calls an electron,[26] and with the fact that the Dirac equation only describes "one" electron, and that electrons could "decide" (or be told) which equation to follow depending upon their speed being near c or not! I also read that as early as 1924, (Louis) de Broglie (1892–1987, 1929 Nobel Prize in Physics) had proposed that "particles" are waves, and have a wavelength: Dirac was then treating the case of an electron, while Schrödinger (1887–1961, 1933 Nobel Prize in Physics) deduced a simpler equation for the "non-relativistic case".[27]

As for the name of Poincaré, which my physics teachers did not mention for (special) relativity, avoiding a mention of de Broglie or a part of the work of Dirac also seems to have political reasons.

When "scientists" behave in this manner, very similar to that of religious fundamentalists, who want to impose the inventions of preceding generations without people using their brain to analyze what they are told, it makes the detective work of mathematicians more difficult to determine *which are the serious problems to solve in order to understand how nature works*, hence *which new mathematical tools should one develop*.

The eighteenth-century point of view is that of classical mechanics, using ODEs and a finite number of degrees of freedom, like for classical particles.

The nineteenth-century point of view is that of continuum mechanics, using PDEs and infinitely many degrees of freedom, like for waves.

Mixing these two points of view is to go backwards, and this is what Lorentz and Poincaré were doing when they tried to put electrons as point singularities

---

[25] How can one expect to describe an interaction (of light and matter) by using a linear framework?

[26] Various waves are called "electrons", and measuring their masses might not give the same number.

[27] I then read that formally letting the speed of light c tend to infinity in the Dirac equation makes the Schrödinger equation appear, so that the Schrödinger equation (in that case) just looks like an approximation where "$c = +\infty$", hence it becomes a logical error to claim that the Schrödinger equation may predict that some effects travel faster than the real speed of light.

in the wave equation or the Maxwell–Heaviside equation. They thought that the mass of an electron could not be entirely of electromagnetic origin, but it was difficult for them to imagine that electrons are waves.

What was done in quantum mechanics also went backwards, when one imagined curious objects which could be sometimes particles and sometimes waves. Actually, starting with a Hamiltonian system is a backward (eighteenth-century) point of view, and it is a better (nineteenth-century) point of view to start from PDE systems. It was difficult to imagine what the twentieth-century point of view would be, as I only perceived it at the end of the twentieth century, and it forces us to go *beyond PDEs*, but in a way which is not so precise at the moment.

The twentieth-century point of view in mechanics or physics is to consider *PDEs with small scales*, as in atomic physics, meteorology, phase transitions, plasticity, turbulence (in alphabetic order).

The mathematical question is to derive *effective equations*, and I call this theory GTH, the *General Theory of Homogenization* [6]. It is absolutely crucial to *avoid probability hypotheses* in homogenization, and it is also better to avoid believing in a world with only one scale, like in the periodically modulated framework, although it was such a work by Évariste Sanchez-Palencia which enabled me to understand the connection of my previous work (in the early 1970s) with François Murat to questions of mixtures in mechanics or physics. Our work, motivated by an academic problem of optimal design, extended the work of Sergio Spagnolo from the late 1960s, which he called G-convergence since it is the convergence of Green kernels, and we called our extension H-convergence; H reminds one of the term homogenization, borrowed from Ivo Babuška, who himself borrowed it from the engineering literature.

Crucial in our approach was the development of our theory of *compensated compactness* (a not-so-well-chosen term, proposed by our common advisor, Jacques-Louis Lions), which I generalized in the late 1980s into the theory of *H-measures*: although I introduced these *microlocal* measures for computing the first (quadratic) correction in *small amplitude homogenization*, I then used them for defining (at last) what a *beam of light* means from a mathematical point of view, and for some (linear) systems of PDEs with adequate conservation laws I derived an ODE which describes the transport of oscillations (and concentration effects) in the solutions presenting high frequency parts.

One should not confuse my ideas with those of Lars Hörmander (1931–2012, 1962 Fields Medal, 1988 Wolf Prize), who introduced in the early 1970s the notion of *microlocal regularity* and the *wave front set* (or *essential singular support*), generalizing the *singular support* of a distribution introduced by Laurent Schwartz (1915–2002, 1950 Fields Medal): the supporters of Lars Hörmander (wrongly) call his results "propagation of singularities", while, as Mike Crandall pointed out to me, they are results of *propagation of microlocal regularity*, of which I do not know any relevance to physics.

It is not that one starts from a Hamiltonian and creates a PDE using the rules of quantum mechanics, but that one should start from special systems of PDEs and derive the simpler equations (probably ODEs with maybe a Hamiltonian structure) which govern the part of the solutions of the system which present large frequencies, since that is the case when one talks about *waves behaving as "particles"*.

However, there is one step further to go, which is that one should extend my results to semi-linear systems, and I conjecture that such an extension will involve questions of geometry.

One such question is related to the *shape of elementary particles*, in the spirit of the guesses of Bostick (1916–1991) concerning electrons or photons. I think that a good candidate for these "elementary particles" and many others is the Dirac equation with no mass term coupled with the Maxwell–Heaviside equation: it is *conformally invariant*, hence it may provide interesting effects at plenty of scales.

Unfortunately, before such a program can be carried out, one needs good existence theorems, which allow for a sequence of initial data to "oscillate", so that one needs to go far beyond the results of Yvonne Bruhat,[28] at least those which I heard around 1980, where she proved global existence for data sufficiently small in some Sobolev space, using a particular conformal transformation introduced by Roger Penrose (1988 Wolf Prize in Physics).

For carrying out such a program, I have introduced ideas of *compensated integrability* and *compensated regularity*, which certainly need to be improved, by finding an adapted geometrical framework.

10.15 Concluding Remarks

In order to progress *towards better mathematical models for physics*, I then propose to teach classical results from continuum mechanics or physics without repeating the errors which were made, and which have become apparent since, and then to see which kind of mathematical tools one should develop.

Unfortunately, it is hard to convince non-mathematicians, since they very often confuse '*A* implies *B*' and '*B* implies *A*': try to make a

---

[28] She signs her articles Yvonne Choquet-Bruhat, using in front of her name that of her husband, Gustave Choquet (1915–2006).

non-mathematician understand an elementary point of logic, that 'if A is false', then 'A implies B' is true. Someone who catches this point may then understand why mathematicians insist that if one finds mistakes in the hypotheses of a game, it is of no use playing such a game, even if it predicts things which are observed, at least if one wants to pretend to be a scientist!

It is unfortunately true that it is difficult to go in a different direction than the majority. I have annoyed many by mentioning that their preferred game is not good physics, that thermodynamics and quantum mechanics are flawed, but I have not said that everything in these theories should be thrown away.

Since Poincaré wrote a note at the Comptes Rendus de l'Académie des Sciences in Paris before Einstein published anything in special relativity, his name should be mentioned.

Some claim that Poincaré was not good enough at physics to see some arguments which Einstein found later, but I strongly disagree with such unethical behaviour of forgetting to quote Poincaré: omitting his name is like claiming that it is OK to steal a car when the owner is not a good driver.

It seems to be that Hilbert wrote on gravitation in a framework of Riemannian manifolds before Einstein published anything in general relativity, and his name should be mentioned.

Einstein seems to be the only one to blame for claiming that light rays are bent by gravitation: light is electromagnetism, described by the Maxwell–Heaviside equation, with a boundary condition at the surface of the sun to be discovered, since the intense electromagnetic effects there are not so well understood.

In the 1970s, Alfvén (1908–1995, 1970 Nobel Prize in Physics) said that some observations in the cosmos seem to result from electromagnetic effects and not gravitational ones, but "specialists of gravitation" prefer to invent dark matter (and other dark things) rather than learn about electromagnetism (a sign that they were not really trained as physicists), or acknowledge that Einstein made a few mistakes!

My explanation of such a behaviour is what I call a Comte complex, resulting from the classification of sciences by (Auguste) Comte (1798–1857): 1- mathematics, 2- astronomy, 3- physics, 4- chemistry, 5- biology. Comte was good at mathematics since he entered the École Polytechnique at a high rank, and for him astronomy must have been what we now call celestial mechanics, probably the applied mathematics of the time, and it may be useful to recall that Gauss (1777–1855), so famous as to have been called the "king of mathematicians", worked at the observatory in Göttingen, and invented quadrature formulas to avoid losing too much time computing the trajectories of planets.

It is not only silly to imagine a complete order among the various sciences, since being creative in mathematics, physics, chemistry, and biology seems to require different kinds of intuition, but one should observe that many mathematicians and physicists do not even behave as scientists, as if they felt superior because their branch appears high in Comte's classification. I read a book of translations (by his daughter) into English of the letters which Born (1882–1970, 1954 Nobel Prize in Physics) wrote to Einstein, and I was surprised that he almost considered that Einstein had a kind of "direct connection to God", despite Einstein's errors in physics which I had already discovered at the time I read these letters. It was not the first time that I had seen physicists show this "Comte complex" of being amazed by their colleagues who used mathematical theories which they ignore, even though any good student could find flaws in the physics.

In a commencement address given at Caltech (California Institute of Technology, Pasadena, CA) in 1974, Feynman noted:

"We have learned a lot from experience about how to handle some of the ways we fool ourselves. One example: Millikan measured the charge on an electron by an experiment with falling oil drops, and got an answer which we now know not to be quite right. It's a little bit off because he had the incorrect value for the viscosity of air. It's interesting to look at the history of measurements of the charge of an electron, after Millikan. If you plot them as a function of time, you find that one is a little bit bigger than Millikan's, and the next one's a little bit bigger than that, and the next one's a little bit bigger than that, until finally they settle down to a number which is higher.

Why didn't they discover the new number was higher right away? It's a thing that scientists are ashamed of – this history – because it's apparent that people did things like this. When they got a number that was too high above Millikan's, they thought something must be wrong – and they would look for and find a reason why something might be wrong. When they got a number close to Millikan's value they didn't look so hard. And so they eliminated the numbers that were too far off, and did other things like that ..."

It is not clear to me whether Millikan (1868–1953, 1923 Nobel Prize in Physics) is to blame, but many experimentalists who followed and did not mention that their measurements were quite different from those of Millikan should be blamed from a purely ethical point of view, but one must realize that they feared for their career if they openly criticized Millikan.

It is not an easy task to have to criticize a famous physicist, and I am certainly not a diplomat in describing some mistakes that Einstein made, but it does not mean that everything he wrote was wrong![29]

At the time I gave the lectures in Hammamet, I had not found the time to read the book *Mécanique quantique: Et si Einstein et de Broglie avaient aussi raison?* [4], which my friend Michel Gondran wrote with his son (Alexandre), and I had only read the few chapters which he had kindly sent me before the book was printed. In their book they describe the Solvay conference of 1927, where the point of view on quantum mechanics which prevailed was that of (Niels) Bohr (1885–1962, 1922 Nobel Prize in Physics), Born, Heisenberg (1901–1976, 1932 Nobel Prize in Physics), and Pauli (1900–1958, 1945 Nobel Prize in Physics), but not that of L. de Broglie, Einstein, and Schrödinger.

One may compare this with what happened at the council of Nicaea in 325, where the trinitarian point of view of Christianity prevailed (and was then imposed as a dogma to all Christians since), to understand my claim that many "scientists" behave like religious fundamentalists.

It does not make sense that a majority of people should accept a dogma concerning the functioning of the Universe if common sense shows a flaw in their arguments: for quantum mechanics, one cannot describe an interaction (here of light and matter) with a linear equation, so why not point out this main flaw of quantum mechanics to new generations of students?

Why present examples of a particle in a square box without saying that a discontinuous electric potential V is unphysical, since the electric field E (which is $-\text{grad}(V)$) should be locally square integrable, and that even a smoother V would evolve because of the Maxwell–Heaviside equation, so that one can explain the game about freezing V because one wants to study a much faster phenomenon than the evolution of V, so that physics is inherently multi-scale!

A few years ago, I heard a (recent) Nobel laureate in Physics give a talk at CMU, and near the end he said that problems in biology are more difficult than problems in physics, since problems in biology have many scales, while problems in physics always have one scale! I wonder if he will issue a corrected statement when he realizes that even students in mathematics are taught how silly he was.

It is quite dangerous for one's academic career to be opposed to the ideas of a ruling majority, but the situation is not as bad as with some religious fundamentalists, who know next to nothing about God but are ready to kill

---

[29] Many point out that Einstein was not so careful in his computations, which I find a quite minor point: since Einstein was not a mathematician, he was not bound to prove what he asserted, and a physicist's ideas are more important than the detail of his computations, but I point out errors in the physics!

those who disagree with the ideas they were brainwashed with! Those who want to trust their religious dogmas concerning creation should learn enough logic to understand that when I write that the question of creation versus no creation is undecidable, it does not contradict what they believe: it just shows that they follow people who postulated things about questions which they did not understand at all!

The question of whether the Universe always existed or was created is mathematically undecidable, as a consequence of the accepted law that energy cannot travel faster than light.

Those who think that the Schrödinger equation (which does not show a finite propagation effect) permits things to travel faster than light should reflect on the point that the Schrödinger equation must be considered a simplified model corresponding to having let the parameter c (the speed of light) tend to infinity, and after that one has $c = +\infty$ in the model. It is not always bad, since many questions show velocities much smaller than c, but inside an atom things do occur at the speed of light, and one must resolve the following paradox: using his methods of diagrams, Feynman claimed that he was solving the Schrödinger equation and he correctly predicted the outcome of an interaction between particles, although the Schrödinger equation cannot be correct in such situations. Had he (without saying it) been working with a kind of Dirac equation, which is relativistic, or is it that his summation of diagrams diverges, but he kept the "correct terms" in order to find the solution of another (better) equation?[30]

Since the 1950s, physicists have boasted that they would soon be able to control thermonuclear fusion. I believe the main reason they failed is that no one knows the laws of matter at millions of degrees, so why be so silly as to play a big-bang game at billions of degrees?

What follows are 13 appendices, and I have already mentioned in the text Appendix A (*Refraction*), Appendix B (*The vibrating string*), Appendix C (*Linearized elasticity*), Appendix D (*Velocity of sound*), Appendix E (*Are electromagnetic waves transverse?*), and Appendix F (*Invariants of the wave equation, and the "Lorentz" group of Poincaré*).

Appendix G (*What is matter?*) contains my conjectures about this important question, remembering that since particles are waves (as L. de Broglie correctly guessed in his 1924 PhD thesis) one must work with systems of PDEs which are hyperbolic (with characteristic velocities $\leq c$), and all the

---

[30] It is a classical exercise that if a real sequence a_n tends to 0, but $\sum_n (a_n)_+ = \sum_n (a_n)_- = +\infty$, then for every $b \in \mathbb{R}$ there exists a bijection f from $\mathbb{N}$ onto itself such that $\sum_n a_{f(n)} = b$.

conjectures of physicists must be shown to follow from such models (or a necessary enlarged class for reasons of homogenization) if one wants to avoid the logical mistake of confusing '*A* implies *B*' with '*B* implies *A*' (i.e. having just chosen a game whose output shows the observed symmetries of so-called elementary particles).

Appendix H (*Why are elementary snowflakes flat?*) and Appendix I (*Are quasi-crystals flat?*) serve to convince us that the question of latent heat has not been correctly defined, and that the rules of thermodynamics may be good only for slow processes, and are wrong for fast processes.

Appendix J (*Non-local effects induced by homogenization*) describes a question that I became interested in during the late 1970s, and is not about the same effects that Évariste Sanchez-Palencia studied in the early 1970s (dependence upon frequency of the dielectric permittivity, and visco-elastic effects), because I wanted to work in a hyperbolic framework: I had guessed that the rules of "spontaneous absorption and emission" used by physicists are just their way to describe some nonlocal effects (in time) induced by homogenization. Actually, there are other non-local effects, in space as I noticed with François Murat, or in space and time as was noticed by Youcef Amirat, Kamel Hamdache, and Hamid Ziani (1949–2004) [1]: their work actually gives an example of a natural power expansion which diverges in the sense of distributions, since all the terms live somewhere, while the correct solution lives elsewhere!

It suggests a way to resolve the paradox of the efficiency of some Feynman diagrams, which give interesting answers by looking at wrong equations.

Maxwell had introduced discrete velocity models, but Appendix K (*On some discrete velocity models in one dimension*) is not about the general case, and it is concerned with a special class of models in one dimension, for which a *compensated integrability* effect permits us to use an approach for existence questions that is not a semi-group approach. I hope that similar ideas could be used for the questions mentioned in Appendix G (a system of Dirac equation without mass term coupled with the Maxwell–Heaviside equation, using Dirac formulas for ρ and j in terms of ψ), going much further than the early 1980s work of Yvonne Bruhat, and my hope lies in a remark that Raoul Bott (1923–2005, 2000 Wolf Prize) made to me in the late 1980s during a visit to CMU (Carnegie Mellon University, Pittsburgh, PA),[31] but I still do not know how to use it.

---

[31] Raoul Bott had been a PhD student at Carnegie Tech (Carnegie Institute of Technology, Pittsburgh, PA) of my colleague Dick Duffin (1909–1996), starting in 1948, when the head of the department of mathematics was Synge (1897–1995), the father of Cathleen Morawetz. In 1967, the Carnegie Institute of Technology merged with the Mellon Institute for Industrial Research to form Carnegie Mellon University.

Appendix L (*Homogenization of elliptic operators*) and Appendix M (*Compensated Compactness*) describe some joint work with François Murat, while Appendix N (*H-measures*) describes my extension for going further in deriving mathematically some conjectures deduced from what physicists say.

I have not included the discussion of my *multi-scales H-measures* [7], since I am not sure yet if these objects are the ones needed to explain formal questions of boundary layers in GTD, the geometric theory of diffraction of Joe Keller, or boundary layers in hydrodynamics, like the triple-deck structure proposed by Stewartson (1925–1983), or to explain a part of what one claims about turbulence.

Some believe that my theorem of propagation of physical quantities along (correctly defined) "beams of light" is the same as classical geometrical optics, but it shows that they do not understand the difference between the quantifiers $\exists$ and $\forall$: in geometrical optics one constructs one particular asymptotic expansion of the type "distorted plane wave", and one has difficulty with caustics, which are points near which the solution of the scalar wave equation considered does not look like a distorted plane-wave; in my proof using H-measures (which I introduced for a question of small amplitude homogenization), I tackle all general sequences of solutions of particular linear systems of PDEs (the scalar wave equation, the Maxwell–Heaviside equation, linearized elasticity, and others) and I describe where energy and other physical quadratic quantities go in the limit. Besides, geometrical optics cannot handle the question of what a "beam of light" is, since it is quite removed from the framework of distorted plane waves, and a few people wrongly talk of "propagation of singularities" when they actually work on "propagation of microlocal regularity" (as I was told by Mike Crandall). Lying has rarely been very productive in science!

Appendix A *Refraction*

In the 1690 argument by Huygens, a plane front propagates downward with speed $\frac{c}{n^+}$ in material of index n^+ and incidence angle i^+: it touches interface ($x_3 = 0$) at $A(t)$, which generates centered waves below the interface, with speed $\frac{c}{n^-}$ in material of index n^-. If

$$n^+ \sin(i^+) < n^-,$$

the envelope of the spherical waves is a plane front, showing refraction angle i^-, and elementary geometry gives

$$\text{law of sines: } n^+ \sin(i^+) = n^- \sin(i^-).$$

The argument is *incomplete*, since $A(t)$ also generates centered waves above the interface, whose envelope is the classical *reflected wave*: how much is reflected, and how much is refracted?

A possible intuition of Descartes.

A general (non-constant) plane-wave propagating at speed v^+ in unit direction e^+ above the interface (with $e_3^+ < 0$) means a function $f\big((x, e^+) - v^+ t\big)$, and if it transforms into a (non-constant) plane-wave propagating at speed v^- in unit direction e^- below the interface (with $e_3^- < 0$), meaning a function $g\big((x, e^-) - v^- t\big)$, then *continuity at the interface* implies

$$f\big((x, e^+) - v^+ t\big) = g\big((x, e^-) - v^- t\big) \text{ for all } (x,t) \text{ with } x_3 = 0.$$

Since f is not constant, it implies

$$\frac{P e^+}{v^+} = \frac{P e^-}{v^-},$$

where P is the projection onto the plane $x_3 = 0$, which is the *law of sines* in disguise, and g is only a rescaling of f, by

$$g(v^- z) = f(v^+ z) \text{ for all } z \in \mathbb{R}.$$

This argument also has the defect of forgetting a reflected wave, but it is valid even for *anisotropic* media, where the velocity of a plane-wave in unit direction e depends upon e, and it is valid for all kinds of equations, so that it seems to be the argument of Descartes which Fermat opposed.

The scalar wave equation (in an anisotropic medium).

The one-dimensional scalar wave equation is

$$\frac{\partial^2 u}{\partial t^2} - w^2 \frac{\partial^2 u}{\partial x^2} = 0,$$

and its general solution is

$$f(x - wt) + g(x + wt) \text{ for arbitrary functions } f, g,$$

so that it describes *waves going in both directions*.

The generalization to an N-dimensional anisotropic medium is

$$\frac{\partial^2 u}{\partial t^2} - \sum_{i,j=1}^{N} A_{i,j} \frac{\partial^2 u}{\partial x_i \partial x_j} = 0, \tag{10.1}$$

where the matrix A is *symmetric and positive definite* (its coefficients have dimension $length^2 time^{-2}$). Equation (10.1) admits plane-wave solutions in every unit direction $e \in \mathbb{R}^N$:

$$f\big((x,e) - w\,t\big) \text{ is a solution if } w^2 = (A\,e,e).$$

The scalar wave equation in the case of variable coefficients is

$$\frac{\partial^2 u}{\partial t^2} - \sum_{i,j=1}^{N} \frac{\partial}{\partial x_i}\left(A_{i,j}\frac{\partial u}{\partial x_j}\right) = 0, \tag{10.2}$$

or more generally

$$\frac{\partial}{\partial t}\left(\rho\,\frac{\partial u}{\partial t}\right) - \sum_{i,j=1}^{N} \frac{\partial}{\partial x_i}\left(A_{i,j}\frac{\partial u}{\partial x_j}\right) = 0, \tag{10.3}$$

where ρ and $A_{i,j}$ may depend upon x (and eventually on t, but in a "smooth" way). Of course, (10.2) or (10.3) are understood *in the sense of distributions*, of Laurent Schwartz but also of Sergei Sobolev (1908–1989), so that at a smooth interface of normal v, $(A\,\mathrm{grad}(u), v)$ has the same trace on both sides, but also $\frac{\partial u}{\partial t}$ has the same trace on both sides.[32]

The construction by Descartes or Huygens does not give a solution of the wave equation: the solution has both a reflected wave and a refracted wave, except for a special incidence angle $i*$,[33] for which there is no reflected wave.

Reflection in an anisotropic material (with matrix A) in front of a mirror uses a *Dirichlet condition* on $(x, v) = 0$. For each unit vector e (with $(e, v) < 0$),

$$f\big((x,e) - w\,t\big) - f\big((x, e + \alpha\,v) - w\,t\big), \text{ with } w^2 = (A\,e,e),$$

satisfies the Dirichlet condition whatever $\alpha \in \mathbb{R}$ is, but for it to satisfy (10.1) one needs

$$\big(A\,(e + \alpha\,v), (e + \alpha\,v)\big) = w^2 = (A\,e,e),$$

and since one wants $\alpha \neq 0$, one deduces that

$$\alpha = -2\frac{(A\,e, v)}{(A\,v, v)}.$$

Some unit vectors e imply $(e + \alpha\,v, v) < 0$ if v is not an eigenvector of A, and it is not what one considers as a reflection!

---

[32] This comes from the compatibility condition $\frac{\partial}{\partial t}\big(\frac{\partial u}{\partial x_j}\big) - \frac{\partial}{\partial x_j}\big(\frac{\partial u}{\partial t}\big) = 0$, which implies the continuity of $\frac{\partial u}{\partial t}\,v_j$ for all j.

[33] The incidence angle $i*$ and refracted angle $r*$ correspond to $n\cos i* = \cos r*$, and it gives $i* + r* = \frac{\pi}{2}$, $\sin i* = \cos r* = \frac{n}{\sqrt{n^2+1}}$, $\cos i* = \sin r* = \frac{1}{\sqrt{n^2+1}}$, $\tan i* = n$, $\tan r* = \frac{1}{n}$.

Refraction: If material in $(x, v) > 0$ has matrix A and material in $(x, v) < 0$ has matrix B, one considers

$$f\big((x, e + \beta v) - w t\big) \text{ in } (x, v) < 0 \, ; \, w^2 = (A e, e),$$

but for it to satisfy (10.1) one needs

$$\big(B (e + \beta v), (e + \beta v)\big) = w^2 = (A e, e),$$

and one only accepts the case

$$(e + \beta v, v) < 0.$$

Isotropic example: $A = c^2 I$ for $x_2 > 0$, $A = \frac{c^2}{n^2} I$ for $x_2 < 0$, with $e = (\sin i, -\cos i)$. One considers the function defined by

$$f(x_1 \sin i - x_2 \cos i - ct) + \lambda f(x_1 \sin i + x_2 \cos i - ct)$$

in $x_2 > 0$, and (remembering the relation $\sin i = n \sin r$)

$$\mu f(n x_1 \sin r - n x_2 \cos r - ct)$$

in $x_2 < 0$. Both satisfy the wave equation outside $x_2 = 0$, and the continuity of u at $x_2 = 0$ gives

$$1 + \lambda = \mu, \tag{10.4}$$

and the value of $c^2 \frac{\partial u}{\partial x_2}$ from above the interface should match the value of $\frac{c^2}{n^2} \frac{\partial u}{\partial x_2}$ from below the interface, giving

$$-\cos i + \lambda \cos i = -\mu \frac{\cos r}{n}, \tag{10.5}$$

and the solution of the linear system (10.4)–(10.5) is

$$\lambda = \frac{n \cos i - \cos r}{n \cos i + \cos r} = \frac{\sin 2i - \sin 2r}{\sin 2i + \sin 2r},$$

$$\mu = \frac{2n \cos i}{n \cos i + \cos r} = \frac{2 \sin 2i}{\sin 2i + \sin 2r}.$$

The case with no refraction: There is a critical angle of refraction r_c corresponding to the case $i = \frac{\pi}{2}$, i.e. $\sin r_c = \frac{1}{n}$. If $r > r_c$, one teaches that all the light is reflected, but an incident wave plus a reflected wave in $x_2 < 0$, and 0 in $x_2 > 0$ is not a solution of the wave equation!

I had heard of *evanescent waves* (introduced by Lord Rayleigh for surface waves in elasticity) which decay exponentially fast away from an interface, so I have tried to use them here.

One considers a given (upward) incident plane-wave

$$f\left(x_1 \sin r + x_2 \cos r - \frac{c}{n} t\right) \text{ in } x_2 < 0,$$

with an unknown (downward) reflected plane-wave

$$g\left(x_1 \sin r - x_2 \cos r - \frac{c}{n}t\right) \text{ in } x_2 < 0,$$

and a transmitted wave

$$U\left(x_1 \sin r - \frac{c}{n}t, x_2\right) \text{ in } x_2 > 0. \tag{10.6}$$

The continuity conditions at $x_2 = 0$ are

$$U(z,0) = f(z) + g(z) \text{ for } z \in \mathbb{R},$$

$$\left.\frac{\partial U}{\partial x_2}\right|_{x_2=0} = \frac{1}{n^2}\left(\cos r f'(z) - \cos r g'(z)\right) \text{ for } z \in \mathbb{R}.$$

The reason for using evanescent waves is that for all $\xi \in \mathbb{R}$

$$u = e^{2i\pi\xi(x_1 \sin r - ct/n)}e^{-2\pi\gamma|\xi|x_2} \text{ solves } \frac{\partial^2 u}{\partial t^2} - c^2\Delta u = 0$$

$$\text{in } x_2 > 0 \text{ if } \gamma = \sqrt{\sin^2 r - \frac{1}{n^2}}.$$

It is natural to use the Fourier transforms of f, g, and U:

$$f(z) = \int_{\mathbb{R}} e^{2i\pi z\xi}\widehat{f}(\xi)\,d\xi \text{ and } g(z) = \int_{\mathbb{R}} e^{2i\pi z\xi}\widehat{g}(\xi)\,d\xi,$$

so that the first part of (10.6) gives

$$U(z,x_2) = \int_{\mathbb{R}} \left(\widehat{f}(\xi) + \widehat{g}(\xi)\right) e^{2i\pi z\xi} e^{-2\pi\gamma|\xi|x_2}\,d\xi$$

and the second part of (10.6) gives

$$-\gamma|\xi|\left(\widehat{f}(\xi) + \widehat{g}(\xi)\right) = \frac{1}{n^2}i\xi\cos r\left(\widehat{f}(\xi) - \widehat{g}(\xi)\right) \text{ for } \xi \in \mathbb{R}.$$

One deduces that

$$\widehat{g}(\xi) = \begin{cases} \delta\widehat{f}(\xi) \text{ for } \xi > 0, \\ \delta\widehat{f}(\xi) = \frac{1}{\delta}\widehat{f}(\xi) \text{ for } \xi < 0, \end{cases} \text{ with } \delta = \frac{i\cos r + n^2\gamma}{i\cos r - n^2\gamma},$$

or

$$\widehat{g}(\xi) = \Re\delta\widehat{f}(\xi) + \Im\delta\, i\, \text{sign}(\xi)\widehat{f}(\xi),$$

i.e.

$$g = \Re\delta f + \Im\delta\, Hf,$$

where H is the *Hilbert transform*, defined by

$$Hf(x) = \frac{1}{\pi}\lim_{\varepsilon\to 0}\int_{|y|>\varepsilon}\frac{f(x-y)}{y}\,dy,$$

for smooth functions with compact support, so that the relation between g and f is not local!

The classical cases of reflection and refraction then involve elementary linear algebra, while the case with no refraction involves more technical mathematics.

Appendix B *The vibrating string*

In what I think is D. Bernoulli's argument, one considers masses $m_1, \ldots, m_{N-1}$ at $0 < x_1(t) < \cdots < x_{N-1}(t) < L$ ($x_0(t) = 0$ and $x_N(t) = L$ are fixed walls). For $j = 1, \ldots, N$, there is a spring with constant $\kappa_j > 0$ and equilibrium length $\ell_j > 0$ between the masses at x_{j-1} and at x_j.

The increase in length of spring j is $x_j - x_{j-1} - \ell_j$, so that it exerts forces $-\kappa_j(x_j - x_{j-1} - \ell_j)$ at x_j and $\kappa_j(x_j - x_{j-1} - \ell_j)$ at x_{j-1}, hence the differential equation for x_j $(j = 1, \ldots, N-1)$

$$m_j \frac{d^2 x_j}{dt^2} + \kappa_j(x_j - x_{j-1} - \ell_j) - \kappa_{j+1}(x_{j+1} - x_j - \ell_{j+1}) = 0.$$

It is not clear if for $t > 0$ one has $x_{j-1}(t) < x_j(t)$ for $j = 1, \ldots, N$.[34] The equilibrium positions y_j of mass j satisfy

$$\kappa_j(y_j - y_{j-1} - \ell_j) - \kappa_{j+1}(y_{j+1} - y_j - \ell_{j+1}) = 0,$$
$$\text{for } j = 1, \ldots, N-1, \text{ with } y_0 = 0, \ y_N = L. \tag{10.7}$$

Since (10.7) is a linear system with as many equations as unknowns, existence is equivalent to uniqueness, but one must then check if $y_{j-1} < y_j$ for $j = 1, \ldots, N$. For proving uniqueness, one considers the homogeneous version of (10.7),

$$\kappa_j(z_j - z_{j-1}) - \kappa_{j+1}(z_{j+1} - z_j) = 0,$$
$$\text{for } j = 1, \ldots, N-1, \text{ with } z_0 = z_N = 0,$$

one multiplies the preceding equation j by z_j, one sums in j, which gives

$$\sum_{j=1}^{N} \kappa_j(z_j - z_{j-1})^2 = 0,$$

hence all z_j are equal, to 0 since $z_0 = 0$. One defines F_L by

$$F_L = \kappa_j(y_j - y_{j-1} - \ell_j) \text{ for } j = 1, \ldots, N, \tag{10.8}$$

---

[34] The law of force for compressing a spring has been linearized, buckling is not taken into account, and only a finite force is necessary to squeeze a spring to zero length!

which uses (10.7): F_L is the force to apply at L for maintaining equilibrium, and (10.8) implies

$$\left(\frac{1}{\kappa_1} + \cdots + \frac{1}{\kappa_N}\right) F_L = L - (\ell_1 + \cdots + \ell_N), \tag{10.9}$$

so that

if $L \geq \ell_1 + \cdots + \ell_N$, then $F_L \geq 0$ and $y_{j-1} < y_j, j = 1, \ldots, N$.

If the springs under no tension have total length larger than L, one wants $y_{j-1} < y_j$, i.e. $\ell_j + \frac{F_L}{\kappa_j} > 0$ for $j = 1, \ldots, N$, or

$$F_L > -\min_{j=1,\ldots,N} \kappa_j \ell_j,$$

and using (10.9) one deduces that

$$\ell_1 + \cdots + \ell_N < L + \left(\frac{1}{\kappa_1} + \cdots + \frac{1}{\kappa_N}\right) \min_{j=1,\ldots,N} \kappa_j \ell_j$$

implies $y_{j-1} < y_j$ for $j = 1, \ldots, N$.

The condition is true if the ℓ_j are equal and the κ_j are equal.

Writing

$$x_j(t) = y_j + z_j(t) \text{ for } j = 0, \ldots, N,$$

one obtains the equations

$$m_j \frac{d^2 z_j}{dt^2} + k_j(z_j - z_{j-1}) - k_{j+1}(z_{j+1} - z_j) = 0, \, j = 1, \ldots, N-1.$$

Multiplying by $\frac{dz_j}{dt}$ and summing in j gives

$$\sum_{j=1}^{N-1} \frac{m_j}{2}\left(\frac{dz_j}{dt}\right)^2 + \sum_{j=1}^{N} \frac{k_j}{2}(z_j - z_{j-1})^2 = \text{constant},$$

which is not always the *total energy*, sum of the *kinetic energy*,

$$\mathcal{K}(t) = \sum_{j=1}^{N-1} \frac{m_j}{2}\left(\frac{dx_j}{dt}\right)^2 = \sum_{j=1}^{N-1} \frac{m_j}{2}\left(\frac{dz_j}{dt}\right)^2,$$

and the *potential energy* (elastic energy inside the springs),

$$\mathcal{P}(t) = \sum_{j=1}^{N} \frac{\kappa_j}{2}(x_j - x_{j-1} - \ell_j)^2,$$

which is

$$\sum_{j=1}^{N} \frac{\kappa_j}{2}(z_j - z_{j-1} + y_j - y_{j-1} - \ell_j)^2$$

$$= \sum_{j=1}^{N} \frac{\kappa_j}{2}\left(z_j - z_{j-1} + \frac{F_L}{\kappa_j}\right)^2$$

$$= \sum_{j=1}^{N} \frac{\kappa_j}{2}(z_j - z_{j-1})^2 + \frac{F_L}{2}\left(L - (\ell_1 + \cdots + \ell_N)\right),$$

using $\sum_{j=1}^{N}(z_j - z_{j-1}) = 0$ and (10.9): the constant is the total energy only if $L = \ell_1 + \cdots + \ell_N$ (hence $F_L = 0$).

One may naively think that if one starts with the springs under no tension, occupying the length $\ell_1 + \cdots + \ell_N$ and one applies the force F_L until the end point is at position L, then the work of the force is $F_L\left(L - (\ell_1 + \cdots + \ell_N)\right)$ and this is not the value $\frac{F_L}{2}\left(L - (\ell_1 + \cdots + \ell_N)\right)$ which appears in the preceding formula, but if one behaves in such a naive way, one will not end up with the mass j at position y_j and with velocity 0, of course.

If one starts with $x_1(0) = \ell_1, x_2(0) = \ell_1 + \ell_2, \cdots, x_N(0) = \ell_1 + \ldots + \ell_N$, with $\frac{dx_j}{dt}(0) = 0$ for $j = 1, \ldots, N$, and one applies a force $F(t)$, then the system is changed and one no longer has $x_N(t) = L$ but $m_N \frac{d^2 x_N}{dt^2} + \kappa_N(x_N - x_{N-1} - \ell_N) = F(t)$, while before there was no mass m_N involved; between time 0 and T the work of the force is going to be $\int_0^T F(t)\frac{dx_N}{dt}\, dt$; multiplying the equation j by $\frac{dx_j}{dt}$ and summing in j one obtains that $\frac{d}{dt}\left[\mathcal{K}(t) + \frac{m_N}{2}\left(\frac{dx_N}{dt}\right)^2 + \mathcal{P}(t)\right] = F(t)\frac{dx_N}{dt}$, and therefore the work done between time 0 and T is $\mathcal{K}(T) + \frac{m_N}{2}\left(\frac{dx_N}{dt}(T)\right)^2 + \mathcal{P}(T)$; if at time T one has succeeded in having $x_N(T) = L$ and $\frac{dx_N}{dt}(T) = 0$, then the work done by the force is exactly the total energy of the system that one had considered from the start.

The existence of $F(t)$ such that (at time T) $x_j(T)$ and $\frac{dx_j}{dt}(T)$ take given values (for $j = 1, \ldots, N$) is a question of *controllability*; there is an algebraic characterization,[35] and it can be checked that the system is controllable if $\kappa_j > 0$ for all j.[36]

---

[35] In the case of a system $\frac{dX}{dt} = AX + Bu$, with X of size M, the necessary and sufficient condition for controllability is that the rank of the matrix with block columns $Y = (B\Lambda B \cdots \Lambda^{M-1}B)$ is M.

[36] $M = 2N$ with $X = \begin{pmatrix} x \\ \frac{dx}{dt} \end{pmatrix}$ and $A = \begin{pmatrix} 0 & I \\ -M_0 & 0 \end{pmatrix}$, where M_0 is a tridiagonal matrix, and $B = e_{2N}$;

AB puts e_N in the span of the columns of Y, and then A^2B puts $M_0 e_N$, which is a combination of e_{2N-1} and e_{2N} if $\kappa_N > 0$, so e_{2N-1} is in the span and A^3B puts e_{N-1} and A^4B puts $M_0 e_{N-1}$, which adds e_{2N-2} in the span if $\kappa_{N-1} > 0$, and so on.

Assuming all m_j equal to m and all κ_j equal to κ, one looks at *periodic solutions* of the form $x(t) = y + e^{i\omega t}a$, where y is the equilibrium solution, and one finds that one must have $M_0 a = \omega^2 a$, where M_0 is the symmetric matrix defined by

$$(M_0)_{i,j} = 0 \text{ for } j \neq i \text{ or } j \neq i \pm 1,$$

$$(M_0)_{i,i} = \frac{2\kappa}{m} \text{ for } i = 1, \ldots, N-1,$$

$$(M_0)_{i,i-1} = (M_0)_{i,i+1} = \frac{-\kappa}{m} \text{ for } i = 1, \ldots, N-1;$$

one either discards the condition for $(M_0)_{1,0}$ and $(M_0)_{N-1,N}$ in this list, or one considers that one must have $a_0 = a_N = 0$.

The eigenvectors and eigenvalues of M_0 are found explicitly: one chooses $p = 1, \ldots, N-1$, and one defines a by

$$a_i = \sin\left(\frac{ip\pi}{N}\right) \text{ for } i = 1, \ldots, N-1,$$

which gives the eigenvalue

$$\omega_p^2 = 2\frac{\kappa}{m}\left[1 - \cos\left(\frac{p\pi}{N}\right)\right] = \frac{4\kappa}{m}\sin^2\left(\frac{p\pi}{2N}\right),$$

so that the corresponding frequencies of vibration of the system are

$$v_p = \frac{\omega_p}{2\pi} = \frac{\sqrt{\kappa}}{\pi\sqrt{m}}\sin\left(\frac{p\pi}{2N}\right),$$

hence for N large the lowest frequency is $v_1 \approx \frac{\sqrt{k}}{2N\sqrt{m}}$.

The next question is to let N tend to infinity, while rescaling correctly m and κ. For rescaling mass, $m = \frac{m_*}{N}$ is natural, where m_* is the mass of the vibrating string that one is trying to model, which one imagines as divided into N equal parts.

The rescaling for κ is less intuitive. Since *force* is *mass* × *acceleration*, the unit for κ is *mass* × *time*$^{-2}$ and if mass is $\frac{m_*}{N}$ one may choose *time* = $\frac{time_*}{N}$, which corresponds to $\kappa = N\kappa_*$, and this corresponds to keeping the lowest frequency almost fixed, which may be how D. Bernoulli thought.

One may think of having a constant (non-zero) speed of propagation so *length* and *time* should be rescaled in the same way, but D. Bernoulli had not derived the one-dimensional wave equation, and he was probably not thinking in this way.

One may ask that forces stay bounded and do not tend to 0, since from the experience of tuning a violin some high tension must be used but certainly not to infinity.

A piece of string of length $\frac{L}{N}$ might increase its length of order $O\left(\frac{1}{N}\right)$ but for a force $O(1)$, hence κ must be $O(N)$.

One may want a bounded kinetic energy and a bounded potential energy. For kinetic energy one has N terms $O\left(\frac{1}{N}\right)$ since m is $O\left(\frac{1}{N}\right)$ and $\frac{dx_j}{dt}$ is $O(1)$; for the potential energy one expects $x_j - x_{j-1} - \ell_j$ to be $O\left(\frac{1}{N}\right)$; for having $\kappa (x_j - x_{j-1} - \ell_j)^2$ of order $O\left(\frac{1}{N}\right)$, one needs κ of order $O(N)$; I do not know if D. Bernoulli thought in terms of potential energy.

All the preceding considerations consist of arguing about physical intuition, but one's physical intuition might be wrong.

For a physical problem that one is not sure to have understood correctly, one should turn to the mathematical side and prove various theorems, under different hypotheses.

One scales $m = \frac{m_*}{N}$, and one uses a sequence $\kappa(N)$ which one lets behave in various ways as $N \to \infty$. One starts from $x_j(0) = y_j = \frac{jL}{N}$, with $\frac{dx_j}{dt} = O(1)$ for $j = 1, \ldots, N-1$, so that $\mathcal{K}(0) = O(1)$ and $\mathcal{P}(0) = 0$; from the solution of the differential system one may construct a function u_N which at any time t is continuous and piecewise affine with $u_N\left(\frac{jL}{N}, t\right) = x_j(t) - y_j$, and one wonders what happens to u_N as $N \to \infty$.

If $\frac{\kappa(N)}{N} \to 0$, one finds that every *weak limit* u_∞ of a *subsequence* extracted from the sequence u_N satisfies $\frac{\partial^2 u}{\partial t^2} = 0$; this is the case where there is only kinetic energy and no elastic behaviour giving rise to a non-zero potential energy.

If $\frac{\kappa(N)}{N} \to \infty$, one finds that every weak limit u_∞ satisfies $\frac{\partial^2 u}{\partial x^2} = 0$: in this case, the string behaves as a rigid body.

If $\frac{\kappa(N)}{N} \to \kappa \neq 0$, one finds that every weak limit u_∞ satisfies $\frac{\partial^2 u}{\partial t^2} - c^2 \frac{\partial^2 u}{\partial x^2} = 0$, with $c^2 = \frac{\kappa L^2}{m_*}$: this case corresponds to a vibrating string (with no dissipation of energy).

These results can be proven with standard *variational techniques*, commonly used in the abstract part of *numerical analysis*, where the problem considered is usually the opposite one: one starts from the wave equation and one wants to approach its solution, but the ideas are the same and one uses the bound on the total energy in a crucial way.

The choice $x_j(0) = y_j$ for all j was chosen as a simplification, but if one wants to start with a non-zero potential energy, then it is better to take it bounded if one wants a subsequence of u_N to converge in a reasonable *functional space*.

Extracting a subsequence converging in a weak topology is only a first step, which uses classical *functional analysis*, but an efficient numerical algorithm must have all the sequences converge (and this part usually follows from uniqueness of a solution), in a strong topology, and as fast as possible.

Appendix C *Linearized elasticity*

The one-dimensional model with masses and springs is about *longitudinal waves*, where the displacements of material points are in the direction of the propagating wave, although the motivation was the motion of a violin string, which correspond to *transverse waves*, where the displacements of material points are in a direction perpendicular to the propagating wave.

It would be better to consider masses moving in a two-dimensional (or three-dimensional) space, but one loses linearity, so that the problem becomes too difficult: if mass i (initially at $i\ell$, with $\ell = \frac{L}{N}$) is at the point (x_i, y_i), the increase in length of the spring i is $\sqrt{(x_i - x_{i-1})^2 + (y_i - y_{i-1})^2} - \ell$.

A limitation of many models used for solids is to consider only nearest neighbour interactions: this is not compatible with the belief that particles (or atoms) attract or repel each other depending upon their distances.

Studying vibrations around an equilibrium position would be like having point i and point j linked by a spring of strength $\kappa_{i,j}$, and a potential energy $\sum_{i \neq j} \frac{\kappa_{i,j}}{2} (x_j - x_i - (j-i)\ell)^2$, and if u was a smooth function and $x_i(t) = u(i\ell, t)$, then besides quantities proportional to $\int_0^L \left| \frac{\partial u}{\partial x} \right|^2 dx$ one could well see quantities of the type $\iint_{(0,L) \times (0,L)} |u(x) - u(y)|^2 w(x, y) \, dx \, dy$ for some weight function w, and one would end with equations which are not the usual PDEs, since non-local terms would appear.

Cauchy studied two-dimensional linearized elasticity by a natural generalization of the idea of D. Bernoulli, considering a square lattice with small masses of size m at the points $(i\ell, j\ell)$ for integers i, j, with springs of strength κ along the horizontal and vertical lines, but also along the two diagonal directions.[37]

The scaling is $\ell = \frac{L}{N}$ and $m = \frac{m_*}{N^2}$, corresponding to a finite density of mass at the limit $N \to \infty$, but $\kappa = \kappa_*$ independent of N, because $\frac{\kappa}{m}$ must have the dimension $\frac{1}{\text{time}^2}$, and time scales as length for having a fixed speed of propagation of waves.

Another way to interpret the scaling for κ is that one does not impose forces $O(1)$ at points of the boundary, but a force per unit of length (i.e. a two-dimensional pressure), and forces are $O\left(\frac{1}{N}\right)$; for a three-dimensional problem, mass scales as $m = \frac{m_*}{N^3}$, κ scales as $\kappa = \frac{\kappa_*}{N}$, forces are $O\left(\frac{1}{N^2}\right)$, and pressure is $O(1)$.

---

[37] Without diagonal springs, the lattice is weak: the infinite lattice has a family of equilibria, where all the squares become lozenges, without increasing the length of any of the springs; with the diagonal springs these equilibria disappear.

The displacement has two components, u^1 and u^2, and one uses notation $u^k_{i,j}$ for $u^k(i\ell, j\ell)$. One assumes u^1 and u^2 smooth for using their Taylor expansion and identifying the PDE governing the motion of masses in the limit $N \to \infty$.

One linearizes by considering the directions of the springs almost fixed, so that only the displacements in the directions of the springs are felt: horizontal forces then contribute a term in $(\partial_{x_1} u^1)^2$ to the potential energy, vertical forces a term in $(\partial_{x_2} u^2)^2$, and diagonal forces terms $\left((\partial_{x_1} \pm \partial_{x_2})(u^1 \pm u^2)\right)^2$, i.e. $(\partial_{x_1} u^1)^2 + (\partial_{x_2} u^2)^2 \pm 2(\partial_{x_1} u^2 + \partial_{x_2} u^1)^2$.

Cauchy's approach only gave the case $\lambda = \mu$ in the general family of isotropic (linearized) elastic materials then proposed by Lamé, where the *Cauchy stress tensor* has the form

$$\sigma_{i,j} = \mu\left(\frac{\partial u_i}{\partial x_j} + \frac{\partial u_j}{\partial x_i}\right) + \lambda \delta_{i,j} \sum_k \frac{\partial u_k}{\partial x_k}, \text{ for all } i,j. \tag{10.10}$$

The general equations of linearized elasticity are

$$\rho \frac{\partial^2 u_i}{\partial t^2} - \sum_j \frac{\partial \sigma_{i,j}}{\partial x_j} = 0 \text{ for all } i, \tag{10.11}$$

which in the isotropic case (10.10) give the *Lamé equations*

$$\rho \frac{\partial^2 u_i}{\partial t^2} - \mu \Delta u_i - (\lambda + \mu)\frac{\partial [\mathrm{div}(u)]}{\partial x_i} = 0 \text{ for all } i. \tag{10.12}$$

(10.12) implies that both $\mathrm{div}(u)$ and $\mathrm{curl}(u)$ satisfy wave equations, but with different speeds of propagation:

$$\rho \frac{\partial^2 [\mathrm{div}(u)]}{\partial t^2} - (2\mu + \lambda) \Delta [\mathrm{div}(u)] = 0, \tag{10.13}$$

$$\rho \frac{\partial^2 [\mathrm{curl}(u)]}{\partial t^2} - \mu \Delta [\mathrm{curl}(u)] = 0. \tag{10.14}$$

(10.13) corresponds to *pressure waves* (or primary waves, or P-waves in seismology), and (10.14) corresponds to the *shear waves* (or secondary waves, or S-waves in seismology). In seismology, one makes the approximation that the ground is linearly elastic (and even isotropic!), and it is useful that P-waves travel faster than S-waves (since most real materials have $\lambda > 0$).

In an earthquake, the P-waves are not so dangerous and arrive first, signaling the coming of the S-waves, which are dangerous for buildings that are not designed carefully enough.[38]

---

[38] There are also tertiary waves used in seismology, but they are surface waves: these are the evanescent waves introduced by Lord Rayleigh, which decay exponentially with depth.

(10.11) is used for anisotropic materials, with (10.10) replaced by

$$\sigma_{i,j} = \sum_{k,\ell} C_{i,j;k,\ell} \varepsilon_{k,\ell} \text{ for all } i,j,$$

$$\text{with } \varepsilon_{k,\ell} = \tfrac{1}{2}\left(\frac{\partial u_k}{\partial x_\ell} + \frac{\partial u_\ell}{\partial x_k}\right), \text{ for all } k,\ell, \tag{10.15}$$

and the elasticity coefficients $C_{i,j;k,\ell}$ enjoy symmetry properties

$$C_{i,j;k,\ell} = C_{j,i;k,\ell} = C_{i,j;\ell,k} = C_{k,\ell;i,j}, \text{ for all } i,j,k,\ell,$$

and for any unit vector e, the *acoustic tensor* $A(e)$ defined by

$$A(e)_{i,k} = \sum_{j,\ell} C_{i,j;k,\ell} e_j e_\ell, \text{ for all } i,k,$$

must be positive definite, i.e. there exists $\alpha > 0$ such that

$$\sum_{i,j,k,\ell} C_{i,j;k,\ell} a_i e_j a_k e_\ell \geq \alpha \,|a|^2 |e|^2 \text{ for all vectors } a,e, \tag{10.16}$$

the strict *Legendre–Hadamard* condition.

(10.16) is that which makes stationary elasticity a *strongly elliptic* system, and it means that

$$\sum_{i,j,k,\ell} C_{i,j;k,\ell} M_{i,j} M_{k,\ell} > 0, \tag{10.17}$$

for all non-zero matrices M of rank ≤ 2, i.e.

$$\text{for all } M = a \otimes b + b \otimes a, \text{ with vectors } a,b \neq 0. \tag{10.18}$$

In the isotropic case (10.17)–(10.18) are $\mu > 0$ and $2\mu + \lambda > 0$.

One also uses a stronger condition,[39] which is that the inequality in (10.17) is true

$$\text{for all non-zero symmetric } M. \tag{10.19}$$

In the isotropic case (10.17)–(10.19) are $\mu > 0$ and $2\mu + 3\lambda > 0$.

Although the condition (10.17)–(10.18) is physical, and expresses the existence of the correct number of waves propagating in any direction (at positive speeds), the stronger condition (10.17)–(10.19) has been imposed for a *mathematical reason*,[40] which is that it permits one to use the Lax–Milgram lemma, which in the symmetric case is just a theorem of F. Riesz (1880–1956).[41]

---

[39] Sometimes called "very strong ellipticity".

[40] Georges Verchery pointed out to me that one can deduce the necessity of (10.19) from considerations of thermodynamics.

[41] Frigyes (Frederic) Riesz, who was Hungarian, introduced the spaces L^p in honour of Lebesgue (1875–1941) and the spaces $\mathcal{H}^p$ in honour of Hardy (1877–1947), but no spaces are named after him; the Riesz operators were introduced by his younger brother Marcel Riesz (1886–1969).

If one looks for plane-waves in the direction e traveling at speed w (velocity $w\,e$) it means using $u\big((x,e)-w\,t\big)$ in the general equation (10.11) with the *constitutive relation* (10.15), which gives

$$\rho\,w^2 u'' - A(e)u'' = 0,$$

which makes the (symmetric) acoustic tensor appear, and explains why one wants it positive definite.

In the isotropic case

$$A(e) = \mu\,I + (\lambda+\mu)\,e\otimes e,$$

which has e as eigenvector with eigenvalues $\lambda+2\mu$ and the plane orthogonal to e as eigenspace for the eigenvalue μ.

Using u parallel to e gives *longitudinal waves*, and using u orthogonal to e gives *transverse waves*. Since $\mathrm{div}(u) = (u',e)$ and $\mathrm{curl}(u) = e \times u'$, one sees that these longitudinal waves satisfy $\mathrm{curl}(u) = 0$, and these transverse waves satisfy $\mathrm{div}(u) = 0$.

The distinction between longitudinal waves and transverse waves is not natural for anisotropic media, since there is no reason then that e or a vector orthogonal to e should be an eigenvector of the acoustic tensor $A(e)$!

It is curious that in seismology one thinks about P-waves and S-waves, since geological materials accumulate in layers, which probably creates transversally isotropic media (often no longer in their original horizontal position). Even if one considers that all rocks are isotropic, there are clear interfaces of discontinuities of coefficients, and in asserting what happens at an interface one must use (10.11) and not (10.13)–(10.14) for discovering which are the reflected and refracted waves!

At a smooth surface of discontinuity with normal v, the displacement u is continuous,[42] so that the tangential derivatives of u are continuous, but equation (10.11) implies that the *traction* is continuous, i.e.

$$\sum_j \sigma_{i,j} v_j \text{ is continuous,}$$

which in the isotropic case means that

$$\mu\left(\sum_j \frac{\partial u_i}{\partial x_j} v_j + \sum_j \frac{\partial u_j}{\partial x_i} v_j\right) + \lambda\,\mathrm{div}(u)\,v_i \text{ is continuous,}$$

so that the combinations of derivatives which match are not related to $\mathrm{div}(u)$ or $\mathrm{curl}(u)$, and (10.13)–(10.14) are not so helpful.

---

[42] Formation of cracks is an important question, not considered here. Once a crack is formed, it may open.

The displacement of the material point at x being $u(x)$, one has to consider the transformation U defined by $U(x) = x + u(x)$, with gradient $F = \nabla U$, and because of rotations it is $F^T F = I + \nabla u + \nabla u^T + \nabla u^T \nabla u$ which measures the stretching.

Linearized elasticity consists of assuming that ∇u is small, or F is near I, and deduce that $\nabla u^T \nabla u$ can be neglected, so that it is $\nabla u + \nabla u^T$ which appears, denoted ε. Even if the solution u of the (linear) elasticity system is small, its gradient may not be small, and linearization should be questioned.

If the domain has corners, both u and ∇u are usually infinite at corners, and cracks are even known to initiate there, so that using linearized elasticity in such situations is problematic.

In a homogeneous material, one argues that some particular results (but not all) may nevertheless be correct because of invariants in (non-linear) elasticity which James Rice called *J-integrals*, and they are sometimes called Rice integrals, although they were introduced by Eshelby (1916–1988) in 1956. It then explains why studying which singularities arise in the solution of linearized elasticity systems in corners may still be useful, although the hypotheses necessary for linearizing are certainly far from being met.

Through the use of the Lax–Milgram lemma, the very strong ellipticity condition (10.17)–(10.19) permits one to prove a general homogenization result, but the conditions of application contradict the requirements for being able to use a linearization: the result applies to situations where the coefficients converge only in a weak topology, so that one may have plenty of interfaces (corresponding to mixing various materials), and it is not reasonable to expect that *in a realistic non-linear setting* the jump conditions will make ∇U near I at all these interfaces.

However, it is a more reasonable expectation if the elastic properties do not change much, and this is the case for what I called *small amplitude homogenization*, for which I introduced the (then) new tool of *H-measures*, which permits one to study the transport of energy (and momentum) along light beams, and similar results hold for other hyperbolic systems, like linearized elasticity or the Maxwell–Heaviside equation.

Since *total energy* is conserved, it is important to find *where it goes*, and avoid the nonphysical rules that many mathematicians seem to believe, that elastic materials "minimize their potential energy", instantaneously (since they also seem to believe in a nonphysical world where time does not exist)!

Appendix D *Velocity of sound*

Linearization may seem reasonable if some quantities are small, but a function may be small without its derivatives being small.

For hyperbolic equations, which are more or less the PDE for which information travels at finite speed, the difference between the linear case and the *quasi-linear* case is important.

For an *elastic string*, with *large deformations*, one may use

$$\frac{\partial^2 w}{\partial t^2} - \frac{\partial}{\partial x}\left(f\left(\frac{\partial w}{\partial x}\right)\right) = 0, \tag{10.20}$$

where w is the *vertical displacement*, $\frac{\partial w}{\partial t}$ the (vertical) *velocity*, $\frac{\partial w}{\partial x}$ the *strain*, and $f\left(\frac{\partial w}{\partial x}\right)$ the *stress*, but f is not affine and satisfies $f' > 0$: $\sqrt{f'}$ is the local *speed of propagation* of perturbations.

The first to study a model like (10.20) was Poisson, around 1807, concerned with a *barotropic* model of *gas dynamics*, where the pressure p is a non-linear function of density ρ:

$$\frac{\partial \rho}{\partial t} + \frac{\partial (\rho u)}{\partial x} = 0,$$

$$\frac{\partial (\rho u)}{\partial t} + \frac{\partial (\rho u^2 + p)}{\partial x} = 0.$$

Newton apparently computed the *speed of sound* in air, without knowing about the wave equation, but knowing about *compressibility of air*, and he found around 200 m/sec.

However, under the usual conditions of temperature and pressure, the real speed is above 300 m/sec. Poisson used a relation $p = c\rho^\gamma$ (possibly suggested by Laplace) with an adapted value of γ, instead of the Boyle–Mariotte law ($PV = constant$ at constant temperature),[43] i.e. $p = c\rho$. Poisson was looking for *rarefaction waves*, and he left his solution in implicit form, which Challis found to be problematic in 1848 (for sinusoidal initial data); Stokes explained that profiles became steeper and steeper, until one had to introduce a discontinuity (*shock*), for which he computed the velocity, by expressing the *conservation of mass* and the *conservation of linear momentum*.

The basic ideas are more easily explained with the equation

$$\frac{\partial u}{\partial t} + u\frac{\partial u}{\partial x} = 0, \tag{10.21}$$

often called the (inviscid) *Burgers equation*, since Burgers (1895–1981) proposed in 1948 the equation

$$\frac{\partial u}{\partial t} + u\frac{\partial u}{\partial x} - \varepsilon\frac{\partial^2 u}{\partial x^2} = 0, \tag{10.22}$$

---

[43] The relation $PV = RT$ appeared after the 1802 work of Gay Lussac: his law states that at fixed volume the pressure is proportional to the *absolute temperature* T (although the notion was not defined yet, and was the temperature in degree C +273), and he mentioned a 1787 unpublished law by Charles (1746–1823), that at constant pressure the volume is proportional to T.

with a viscosity term, which he believed to be a one-dimensional model of turbulence: Eberhard Hopf (1902–1983) immediately pointed out that turbulence is something very different, and he was able to study the limiting case $\varepsilon \to 0$, by using a non-linear transformation which changes the equation into the linear heat equation, now known as the *Hopf–Cole transform*, since Julian Cole (1925–1999) found it independently. However, as I learned from Peter Lax, (10.22) was used in 1915 by Bateman (1882–1946), with the limiting case (10.21). Actually, Forsyth (1858–1942) had introduced (10.22) ten years earlier, in connection with the "Hopf–Cole" transform.[44]

If one uses a smooth initial datum v for (10.21), then the method of *characteristic curves* (going back at least to Cauchy) gives the solution as long as it stays smooth:

$$\text{if } \frac{dx(t)}{dt} = u(x(t),t); x(0) = \xi, \text{ then } \frac{d}{dt}\big(u(x(t),t)\big) = 0,$$
$$\text{so that } u(x(t),t) = v(\xi),$$

i.e.

$$\text{on the line } x(t) = \xi + t v(\xi), \text{ one has } u(x(t),t) = v(\xi). \tag{10.23}$$

The implicit formula of Poisson is like replacing (10.23) by

$$v\big(x - t u(x,t)\big) = u(x,t) \text{ for all } x. \tag{10.24}$$

(10.24) is not as useful as (10.23), which shows that, except for the case where v is non-decreasing, the solution of (10.21) with initial data v is only smooth up to T_c, given by[45]

$$-\frac{1}{T} = \inf_{\xi \in \mathbb{R}} v'(\xi).$$

Since the solution cannot be smooth all the time, one writes (10.21) in *conservative form*

$$\frac{\partial u}{\partial t} + \frac{\partial(\frac{u^2}{2})}{\partial x} = 0, \tag{10.25}$$

which is understood *in the sense of distributions*.[46]

---

[44] If U solves $\frac{\partial U}{\partial t} - \varepsilon \frac{\partial^2 U}{\partial x^2} + \frac{1}{2}\big(\frac{\partial U}{\partial x}\big)^2 = 0$, then $u = \frac{\partial U}{\partial x}$ solves (10.22). If $\frac{\partial w}{\partial t} - \varepsilon \frac{\partial^2 w}{\partial x^2} = 0$, then $\frac{\partial \varphi(w)}{\partial t} - \varepsilon \frac{\partial^2 \varphi(w)}{\partial x^2} = -\varepsilon \varphi''(w)\big(\frac{\partial w}{\partial x}\big)^2$. One chooses φ such that $\varepsilon \varphi''(w) = \frac{1}{2}(\varphi')^2$.

[45] In (10.21) u is a velocity, so that u_x is the inverse of a time.

[46] (10.25) only makes sense if u and u^2 are distributions: practically, it means $u \in L^2_{\text{loc}}$.

From the moment one accepts discontinuous functions, one should be careful, because of the apparent paradox,

$$v = a \, \text{sign}(x) \text{ satisfies } v^2 = a^2 \text{ and } v^3 = a^2 v, \text{ but if } a \neq 0$$

$$2a^3 \delta_0 = \frac{d(v^3)}{dx} \neq 3v^2 \frac{dv}{dx} = 6a^3 \delta_0,$$

which reminds one that multiplication of distributions is not defined, and the formula $\frac{d(fg)}{dx} = f\frac{dg}{dx} + g\frac{df}{dx}$ needs smoothness hypotheses on f and g: for example

$$\frac{d(fg)}{dx} = f\frac{dg}{dx} + g\frac{df}{dx} \text{ is true if } f, g \in C^0(\mathbb{R}) \cap BV_{\text{loc}}(\mathbb{R}),$$

it is not always true if $f, g \in L^\infty_{\text{loc}}(\mathbb{R}) \cap BV_{\text{loc}}(\mathbb{R})$.

As was correctly found by Stokes, there is a formula for computing the velocity of a discontinuity, corresponding to

$$f(x,t) = \begin{cases} a_- \text{ for } x < s(t), \\ a_+ \text{ for } s(t) < x, \end{cases} \qquad g(x,t) = \begin{cases} b_- \text{ for } x < s(t), \\ b_+ \text{ for } s(t) < x, \end{cases}$$

which satisfy $\dfrac{\partial f}{\partial t} + \dfrac{\partial g}{\partial x} = 0$ if and only if $(a_+ - a_-)s'(t) = b_+ - b_-$.

If the initial datum for (10.21) has the form

$$v(x) = a_- \text{ for } x < \xi, v(x) = a_+ \text{ for } \xi < x, \text{ and } a_- < a_+,$$

then *the solution* of (10.21) is the *rarefaction wave*

$$u(x,t) = \begin{cases} a_- \text{ for } x < \xi + a_- t, \\ \frac{x-\xi}{t} \text{ for } \xi + a_- t < x < \xi + a_+ t, \\ a_+ \text{ for } \xi + a_+ t < x. \end{cases}$$

If the initial datum for (10.21) has the form

$$v(x) = a_- \text{ for } x < \xi, v(x) = a_+ \text{ for } \xi < x, \text{ and } a_- > a_+,$$

then *the solution* of (10.21) is the *shock*

$$u(x,t) = \begin{cases} a_- \text{ for } x < \xi + st, \\ a_+ \text{ for } \xi + st < x \end{cases} \qquad \text{with } s = \frac{a_- + a_+}{2}.$$

Since (10.21) has plenty of solutions in the sense of distributions, one then needs a natural way of selecting the correct solution, for example by considering the limit as ε tends to 0 of the solution of (10.22), as first done by Eberhard Hopf.

More general questions were then considered by Peter Lax, in particular an extension to systems of various examples considered by Stokes, Riemann (who studied in his 1860 thesis the system of gas dynamics, but with a wrong choice of isentropic solutions, although the term entropy had not been coined yet), Rankine and Hugoniot after whom the jump conditions are now called, and many others who worked at finding a correct formulation of some thermodynamical questions.

Peter Lax had told me that part of his motivation had been to develop a mathematical framework for teaching a course out of a book by Courant (1888–1972) and Friedrichs (1901–1982), who summarized the content of many technical reports, possibly written during World War II.

Similar efforts occurred in the Soviet Union, with a different policy about publications: Sergei Godunov found (without publishing them) many results published by Peter Lax.

The Burgers equation is Galilean invariant,[47] but academic equations that are not Galilean invariant were studied:

$$\frac{\partial u}{\partial t} + \frac{\partial \big(f(u)\big)}{\partial x} = 0 \tag{10.26}$$

is only Galilean invariant if $f(u) = \alpha u^2 + \beta u + \gamma$.

The correct notion of the solution for (10.26) was found by Olga Oleinik (1925–2001), and her condition was reformulated by Eberhard Hopf and by Krushkov (1936–1997): for a function φ (called an "entropy" by Peter Lax) one defines its "entropy flux" ψ by

$$\psi' = \varphi' f'$$

and then the *Oleinik condition* is that

$$\frac{\partial \big(\varphi(u)\big)}{\partial t} + \frac{\partial \big(\psi(u)\big)}{\partial x} \le 0 \text{ for all convex } \varphi. \tag{10.27}$$

The idea of Eberhard Hopf was to consider the approximation

$$\frac{\partial u_\varepsilon}{\partial t} + \frac{\partial f(u_\varepsilon)}{\partial x} - \varepsilon \frac{\partial^2 u_\varepsilon}{\partial x^2} = 0$$

and to multiply it by $\varphi'(u_\varepsilon)$; one obtains

$$\frac{\partial \varphi(u_\varepsilon)}{\partial t} + \frac{\partial \psi(u_\varepsilon)}{\partial x} - \varepsilon \frac{\partial^2 \varphi(u_\varepsilon)}{\partial x^2} + \varphi''(u_\varepsilon) \left(\frac{\partial u_\varepsilon}{\partial x} \right)^2 = 0,$$

---

[47] If u is a solution and $a \in \mathbb{R}$, $a + u(x - at, t)$ is a solution.

so that if u_ε converges a.e. to u, then (10.27) is true. Peter Lax called (10.27) an "entropy condition", but it is better to follow Costas Dafermos and call it an E-condition.

Krushkov showed uniqueness of the solution satisfying (10.27) for a more general equation than (10.26) (using $x \in \mathbb{R}^N$), and using the family of "entropies/entropy flux" φ_k, ψ_k defined by

$$\varphi_k(v) = (v - k)_+, \psi_k(v) = \begin{cases} 0 \text{ for } v < k, \\ f(v) - f(k) \text{ for } k < v. \end{cases}$$

Peter Lax considered systems in conservative form

$$\frac{\partial U}{\partial t} + \frac{\partial (F(U))}{\partial x} = 0, U(x, t) \in \mathbb{R}^p,$$

which are *strictly hyperbolic*, in the sense that

(the matrix) $F'(U)$ has p distinct real eigenvalues,

and an "entropy" φ and its "entropy flux" are defined by

$$\psi' = \varphi' F',$$

but unlike the scalar case where every function is an entropy, the conditions for a system are quite constraining, besides the "trivial entropies" $\varphi_j = U_j, \psi_j = F(U)_j$.

If one writes the equation for a non-linear string as a system

$$\begin{cases} \dfrac{\partial u}{\partial t} - \dfrac{\partial v}{\partial x} = 0, \\ \dfrac{\partial v}{\partial t} - \dfrac{\partial f(u)}{\partial x} = 0 \end{cases} \tag{10.28}$$

(u the strain, v the velocity, $\sigma = f(u)$ the stress), then

(10.28) is strictly hyperbolic if and only if $f'(u) > 0$ for all u.

An entropy/entropy flux pair $\big(\varphi(u, v), \psi(u, v)\big)$ means

$$\frac{\partial \psi}{\partial v} = -\frac{\partial \varphi}{\partial u},$$

$$\frac{\partial \psi}{\partial u} = -f'(u)\frac{\partial \varphi}{\partial v},$$

so that φ must satisfy the compatibility condition

$$\frac{\partial^2 \varphi}{\partial u^2} = f'(u)\frac{\partial^2 \varphi}{\partial v^2}. \tag{10.29}$$

Besides the trivial cases $\varphi_1 = u, \psi_1 = -v$, $\varphi_2 = v, \psi_2 = -f(u)$, there are infinitely many other solutions of (10.29), but two are special:

$$\text{the } \textit{total energy } \varphi = \frac{v^2}{2} + F(u); \quad \psi = v f(u),$$

$$\text{with } F(s) = \int^s f(z) \, dz, \tag{10.30}$$

$$\varphi = uv; \quad \psi = -\frac{v^2}{2} - u f(u) + F(u). \tag{10.31}$$

One sees why calling φ an "entropy" may be misleading, since in (10.30) φ is the (density of) total energy, sum of the density of kinetic energy $\frac{v^2}{2}$ (since v is the vertical velocity $\frac{\partial w}{\partial t}$) and a density of potential energy $F(u)$ (i.e. $F\left(\frac{\partial w}{\partial x}\right)$).

Since (10.20) has a Hamiltonian structure, the conservation of the total energy is a consequence of the invariance of the equation by translations in t.

The invariance of (10.20) by translations in x is related to the conservation of *linear momentum*, so that in (10.31) the function φ may be considered as a density of linear momentum.

My interest in the genesis of ideas in continuum mechanics was initiated by a historical section in the book by Courant and Friedrichs.[48] Many references are hard to find, but there is a copy of the complete works of Stokes (which he edited) in the library of CMU (Carnegie Mellon University), which is not so surprising since CarnegieTech (Carnegie Institute of Technology) was an engineering school in the days when Pittsburgh was famous for its steel industry. I was surprised to read that Stokes had not repeated his 1848 derivation of the jump conditions: he was (wrongly) convinced by Lord Rayleigh and Thomson (not yet Lord Kelvin) that his solution must be wrong since it does not conserve energy: I thought that the intuition of internal energy is clear, although I found the rules of thermodynamics curious.

Poisson corrected the discrepancy about the speed of sound by introducing an *ad hoc* parameter γ, so that he had an *engineering model*, good enough for making predictions[49] but not for providing a scientific explanation of the phenomenon.

I consider thermodynamics to be not so scientific, since it consists of saying that one can talk about the macroscopic behaviour of materials near their equilibrium in temperature without really knowing what happens at

---

[48] In 1997, Cathleen Morawetz told me that her advisor (Friedrichs) had asked her to proofread his book: many other historical references were suppressed, since the publisher was finding the book too long.

[49] Saint Venant (1797–1886) mentioned one motivation: to compute the exit velocity of a shell out of the barrel of a gun, to understand whether it would be higher by changing the length of the barrel!

mesoscopic scales, and it says nothing about what happens during transitory periods. The explanation given for the discrepancy in the speed of sound is that propagation occurs too fast for leaving time to arrive at an equilibrium in temperature, so that the process is *adiabatic* (no heat exchange), or equivalently *isentropic*,[50] and other rules apply.

Stokes even argued against his 1848 discontinuous solutions by saying that even a small viscosity forces solutions to be continuous. However, viscosity means dissipation of energy, contradicting also the conservation of energy!

Thermodynamics "solved" the problem by inventing names, *heat* or *internal energy*, but being unable to say much about how it evolves, apart from some inequalities, since it failed to describe how energy finds a way to hide at mesoscopic scales.

I knew in the mid 1970s that the flaws of thermodynamics could only be repaired by improving the theory of *homogenization*, which explained without probabilities what happens: the difference between weak convergence and strong convergence.

There is a belief in the theory of shock waves that only the values on each side of a shock are important, although there was an extensive study of the *internal structure of shocks* by Charles Conley (1933–1984) and Joel Smoller, and it depends upon the form of viscosity which is used.

Besides, viscosity has already postulated a kind of dissipation, and there is no reason to believe that the guessed reason for dissipation is that which is used in reality: one is so used to playing with such "viscous" models that one forgets that they are just engineering models, which mimic something observed but cannot give any scientific understanding about why something like dissipation happens!

Appendix E *Are electromagnetic waves transverse?*

Nowadays, one mentions the *dielectric permittivity* ε_0 and the *magnetic susceptibility* μ_0 of the *vacuum*, related by $\varepsilon_0 \mu_0 c^2 = 1$.

No one cares about calling the vacuum solid, liquid, or gas,[51] but in the nineteenth century, physicists talked about aether (where electromagnetic waves propagate) as if it was "solid": they thought that light propagates by transverse waves.

---

[50] Adiabatic means $\delta Q = 0$, but $\delta Q = \theta \, ds$ for θ the absolute temperature and s the entropy, so that $ds = 0$ (since $\theta > 0$).

[51] These are only intuitive notions in usual circumstances: in a container, a liquid slowly goes down until it is below a horizontal interface, a gas fills all the container, and a solid keeps its shape. However, near the *triple point*, one goes without discontinuity from liquid to gas, with no *latent heat*.

Fresnel had used such an idea before Cauchy and Lamé had written a system of PDEs for elasticity, and (linearized) elastic waves are either longitudinal or transverse in an isotropic medium, but not in an anisotropic medium!

I have not heard of any definition for what longitudinal or transverse waves could mean for a general hyperbolic system.

Since what Maxwell wrote was encumbered with ideas about aether, the form one calls the Maxwell equation is that resulting from the simplifying work of Heaviside, so that I prefer to call it the *Maxwell–Heaviside equation*: it has the form

$$\frac{\partial B}{\partial t} + \mathrm{curl}(E) = 0; \; \mathrm{div}(B) = 0$$

and

$$-\frac{\partial D}{\partial t} + \mathrm{curl}(H) = j; \; \mathrm{div}(D) = \rho,$$

which implies the *conservation of (electric) charge*

$$\frac{\partial \rho}{\partial t} + \mathrm{div}(j) = 0,$$

and I learned from Joel Robbin how to write this with differential forms. The constitutive relation for the vacuum is

$$D = \varepsilon_0 E; \; B = \mu_0 H.$$

If one looks for electromagnetic waves propagating in the direction of a unit vector $e \in \mathbb{R}^3$ in the vacuum, without matter (i.e. $\rho = 0$ and $j = 0$), it means that one uses

$$E(x,t) = E_0\big((x,e) - ct\big); \; H(x,t) = H_0\big((x,e) - ct\big)$$

in the Maxwell–Heaviside equation, and since

$$\mathrm{div}(B) = 0 \text{ means } H_0' \perp e, \quad \mathrm{div}(D) = 0 \text{ means } E_0' \perp e,$$

$$\frac{\partial B}{\partial t} + \mathrm{curl}(E) = 0 \text{ means } -\mu_0 c H_0' + e \times E_0' = 0, \tag{10.32}$$

$$-\frac{\partial D}{\partial t} + \mathrm{curl}(H) = 0 \text{ means } \varepsilon_0 c E_0' + e \times H_0' = 0,$$

E_0' and H_0' are orthogonal to e and to each other. Assuming non-constant fields ($E_0' \neq 0$, $H_0' \neq 0$), (10.32) implies

$$\varepsilon_0 \mu_0 c^2 = 1,$$

which defines the speed of light c in terms of ε_0 and μ_0.

Although the electric and magnetic fields may be orthogonal to e, it does not mean that a charged particle moves in a plane orthogonal to e while the electromagnetic wave passes by.

For simplification, let e^1, e^2, e^3 be an orthonormal basis, and for $e = e^3$ assume that the electric field keeps a constant direction e^1 (so that the magnetic field keeps a constant direction e^2), and that the electromagnetic wave has the form

$$E(x,t) = \psi(x_3 - ct)\,e^1; H(x,t) = \varepsilon_0 c\,\psi(x_3 - ct)\,e^2. \tag{10.33}$$

For $\psi \in L^2_{loc}(\mathbb{R})$, the plane-wave (10.33) has a locally integrable

$$density\ of\ electromagnetic\ energy\ \frac{1}{2}\,(\varepsilon_0|E|^2 + \mu_0|H|^2). \tag{10.34}$$

A charged particle of charge q (and mass m) feels the

$$\text{``Lorentz'' force } q\,(E + v \times B), \tag{10.35}$$

which Lorentz introduced only in 1892, but it had appeared in an 1862 article by Maxwell. Before knowing about (10.35), physicists may have thought that the force is qE, which was probably the intuition behind *Ohm's law*.

Before studying the ODE governing how the particle moves when the electromagnetic wave passes by, there is a philosophical question to address: one has argued about which equations to use when there is no matter, and then one uses them in the presence of a charged particle, which is matter!

One is discussing an *interaction between light* (meaning electromagnetic waves) *and matter*, and light is described by a PDE, the Maxwell–Heaviside equation, but there is no PDE used for describing matter.

Since *one cannot expect to describe an interaction with a linear equation*, it only looks like a *first level of approximation*.

Maybe one could start with a system of PDEs (not shown) describing both light and matter, and for deriving a formula like the "Lorentz" force (10.35) one would fix matter, maybe by considering a fixed small obstacle and computing the electromagnetic field outside it with adequate boundary conditions related to a total charge q, and then compute the forces acting on the boundary of this small obstacle.

Since the "Lorentz" force implies a change in the linear momentum, the work of Poincaré of studying which waves are created in the electromagnetic field for conserving momentum was like a *second level of approximation*, which made him discover in 1900 that the density of electromagnetic energy (10.34) is equivalent to a density of mass, with the formula $e = mc^2$.

Of course, even if one knew from which interacting system of PDEs to start, and one imagined a list of successive levels of approximation, it might not necessarily be a convergent scheme: Feynman later described his method of using diagrams, which create a book-keeping problem of all the terms, and a question of how to "sum" and avoid the divergence of the scheme.

Here, using $B = \mu_0 H$, the position $x(t)$ of the particle satisfies

$$m \frac{d^2 x}{dt^2} = q \left(E + \frac{dx}{dt} \times \mu_0 H \right).$$

In the particular case (10.33), using $'$ instead of $\frac{d}{dt}$, it gives

$$mx_1'' = q\psi(x_3 - ct)\left(1 - \frac{x_3'}{c}\right),$$

$$mx_2'' = 0, \tag{10.36}$$

$$mx_3'' = q\psi(x_3 - ct)\frac{x_1'}{c}.$$

If the particle starts *at rest* at a, it stays in the plane $x_2 = a_2$, but unless $\psi = 0$ its component x_3 changes, so that *the transverse character of the electromagnetic waves is not exact.*

Multiplying the first equation of (10.36) by x_1', one has

$$mx_1''x_1' = q\psi(x_3 - ct)\left(1 - \frac{x_3'}{c}\right)x_1' = mcx_3''\left(1 - \frac{x_3'}{c}\right), \tag{10.37}$$

so that if ψ has compact support

$$(x_1')^2 + (x_3')^2 - 2cx_3' = 0, \tag{10.38}$$

since it must be constant by integrating (10.37), and the constant is 0 for large negative time (before the wave arrives).

This relation is independent of ψ, and expresses that (x_1', x_3') lies on a circle centered at $(0, c)$ and with radius c.

One defines Ψ by

$$\Psi(s) = -\int_s^\infty \psi(\sigma)\,d\sigma \quad \text{for } s \in \mathbb{R},$$

so that $\Psi' = \psi$, and

$$x_1' = -\frac{q}{mc}\Psi(x_3 - ct) \quad \text{for all } t \in \mathbb{R},$$

since the derivatives (in t) are the same by the first equation of (10.36), and both sides are 0 for large negative t. By (10.38)

$$(x_3')^2 - 2cx_3' = -\frac{q^2}{m^2c^2}\Psi^2(x_3 - ct) \quad \text{for all } t \in \mathbb{R}.$$

If the electromagnetic wave is *strong enough*, in the sense that

$$|\Psi(z)| > \frac{mc^2}{q} \text{ for some } z \in \mathbb{R}, \tag{10.39}$$

then

$$x_3(t) > z + ct \text{ for all } t \in \mathbb{R}, \tag{10.40}$$

since the inequality is true for large negative t, and it is impossible to have $x_3(t_0) = z + ct_0$ for some $t_0 \in \mathbb{R}$, since (10.39)–(10.40) would imply $|x_1'(t_0)| > c$, contradicting (10.38).

It means that every non-zero ψ gives an electromagnetic wave which sweeps all the charged particles with a ratio $\frac{m}{q}$ small enough (for (10.39) to be valid): how can one call such electromagnetic waves transverse?

Of course, the preceding computations were carried out in a classical (non-relativistic) way. Since (10.40) implies that x_3' must take values $\geq c - \varepsilon$ for all $\varepsilon > 0$, which implies that $|x_1'(s)|$ takes values near c, it means that $|x'|$ takes values near $\sqrt{2}c$, corresponding to a kinetic energy near mc^2, but relativistic corrections preclude this to happen, of course!

Appendix F *Invariants of the wave equation, and the "Lorentz" group of Poincaré*

In my graduate course in Madison (WI) in 1974–75, I wanted to explain how to use invariants of the wave equation to prove some local decay of the energy of solutions of the wave equation $u_{tt} - \Delta u = 0$ outside a star-shaped obstacle,[52] according to articles by Cathleen Morawetz, Peter Lax, and Walter Strauss. Since I was reading about three (quadratic) invariants using $u, u_t, u_r, |\text{grad}(u)|^2$, with coefficients depending only on t, r, I decided to discover all the quadratic invariants in u and its partial derivatives in x and t, with coefficients depending upon x and t. I was helped by discussions with Joel Robbin, which in some way made me (re)discover an idea which must go back to Lie (1842–1899) that non-linear questions on a Lie group may become linear questions on its Lie algebra, and (re)discover an idea attributed to (Emmy) Noether (1882–1935), that groups of symmetries are related to conservation laws.

Actually, one difference is that I worked with PDEs, and not with Lagrangians, which I do not consider good physics.[53]

---

[52] A common "mistake" among mathematicians was to use a characteristic velocity equal to 1.
[53] The physics is like a Cauchy problem, with $u(0)$ and $u_t(0)$ given, and not with $u(0)$ and $u(T)$ given.

It is traditional to say that solutions of the *wave equation*

$$\frac{\partial^2 u}{\partial t^2} - c^2 \Delta u = 0 \tag{10.41}$$

satisfy

$$\int \frac{1}{2} \left(\left| \frac{\partial u}{\partial t} \right|^2 + c^2 |\mathrm{grad}(u)|^2 \right) dx = \text{constant}, \tag{10.42}$$

which is *conservation of total energy*, but this requires some hypotheses, like

$$u\Big|_{t=0} = u_0 \in H^1; \frac{\partial u}{\partial t}\Big|_{t=0} = u_1 \in L^2, \tag{10.43}$$

which ensures that (10.41) has a unique solution (in the sense of distributions) with such initial data, satisfying

$$u \in C(\mathbb{R}; H^1); \frac{\partial u}{\partial t} \in C(\mathbb{R}; L^2),$$

so that (10.43) makes sense, and is true. Moreover, there is a finite speed c of propagation of information, i.e.

$$\text{if support } (u_0), \text{ support } (u_1) \subset K, \text{ then}$$
$$u(x,t) = 0 \text{ if } \mathrm{dist}(x;K) > c\,|t|. \tag{10.44}$$

(10.44) permits one to generalize (10.42) to situations where (10.41) and (10.43) are not necessarily valid, but u is smooth enough:

$$\frac{\partial u}{\partial t} \left(\frac{\partial^2 u}{\partial t^2} - c^2 \Delta u \right)$$
$$= \frac{\partial}{\partial t} \left(\frac{1}{2} \left| \frac{\partial u}{\partial t} \right|^2 + \frac{c^2}{2} |\mathrm{grad}(u)|^2 \right) - \sum_j \frac{\partial}{\partial x_j} \left(c^2 \frac{\partial u}{\partial t} \frac{\partial u}{\partial x_j} \right). \tag{10.45}$$

Similarly, one has (for all k)

$$\frac{\partial u}{\partial x_k} \left(\frac{\partial^2 u}{\partial t^2} - c^2 \Delta u \right) = \frac{\partial}{\partial t} \left(\frac{\partial u}{\partial t} \frac{\partial u}{\partial x_k} \right) - \sum_j \frac{\partial}{\partial x_j} \left(c^2 \frac{\partial u}{\partial x_k} \frac{\partial u}{\partial x_j} \right) - \frac{\partial (\text{action})}{\partial x_k},$$

$$\text{with } action = \frac{1}{2} \left| \frac{\partial u}{\partial t} \right|^2 - \frac{c^2}{2} |\mathrm{grad}(u)|^2. \tag{10.46}$$

A first natural question then is to find which combinations

$$v = \alpha \frac{\partial u}{\partial t} + \sum_k \beta_k \frac{\partial u}{\partial x_k} + \gamma u \tag{10.47}$$

(with α, β_k for all k, and γ functions of (x,t)) are such that

$$v \left(\frac{\partial^2 u}{\partial t^2} - c^2 \Delta u \right) = \frac{\partial A}{\partial t} + \sum_j \frac{\partial B_j}{\partial x_j}, \tag{10.48}$$

a *conservative form* having A and all the B_j quadratic (in u, $\frac{\partial u}{\partial t}$, and $\mathrm{grad}(u)$), with coefficients depending upon (x, t).

Besides (10.45) and (10.46), one also needs to use

$$u\left(\frac{\partial^2 u}{\partial t^2} - c^2 \Delta u\right) =$$
$$\frac{\partial}{\partial t}\left(u\frac{\partial u}{\partial t}\right) - \sum_j \frac{\partial}{\partial x_j}\left(c^2 u\frac{\partial u}{\partial x_j}\right) - 2\,\text{action}.$$

One finds

$$A = \alpha\left(\frac{1}{2}\left|\frac{\partial u}{\partial t}\right|^2 + \frac{c^2}{2}|\mathrm{grad}(u)|^2\right) + \sum_k \beta_k \frac{\partial u}{\partial t}\frac{\partial u}{\partial x_k} + \gamma\, u\frac{\partial u}{\partial t} - \frac{1}{2}\frac{\partial \gamma}{\partial t}u^2,$$

$$B_j = -\alpha c^2 \frac{\partial u}{\partial t}\frac{\partial u}{\partial x_j} - \beta_j \,\text{action} - \sum_k \beta_k c^2 \frac{\partial u}{\partial x_k}\frac{\partial u}{\partial x_j} \qquad (10.49)$$
$$\qquad - \gamma c^2 u\frac{\partial u}{\partial x_j} + \frac{c^2}{2}\frac{\partial \gamma}{\partial x_j}u^2, \text{ for all } j,$$

and a list of *linear* differential constraints on the coefficients (α, β_k for all k, and γ): terms in $\left|\frac{\partial u}{\partial t}\right|^2$ give

$$-\frac{1}{2}\frac{\partial \alpha}{\partial t} + \frac{1}{2}\sum_k \frac{\partial \beta_k}{\partial x_k} - \gamma = 0,$$

terms in $\frac{\partial u}{\partial t}\frac{\partial u}{\partial x_j}$ give (for all j)

$$c^2\frac{\partial \alpha}{\partial x_j} - \frac{\partial \beta_j}{\partial t} = 0,$$

terms in $\frac{\partial u}{\partial x_j}\frac{\partial u}{\partial x_k}$ (for all $j \neq k$) give

$$c^2\left(\frac{\partial \beta_j}{\partial x_k} + \frac{\partial \beta_k}{\partial x_j}\right) = 0,$$

terms in $\left(\frac{\partial u}{\partial x_j}\right)^2$ (for all j) give

$$-\frac{c^2}{2}\frac{\partial \alpha}{\partial t} + \frac{c^2}{2}\frac{\partial \beta_j}{\partial x_j} + \gamma = 0,$$

and terms in u^2 give

$$\frac{1}{2}\left(\frac{\partial^2 \gamma}{\partial t^2} - c^2 \Delta \gamma\right) = 0.$$

A second natural question is to find when (10.47) gives

$$\frac{\partial^2 u}{\partial t^2} - c^2 \Delta u = 0 \text{ implies } \frac{\partial^2 v}{\partial t^2} - c^2 \Delta v = 0,$$

in order to deduce a conservative form by using the identity

$$v\left(\frac{\partial^2 u}{\partial t^2} - c^2 \Delta u\right) - u\left(\frac{\partial^2 v}{\partial t^2} - c^2 \Delta v\right) = \frac{\partial}{\partial t}\left(v\frac{\partial u}{\partial t} - u\frac{\partial v}{\partial t}\right)$$
$$- c^2 \sum_j \frac{\partial}{\partial x_j}\left(v\frac{\partial u}{\partial x_j} - u\frac{\partial v}{\partial x_j}\right).$$

This amounts to a question of commutator:[54]

$$\left[\frac{\partial^2}{\partial t^2} - c^2\Delta, \; \alpha\frac{\partial}{\partial t} + \sum_k \beta_k\frac{\partial}{\partial x_k} + \gamma\right] = d\left(\frac{\partial^2}{\partial t^2} - c^2\Delta\right), \qquad (10.50)$$

for a function d, giving a (different) list of *linear* differential constraints on α, β_k for all k, and γ, involving d.

Identifying the coefficients of $\frac{\partial^2}{\partial t^2}$ in (10.50) gives

$$2\frac{\partial\alpha}{\partial t} = d,$$

the coefficients of $\frac{\partial^2}{\partial t\partial x_j}$ for all j give

$$-2c^2\frac{\partial\alpha}{\partial x_j} + 2\frac{\partial\beta_j}{\partial t} = 0,$$

the coefficients of $\frac{\partial^2}{\partial x_j\partial x_k}$ for all $j \neq k$ give

$$-2c^2\left(\frac{\partial\beta_j}{\partial x_k} + \frac{\partial\beta_k}{\partial x_j}\right) = 0,$$

the coefficients of $\frac{\partial^2}{\partial x_j^2}$ for all j give

$$-2c^2\frac{\partial\beta_j}{\partial x_j} = -d,$$

the coefficients of $\frac{\partial}{\partial t}$ give

$$\frac{\partial\alpha}{\partial t^2} - c^2\Delta\alpha + 2\frac{\partial\gamma}{\partial t} = 0,$$

the coefficients of $\frac{\partial}{\partial x_k}$ for all k give

$$\frac{\partial\beta_k}{\partial t^2} - c^2\Delta\beta_k - 2c^2\frac{\partial\gamma}{\partial x_k} = 0,$$

and the constant coefficient gives

$$\frac{\partial\gamma}{\partial t^2} - c^2\Delta\gamma = 0.$$

[54] For two operators L_1, L_2, $[L_1, L_2]$ means $L_1 L_2 - L_2 L_1$.

One finds all the solutions easily, and in three space dimensions, five parameters correspond to $d \neq 0$ (with d a polynomial of degree ≤ 1),[55] and ten parameters correspond to $d = 0$.

Among these ten, four correspond to $\frac{\partial}{\partial t}$ and $\frac{\partial}{\partial x_j}$, and are related to the invariance of the wave equation by translations; three correspond to $x_j \frac{\partial}{\partial x_k} - x_k \frac{\partial}{\partial x_j}$ for $j \neq k$ and are related to the invariance of the wave equation by rotations; the other three correspond to

$$x_j \frac{\partial}{\partial t} + c^2 t \frac{\partial}{\partial x_j}, \tag{10.51}$$

and are related to the invariance of the wave equation by transformations introduced by Lorentz.

For $x \in \mathbb{R}$, one looks at transformations

$$U(x,t) = u(\alpha x + \beta t, \gamma x + \delta t), \tag{10.52}$$

and one wants that

$$U_{tt} - c^2 U_{xx} = u_{tt} - c^2 u_{xx}, \tag{10.53}$$

the left side being evaluated at (x,t), and the right side being evaluated at $(\alpha x + \beta t, \gamma x + \delta t)$.

Expressing U_{tt} and U_{xx} in terms of u_{tt}, u_{tx}, u_{xx} (10.53) requires

$$\delta^2 - c^2 \gamma^2 = 1; \quad \beta \delta - c^2 \alpha \gamma = 0; \quad \beta^2 - c^2 \alpha^2 = -c^2,$$

from which one deduces $\beta^2 = c^4 \gamma^2$. If one looks for the branch near $(1,0,0,1)$, one uses $c\gamma = \sinh \lambda$, and one deduces easily

$$\alpha = \delta = \cosh \lambda; \quad \beta = c \sinh \lambda; \quad \gamma = \frac{1}{c} \sinh \lambda, \text{ for } \lambda \in \mathbb{R}. \tag{10.54}$$

Then the combination (10.51) appears by taking the derivative of $u(\alpha x + \beta t, \gamma x + \delta t)$ with respect to λ, at $\lambda = 0$.

Galilean invariance usually means that, moving at velocity w, the function $u(x - wt, t)$ satisfies the same equation as u,[56] but the wave equation is not Galilean invariant, and one should use (10.52) with (10.54) instead.

---

[55] $t \frac{\partial}{\partial t} + \sum_j x_j \frac{\partial}{\partial x_j}$ corresponds to $d = 2$, and it is related to the fact that if u solves the (homogeneous) wave equation, then for any $\lambda \neq 0$, $u(\lambda x, \lambda t)$ also solves the wave equation.

[56] If u is a velocity, one must consider $w + u(x - wt, t)$.

It is $\frac{\beta}{\alpha}$ which plays the role of $-w$, so that $\frac{w}{c} = -\tanh\lambda$, hence $\cosh\lambda = \frac{1}{\sqrt{1-w^2/c^2}}$ and $\sinh\lambda = \frac{-w}{c\sqrt{1-w^2/c^2}}$, and the *Lorentz transformation* is given by

$$u\left(\frac{x-wt}{\sqrt{1-w^2/c^2}}, \frac{-wx+c^2t}{c^2\sqrt{1-w^2/c^2}}\right).$$

The factor $\sqrt{1-w^2/c^2}$ is called the *Fitzgerald contraction*, since Fitzgerald had first proposed that matter contracts in its direction of motion, as a way to "explain" the measurements of Michelson and Morley, that the speed of light is c in every direction.

Without a change of time, introduced later by Lorentz, there would be a contradiction,[57] but there is no contradiction in having a group of matrices $M(s)$ all having diagonal elements ≥ 1:

$$M(s) = \begin{pmatrix} \cosh s & c\sinh s \\ c^{-1}\sinh s & \cosh s \end{pmatrix} \text{ satisfies}$$

$$M(s)M(\sigma) = M(s+\sigma) \text{ for all } s,\sigma \in \mathbb{R},$$

and $M(s)$ is actually an exponential:

$$M(s) = e^{sJ} \text{ for } s \in \mathbb{R}, \text{ with } J = \begin{pmatrix} 0 & c \\ c^{-1} & 0 \end{pmatrix}.$$

The Maxwell–Heaviside equation in the vacuum is

$$\mu_0\frac{\partial H}{\partial t} + \mathrm{curl}(E) = 0,$$

$$\mu_0\mathrm{div}(H) = 0,$$

$$-\varepsilon_0\frac{\partial E}{\partial t} + \mathrm{curl}(H) = j,$$

$$\varepsilon_0\mathrm{div}(E) = \rho,$$

which imply

$$\frac{\partial^2 H}{\partial t^2} - c^2\Delta H = c^2\mathrm{curl}(j),$$

$$\frac{\partial^2 E}{\partial t^2} - c^2\Delta E = -\frac{c^2}{\varepsilon_0}\mathrm{grad}(\rho) - \mu_0 c^2\frac{\partial j}{\partial t},$$

since $\varepsilon_0\mu_0 c^2 = 1$, and $\mathrm{curl}\big(\mathrm{curl}(V)\big) - \mathrm{grad}\big(\mathrm{div}(V)\big) = -\Delta V$.

Although each component of E and H satisfies a homogeneous wave equation if there is no matter ($\rho = 0$ and $j = 0$), it is useful to work directly on

---

[57] For two observers in frames moving at constant velocity, one could not observe a contraction in both frames.

the Maxwell–Heaviside equation, as for the *conservation of electromagnetic energy*

$$\frac{\partial}{\partial t}\left(\frac{\mu_0}{2}|H|^2 + \frac{\varepsilon_0}{2}|E|^2\right) + \text{div}(E \times H) = 0,$$

where $E \times H$ is the *Poynting vector*, since

$$(\text{curl}(E),H) - (\text{curl}(H),E) = \sum_i\left(\sum_{j,k}\varepsilon_{i,j,k}\frac{\partial E_k}{\partial x_j}\right)H_i - \sum_k\left(\sum_{j,i}\varepsilon_{k,j,i}\frac{\partial H_i}{\partial x_j}\right)E_k$$

$$= \sum_{i,j,k}\varepsilon_{i,j,k}\frac{\partial(E_kH_i)}{\partial x_j} = \sum_j\frac{\partial}{\partial x_j}\left(\sum_{i,k}\varepsilon_{j,k,i}E_kH_i\right).$$

Inside matter one uses the constitutive relation

$$D = \varepsilon E; \ B = \mu H,$$

with the *dielectric permittivity* ε and the *magnetic susceptibility* μ being symmetric positive definite matrices, and the density of electromagnetic energy is

$$\frac{(\varepsilon E,E)}{2} + \frac{(\mu H,H)}{2}, \text{ i.e. } \frac{(D,E)}{2} + \frac{(B,H)}{2}.$$

If the matrices ε and μ are discontinuous on a smooth interface,

the tangential components of E,H are continuous,

the normal components of D,B are continuous.

On the boundary of a *perfect conductor*, the tangential component of E is 0. On the boundary of a *perfect insulator*, the normal component of D is 0. In general, boundary conditions in electromagnetism are not as simple as Dirichlet or Neumann conditions, and they tend to be frequency dependent.

The plane waves in a unit direction e, propagating with velocity v, are non-constant solutions of the form

$$E\big((x,e) - vt\big); \ H\big((x,e) - vt\big),$$

and the (non-zero) derivatives E' and H' must satisfy

$$-v\mu H' + e \times E' = 0; \ (\mu H',e) = 0,$$
$$v\varepsilon E' + e \times H' = 0; \ (\varepsilon E',e) = 0,$$

(10.55)

and since $v \neq 0$ (because ε and μ are positive definite), the second equation of each line is a consequence of the first equation.

One makes the change of notation

$$M = \varepsilon^{-1/2}\mu\varepsilon^{-1/2}; \ E' = \varepsilon^{-1/2}\widetilde{E}; \ H' = \varepsilon^{-1/2}\widetilde{H},$$

and (10.55) becomes

$$M\tilde{H} - \eta \times \tilde{E} = 0; \ \tilde{E} + \eta \times \tilde{H} = 0,$$

with

$$\eta = \frac{1}{v\sqrt{\det(\varepsilon)}} \varepsilon^{1/2} e, \tag{10.56}$$

and one has used the formula $A^{\mathrm{T}}(A a \times A b) = \det(A)\,(a \times b)$ for every matrix A and vectors a, b (all this being done in $\mathbb{R}^3$).

The equation for $\tilde{H}$ is then

$$M\tilde{H} + \eta \times (\eta \times \tilde{H}) = 0,$$

and using the formula $a \times (b \times c) = (a, c)\,b - (a, b)\,c$, it becomes

$$(M + \eta \otimes \eta - |\eta|^2 I)\tilde{H} = 0,$$

and for a non-zero $\tilde{H}$ to exist one must have

$$\det(M + \eta \otimes \eta - |\eta|^2 I) = 0.$$

One has the following identity:[58]

$$\det(M + a \otimes a - |a|^2 I)$$
$$= \det(M) - \det(M)\,\mathrm{tr}(M^{-1})|a|^2 + \det(M)\,(M^{-1}a, a) + |a|^2 (Ma, a).$$

The next step is to show that there are waves propagating in every direction, with two different speeds of propagation except for very special directions with only one speed.

If $M = mI$ (corresponding to $\mu = m\varepsilon$), the equation becomes $m\,(m - |\eta|^2)^2 = 0$, and by (10.56) there is only one speed in direction e, which is

$$v = \frac{|\varepsilon^{1/2} e|}{\sqrt{m \det(\varepsilon)}} \quad (\text{in the case } \mu = m\varepsilon).$$

In the general case, one chooses a basis of eigenvectors for M, ordering eigenvalues as $\lambda_1 \leq \lambda_2 \leq \lambda_3$, and one calls f the unit vector in the direction of η given by (10.56), with components f_1, f_2, f_3, and one applies the formula to $a = \sigma f$, and it gives a quadratic equation for $s = \sigma^2$:

$$s^2 \sum_j \theta_j \lambda_j + s\lambda_1\lambda_2\lambda_3 \sum_j \frac{\theta_j - 1}{\lambda_j} + \lambda_1\lambda_2\lambda_3 = 0, \tag{10.57}$$

---

[58] In a basis with first vector parallel to a, the determinant is $\det(M) - |a|^2 (m_{11}m_{33} - m_{13}m_{31}) - |a|^2 (m_{11}m_{22} - m_{12}m_{21}) + |a|^4 m_{11}$. One notices that $(Ma, a) = |a|^2 m_{11}$ and that the diagonal coefficients of $\det(M)M^{-1}$ are $(m_{22}m_{33} - m_{23}m_{32})$, $(m_{11}m_{33} - m_{13}m_{31})$ and $(m_{11}m_{22} - m_{12}m_{21})$.

where θ_j denotes f_j^2 (with sum 1 in j). If one denotes

$$X = \sum_j \theta_j \lambda_j; \quad Y = \sum_j \frac{\theta_j}{\lambda_j},$$

the point (X, Y) belongs to the triangle of vertices $\left(\lambda_j, \frac{1}{\lambda_j}\right)$, so that it belongs to the convex hull of the piece of hyperbola $\left(\lambda, \frac{1}{\lambda}\right)$ for $\lambda_1 \leq \lambda \leq \lambda_3$, hence X and Y satisfy

$$\lambda_1 \leq X \leq \lambda_3; \quad \frac{1}{X} \leq Y \leq \frac{\lambda_1 + \lambda_3 - X}{\lambda_1 \lambda_3}.$$

The discriminant of the quadratic equation (10.57) is

$$(\lambda_1 \lambda_2 \lambda_3)^2 \left(\sum_j \frac{\theta_j - 1}{\lambda_j}\right)^2 - 4\lambda_1 \lambda_2 \lambda_3 \sum_j \theta_j \lambda_j,$$

and for X given it is a minimum for Y at its maximum value, i.e. $Y = \frac{\lambda_1 + \lambda_3 - X}{\lambda_1 \lambda_3}$, when the value of the discriminant is

$$(\lambda_1 \lambda_2 \lambda_3)^2 \left(\frac{X}{\lambda_1 \lambda_3} + \frac{1}{\lambda_2}\right)^2 - 4\lambda_1 \lambda_2 \lambda_3 X = (\lambda_1 \lambda_2 \lambda_3)^2 \left(\frac{X}{\lambda_1 \lambda_3} - \frac{1}{\lambda_2}\right)^2.$$

There are solutions for every direction e, with two different speeds of propagation, except when f satisfies

$$\sum_j |f_j|^2 \lambda_j = \frac{\lambda_1 \lambda_3}{\lambda_2},$$
$$\sum_j \frac{|f_j|^2}{\lambda_j} = \frac{1}{\lambda_1} + \frac{1}{\lambda_3} - \frac{1}{\lambda_2}. \tag{10.58}$$

In the case $\lambda_1 < \lambda_2 < \lambda_3$, (10.58) only happens if $f_2 = 0$, and one finds exactly four special directions.[59]

If two eigenvalues coincide (different from the third one), (10.58) defines the eigen-direction of the simple eigenvalue.

Since the speed of propagation of a plane wave is not so simple a question for the Maxwell–Heaviside equation in an anisotropic media, one can guess that the question of reflection and refraction of such waves at an interface may lead to quite intricate problems, of elementary geometry or linear algebra.

Appendix G *What is matter?*

Even before I entered École Polytechnique (where I studied from 1965 to 1967), I was taught that in the beginning of the twentieth century, physicists "understood" that "elementary particles" could not be just points, and the

---

[59] The linear system gives $f_1^2 = \frac{\lambda_3(\lambda_2 - \lambda_1)}{\lambda_2(\lambda_3 - \lambda_1)}$, $f_3^2 = \frac{\lambda_1(\lambda_3 - \lambda_2)}{\lambda_2(\lambda_3 - \lambda_1)}$.

explanation I was given is that "a point would radiate energy", whatever that means!

During my first week in the class of *mathématiques supérieures* (in the Fall of 1963), I had been taught (by my mathematics teacher, obviously) that if a proposition A is false, then the statement 'A implies B' is true, whatever B is.

I then wondered for many years why, when one tells a "physicist" that one of his/her hypotheses A is flawed, he/she usually answers something like 'but it shows B, which is observed', a sign that he/she does not understand much about logical thinking, and confuses 'A implies B' with 'B implies A'.

At École Polytechnique, I was taught the curious dogmas of quantum mechanics, and it was my first hint that some "physicists" behave like religious fundamentalists, forcing people to adhere to silly dogmas which earlier generations invented.

Later (when I was already a university professor), I wondered why one loses time arguing for or against a "big-bang": if matter cannot travel faster than the speed of light c, then a (mathematical) consequence is that the question 'creation versus non-creation' becomes undecidable.

However, there is a social reason for criticizing the hypotheses of the "physicists" who argue for the big-bang, since they mention temperatures of billions of degrees, and there is ample evidence that one does not yet know the laws that matter follows at millions of degrees:[60] since the mid 1950s, physicists have boasted that they were going to control thermonuclear fusion, and billions of dollars have been spent, but if one builds huge lasers (the size of buildings) to shoot at little pellets filled with deuterium or tritium, is not the reason that one knows the postulated laws to be incorrect, and one needs experimental data to get a better understanding of what happens?

One claims that *matter cannot move faster than the speed of light c*, without defining what matter is! Lorentz and Poincaré thought that the mass of an electron cannot be only of electromagnetic origin, but the Maxwell–Heaviside equation describes "light" (meaning electromagnetic waves), and not matter (since ρ and j must be given in the Maxwell–Heaviside equation), and an electron is "matter" (whatever that is), so they must have argued from some intuitive model.

There is a difference between what physicists call a *phase velocity* and a *group velocity*.

Phase velocity is the velocity of plane waves, and electromagnetic plane waves in the vacuum *move at the speed of light* in every direction; to show that

---

[60] Some laws were postulated, maybe using the dogmas of statistical mechanics, but even if all the physicists in the world accept the use of a particular law, would it force nature to use it?

group velocity is $\leq c$, one should prove that if initial data have support in K at $t = 0$, then at $t = T$ the support is included in $\{x \mid dist(x;K) \leq c|T|\}$.

In the case of a homogeneous medium, I was taught a proof using an *elementary solution*, but the concept does not apply to variable coefficients (although one may talk of a Green kernel), so I proved such an extension in my 1974–75 course.

I consider a scalar wave equation

$$\rho \frac{\partial^2 u}{\partial t^2} - \operatorname{div}\left(A \operatorname{grad}(u)\right) = 0 \text{ in } \mathbb{R}^N \times (0, T), \qquad (10.59)$$

with ρ and A depending only upon x, and

$$\rho \in L^\infty(\mathbb{R}^N), \ \rho(x) \geq \rho_- > 0 \text{ a.e. } x,$$
$$A \in L^\infty\left(\mathbb{R}^N; \mathcal{L}_{\mathrm{sym}}(\mathbb{R}^N, \mathbb{R}^N)\right), \ A(x) \geq \alpha I \text{ a.e. } x; \ \alpha > 0.$$

Existence is known by methods of functional analysis, with initial data in $V = H^1(\mathbb{R}^N)$ for u, and in $H = L^2(\mathbb{R}^N)$ for $\frac{\partial u}{\partial t}$.

If the initial data have support in a compact set K:

if $(A(x)\xi, \xi) \leq \gamma^2 \rho(x) |\xi|^2$ for all $\xi \in \mathbb{R}^N$ a.e. x, then

the solution has support in $\{(x, t) \mid dist(x;K) \leq \gamma |t|\}$.

I wrote a more precise statement, as well as extensions to the Maxwell–Heaviside equation and to the system of linearized elasticity (for proving bounds in homogenization) in 1997 on the occasion of a conference in memory of Ennio De Giorgi (1928–1996, 1990 Wolf Prize).

The more precise statement uses (an upper bound of) the speed of propagation in the direction of a unit vector ξ:

if $K \subset \{x \mid a_- \leq (x, \xi) \leq a_+\}$, if $(A(x)\xi, \xi) \leq \gamma^2(\xi) \rho(x)$ a.e. $x \in \mathbb{R}^N$, then

the solution has support in $\{(x, t) \mid a_- - \gamma(\xi)|t| \leq (x, \xi) \leq a_+ + \gamma(\xi)|t|\}$.

One multiplies (10.59) by $\varphi \frac{\partial u}{\partial t}$, for φ smooth in (x, t):

$$\frac{\partial}{\partial t}\left(\frac{\varphi \rho}{2}\left|\frac{\partial u}{\partial t}\right|^2 + \frac{\varphi}{2}\left(A \operatorname{grad}(u), \operatorname{grad}(u)\right)\right) - \operatorname{div}\left(\varphi \frac{\partial u}{\partial t} A \operatorname{grad}(u)\right)$$
$$= \frac{\partial \varphi}{\partial t}\left(\frac{\rho}{2}\left|\frac{\partial u}{\partial t}\right|^2 + \frac{1}{2}\left(A \operatorname{grad}(u), \operatorname{grad}(u)\right)\right) - \frac{\partial u}{\partial t}\left(A \operatorname{grad}(u), \operatorname{grad}(\varphi)\right).$$
$$(10.60)$$

One makes the right side of (10.60) ≤ 0 by choosing φ such that

$$\frac{\partial \varphi}{\partial t} \leq 0, \text{ and } \left(A \operatorname{grad}(\varphi), \operatorname{grad}(\varphi)\right) \leq \rho \left|\frac{\partial \varphi}{\partial t}\right|^2. \qquad (10.61)$$

Choosing

$$\varphi(x,t) = \varphi_0\big((x,\xi) - \kappa\,t\big), \text{ with } \kappa > 0,$$

(10.61) becomes

$$\varphi_0' \geq 0, \text{ and}\big(A\xi,\xi\big) \leq \rho\kappa^2 \text{ where } \varphi_0' \neq 0.$$

Taking $\varphi_0(z) = 0$ for $z \leq a_+$ and $0 < \varphi_0(z) \leq 1$ and φ_0 increasing for $z > a_+$, one deduces that for $t > 0$ one has

$$\int_{R^N} \varphi\left(\frac{\rho}{2}\left|\frac{\partial u}{\partial t}\right|^2 + \frac{1}{2}\big(A\,\mathrm{grad}(u),\mathrm{grad}(u)\big)\right)dx \leq 0,$$

so that $u(x,t) = 0$ for $(x,\xi) - \kappa\,t > a_+$. Similarly with a function $\varphi_0\big((x,\xi)+\kappa\,t\big)$.

Of course, the estimate for the wave equation is first derived for smooth coefficients and smooth initial data: the solution is more regular and all integrations by parts are valid; the result obtained is kept at the limit for discontinuous coefficients.

For the Maxwell–Heaviside equation, one multiplies equations by φE and $-\varphi H$ for a smooth function φ in (x,t), and (assuming ρ and j equal to 0) one obtains

$$\begin{aligned}
0 &= \varphi\left(\frac{\partial B}{\partial t},H\right) + \varphi\,(\mathrm{curl}(E),H) + \varphi\left(\frac{\partial D}{\partial t},H\right) - \varphi\,(\mathrm{curl}(H),E) \\
&= \frac{\partial}{\partial t}\left(\frac{\varphi\,(B,H)}{2} + \frac{\varphi\,(D,E)}{2}\right) - \frac{1}{2}\frac{\partial\varphi}{\partial t}\big((B,H) + (D,E)\big) \\
&\quad + \mathrm{div}\big(\varphi\,(E \times H)\big) - (\mathrm{grad}(\varphi),E \times H).
\end{aligned}$$

One wants to have

$$\frac{\partial\varphi}{\partial t}\big((\mu H,H) + (\varepsilon E,E)\big) + 2(\mathrm{grad}(\varphi),E \times H) \geq 0$$

for all $E,H \in \mathbb{R}^3$, and using functions $\varphi\big((x,\xi) - \kappa\,t\big)$, one deduces that

$$\gamma^2(\xi) = \sup_{x,E\neq 0,H\neq 0} \frac{2(\xi,E \times H)}{(\mu H,H) + (\varepsilon E,E)}.$$

These techniques serve to find bounds for the group velocities of general electromagnetic waves, but nothing comparable will be possible for matter as long as matter is not defined by a particular system of PDEs of hyperbolic nature.

What does "an elementary particle cannot be a point" mean?

The one-dimension wave equation $\frac{\partial^2 u}{\partial t^2} - c^2\frac{\partial^2 u}{\partial x^2} = 0$ has solutions $f(x - ct)$ for any smooth function f. If a sequence f^n converges to the Dirac mass δ_0 at 0, in the sense of Radon measures, or in the sense of distributions, one deduces at the limit that a Dirac mass moving at velocity c is a solution of the one-dimension wave equation.

It is not the case for the N-dimensional wave equation $\frac{\partial^2 u}{\partial t^2} - c^2 \Delta u = 0$ for $N \geq 2$, which does not have solutions which are combinations of moving Dirac masses or their derivatives: using the partial Fourier transform $\mathcal{F}_x$ (extended by Laurent Schwartz to tempered distributions), the equation becomes

$$\frac{\partial^2 (\mathcal{F}_x u)}{\partial t^2} + 4\pi^2 c^2 |\xi|^2 \mathcal{F}_x u = 0,$$

which does not have solutions of the form

$$\sum_{\text{finite}} a_j(t) P_j(\xi) e^{2i\pi(\xi, b(t))},$$

with polynomials P_j. This effect holds for many systems of PDEs, and comes with the name *dispersion*: since plane-waves travel at different (phase) velocities in different directions, they can hardly collaborate to make solutions of certain types.

Another difficulty that physicists (and mathematicians like Poincaré) faced is that if an "electron" were a point, it would correspond to a singularity in the Maxwell–Heaviside equation (i.e. ρ and j would contain Dirac masses), creating a quite singular electromagnetic field, and since the "electron" is supposed to feel the "Lorentz" force, would it feel that force coming from its own electromagnetic effect?

If "elementary particles" exist, and if they are not points, I wonder why Poincaré did not think that they are waves, and since such waves would not be supposed to travel faster than the speed of light, why not have thought that one needs a hyperbolic system of PDEs having characteristic speeds $\leq c$?

(Louis) de Broglie proposed in his 1924 thesis that electrons are waves, and associated a wavelength with them, but I do not think that he wrote a precise hyperbolic system of PDEs.

Dirac wrote a system of PDEs for a "relativistic electron" in 1928. Although physicists agree that Dirac's equation not only has relativistic invariance (which Dirac did on purpose), but also explains the "spin of the electron" and the "positron" (whatever they are), they often do not mention Dirac's equation, maybe because Dirac had to bend the rules of quantum mechanics to arrive at it.

Dirac's equation has a few other good points, and some defects.

It seems curious to expect to describe an interaction (of "light" and "matter") with linear equations, but Dirac's equation is a semi-linear system: the *Planck constant h* appears in coupling the part about "matter" (Dirac's equation, with $\psi \in \mathbb{C}^4$) and the part about "light" (the Maxwell–Heaviside equation, with the scalar potential V and the vector potential A).

Using the (then new) tool of H-measures (which I introduced in the late 1980s), I proved transport properties, like for energy and momentum along light beams, and it is the derivatives of the coefficients that appear, and not the coefficients themselves. It is therefore worth noting that V and A appear in Dirac's equation, and not the electric field E and the magnetic field H, so that it leaves room for the "Lorentz" force to appear at some scale. As suggested to me by Patrick Gérard (who may have just repeated an intuition of physicists), the "Lorentz" force does not exist at the microscopic scale (that of atoms) but starts appearing at a particular mesoscopic scale.[61]

If one formally lets c tend to ∞ in Dirac's equation, only two components of ψ stay relevant, and they are related to the Schrödinger equation, and A has disappeared at the limit but V remains. One deduces that for questions that take place at small velocities compared to the speed of light c, it may be a good approximation to play with the Schrödinger equation, but since the Schrödinger equation is a model in which $c = +\infty$, it becomes an error to use the Schrödinger equation and pretend that some effects can happen faster than the real speed of light!

Since V is governed by the Maxwell–Heaviside equation, a non-constant V starts evolving, and a one-dimensional example with V discontinuous is unphysical, since E then contains Dirac masses, hence does not belong to L^2_{loc}: the Schrödinger equation is only an approximation for effects occurring much faster than the time scale at which V changes.

Dirac considered that his equation describes "one" "relativistic electron". That it only describes one particle comes from a rule of quantum mechanics that I believe to be wrong, since it is not when there are m "particles" that one should work in $\mathbb{R}^{3m}$ (or $\mathbb{R}^{6m}$) but when there are m scales present.

If there are electrons they should all be relativistic, since I find nonphysical the scenario of an electron needing to be told if an experiment is going to give it a relativistic speed or not in order for it to choose which equation it should satisfy!

Dirac added a (mass) term to his equation, containing the "rest mass of the electron", in order to have a dispersion relation analogous to that of a previous (scalar) model for an electron, the Klein–Gordon equation.

Some authors mention "photons" when dealing with the scalar wave equation, but "real photons" are not even related to the Maxwell–Heaviside

---

[61] Patrick Gérard used his semi-classical measures, which are just the particular case of 1-scale H-measures of the more general multi-scales H-measures that I introduced recently.

equation, since they are created by interaction of light and matter, which I think the Dirac equation without a mass term could describe. In one of his lectures (for a general audience), Feynman used the term photon in the scalar case, but added that they were not "real photons", which have spin 1,[62] so instead he called them "spin-0 photons". In a similar way, I do not think that the Klein–Gordon equation can explain that electrons have spin 1/2.

I disagree with the introduction by Dirac of the mass term of the electron in his equation, and I think that one should work with Dirac's equation without a mass term and leave it to appear by a homogenization effect similar to that for which François Murat and Doina Cioranescu wrote a general theory in an elliptic framework, analogous to H-convergence:[63] although they talked of "a strange term coming from nowhere", it would be explained as the electromagnetic energy stored inside the wave which one calls an electron!

Around 1980, I heard a talk by Yvonne Bruhat on the Dirac equation without a mass term, but she did not mention why Dirac introduced it (which I only learned in 1983), and she used the fact that it is conformally invariant,[64] and extended the definition to Riemannian manifolds she proved a local existence for small data in an adapted Sobolev space, and thanks to a conformal transformation introduced by Roger Penrose, she deduced a global existence theorem for small data.

In 1983, when I read about Dirac's introduction of his equation, I immediately thought that the mass term should not be there and should appear by a homogenization process. When at the end of the year I described a program of studying the "oscillating solutions" of (semi-linear) hyperbolic systems, I forgot to explain what I had in mind, which Robin Knops asked about, and I then mentioned to him Dirac's equation with no mass term coupled with the Maxwell–Heaviside equation.

Early in 1985, I read about Bostick's idea for a possible toroidal shape of an electron,[65] and I thought that it could be transposed to the setting I had in mind: I conjectured that the Dirac equation without mass term coupled with the Maxwell–Heaviside equation could describe many other "particles".

I later developed *H-measures* (for small amplitude homogenization) and applied them to the transport of energy and momentum along light beams, but the extension to semi-linear problems seems to require a better understanding of geometry.

---

[62] The unit for spin is $\hbar$ (pronounced h-bar), which is $\frac{h}{2\pi}$.

[63] The term used by François Murat and myself for our generalization of the G-convergence of Sergio Spagnolo.

[64] The Dirac equation with a non-zero mass term is not conformally invariant.

[65] I did not catch his idea about (two-dimensional) photons.

For a semi-linear equation in three-space dimensions with a quadratic non-linearity, one has to consider special cubic quantities, but one has to improve bounds on solutions, since $W^{1,p}$ regularity for solutions of a scalar wave equation does not always hold for $p \neq 2$, as was observed by Walter Littman.

For a quite different class of semi-linear models in one-dimensional kinetic theory, I developed a method based on *integrability by compensation*, and the use of functional spaces in (x, t), to bypass the defects of the semi-group approach.

I hope that something similar could hold for systems like the Dirac equation without mass term coupled with the Maxwell–Heaviside equation, using an interesting hint mentioned to me by Raoul Bott in the late 1980s.

Raoul Bott mentioned that physicists play with models in two space dimensions (plus time) with a cubic nonlinearity, instead of models in three space dimensions (plus time) with a quadratic nonlinearity: the scaling is related to the Sobolev embedding. In three-space-time, $H^1(\mathbb{R}^3)$ is embedded in $L^6(\mathbb{R}^3)$, so that a cubic nonlinearity belongs to $L^2(\mathbb{R}^3)$; in four-space-time, $H^1(\mathbb{R}^4)$ is embedded in $L^4(\mathbb{R}^4)$, so that a quadratic nonlinearity belongs to $L^2(\mathbb{R}^4)$. This then suggests one should aim to prove existence theorems in a space like H^1 in space-time!

In the late 1970s, I decided to investigate questions of appearance of memory effects by homogenization, for different reasons than Évariste Sanchez-Palencia, who (using the Laplace transform) had explained viscoelasticity,[66] and the dependence of the dielectric permittivity ε with respect to frequency.[67]

My idea was that the strange laws of spontaneous absorption and emission that physicists use to explain experiments of spectroscopy are their way to express that an effective equation contains memory effects: no probabilities are needed, and playing probabilistic games is just one way to make such equations appear, and it is a typical error of logical thinking to believe that nature has no other choice than to play such games.

Spectroscopy is about sending light inside a gas, and it is natural to expect resonances at some wavelengths corresponding to some characteristic lengths occurring in the gas.

Although physicists tell us how atoms are supposed to look, and chemists tell us how molecules are supposed to look, the point of view of the

---

[66] In a periodic framework, he used a fluid (governed by the dissipative Stokes's equation) trapped in a linearized elastic material (governed by the conservative Lamé's equation).

[67] He used Ohm's law $j = \sigma E$, and the variations of ε and σ induce a frequency dependence. I later asked Joe Keller if there was a similar hint for the dependence of the magnetic susceptibility μ with respect to frequency, and he said that it is difficult to explain magnetism without quantum mechanics.

mathematician should be to consider that these are just hypotheses which one should check, and for doing that one must develop a mathematical theory. Of course, it will be developed as a part of GTH, the general theory of homogenization which I have tried to expand [6], since the question is about homogenization of hyperbolic equations, but because turbulence is precisely a question of this type, it is not considered an easy question, and only simple examples are understood.

However, one does not need to wait for the completed theory to observe some defects in the rules that physicists invented. It is unfortunate that they were transformed into dogma, since people are rarely trained to criticize dogmas.

The first impression was that one observed "lines".

In 1862, for a gas of hydrogen, Ångström (1814–1874) observed four lines in the visible spectrum, and in 1885 Balmer (1825–1898) noticed that they follow (Balmer's formula)

$$\frac{1}{\lambda} = R_H\left(\frac{1}{4} - \frac{1}{n^2}\right) \text{ for } n = 3,4,5,6,$$

where R_H is the Rydberg constant ($10\,967\,760$ m^{-1}).

In 1888, Rydberg (1854–1919) and Ritz (1878–1909) proposed a generalization

$$\frac{1}{\lambda} = R_H\left(\frac{1}{m^2} - \frac{1}{n^2}\right) \text{ for } m < n,$$

called the Rydberg–Ritz formula, so that Ångström had observed the case $m = 2$ (Balmer series); the case $m = 1$ (Lyman series) was observed in the UV (ultraviolet) by Lyman (1874–1954), who studied the ultraviolet spectrum of electrically excited hydrogen gas between 1906 and 1914; the case $m = 3$ (Paschen series) was observed in the IR (infrared) in 1908 by Paschen (1865–1947); the case $m = 4$ (Brackett series) was observed (in the infrared) in 1922 by Brackett (1896–1988); the case $m = 5$ (Pfund series) was observed (in the infrared) in 1924 by Pfund (1879–1949); the case $m = 6$ (Humphreys series) was observed (in the infrared) in 1953 by Humphreys (1898–1986).

In 1896, Zeeman (1865–1943, 1902 Nobel Prize in Physics) discovered that a magnetic field splits spectral lines into several components; if it is strong enough (or for non-zero spin) the Zeeman effect is called the Paschen–Back effect, and Back (1881–1959) studied this effect in his 1913 PhD thesis.

In 1913, Stark (1874–1957, 1919 Nobel Prize in Physics) discovered that spectral lines are split into several components in the presence of an electric field, but the "Stark" effect was also found independently by Lo Surdo (1880–1949), so in Italy it is often called the Stark–Lo Surdo effect.

I have not discovered who first observed that spectral "lines" have a shape, and various shapes are now used, like Lorentzian (rescaling of $\frac{1}{1+x^2}$), Gaussian (rescaling of e^{-x^2}), and Voigt functions (convolution of a Lorentzian and a Gaussian): various explanations are given, like a Doppler effect, and broadening corresponding to collision and proximity of molecules in a gas.

Maybe physicists prefer to avoid questioning the basic dogma of quantum mechanics, that it deals with the spectrum of an operator in a Hilbert space, and at a conference at the end of 1983 I heard Jean Leray (1906–1998, 1979 Wolf Prize) say that physicists never describe the spectrum of the helium atom ($Z = 2$, $A = 4$), probably because the measured values do not fit with their theories!

For hydrogen, why attribute all the "lines" to "the electron"?

Under the usual conditions, a mole of a gas (2 grams for hydrogen H_2) occupies 22.4 liters,[68] and the number of molecules in it is the Avogadro number, about $6.022 \ 10^{23}$.

If one believes in the kinetic theory of gases of Maxwell and Boltzmann, these molecules move with a Maxwellian distribution in velocities (depending upon temperature and pressure), but even if they are fixed, their average distance is about 3 nanometers (30 Ångströms), and if some special resonance effect occurs when the wavelength (of the light sent through the gas) is 4102 Ångströms (for example), there must be a complicated mechanism of multiple scattering at play.

I find it amazing that anyone accepted the explanation of a game putting the blame on "the electron" alone. After all, each atom of hydrogen is supposed to contain an electron, and a proton which interacts with it, and moreover two atoms of hydrogen interact to form quite stable molecules, and since they are at an average distance of less than one per cent of the wavelength of the light, there must be a complicated multi-scales picture underneath.

What else besides the spectrum of an operator can deliver an infinite list of numbers? Using the usual error in logical thinking, one must have thought that exhibiting a game showing a correct list of numbers would force nature to play it!

Then one observed that these numbers (the positions of the "lines") do not exist, since lines have a shape, and if the shape is Lorentzian, it looks like the real (or imaginary) part of a meromorphic function with poles near an axis.

Of course, all this was done without a clear idea of what an electron really is, apart from being a wave, as L. de Broglie first proposed in his 1924 PhD thesis.

---

[68] A liter is the volume of a cube of side 10 cm, i.e. $10^{-3} \ m^3$.

The introduction of "particles" playing funny games was a backward move to eighteenth-century ideas of classical mechanics, mixed with a wrong use of probabilities.

The nineteenth-century idea of continuum mechanics was already to consider waves, but waves without small scales. It would be strange to study an interaction (here of light and matter) in a linear framework, and this is a main flaw of the (classical) ideas of quantum mechanics, so that one has to consider PDEs that are not linear! However, why not go *beyond PDEs*?

Homogenization (when it is correctly understood, without a periodic framework and without probabilities) shows that giving statistical information on the components of a mixture is not enough to compute its effective behaviour. Since it is unrealistic to expect to know everything at all scales, one may need to identify a hierarchy of problems, and since even simple examples show the need to go beyond PDEs, like when using memory effects, one may need (for a mathematical proof) to know everything from the past to determine the future.

Actually, Bernard Coleman gave me (in the late 1980s) an interesting idea after I had shown him my results on the appearance of memory effects by homogenization (on a simple model): in a situation where the initial datum $u(0)$ is not sufficient for predicting the future of u, the information on $u(t)$ for $t \in (0, T)$ could be used to filter out an estimate of hidden (internal) variables, and limit the possible future.

Since good equations for describing the interaction of light and matter cannot be linear, gathering all the information from spectroscopy (i.e. sending various monochromatic waves and measuring what happens) should not be sufficient for predicting the output when the input is a general plane-wave (mixing frequencies): one could decompose it (by Fourier transform) into a sum of monochromatic waves, but there is no reason why the output should be the sum of the outputs of these monochromatic waves, and a "line" from one of these outputs could well be shifted because of a nonlinear interaction![69]

Feynman mentioned in a lecture for a general audience that he visualized an electron going near the speed of light as a pancake (because of Fitzgerald contraction), a sign that at rest he visualized it as a ball of dough.

Bostick expressed a different intuition, which I prefer, since he visualized an electron as a doughnut, and the hole is crucial for having magnetic lines go through it, and avoiding some infinities that have plagued physicists (and Poincaré).

---

[69] So that a red shift might not always result from a Doppler effect.

After reading about Bostick's idea (in the beginning of 1985), I adopted it as a conjecture about special solutions of the system made of Dirac's equation (without the mass term) coupled with the Maxwell–Heaviside equation, not only about electrons but other "elementary particles". The proposed toroidal structure imagined by Bostick does not have cylindrical symmetry, since he thought about a current showing a swirl, and I later thought that it also gives an idea of what the two spins are, and why such "electrons" are Fermions.

Since Bostick's idea about "photons" is two-dimensional, I could not see how to use it for my system, but a few years ago, I saw a drawing which made me understand a conjecture about "photons", and how their somewhat two-dimensional nature permits them to be stacked, explaining that such "photons" are Bosons.

However, I thought again about why "electrons" satisfy the Pauli principle, by considering a sequence of initial data consisting of two "electrons" with the same spin in nearby points (and parallel planes) and how the limit could be a solution consisting only of light: it would not be a two-dimensional (linearly polarized) "photon", but another kind of "photon", also two-dimensional but circularly polarized (and maybe with no spin) and going away in all the directions of its plane.

In Bostick's "electron", one starts from a circular loop in a plane with a current going through it, and if one started from a general loop (not making a knot) I guess that the electric forces would make it become circular and planar.

Starting with a knot with a current through it, a question is to identify the geometrical shapes which should be preferred, so that it is not a question of topology but of geometry.

The next step is that the circle is just the seed for toroidal surfaces around it, where the current is tangential and goes around the torus with a swirl, but I do not remember if Bostick's article has this idea and if he mentioned how much the swirl should be, because it seems to me now that the swirl should make exactly one turn, so that there are two choices (with different spin): if it makes an even number of turns it should be unstable and disintegrate completely into light, and if it makes an odd number of turns it should partially disintegrate into light and leave an electron (with a single turn).

For other "particles" built over a different geometrical shape, I wonder if the manifolds named after Eugenio Calabi and Shing-Tung Yau appear in a natural way.

In the Maxwell–Heaviside equation, Dirac used

$$\rho = \left(A_0\psi, \overline{\psi}\right), \; j_k = \left(A_k\psi, \overline{\psi}\right) \text{ for } k = 1,2,3, \qquad (10.62)$$

where A_0, A_1, A_2, and A_3 are 4×4 complex matrices such that

$$L = A_0 \frac{\partial}{\partial t} + \sum_j A_j \frac{\partial}{\partial x_j} \text{ satisfies } L^2 = I \left(\frac{\partial^2}{\partial t^2} - c^2 \Delta \right), \tag{10.63}$$

and he constructed them by blocks using 2×2 complex Pauli matrices. (10.62) implies the conservation of electric charge

$$\frac{\partial \rho}{\partial t} + \text{div}(j) = 0 \text{ if } L\psi = 0,$$

because the matrices A_0, A_1, A_2, and A_3 are Hermitian symmetric, but Dirac's equation is not $L\psi = 0$, since the matter field $\psi \in \mathbb{C}^4$ is coupled to the electromagnetic field, and Dirac's equation without a mass term has the form

$$L\psi + \frac{1}{h} B(\psi; V, A) = 0 :$$

the nonlinearity is bilinear,[70] linear in ψ and in the scalar and vector potentials V, A, but the particular form of B does not destroy the conservation of charge.

The operator L in (10.63) is reminiscent of the quaternions of Hamilton, or of the construction of Clifford algebras, themselves extensions of the algebraic question of creating $\mathbb{C}$ by adding a square root of -1 to $\mathbb{R}$: the question of adding roots that do not exist in a given field has been well studied in algebra, and is taught as a first step to Galois theory.

I have wondered if more than quadratic roots are necessary to understand the "elementary particles" created by nature, but before discussing this, I must insist that my program is about the system of PDEs, with the difficulty that the first step is to prove the existence of solutions, a question which already seems to require introducing new mathematical ideas.

The question of finding special explicit solutions, like the "electron" or the "photon" as imagined by Bostick, should be easier, although it is not clear that they exist, since they could simply be data presenting a small rate of decay (enabling their possible observation). If such special solutions exist, assessing their stability might be an easier question of analysis.

The question of interaction of such special solutions may require difficult proofs, but guesses may come from the study of invariants, and those defined by integrals should be easy to determine, while the study of special solutions might point to others: although these questions look more like ones in geometry than topology, it might be like selecting one special solution in an equivalence class defined by the value of topological invariants.

---

[70] The constants appearing in B are the speed of light c and the elementary charge e.

One may seek a solution of a PDE system by a power expansion: if it is divergent, one may propose a special way to sum various terms, which by a "miraculous coincidence" may provide numbers similar to those one observes in an experiment.

It is not scientific to say that one understands the question, and one may have actually corrected some defect of the equation used, and the proposed scheme solves a better model: Feynman used methods of diagrams to solve the Schrödinger equation, which is not a good model since it corresponds to an approximation $c = +\infty$, but his selection of diagrams gave good results for interaction of "particles".

However, what happens "inside particles" takes place at the speed of light, so I wondered which corrected equation is behind his scheme, and I asked Jim Glimm (in the late 1980s) what to read to understand Feynman diagrams: his answer was "it is a state of mind", and I guess it will be some time before a mathematician explains what is going on. A crucial question is to understand in what way a sum of Feynmann diagrams converges, since one cannot define something using sums that do not converge: a mathematical question where something similar occurs is for the appearance of nonlocal effects by homogenization, in a particular case solved by Youcef Amirat, Kamel Hamdache, and Hamid Ziani, a natural power expansion diverges in the sense of distributions, since all the terms live somewhere, while the correct solution lives elsewhere!

A proton has electric charge $+e$, so that two protons repel each other, but they seem to coexist inside the nucleus of an atom for $Z \geq 2$, and physicists explain this by an attracting nuclear force at short range, which they consider consists of an exchange of mesons.

However, this seems to use the poor guess that a proton is like a ball. If one uses Bostick's guess that an electron is like a torus, why not imagine a proton as some kind of knot, containing much more electromagnetic energy than an electron? If two knots are entangled, they may choose a particular geometrical position which has not much to do with two copies of the geometrical position a single knot likes to be in.

What would mesons be in such a case?

Adding a square root to a field creates an extension of order 2; adding a cubic root creates an extension of order 3; adding both creates an extension of order 6.

If Dirac did the analog of adding a square root, was Gell-Mann doing the analogue of also adding a cubic root when he introduced quarks? (George) Zweig independently introduced something similar to quarks, which he called aces, since he initially thought there were four new particles.

My program of describing oscillations (and concentration effects) in the coupled system of Dirac's equation without a mass term and the Maxwell–Heaviside equation might then be just a first step, and one may need to consider more general coupled systems afterwards, but the purpose is to start from a PDE and explain all the small scales that appear in solutions, since it is when there are small scales in a solution that one uses the language of "particles".

For linear equations, the transport of small scales can be done using H-measures (which I introduced in the late 1980s), but for semi-linear systems the more general multi-scales H-measures (which I introduced recently [7]) might be helpful, but they are not powerful enough.

Only after having found adapted mathematical tools for semi-linear systems can one expect to be able to say which of the guesses of physicists were correct and which were not.

Appendix H *Why are elementary snowflakes flat?*

Elementary snowflakes are flat and usually present a symmetry of order 3 or even 6, whose details vary. They get entangled during their fall, so that they trap air, which may play a role in the quietness when snow falls without wind, since acoustic waves are quickly damped, all the "noise" is suppressed.

Elementary snowflakes have access to three dimensions when they form (from the humidity in high cold parts of the atmosphere). Why don't they create "beautiful" three-dimensional structures?

Do they try, and something disturbs them?

I give a "personality" to snowflakes to argue in the spirit of Poincaré's ideas in relativity: "particles" do not know what they do, and in particular which structure they help to build; they do not care if these structures are artistic or not:[71] there are no "particles", hence no "thinking" or "making choices".

One cannot answer the question by a PDE game in the plane!

In high school, I was taught that one should not put in the hypothesis what one wants to find in the conclusion! A crucial problem for the correct evolution of scientific knowledge is that too many "scientists" do exactly that.

For this problem, isn't it silly to play a two-dimensional game with a density of surface energy depending upon the normal to the surface and presenting a symmetry of order 3 or 6?

That "particles" approaching a structure know how to distinguish the "normal" suggests "thinking particles", although one should consider this

---

[71] There is no definition for which structures are artistic, and which structures are not.

further if there are enough fields that a "particle" could feel. If there are two elementary snowflakes in formation in the same plane, and a "particle" moves in the plane in their vicinity, towards which one would it move?[72]

In a plane model with a density of surface energy depending upon the normal, one has selected a center and three lines making angles of $\frac{\pi}{3}$. Why would the plane be so anisotropic?

One may assume that in the high atmosphere there is a particular direction imposed by gravity, and that elementary snowflakes are formed in a horizontal position, but apart from introducing another field (electric, magnetic, or other), how could three special directions making angles of $\frac{\pi}{3}$ appear in a natural way?

Why not say that there is only one elementary snowflake in formation, with others sufficiently far away, so that one understands how unrealistic this problem is?

If a rigid solid falls inside a fluid filling the whole space,[73] one may choose a frame attached to the solid and consider that the fluid turns around the solid, taking into account the "Coriolis force", which de Coriolis (1792–1843) introduced only in 1835, while Laplace used it in 1803,[74] but if two or more rigid bodies fall in the fluid, such a trick of using a fixed domain does not work.

The laws of physics are the same everywhere, and "particles" near two forming elementary snowflakes do not know their orientation: they are as blind as "particles" already attached to one snowflake. The difference between them is that the latter has exchanged most of its kinetic energy for some stability.[75]

That mathematicians often adopt a caricature of physics may result from physicists failing in their duty to discover what physical laws are (at microscopic level): one reason for their failure is that they often adopt the point of view of engineers, who can control phenomena even though only erroneous equations are known about them. Also, having been too accustomed to situations near equilibrium, many forget that time and motion exist: the physical laws which I mention are the *dynamical laws*, and not just the truncated versions that describe equilibria, or situations near these equilibria.

Some physicists complain that Dirac's arguments were too mathematical, but since Galileo physical laws have been expressed in a language called

---

[72] A "particle" is blind: it cannot identify elementary flakes in formation around it, and choose which structure to attach itself to.

[73] The argument needs the domain occupied by the fluid to be invariant by translations and by rotations.

[74] Laplace deduced that falling bodies are slightly pushed eastward, but this had been considered by Borelli (1608–1679) in the 1660s, although with some errors in his analysis.

[75] This should be understood in terms of solutions of PDEs, and not in terms of nonexistent "particles", which only serve as a language adapted to some high frequency limits.

mathematics,[76] even though he himself used only elementary mathematics: since Newton, one considers that this language must be enriched at crucial times, when the need appears to develop new mathematical tools to express a "new law" of nature, i.e. one that no mathematician or physicist perceived before.

One should criticize "well-known" notions, like kinetic energy and mass, and try to perceive what it means in our real world where there are no particles but only waves.

If the difference between a "particle" already settled in an elementary snowflake in formation and another one that still "hesitates" between two crystalline orientations lies in their kinetic energy, one should remember that when energy is localized at a mesoscopic level it is a part of internal energy (called heat), and internal energy moving around is called heat flux.

One makes sherbet by freezing a blend of fruit juice, water, and sugar, and stirring the mixture to avoid the formation of flat pieces of ice. As for the formation of elementary snowflakes, which surprisingly end up being flat, there are flat pieces of ice which form when an aqueous mixture starts freezing!

One calorie makes the temperature of one gram of water increase by one degree C (= Celsius).

The *latent heat*[77] necessary to melt one gram of ice at $0°$ C (into one gram of water at $0°$ C) is around 80 calories, so the heat released by one gram of water at $0°$ C turning into ice could warm another gram of water from $0°$ C to $80°$ C,[78] hence it is crucial to evacuate all the heat released by the part of the water which has already frozen.

If some water finds itself surrounded by ice, it is probable that the heat released when the water freezes may then melt a part of the surrounding ice, so the geometry of ice should adapt well to evacuating heat.[79]

---

[76] Hence part of mathematics is science. Not all mathematicians do science, since many do their best to choose areas in which no applications are known!

[77] Part of my argument is that the notion of "latent heat" is ill defined: it should depend upon the path (into the space of mesostructures) one uses from a first phase to a second phase.

[78] Water boils at 100 degrees C at usual atmospheric pressure, with "latent heat" of vaporization of about 537 calories.

[79] According to a friend who had lived in Alaska they have a saying there that when caught in a blizzard one should lie down in the snow and let oneself be covered by snow: one will be insulated from the cold and not feel the wind chill effect. When he went to Antarctica, he was warned that it does not work there and one should avoid being caught in a blizzard: the ground is frozen solid at a very low temperature. In my first version, I mistakenly wrote that ice is less conducting than water: another friend pointed out that the converse holds; it is true for solid ice, or for compressed snow if not much air is left inside so that it becomes a conductor, which then explains the difference between Alaska and Antarctica. Actually, my argument only requires water and ice to have different heat conductivities.

A simple computation in homogenization gives a bound of the effective conductivity of a mixture made of m conducting materials with conductivity tensors $M_1, \ldots, M_m$, used in proportions $\theta_1, \ldots, \theta_m$, for arbitrary orientations: the highest conductivity in a direction e is $\sum_{i=1}^{m} \theta_i \lambda_i^{\max}$, where $\lambda_i^{\max}$ is the largest eigenvalue of M_i, each M_i being turned so that e becomes an eigenvector for the eigenvalue $\lambda_i^{\max}$: when one mixes isotropic conducting materials, any microstructure using parallel planes is optimal for evacuating heat in the directions of the plane.[80]

Water and bismuth are the only materials, to my knowledge, whose volume increases while freezing,[81] and this change in volume creates elastic tensions: if a geometry is good for evacuating heat but is easily deformed by the elastic tensions, it will probably lose its "optimality", so looking for a better geometry should lead to a compromise between being efficient at heat evacuation and being stable under elastic constraints.

My analysis then is that in questions of phase transitions one should observe microstructures that have good properties of evacuation of heat and of elastic stability, so I differ on this question from Jerry Ericksen, who proposed to explain some observed microstructures by elasticity only.[82]

I think that elementary snowflakes are flat because this geometry is well adapted to evacuating heat, and that drafts of snowflakes which are not two-dimensional are bothered and partially destroyed by the warm plane currents that seek to evacuate the heat liberated when the local humidity freezes.

I think that the symmetry of order 3 is then chosen for a question of *two-dimensional elasticity*, to create an effective isotropic structure, but maybe also to create a dendritic plane structure, adapted again to evacuating heat around the elementary snowflake, but in its own plane.

In making sherbet, one destroys the microstructures that appear naturally: one forces the mixture of water and ice to go through isotropic microstructures from pure water to pure ice.

---

[80] For evacuating heat in an optimal way in a direction e, any microstructure with cylindrical repartition with axis e is optimal, and I saw (in the Carnegie Museum in Pittsburgh) crystalline formations showing parallel needles, which looked very fragile: either because the direction of the heat flux is constant, or because elastic effects come into play, one should rarely observe such microstructures.

[81] In most materials that do not show this unusual change in volume while freezing, it is not two-dimensional elasticity that comes into play but three-dimensional elasticity. However, fluids still have another way of evacuating heat, by convection, like in Rayleigh–Bénard instability.

[82] I wonder if the microstructures shown by his followers for an austenite to martensite transition correspond to a case where the latent heat is small, which would then make the heat evacuation process less crucial.

Since I was taught that to measure the latent heat one stirs the mixture (of two phases of the same material) while it undergoes the transition, it seems that one should *reject the notion of "latent heat" independent of the path chosen*, and the amount of energy stored or released depends upon which path is used while the phase transition takes place.[83]

In the *new thermodynamics* which should be developed to correct the defect of the actual thermodynamics (which is not about dynamics!) one should introduce geometrical parameters (maybe the moments of H-measures or variants) to better describe what kind of microstructure one has at a particular moment during the phase transition.[84]

Appendix I *Are quasi-crystals flat?*

In 1984, Schechtman (2011 Nobel Prize in Chemistry) and his coworkers experimentally discovered what they called *quasi-crystals*. They may have used various techniques, but according to what I read in the mid 1980s one was to create metallic ribbons with interesting magnetic properties, by heating a metallic ribbon above its Curie point (the temperature above which it loses magnetic properties), then orienting the individual "spins" of atoms by applying a transverse magnetic field, and finally quickly cooling the ribbon by pulling on it. To check that the magnetic properties had stayed as they hoped (despite the tempering of the ribbon), they used x-ray diffraction, and they were quite surprised to observe a diffraction pattern with a symmetry of order 5 (or 10), which is impossible for a crystalline structure, hence the chosen name.

This term was used earlier by Yves Meyer for a sum of Dirac masses whose Fourier transform is also a sum of Dirac masses.

In 1973, Roger Penrose had discovered tilings of the plane with two types of rhombi which show a *macroscopic symmetry* of order 5.[85] Although *it obviously has nothing to do with the experiment with the metallic ribbon*, physicists started studying Penrose's tilings!

I find it sad that experienced physicists fail to warn the young generations of researchers against this type of nonsense. There is nothing more upsetting for a mathematician than to hear examples of pseudo-logic of the type "since

---

[83] Since phase transitions are obviously not instantaneous!

[84] It should be spatially dependent, since there is no reason why the evolution of the transition at two different points should follow the same path, or should happen with the same timing if they follow the same path.

[85] The ratio of thick rhombi (with one angle of 72 degrees) to thin rhombi (with one angle of 36 degrees) is the golden ratio.

A implies *B* and one observes something which looks like *B*, then *A* must be true" and "this explains the experiment".[86]

A few physicists do not understand the criticisms of mathematicians, since by their usual answers of the type "but it explains the measured values in some experiments", one deduces that one should start by teaching them the quite elementary observation, that '*A* implies *B*' does not mean '*B* implies *A*'! We should then look for a physical explanation of the experiment on the metallic ribbon, since it obviously has nothing to do with Penrose's tilings.

One has the ingredients in the discussion about elementary snowflakes above: one must evacuate heat (since one starts above the Curie point, 770 degrees C for iron) and there is some elasticity (since one cools the ribbon by pulling on it).

Experimentalists knew that it could change the microstructures inside the ribbon, since it was for checking this that they did an x-ray diffraction!

Why is the result so different from the case of the snowflakes?

A ribbon is a three-dimensional structure: the distance between atoms in a solid is of the order of an Ångström: for a ribbon which is a tenth of a millimeter thick, there are about a million atoms in the small direction, and since this is not negligible it is silly to pretend that it is a two-dimensional problem![87]

Actually, heat must quickly find its way out of the ribbon along the small direction, so it is already quite different from the case of the snowflakes, and one must then understand how the atoms rearrange themselves because of the elastic constraints.

Since there are no "particles", there are no "atoms" either, but as usual one uses classical language to obtain an intuitive understanding of what is going on, and for the sake of simplicity I omit quotes.

The theory of x-ray diffraction was developed for crystals, in order to use the notion of Bragg angles and interpret the diffraction pattern in terms of the crystalline network.

Since heat is evacuated in one direction, it could favour microstructures with cylindrical symmetry, but such microstructures are not efficient from an elastic

---

[86] Besides comforting naive people in their bad habit of repeating something false (since '*A* implies *B*' does not imply '*B* implies *A*'), it shows an astonishing lack of creativity, since one should be able to imagine a few other hypotheses which also imply *B*.

[87] The atoms inside the ribbon must adapt to the constraints, and it does not matter that the ribbon is thin for us!

point of view: in 1984, Jerry Ericksen mentioned to me that one can create in the laboratory a mono-crystal of aluminum in the form of a straight-edge, and one can then bend it easily by hand, but the bent object is so hard that it is impossible to straighten it by hand.

Strain hardening and stress hardening are about the material changing its mesostructure to adapt to the constraints, adopting one quite resistant to tension and bending.

Therefore, a metallic ribbon cannot be a crystal! However, I do not think physicists developed a general theory to explain x-ray diffraction for solids without a crystalline structure.

I first used the intuition coming from my theory of *H-measures*, which are *not adapted to the question*, since *they lack a characteristic length*. As a second approximation, one could use my variant of H-measures with one characteristic length, almost identical to the semi-classical measures introduced by Patrick Gérard, and later Pierre-Louis Lions (1994 Fields Medal) and Thierry Paul, who found a different way to introduce them, using the Wigner transform (so that they called them Wigner measures), but the idea of limiting myself to only one characteristic length hurts my physical intuition: a few years ago, I heard a talk by a physicist (who had won the Nobel Prize) who startled my collaborator Amit Acharya and me by saying that biology is more difficult than physics because it involves many scales, while in physics there is always one scale!

In my opinion the specialists of material science talk about a few different mesoscopic scales (keeping the term microscopic for the level of atoms), and I have spent most of my career trying to develop mathematical tools to understand the effects of multi-scale phenomena in what I call GTH (the general theory of homogenization) [6],[88] and I recently defined a larger class of microlocal objects, *multi-scales H-measures* [7].

It is not clear what "macroscopic spins" are, which orient easily with the magnetic field above the Curie point, and hardly move below the Curie point.[89] Not knowing how to treat magnetism, I thought that understanding which microstructures are adapted to elasticity would already be a success, but homogenization in non-linear elasticity is not yet understood.

---

[88] It is important to repeat that a general theory should have no limitations of periodicity or using probabilities. Also, one cannot explain why nature likes (almost) crystalline structures by only studying (exact) crystalline structures!

[89] I imagine that it is like the difficulty of turning around a long stick inside a narrow cell (and heat pushes the walls away); it seems linked to questions of elasticity, like magneto-striction, which couples elasticity and magnetism (while piezo-electricity couples elasticity and electricity).

In 1988, Owen Richmond (1928–2001) told me that he conjectured a non-elastic behaviour for the effective material, which may be like the behaviour which Eckard Salje calls ferro-elastic or co-elastic, which I heard about 10 years later.

Actually, it is not even clear that one may talk about elasticity, even non-linear, since having heated the ribbon above the Curie point not only permits the "macroscopic spins" to move, but also the atoms themselves, and it looks like plasticity.

Homogenization in linearized elasticity is not so physical, but I thought (as a first step) of finding which microstructures are adapted to the case of mixtures of elastic materials with nearby properties: it becomes a problem of small amplitude homogenization for which I initially developed H-measures.

I knew how to associate this question with a characterization of moments, because of calculations which I had done (with an error), similar to the calculations of Gilles Francfort and François Murat (without error), and I then convinced them to join me for characterizing moments of order 4 of a non-negative (H-)measure living on $S^2 \subset \mathbb{R}^3$ (and not on $S^1 \subset \mathbb{R}^2$ as for a two-dimensional problem). In other terms, quasi-crystals in a metallic ribbon are not really flat, and it is a mistake to analyze the question as if it was a problem in a plane.[90]

It would be too long to explain the appearance of homogenization in questions of optimal design, which I discovered with François Murat in the early 1970s: in 1983, I heard Michael Renardy talk about extrusion of molten polymers and introducing a little water in pipelines for a lubrication purpose. In this joint work with Dan Joseph (1929–2011), they imagined an optimality principle for explaining the observations. I was not convinced,[91] but it made me work again on the problems which I had studied with François Murat almost 10 years earlier.

The idea that nature chooses "optimal" configurations is then not mine, and I imagined that nature could like turbulent flows because they are "optimal" in some sense; I then found in a book of Dan Joseph that such an idea was used by Busse, maybe only for questions of turbulent boundary layers, but without reading Busse's work, I thought that a turbulent boundary layer serves to create a gradient of temperature in a direction quite different from that of transport of mass, and Olivier Pironneau confirmed that heat transfer in turbulent flows may be unusual, and it seems different from the property

---

[90] Although turbulence is a three-dimensional effect, meteorologists talk of "two-dimensional turbulence": turbulent structures are localized in altitude but extended in horizontal directions.

[91] In a non-circular geometry, their principle may imply a mixture, so that one cannot limit attention to Poiseuille flows.

mentioned by Jean Mandel (1907–1982) in his course on continuum mechanics at École Polytechnique, that at the surface of a sea of constant depth linear momentum may be transported without transport of mass, at a phase velocity which depends upon the depth.

For the moment, there is no mathematical theory for studying the evolution of microstructures.

My idea for a first step was to consider a finite number of effective properties of a mixture (like its elastic, electric, magnetic, and thermal properties) together with the statistical information given by Young measures, introduced by Laurence Young (1905–2000), and identify the set of admissible effective coefficients, which one does not yet know how to do.

I imagined that the observed microstructures would often correspond to points on the boundary of such a set, hence giving a way to guess an evolution equation on the boundary. Nevertheless, I was ready to admit the possibility that the evolution could push some boundary points towards the interior of the admissible region, maybe for a quick evolution for attaining another point of the boundary not so near to the first.

Reality is a little more intricate than this first idea.

Under the simplifying hypothesis that the materials one mixes have nearby properties, effective properties can be computed at second order with H-measures; in this approximation, the admissible set is described by a list of integrals of H-measures.

If one perturbs an isotropic medium, these integrals are moments (i.e. integrals of polynomials) of degree 2 for electric, magnetic, and thermal properties, and of degree 4 for elastic properties. The characterization of moments of order 2 of non-negative measures is easy, but it is not so for moments of order 4 in three dimensions, since the measures live on S^2, and all the classical results for moments were done for $\mathbb{R}$ or for S^1.

I had linked the knowledge of effective coefficients in linearized elasticity to the characterization of the closed convex cone $K \subset \mathbb{R}^{15}$ formed by all the moments $\int_{S^2} x_i x_j x_k x_\ell \, d\mu$ for a measure $\mu \geq 0$ living on S^2. With Gilles Francfort and François Murat, we proved that any point $k \in K$ may be obtained by a measure ν_k which is a combination of at most six Dirac masses, and that $k \in \partial K$ if and only if one can find a combination ν_k of at most five Dirac masses having these moments. Moreover, if $k \in \partial K$ is not obtained by a combination of at most four Dirac masses, there then exists a one-parameter family of measures with five Dirac masses having moments k.

In the case where $k \in \partial K$ is obtained for a *transversely isotropic* distribution, then each of these measures has the same weight for its five Dirac masses, which are at the vertices of a regular pentagon (and the one-parameter family consists of turning this pentagon around its axis).

I have not said that the moments permit me to find μ, and I only suggested a conjecture concerning the formation of quasi-crystals, that the change in microstructure imposed by the constraints (of evacuating heat in the small direction of the ribbon and adapting to the elastic constraints due to cooling by pulling on the ribbon) favour an effective material which is transversely isotropic with an H-measure with five Dirac masses with the same weight at the vertices of a regular pentagon.

I only have a vague idea about a mechanism favouring H-measures with few Dirac masses.

Quite a few years ago, I heard a very interesting talk by "Raj" Rajagopal: he took some paste out of a jar, and started to mold it into a ball, while mentioning that one may consider it a liquid, possibly visco-elastic, because in a bowl it would flow slowly toward the bottom; when the ball was warm enough he showed that it bounced back like a good rubber ball, with no apparent dissipation of energy, so that one could consider it an elastic solid; he then threw it as fast as he could on the blackboard, and everyone in the room ducked (since one expected the ball to bounce back into the room), but the ball just splashed onto the blackboard as if it was made of jelly! He then mentioned that there was no good modeling for such a material which reacted so differently to slow variations or to fast variations.

He added that, if he had not warmed it and had hit it hard with a hammer, it would have broken into fine pieces, like a very brittle solid, and that would not have been a good idea since the material is slightly corrosive, and he went off to wash his hands before continuing his talk.

The formation of snowflakes and the formation of quasi-crystals are quite different: snowflakes are the result of a slow evolution, while quasi-crystals are the result of a fast evolution.

For slow evolution, the scenario which I described (of moving on the boundary of the set of admissible effective parameters) looks reasonable, and one can probably make this evolution compatible with classical ideas in thermodynamics.

On the contrary, fast evolution seems to require objects more general than H-measures, since they do not use any characteristic length, hence cannot predict what the result of an x-ray diffraction experiment will be. Since realistic problems often involve a few mesoscopic scales, I hope that the *multi-scales H-measures* which I introduced [7] will permit us to attain a better understanding of these questions.

It is usual for physicists to perform computations in a periodic setting and then apply some results to a general setting without periodicity, and for a mathematician this is only a way to state conjectures. Before my introduction of H-measures there was no mathematical method for proving conjectures of this type, but for the more difficult problems which involve a few scales, one will have to check if ℓ-scale H-measures help, or if one must create another type of mathematical object.

If one succeeds in creating an object (using a few characteristic lengths) which explains the x-ray diffraction through general materials, and if one can also follow the evolution of such an object, it will be useful to ascertain if a quick evolution can indeed produce materials exhibiting a macroscopic symmetry of order 5, and then be able to address a quantitative question: for each value of $h \in [0, 1)$ there is a measure with five Dirac masses at the vertices of a regular pentagon in the plane $\xi_3 = h$, giving moments $k^h \in \partial K$, and one should wonder if the experiment creates quasi-crystals with a particular value of h, or if this value depends upon the experiment, or if h evolves with time.

Appendix J *Non-local effects induced by homogenization*

The earliest physical situations which I knew with homogenization making non-local effects appear were works by Évariste Sanchez-Palencia: he treated (in a periodic setting) examples from acoustics, viscoelasticity, and electromagnetism.

I already knew that one attacks such problems with the Laplace transform, because Jacques-Louis Lions had made a similar remark in one of his lectures, on an academic example

$$\frac{d^2}{dt^2}(A_\varepsilon u_n) + B_\varepsilon u_n = f; \ u_n(0) = v; \ \frac{du_n}{dt}(0) = w,$$

where A_ε and B_ε are second-order elliptic operators with periodically oscillating coefficients; applying the Laplace transform $\mathcal{L}$ (in t) gives a one-parameter homogenization problem

$$(p^2 A_\varepsilon + B_\varepsilon)\mathcal{L}u_n(p) = \mathcal{L}f(p) + w + pv,$$

so that the limit u satisfies an equation of the type

$$C(p)\mathcal{L}u(p) = \mathcal{L}f(p) + w + pv,$$

where $C(p)$ denoted the homogenization limit of $p^2 A_\varepsilon + B_\varepsilon$.

Jacques-Louis Lions noticed that, since $C(p)$ is usually not a polynomial in p, it gives a "pseudo-differential" operator after applying the inverse Laplace transform.

Talking about pseudo-differential operators is a fancy way to say that the limiting equation shows non-local effects.

In the late 1970s, I thought that the reason why physicists invented laws of spontaneous emission and absorption is that an effective equation contains memory effects,[92] and I started studying the appearance of memory effects by homogenization.

A wave equation with variable coefficients in (x, t) is difficult to analyze, and I thought of considering (hyperbolic) first-order scalar equations, but I noticed that turbulence is precisely the question of identifying the class of effective equations for

$$\frac{\partial}{\partial t} + v^n \cdot \operatorname{grad}_x, \text{ usually with } \operatorname{div}(v^n) = 0,$$

when the sequence v^n converges weakly. I chose a simpler case, with fixed characteristic curves, but with damping:

$$\frac{\partial u_n}{\partial t} + a_n u_n = f, \text{ with } u_n\Big|_{t=0} = v. \qquad (10.64)$$

With a_n depending only upon x, I expected a limit equation[93]

$$\frac{\partial u}{\partial t} + a_\infty u - I_{\text{eff}} = f, \text{ with } u\Big|_{t=0} = v,$$

$$\text{and } I_{\text{eff}}(x, t) = \int_0^t K_{\text{eff}}(x, t - s) u(x, s) \, ds \text{ a.e. } (x, t). \qquad (10.65)$$

Using the Laplace transform, (10.64) and (10.65) mean

$$(p + a_n) \mathcal{L}u_n = \mathcal{L}f + v, \text{ i.e. } \mathcal{L}u_n = \frac{\mathcal{L}f + v}{p + a_n},$$

$$(p + a_\infty - \mathcal{L}K_{\text{eff}}) \mathcal{L}u = \mathcal{L}f + v, \text{ i.e. } \mathcal{L}u = \frac{\mathcal{L}f + v}{p + a_\infty - \mathcal{L}K_{\text{eff}}},$$

so that, if the conjecture is right, K_{eff} is "defined" by

$$\frac{1}{p + a_\infty - \mathcal{L}K_{\text{eff}}} = \text{weak limit of } \frac{1}{p + a_n}.$$

---

[92] A mathematician should not confuse an equation with a proof of existence of its solutions: using probabilities as a way to prove a result (in case of memory effects) does not mean that is the only way to prove it.

[93] Since $f, v \geq 0$ implies $u \geq 0$, I conjectured $K_{\text{eff}} \geq 0$: it is a sufficient condition for having this property.

Since I expected K_{eff} to be ≥ 0, I used a theorem of (Sergei) Bernstein (1880–1968), characterizing the Laplace transform of a non-negative measure,[94] which I had heard, mixed with questions on Padé approximants, in a 1977 talk by Daniel Bessis.

Then I found a different proof, based on properties of Pick functions, which I had heard (also mixed with questions on Padé approximants) a few years later in talks by two other physicists, David Bergman and Graeme Milton.

One has implicitly assumed that a_n is bounded in L^∞, specifically

$$-\infty < a_- \leq a_n(x) \leq a_+ < +\infty \text{ a.e. } x \text{ for all } n, \qquad (10.66)$$

and one extracts a subsequence for which there is a Young measure, i.e. a family of non-negative Radon measures of total mass 1 (i.e. probabilities) $dv_x(a)$ with support in the interval $[a_-,a_+]$ such that for every real continuous function G one has

$$G(a_n) \to g \text{ in } L^\infty \text{ weak } \star, \text{ with } g(x) = \int G(a)\,dv_x(a) \text{ a.e. } x.$$

Using the Young measure, (10.48) gives

$$\mathcal{L}K_{\text{eff}} = p + a_\infty - \left(\int_x \frac{dv_x(a)}{p+a}\right)^{-1}.$$

The useful information is then that $\left(\int_x \frac{dv_x(a)}{p+a}\right)^{-1}$ is a Pick function, i.e. for $p \in \mathbb{C} \setminus \mathbb{R}$ it maps the upper half (complex) space into itself, and there is a representation for such functions: in the simpler case where a Pick function is holomorphic outside a compact interval K of the real axis, then it is equal to

$$\alpha z + \beta - \int_K \frac{d\mu}{z-k}, \text{ for all } z \notin \mathbb{R},$$

for some $\alpha \geq 0$, $\beta \in \mathbb{R}$, and a Radon measure $d\mu \geq 0$ on K.

Using a Taylor expansion at ∞, one deduces that

$$\left(\int_x \frac{dv_x(a)}{z+a}\right)^{-1} = z + a_\infty - \int_{[-a_+,-a_-]} \frac{d\mu_x}{z-k}, \qquad (10.67)$$

and a_∞ is given by

$$a_\infty(x) = \int a\,dv_x(a), \text{ i.e. it is the weak } \star \text{ limit of } a_n.$$

---

[94] A C^∞ function $g(p)$ defined on $(0,+\infty)$ is the Laplace transform of a non-negative measure μ if and only if $g \geq 0$ and $(-1)^m \frac{d^m g}{dp^m} \geq 0$ in $(0,\infty)$ for all $m \geq 1$.

From (10.49), one deduces that

$$\mathcal{L}K_{\text{eff}}(x;p) = \int_{[-a_+,-a_-]} \frac{d\mu_x}{p-k}$$

so that by inverse Laplace transform one has

$$K_{\text{eff}}(x;t) = \int_{[-a_+,-a_-]} e^{kt} d\mu_x.$$

It looks like I have imposed the form (10.65) and then identified the only possible candidate for K_{eff}.

One can argue in another way, and observe that

$$u_n(x,t) = v(x)e^{-a_n(x)t} + \int_0^t e^{-a_n(x)(t-s)}f(x,s)\,dx \text{ implies}$$

$$u(x,t) = v(x)\int e^{-at}\,dv_x(a) + \int_0^t \left(\int e^{-a(t-s)}\,dv_x(a)\right)ds,$$

so that one faces an unnatural question: knowing u, what equation does it satisfy? One then observes that

$$\frac{\partial}{\partial t} + a_n \text{ is linear and invariant by translations in } t,$$

so that it is natural to restrict attention to the class of operators which are linear and invariant by translations in t.

A theorem of Laurent Schwartz says (with minimal hypotheses) that such an operator is a convolution in t with a kernel which is a distribution. In this class, only one operator does the job of always selecting u.

I asked Luísa Mascarenhas to study the case where a_n also depends on t: I expected an effective equation

$$\frac{\partial u}{\partial t} + a_\infty u - I_{\text{eff}} = f, \text{ with } u\Big|_{t=0} = v,$$
$$\text{and } I_{\text{eff}}(x,t) = \int_0^t K_{\text{eff}}(x,t,s)\,u(x,s)\,ds \text{ a.e. } (x,t), \tag{10.68}$$

with a kernel which is no longer a convolution (since my argument of invariance does not apply). Assuming a uniform bound on $\frac{\partial a_n}{\partial t}$, she used a discretization in time as a way to select a particular kernel. The sign of K_{eff} may change.

Youcef Amirat, Kamel Hamdache, and Hamid Ziani applied my method using Pick functions to the case

$$\frac{\partial u_n(x,y,t)}{\partial t} + a_n(y)\frac{\partial u_n(x,y,t)}{\partial x} = 0; \; u_n\Big|_{t=0}(x,y) = v(x,y), \tag{10.69}$$

which is invariant by translation in (x,t), so that they expected a convolution in (x,t). In [1], they adapted my method by using also a partial Fourier transform $\mathcal{F}_x$ in x:

$$\mathcal{L}\mathcal{F}_x u_n(\xi,y,p) = \frac{\mathcal{F}_x v(\xi,y)}{p + 2i\pi\,\xi\,a_n(y)},$$

which is valid for $\Re p > 0$; with a bound (10.66) for a_n, assuming a Young measure dv_y, the weak limit u satisfies

$$\mathcal{L}\mathcal{F}_x u(\xi,y,p) = \mathcal{F}_x v(\xi,p)\int \frac{dv_y(a)}{p + 2i\pi\,\xi\,a}.$$

The function $H(\xi,p)$ defined by

$$H(\xi,p) = \int \frac{dv_y(a)}{p + 2i\pi\,\xi\,a}$$

is holomorphic in a strip $|\Im\xi| < \eta_0$ if $2\pi\eta_0 a_\pm < \Re p$, and formula (10.67) can be used by analytic continuation, the analog being

$$\frac{1}{\int \frac{dv_y(a)}{q+a}} = q + a_\infty(y) - \int \frac{d\mu_y(a)}{q+a} \text{ for } q \notin [-a_+, -a_-],$$

and by choosing $q = \frac{p}{2i\pi\,\xi\,p}$ one obtains

$$\frac{1}{\int \frac{dv_y(a)}{p+2i\pi\,\xi\,a}} = p + 2i\pi\,\xi\,a_\infty(y) - (2i\pi\,\xi)^2\int \frac{d\mu_y(a)}{p+2i\pi\,\xi\,a} \text{ if } \Re p > 0.$$

The inverse Laplace transform gives a delay equation for $\mathcal{F}_x u$:

$$\frac{\partial \mathcal{F}_x u(\xi,y,t)}{\partial t} + 2i\pi\,\xi\,a_\infty(y)\mathcal{F}_x u(\xi,y,t) - \int_0^t K(\xi,y,t-s)\mathcal{F}_x u(\xi,y,s)\,ds = 0,$$

$$\mathcal{F}_x u(\xi,y,0) = \mathcal{F}_x v(\xi,y) \text{ with } K(\xi,y,t) = (2i\pi\,\xi)^2\int e^{-2i\pi\,\xi\,a\,t}\,d\mu_y(a),$$

which, after applying the inverse Fourier transform, gives

$$\frac{\partial u(x,y,t)}{\partial t} + a_\infty(y)\frac{\partial u(x,y,t)}{\partial x}$$
$$- \int_0^t\int \frac{\partial^2 u(x-a(t-s),y,s)}{\partial x^2}\,d\mu_y(a)\,ds = 0;\ u\Big|_{t=0} = v. \qquad (10.70)$$

In [1], Youcef Amirat, Kamel Hamdache, and Hamid Ziani also studied directly the finite speed propagation of equations like (10.70), and wrote it as a system from kinetic theory,

$$\frac{\partial u(x,y,t)}{\partial t} + a_\infty(y)\frac{\partial u(x,y,t)}{\partial x} = \frac{\partial}{\partial x}\left(\int F(x,y,t;a)\,d\mu_y(a)\right),\ u\Big|_{t=0} = v,$$

$$\frac{\partial F(x,y,t;a)}{\partial t} + a\frac{\partial F(x,y,t;a)}{\partial x} = \frac{\partial u(x,y,t)}{\partial x},\ F\Big|_{t=0} = 0,$$

where one has defined

$$F(x,y,t;a) = \int_0^t \frac{\partial u(x-a(t-s),y,s)}{\partial x}\,ds.$$

It is important to notice that if the sequence a_n takes m different values, then the measures dv are combinations of m Dirac masses, but by (10.67) the measures $d\mu$ are combinations of $m-1$ Dirac masses, at intermediate values which are the roots of a polynomial of degree $m-1$.[95]

In the late 1980s, I noticed that one can obtain the result of Luísa Mascarenhas (10.68) with weaker hypotheses (uniform equicontinuity in t) by a power expansion in a parameter γ:

$$\frac{\partial U_n(x,t;\gamma)}{\partial t} + \left(a_\infty(x,t) + \gamma\,b_n(x,t)\right)U_n(x,t;\gamma) = f,$$

$$\text{with } b_n(x,t) = a_n(x,t) - a_\infty(x,t);\ U_n\Big|_{t=0} = v,$$

which has for solution

$$U_n(x,t;\gamma) = V_0(x,t) + \sum_{k=1}^\infty \gamma^k V_{k,n}(x,t), \qquad (10.71)$$

with V_0 (independent of n) a solution of

$$\frac{\partial V_0(x,t)}{\partial t} + a_\infty(x,t)\,V_0(x,t) = f;\ V_0\Big|_{t=0} = v,$$

and $V_{k,n}$ defined by induction for $k \geq 1$ by

$$\frac{\partial V_{k,n}(x,t)}{\partial t} + a_\infty(x,t)V_{k,n}(x,t) + b_n(x,t)V_{k-1,n}(x,t) = 0;\ V_{k,n}\Big|_{t=0} = 0. \qquad (10.72)$$

One easily finds bounds for the $V_{k,n}$ showing that the sum in (10.71) has an infinite radius of convergence.

---

[95] From an incomplete 1799 "proof" by Ruffini (1765–1822), which Abel (1802–1829) completed in 1823, and which Galois (1811–1832) found independently, there are polynomials of degree 5 (or more) whose roots cannot be expressed by radicals.

By letting n tend to ∞, one extracts a subsequence such that for each $m \geq 2$,[96] and all times $s_1, \ldots, s_m \in (0, T)$,

$$b_n(\cdot, s_1) \ldots b_n(\cdot, s_m) \rightharpoonup M_m(s_1, \ldots, s_m) \text{ in } L^\infty \text{weak} \star, \tag{10.73}$$

which is possible due to the equicontinuity hypothesis. One notices that although $M_2 \geq 0$ if b_n is independent of t, it is not always true in the general case. Since by (10.72)

$$V_{k,n}(x,t) = -\int_0^t e^{-\int_s^t a_\infty(x,\sigma)\,d\sigma} b_n(x,s) V_{k-1,n}(x,s)\,ds,$$

for $k \geq 1$, one deduces that

$$V_{k,n}(x,t) = (-1)^k \int_\Delta b_n(x,s_1) \cdots b_n(x,s_k) I(x,s_1,t)\,ds_1 \cdots ds_k,$$

with $I(x, s_1, t) = e^{-\int_{s_1}^t a_\infty(x,\sigma)\,d\sigma} U_0(x, s_1)$,

and Δ defined by $0 \leq s_1 \leq \ldots \leq s_k \leq t$.

One sees why the definition (10.73) is natural for dealing with limits: $V_{k,n}$ converges to $V_{k,\infty}$ in L^∞ weak $\star$, with

$$V_{k,\infty}(x,t) = (-1)^k \int_\Delta M_k(x, s_1, \ldots, s_k) I(x, s_1, t)\,ds_1 \cdots ds_k,$$

and $U_\infty(x, t; \gamma) = V_0(x,t) + \sum_{k=1}^\infty \gamma^k V_{k,\infty}(x,t)$ satisfies

$$\frac{\partial U_\infty(x, t; \gamma)}{\partial t} + a_\infty(x,t) U_\infty(x, t; \gamma) - \int_0^t K(x,t,s; \gamma) U_\infty(x,s; \gamma)\,ds = f(x,t),$$

with the kernel $K(x, t, s; \gamma)$ given by

$$K(x,t,s;\gamma) = \sum_{k=2}^\infty \gamma^k K_k(x,t,s), \text{ with } K_2(x,t,s) = M_2(x,t,s) e^{-\int_s^t a_\infty(x,\sigma)\,d\sigma},$$

and K_k obtained by induction for $k \geq 3$: integrals of M_m on various sets appear, and uniform bounds are easy to obtain, which show that the radius of convergence is infinite, and there is then no difficulty using the power series with $\gamma = 1$.

When a_n does not depend upon t, the preceding algorithm for computing the convolution kernel $K(x,t,s;1)$ differs from the one using the representation of Pick functions.

I tried this method for a non-linear equation:

$$\frac{\partial u_n(x,t)}{\partial t} + a_n(x,t) u_n(x,t)^2 = f(x,t); \; u_n(x)\Big|_{t=0} = v, \tag{10.74}$$

[96] For $m = 1$, b_n converges to 0 in L^∞ weak $\star$.

with

$$a_n \geq 0, \; f(x,t) \geq 0, \; v(x) \geq 0,$$

to avoid questions of blow-up on $(0, T)$. The equation

$$\frac{\partial U_n}{\partial t} + (a_\infty + \gamma \, b_n) U_n^2 = f; \; U_n \Big|_{t=0} = v,$$

gives a power expansion in γ, but with two difficulties, one of book-keeping since many more terms appear, and one of convergence since the radius of convergence is limited.[97]

The class of non-linear operators which commute with translations is too big,[98] and one should imagine other ways than power series for studying (10.74).

Power series are still useful, in that they may permit us to understand what a more natural (and smaller) class could be.

Also, such academic examples may help us understand how to deal with the algebraic complexity, like the use of diagrams, and the convergence of summation rules.

It is important to observe that for (10.69) the method of power series *diverges in the sense of distributions*, except if a_n converges strongly (the Young measure is 1 Dirac mass): all the terms of the expansion live on the characteristic curve moving at velocity a_∞, but the limit does not live there.

However, the series may converge in the weak sense of analytic functionals, for which the notion of support is not defined!

There is a classical counter-example that if a real sequence tends to 0, and both the series of positive parts and the series of negative parts diverge, then whatever ℓ is one may reorder the terms so that the sum tends to ℓ, hence if for a divergent series one finds a summation procedure giving a finite sum, it is not so clear if the numerical value is relevant.

Since the Schrödinger equation is not such a good model (because it corresponds to $c = +\infty$), and it is not so useful for studying an interaction of particles which takes place (at the microscopic level) at the real speed of light, one should be surprised that a method of diagrams "based on the Schrödinger equation" could give a numerical result similar to an experimental value.

I see a possibility that the summation procedure actually gives the solution of another equation which is a better model!

---

[97] If the coefficient a_n changes sign, there may be blow-up at some points before time T.

[98] Convolution operators $u \mapsto K \star u$ and pointwise operators $u \mapsto \psi(u)$ do not commute, and generate something too big.

Appendix K *On some discrete velocity models in one dimension*

Partial differential equations should be stable with respect to weak convergence if they describe the behaviour at the macroscopic level of extensive quantities or coefficients of differential forms. Although introduced by Maxwell, discrete velocity models are not so physical, but I used them as a training ground, and I discovered something useful.

For F_i locally Lipschitz continuous, $i = 1,\ldots,m$, one can solve the system

$$\frac{\partial u_i}{\partial t} + C^i \cdot \mathrm{grad}(u_i) = F_i(u_1,\ldots,u_m) \text{ in } \mathbb{R}^N \times (0,T),$$

$$\text{and } u_i(\cdot,0) = v_i \text{ in } \mathbb{R}^N, \; i = 1,\ldots,m,$$

locally in time for bounded data, and I wondered if

$$v_i^n \rightharpoonup v_i^\infty \text{ in } L^\infty(\mathbb{R}^N) \text{ weak } \star, \; i = 1,\ldots,m \text{ implies}$$

$$u_j^n \rightharpoonup u_j^\infty \text{ in } L^\infty\big(\mathbb{R}^N \times (0,T)\big) \text{ weak } \star, \; j = 1,\ldots,m.$$

I found that either $N \geq 2$ and all F_i are affine, or $N = 1$ and

$$F_i(u) = \mathrm{affine}_i(u) + \sum_{jk} A_{ijk} u_j u_k, \; i = 1,\ldots,m, \text{ with } A_{ijk} = A_{ikj} \text{ for all } i,j,k,$$

satisfying (S) $C^j \neq C^k$ implies $A_{ijk} = 0, i = 1,\ldots,m$.

This condition (S) is crucial for the analysis.

If $N = 1$ and $C^j \neq C^k$, it is related to *compensated compactness*

$$\text{if } \frac{\partial u_j^n}{\partial t} + C^j \frac{\partial u_j^n}{\partial x} \text{ is bounded in } L^\infty\big(\mathbb{R} \times (0,T)\big),$$

$$\text{and } u_j^n \rightharpoonup u_j^\infty \text{ in } L^\infty\big(\mathbb{R} \times (0,T)\big) \text{ weak } \star,$$

$$\text{if } \frac{\partial u_k^n}{\partial t} + C^k \frac{\partial u_k^n}{\partial x} \text{ is bounded in } L^\infty\big(\mathbb{R} \times (0,T)\big),$$

$$\text{and } u_k^n \rightharpoonup u_k^\infty \text{ in } L^\infty\big(\mathbb{R} \times (0,T)\big) \text{ weak } \star,$$

$$\text{then } u_j^n u_k^n \rightharpoonup u_j^\infty u_k^\infty \text{ in } L^\infty\big(\mathbb{R} \times (0,T)\big) \text{ weak } \star.$$

If (S) holds, $v_1,\ldots,v_m \in L^1(\mathbb{R})$, there exists $T > 0$ such that

$$\frac{\partial u_i}{\partial t} + C^i \frac{\partial u_i}{\partial x} = \sum_{jk} A_{ijk} u_j u_k \text{ in } \mathbb{R} \times (-T,+T), \text{ and } u_i(\cdot,0) = v_i \text{ in } \mathbb{R}, \; i = 1,\ldots,m,$$

$$(10.75)$$

has a unique solution satisfying

$$A_{ijk} u_j u_k \in L^1\big(\mathbb{R} \times (-T,+T)\big), \; i,j,k = 1,\ldots,m. \qquad (10.76)$$

This is related to *compensated integrability*:

$$\text{if } C^j \neq C^k, \; f_j, f_k \in L^1\big(\mathbb{R} \times (0,T)\big), v_j, v_k \in L^1(\mathbb{R}),$$

$$\frac{\partial u_j}{\partial t} + C^j \frac{\partial u_j}{\partial x} = f_j, \; \frac{\partial u_k}{\partial t} + C^k \frac{\partial u_k}{\partial x} = f_k \text{ in } \mathbb{R} \times (0,T),$$

$$\text{and } u_j(\cdot,0) = v_j, u_k(\cdot,0) = v_k \text{ in } \mathbb{R} \text{ imply } u_j u_k \in L^1\big(\mathbb{R} \times (0,T)\big) \quad (10.77)$$

$$\text{with } ||u_j u_k||_{L^1} \leq \frac{(||f_j||_{L^1} + ||v_j||_{L^1})(||f_k||_{L^1} + ||v_k||_{L^1})}{|C^j - C^k|}.$$

One finds a better integrability because of the information on derivatives: if $N \geq 2$, one has the following result, used by Emilio Gagliardo (1930–2008) and by Louis Nirenberg (1982 Crafoord Prize, 2015 Abel Prize),[99,100]

$$\text{if } u_i \in L^{N-1}_{\text{loc}}(\mathbb{R}^N) \text{ and } \frac{\partial u_i}{\partial x_i} = 0, \; i = 1,\ldots,N,$$

$$\text{then } \prod_j u_j \in L^1_{\text{loc}}(\mathbb{R}^N), \quad (10.78)$$

but for $N \geq 3$, it is not related to compensated compactness.

The time of existence T given by my proof of (10.75) is not just a function of the norm of the data, and I deduced it from a global existence theorem for small data:

$$\text{if (S) holds, there exists an } \varepsilon_0 > 0 \text{ such that } \sum_j ||v_j||_{L^1(\mathbb{R})} \leq \varepsilon_0$$

$$\text{implies that one may take } T = +\infty \text{ in (10.75) and (10.76).} \quad (10.79)$$

For deducing the local existence result from (10.79), one notices that for arbitrary data in $L^1(\mathbb{R})$, there exists $r > 0$ with

$$\sum_j \int_y^{y+r} |v_j(x)| \, dx \leq \varepsilon_0 \text{ for all } y \in \mathbb{R}.$$

For a given $y \in \mathbb{R}$, one replaces the data by 0 outside $[y, y+r]$ and one applies the global existence result (10.79), and the only piece useful is in the two triangles of dependence

$$\{(x,t) \mid t > 0, \; y \leq x - t \max_j C^j \text{ and } x - t \min_j C^j \leq y + r\},$$

$$\{(x,t) \mid t < 0, \; y \leq x - t \min_j C^j \text{ and } x - t \max_j C^j \leq y + r\},$$

---

[99] They used (the global version) for a new proof of the Sobolev embedding theorem, valid even for $p = 1$.

[100] I used (10.78) for proving global existence of solutions of the two-dimensional "Broadwell" model (which arises from a model by Maxwell), for small non-negative data in $L^2(\mathbb{R}^2)$. When I mentioned it to Takaaki Nishida in 1985, he told me that he had already made a similar observation.

which by gluing together all the restrictions to these triangles when one varies y (the compatibility resulting from the uniqueness for small data) gives

$$T = \frac{r}{\max_j C^j - \min_j C^j}.$$

I had first used this trick with Mike Crandall in 1975, for transforming a global existence result for the "Broadwell" model, obtained by Takaaki Nishida and Mimura, who proved that for non-negative bounded data with small norm in L^1 the L^∞ norm is at most multiplied by C.

By the preceding argument, we deduced that the L^∞ norm is at most multiplied by C for an interval of time of the order of r, and for checking that r would not deteriorate too fast with time, we took advantage of the entropy inequality.

My proof of (10.79) results from the classical fixed point argument for strict contractions, which one usually attributes to Banach (1892–1945), although Picard (1856–1941) seems to have used it much earlier. An interesting point is that I used function spaces of functions in (x,t), and not a semi-group approach.

For $a \in \mathbb{R}$, I introduced two function spaces:

$$W_a = \left\{ f(x,t) \mid \frac{\partial f}{\partial t} + a\frac{\partial f}{\partial x} \in L^1(\mathbb{R} \times \mathbb{R}), f\big|_{t=0} \in L^1(\mathbb{R}) \right\},$$
$$V_a = \left\{ f(x,t) \mid |f(x,t)| \le g(x - at), g \in L^1(\mathbb{R}) \right\},$$

with obvious choices of norms, implying $W_a \subset V_a$ with the injection of norm ≤ 1, which proves (10.77) by integrating a product $f_1(x - a_1 t)f_2(x - a_2 t)$ in (x,t) for $a_1 \ne a_2$.

Although the fixed point argument is done in $V_{C^1} \times \cdots \times V_{C^m}$, one finds the solution u_j in W_{C^j}, and this function space has the advantage that if one travels at velocity C^j, then the profile converges to limits in $L^1(\mathbb{R})$ as t tends to $-\infty$ or to $+\infty$, and one deduces results of scattering.

The mapping for which one looks for a fixed point consists of

$$\frac{\partial U_i}{\partial t} + C^i\frac{\partial U_i}{\partial x} = \sum_{jk} A_{ijk} u_j u_k \text{ in } \mathbb{R} \times \mathbb{R}, \text{ and } U_i(\cdot,0) = v_i \text{ in } \mathbb{R}, i = 1,\dots,m.$$

Condition (S) ensures that if $u_j \in V_{C^j}$ for all j, then each term $A_{ijk} u_j u_k$ belongs to $L^1(\mathbb{R} \times \mathbb{R})$, hence $U_j \in W_{C^j}$ for all j:

$$\text{if } \sum_j \|v_j\|_{L^1} \le \varepsilon, \sum_j \|u_j\|_{V_{C^j}} \le \sigma,$$

then

$$\sum_i \|U_i\|_{VC^i} \le \varepsilon + \sum_{ijk} \frac{|A_{ijk}|}{|C^j - C^k|} \|u_j\|_{VC^j} \|u_k\|_{VC^k}, \tag{10.80}$$

where one omits the terms for which $C^j = C^k$. If one denotes

$$M = \max_{j,k}\left(\sum_i \frac{|A_{ijk}|}{|C^j - C^k|}\right),$$

then (10.80) implies

$$\sum_i \|U_i\|_{VC^i} \le \varepsilon + M\sigma^2.$$

For having a ball $\sum_i \|u_i\|_{VC^i} \le \sigma$ sent into itself, it is enough to have $\varepsilon + M\sigma^2 \le \sigma$, and for having a strict contraction there, it is enough to have $2M\sigma < 1$, and one may choose

$$\varepsilon_0 < \frac{1}{4M}, \quad \sigma \le 2\varepsilon_0.$$

The "Broadwell" model does not satisfy condition (S): it is[101]

$$u_t + u_x + uv - w^2 = 0,$$
$$v_t - v_x + uv - w^2 = 0,$$
$$w_t - uv + w^2 = 0,$$

and one cannot get a bound on the integral of w^2 by (10.77). Since non-negative initial data give non-negative solutions for $t > 0$, one obtains this bound from a bound on the integral of uv by integration on a strip $(0 \le t \le T)$. For small total mass, one obtains $u \in W_1, v \in W_{-1}, w \in W_0$, so that for large $t > 0$, u looks like $u_\infty(x - t)$, v looks like $v_\infty(x + t)$, and w looks like $w_\infty(x)$, but since w^2 is integrable (in (x,t)) one has $w_\infty = 0$.

Instead of conservation of mass (which is $u + v + 2w$) and of momentum in x (which is $u - v$), one may instead use the conservation of $u + w$ and $v + w$, and their integrals are precisely the integral of u_∞ (the mass which eventually goes to $+\infty$), and the integral of v_∞ (the mass which eventually goes to $-\infty$):

$$(u + w)_t + u_x = 0,$$
$$(v + w)_t - v_x = 0, \tag{10.81}$$

---

[101] One starts from a plane model due to Maxwell, with speeds 1 in the directions of the axes, and there would be an interaction term in $\pm(u_1 u_2 - u_3 u_4)$, but for data independent of x_2 one may start with $u_3 = u_4$ and it stays that way. The model has no temperature, since all the speeds are the same, but has an entropy: the integral of $u \log(u) + v \log(v) + 2w \log(w)$ is non-increasing in time.

and it is useful to introduce two potentials

$$U(x,t) = \int_{-\infty}^{x} \big(u(\xi,t)+w(\xi,t)\big)\,dx,$$

$$V(x,t) = \int_{x}^{+\infty} \big(v(\xi,t)+w(\xi,t)\big)\,dx,$$

which satisfy

$$U_x = u+w,\, U_t = -u,$$
$$V_x = -(v+w),\, V_t = -v.$$
(10.82)

An argument of compensated compactness applied to (10.81) would make the quantity $u(v+w)+v(u+w)$ appear, but there is again an argument of compensated integrability, giving an estimate of its integral. Multiplying the second equation of (10.81) by U, and using (10.82), one obtains

$$\big(U(v+w)\big)_t - (Uv)_x + u(v+w) + v(u+w) = 0,$$
(10.83)

from which one deduces an interesting estimate, which I heard attributed to Raghu Varadhan (2007 Abel Prize):

$$\int_{\mathbb{R}} U(x,t)\big(v(x,t)+w(x,t)\big)\,dx \text{ is non-increasing in time,}$$
(10.84)

and it can be written as

$$I(t) = \int\int_{y<x} \big(u(y,t)+w(y,t)\big)\big(v(x,t)+w(x,t)\big)\,dx\,dy.$$

The mass $u(y,t)+w(y,t)$ at y is bound to go to $+\infty$, and the mass $v(x,t)+w(x,t)$ at x is bound to go to $-\infty$, and since $y < x$ they will interact later, hence $I(t)$ may be interpreted as an amount of future interaction at time t; the second one is

$$\int_0^T \int_{\mathbb{R}} \big(u(v+w)+v(u+w)\big)\,dx\,dt = I(0) - I(T) \le I(0),$$

obtained by integrating (10.83) on the strip $0 < t < T$: one estimates the integral of uv without a hypothesis that the mass of the initial data is small. Of course, by (10.84) one has

$$I(0) \le \Big(\int_{\mathbb{R}} \big(u(x,0)+w(x,0)\big)\,dx\Big)\Big(\int_{\mathbb{R}} \big(v(x,0)+w(x,0)\big)\,dx\Big).$$

Appendix L *Homogenization of Elliptic Operators*

When Sergio Spagnolo worked in the late 1960s on what he called G-convergence, the convergence of Green kernels, I assume that he was answering a question of Ennio De Giorgi concerning scalar second-order elliptic equations,

$$-\sum_{i,j} \frac{\partial}{\partial x_i} \left(A_{ij} \frac{\partial u}{\partial x_j} \right) = f, \tag{10.85}$$

in a bounded open domain $\Omega \subset \mathbb{R}^N$ with coefficients

$$A_{ij} \in L^\infty(\Omega), \sum_{ij} A_{ij}(x)\,\xi_i\xi_j \geq \alpha\,|\xi|^2 \text{ for all } \xi \in \mathbb{R}^N \text{ a.e. } x: \tag{10.86}$$

he had proved some $C^{0,\alpha}$ regularity (outside the diagonal) for the Green kernels (say with Dirichlet conditions), and if one considers a sequence A^n satisfying (10.86) it implies that a subsequence G_n of Green kernels converges (pointwise outside the diagonal) to a function G_∞, and the question was to show that G_∞ is the Green kernel of an operator of the type (10.85) but not for a classical limit of A^n, and there is a different topology on the matrices of coefficients (in dimension $N \geq 2$).[102]

Sergio Spagnolo proved such a result, under symmetry

$$A^T(x) = A(x) \text{ a.e. } x, \tag{10.87}$$

and a uniform bound

$$\sum_{ij} A_{ij}(x)\,\xi_i\xi_j \leq \beta\,|\xi|^2 \text{ for all } \xi \in \mathbb{R}^N \text{ a.e. } x. \tag{10.88}$$

He also proved that G-convergence is local: the result for the Dirichlet condition serves for other boundary conditions as well.

For an academic question of optimization, François Murat and I proved the same result in the early 1970s, without knowing about the previous work, but by a slightly different method. When we extended our method to the non (necessarily) symmetric case, i.e. without (10.87), we changed (10.88) into

$$\big(A(x)\xi,\xi\big) \geq \frac{1}{\beta}\,|A(x)\xi|^2 \text{ for all } \xi \in \mathbb{R}^N \text{ a.e. } x, \tag{10.89}$$

which under (10.87) is equivalent to (10.88). Since it is not about the convergence of Green kernels,[103] we adopted the term H-convergence, with H related to the term homogenization.[104]

---

[102] For $N = 1$, it is the weak $\star$ limit of $\frac{1}{a_n}$ which is needed.

[103] The Green kernel cannot tell what $A(x)$ is, since the addition of a constant skew-symmetric matrix to $A(x)$ does not change solutions, hence does not change the Green kernel.

[104] I borrowed the term from an article of Ivo Babuška, who himself had borrowed it from the engineering literature.

I then learned that our work is related to the notion of effective properties of mixtures, thanks to some work of Évariste Sanchez-Palencia, who worked in a periodic framework, although it is important in GTH (the general theory of homogenization) to avoid a periodic, or a probabilistic framework.[105]

Since (in dimension ≥ 2) effective properties cannot be computed from Young measures (i.e. one point statistics), the limitations of some laws used in parts of physics (like thermodynamics) should be discussed: *constitutive relations or equations of state are only first approximations*!

Starting from a sequence

$$A^n \in \mathcal{M}(\alpha, \beta; \Omega), \text{ i.e. satisfying (10.86) and (10.89) in } \Omega,$$

defining operators

$$\mathcal{A}^n = -\mathrm{div}\big(A^n \mathrm{grad}(\cdot)\big) \in \mathcal{L}(V; V'), \text{ for } V = H_0^1(\Omega),$$

and using the separability of $V' = H^{-1}(\Omega)$, and a Cantor diagonal subsequence, one extracts a subsequence A^m such that

$$\text{for all } f \in V, \big(\mathcal{A}^m\big)^{-1} f \rightharpoonup \big(\mathcal{A}^{\mathrm{eff}}\big)^{-1} f \text{ in } V \text{ weak,}$$

and moreover

$$-\mathrm{div}\big(A^m \mathrm{grad}(u_m)\big) = f \text{ implies}$$
$$u_m \rightharpoonup u_\infty \text{ in } V \text{ weak, and } A^m \mathrm{grad}(u_m) \rightharpoonup C(u_\infty) \text{ in } L^2(\Omega; \mathbb{R}^N) \text{ weak,}$$

and one wants to show that

$$\text{there exists } A^{\mathrm{eff}} \in \mathcal{M}(\alpha, \beta; \Omega) : C(u_\infty) = A^{\mathrm{eff}} \mathrm{grad}(u_\infty) \text{ for all } f.$$

Even if A^m converges weakly to A^∞ in $L^\infty(\Omega)^{N^2}$ weak $\star$, $\mathrm{grad}(u_m)$ only converges weakly in $L^2(\Omega)^N$ and the product $A^m \mathrm{grad}(u_m)$ has no reason to converge to $A^\infty \mathrm{grad}(u_\infty)$, and actually the limit is $A^{\mathrm{eff}} \mathrm{grad}(u_\infty)$, and $A^{\mathrm{eff}} \neq A^\infty$ in general.

We observed that

$$\big(A^m \mathrm{grad}(u_m), \mathrm{grad}(u_m)\big) \rightharpoonup \big(C(u_\infty), \mathrm{grad}(u_\infty)\big),$$

in the sense of Radon measures (i.e. with test functions in $C_c(\Omega)$).[106] Since we started with the symmetric case, we deduced a good localization for $C(u_\infty)$ by

---

[105] Also, Γ-convergence is not homogenization.
[106] It can be improved with a result of Meyers which we learned afterwards in the work of Sergio Spagnolo.

convexity inequalities:

> if $\alpha_n(x)I \leq A^n(x) \leq \beta_n(x)I$ a.e. x,
> if $\frac{1}{\alpha_n} \rightharpoonup \frac{1}{\alpha_\infty}, \beta_n \rightharpoonup \beta_\infty$ in $L^\infty(\Omega)$ weak $\star$, then $\qquad\qquad$ (10.90)
> $\big(C(u_\infty) - \alpha_\infty\mathrm{grad}(u_\infty), C(u_\infty) - \beta_\infty\mathrm{grad}(u_\infty)\big) \leq 0,$

i.e. $C(u_\infty)$ belongs to a sphere, with two ends of a diameter being $\alpha_\infty\mathrm{grad}(u_\infty)$ and $\beta_\infty\mathrm{grad}(u_\infty)$, so that C is a local operator in $\mathrm{grad}(u)$, hence a matrix A^{eff}. When we extended our result to the non-necessarily symmetric case, we used oscillating test functions, solutions of the transposed operators.

As François Murat noticed in 1970, if the A^n depend upon only one coordinate (x, e), there are explicit formulas for A^{eff} using only weak limits of some non-linear functions of A^n.

In the spring of 1975, Louis Nirenberg showed me a preprint of McConnell, who had computed the corresponding formulas for linearized elasticity, and since in the late 1970s I wanted to explain this to an engineer who needed the extension to sandwiches of steel and rubber (in non-linear elasticity), I had to find the general algorithm.

An application of the div–curl lemma (which François Murat and I noticed in the spring of 1974) shows that

> if $D^n \rightharpoonup D^\infty$ in $L^2(\Omega)^N$ weak,
>
> if $\mathrm{div}(D^n)$ stays in a compact of $H^{-1}(\Omega)$ strong,
>
> if $\psi_n(x_1) \rightharpoonup \psi_\infty(x_1)$ in $L^\infty(\mathbb{R})$ weak $\star$,
>
> then $\psi_n(x_1)D_1^n \rightharpoonup \psi_\infty(x_1)D_1^\infty$ in $L^2(\Omega)$ weak,

and I say that such a D_1^n "does not oscillate in x_1". Denoting $E^n = \mathrm{grad}(u_n)$, instead of $D^n = A^n E^n$ one should use

$$\begin{pmatrix} E_1^n \\ D_2^n \\ \vdots \\ D_N^n \end{pmatrix} = \Phi(A^n) \begin{pmatrix} D_1^n \\ E_2^n \\ \vdots \\ E_N^n \end{pmatrix},$$

and since $E_2^n, \dots, E_N^n$ do not oscillate in x_1,[107] one finds that

$$\Phi(A^n) \rightharpoonup \Phi(A^{\mathit{eff}}) \text{ in } L^\infty(\Omega)^{N^2} \text{ weak } \star. \qquad (10.91)$$

For computing A^{eff}, one notices then that

$$A^{\mathrm{eff}} = \Phi\big(\Phi(A^{\mathrm{eff}})\big),$$

---

[107] Because of the compatibility conditions $\frac{\partial E_j^n}{\partial x_1} = \frac{\partial E_1^n}{\partial x_j}$.

but practically (10.91) means

$$\frac{1}{A_{11}^n} \rightharpoonup \frac{1}{A_{11}^{\text{eff}}} \text{ in } L^\infty(\Omega) \text{ weak } \star,$$

$$\frac{A_{1j}^n}{A_{11}^n} \rightharpoonup \frac{A_{1j}^{\text{eff}}}{A_{11}^{\text{eff}}}, \frac{A_{j1}^n}{A_{11}^n} \rightharpoonup \frac{A_{j1}^{\text{eff}}}{A_{11}^{\text{eff}}} \text{ in } L^\infty(\Omega) \text{ weak } \star, j \geq 2, \qquad (10.92)$$

$$A_{ij}^n - \frac{A_{i1}^n A_{1j}^n}{A_{11}^n} \rightharpoonup A_{ij}^{\text{eff}} - \frac{A_{i1}^{\text{eff}} A_{1j}^{\text{eff}}}{A_{11}^{\text{eff}}} \text{ in } L^\infty(\Omega) \text{ weak } \star, i,j \geq 2.$$

Sergio Spagnolo and Antonio Marino showed that in order to obtain all the materials with matrices satisfying (10.86)–(10.88), it suffices to take sequences $c_n I$ with c_n taking only two values, c_-, c_+, but François Murat and I wanted to find a more precise characterization, the set of H-limits for

$$A^n = \big(\alpha \chi_n + \beta (1 - \chi_n)\big) I, \text{ for a sequence of characteristic functions}$$

satisfying $\chi_n \rightharpoonup \theta$ in $L^\infty(\Omega)$ weak $\star$.

For finding more constraints than (10.92) concerning A^{eff}, I improved our method by using more of our *compensated compactness theory*, and I applied it for $A^{\text{eff}} = cI$ in June 1980, while I was visiting NYU (New York University).

George Papanicolaou told me to compare my bounds with "classical" ones, which I had never heard of, those of Zvi Hashin and Shtrikman (1930–2003): since their argument was incomplete, I then was the first to prove these bounds.

However, their construction using coated spheres made sense, so that the bounds (when A^{eff} is isotropic) are optimal.

François Murat thought that the same method should work for the general case, but it took us a few months to compute a construction with coated confocal ellipsoids, and I described our result at a meeting in June 1981 at NYU: (10.92) says

$$\lambda_-(\theta) I \leq A^{\text{eff}} \leq \lambda_+(\theta) I \text{ a.e. in } \Omega, \text{ with}$$
$$\frac{1}{\lambda_-(\theta)} = \frac{\theta}{\alpha} + \frac{1-\theta}{\beta},$$
$$\lambda_+(\theta) = \theta \alpha + (1 - \theta) \beta,$$

which in terms of the eigenvalues λ_j of A^{eff} is

$$\lambda_-(\theta) I \leq \lambda_j \leq \lambda_+(\theta) I \text{ a.e. in } \Omega \text{ for all } j, \qquad (10.93)$$

but the characterization requires also

$$\sum_{j=1}^N \frac{1}{\lambda_j - \alpha} \leq \frac{1}{\lambda_-(\theta) - \alpha} + \frac{N-1}{\lambda_-(\theta) - \alpha}, \text{ a.e. in } \Omega,$$

$$\sum_{j=1}^N \frac{1}{\beta - \lambda_j} \leq \frac{1}{\beta - \lambda_-(\theta)} + \frac{N-1}{\beta - \lambda_-(\theta)}, \text{ a.e. in } \Omega. \qquad (10.94)$$

Since θ may vary with x, one must consider that $A^{\text{eff}} = \alpha I$ where $\theta = 1$, $A^{\text{eff}} = \beta I$ where $\theta = 0$, and one is left with a set where $0 < \theta < 1$, so that (10.94) makes sense there because the λ_j cannot take the values α or β by (10.93).

For layers in parallel planes with normal e, the analog of (10.92) is that A^{eff} has e as eigenvector for the value $\lambda_-(\theta)$, and the hyperplane orthogonal to e as eigenspace for the eigenvalue $\lambda_+(\theta)$, so that both inequalities of (10.94) are equalities.

I naively thought that our construction of coated confocal ellipsoids could not be avoided, and, as an exam for two students of École Polytechnique who followed a set of lectures I gave in the spring of 1982, I asked them to check that the set attainable by repeating the formula (10.92) for layering is smaller than the set described by (10.93)–(10.94), but Philippe Braidy and Didier Pouilloux first reported that numerically the two sets looked equal, and a few days after that they had a simple proof that they coincide.

Since they used layerings in orthogonal directions which were eigenvectors of the matrices used, I thought that it could be useful to write (10.92) in an intrinsic way for the general case:[108] if one layers A with proportion θ and B with proportion $1 - \theta$ using planes orthogonal to a unit vector e, one has

$$A^{\text{eff}} = \theta A + (1-\theta) B - \frac{\theta (1-\theta) (B-A) e \otimes e (B-A)}{(1-\theta) (A e, e) + \theta (B e, e)}.$$

There are "explicit" formulas in the periodic setting, which one should name differently: one needs to solve a PDE on a unit period (with periodic boundary conditions), which hardly makes sense if one does not have a numerical code at one's disposal.

Also, the number of operations to perform grows fast when one requires a precise estimate of these "explicit" coefficients.

In 1974, I had found in the French translation of a book by (Lev) Landau (1908–1968, 1962 Nobel Prize in Physics) and Lifschitz (1915–1985) a section about the conductivity of a mixture: if one grinds two conductors into fine powder, which one pours into a container in proportions θ and $1 - \theta$, if one mixes by shaking well, if one compresses it, what is the conductivity of the resulting mixture?

I guessed that shaking meant that one expected the result to be an isotropic conductor, and that compressing was to avoid air as a third component, but without a better understanding of grinding and shaking, how could they expect

---

[108] I did this computation while I visited MSRI (Mathematical Sciences Research Institute) in Berkeley, in the spring of 1983, and I actually deduced an ODE for moving around in the spaces of (not necessarily symmetric) matrices.

to answer with a number? We did not know exactly what the correct interval is, but we knew that it is not reduced to a point!

I did not try to check exactly what they were saying, but there was no indication that θ or $1 - \theta$ should be small, and their formula for $\theta = \frac{1}{2}$ was not symmetric in α and β.

There was no mention of experimental measurements either.

In my opinion these authors did not know what they were talking about.

In the 1980s, George Papanicolaou mentioned that physicists do not like the Hashin–Shtrikman bounds because they do not show a percolation effect. It then seems that physicists do not perceive that an effective conductivity is more than proportions used, and that details at mesoscopic scales are needed, hence the method for creating a mixture is important.

Percolation is about mixing a conductor and an insulator, so that it is natural that the Hashin–Shtrikman bounds are not so informative: an efficient use of a small amount of insulator can result in almost no current going through, and a large amount of insulator can be used adequately for having little impact on the current going through the conductor part.

It would be useful to develop a mathematical theory about the evolution of mixtures, for asserting the cost of various methods of mixing, and deduce the total cost of creating a mixture.

In 1986, after learning more physics and how physicists think, I checked the formula in the book of L. Landau and Lifschitz: it was one of the two Hashin–Shtrikman bounds, deduced by a hand-waving argument involving a material filling the space except for a sphere of the other material embedded into it.

Reading further on, there was an interesting guess

$$a^{\text{eff}} \approx \overline{a} - \frac{\overline{(a - \overline{a})^2}}{3\overline{a}} \text{ (in dimension 3),} \qquad (10.95)$$

where (as usual) $\overline{b}$ is an intuitive local average of b: I interpreted (10.95) as a guess for a *small amplitude homogenization* problem:

$$\text{if } a^n = a_* + \gamma\, b_n, \text{with } b_n \rightharpoonup 0 \text{ in } L^\infty(\Omega) \text{ weak } \star \text{ and}$$
$$a^n I \text{ H-converges to } a_* I + \sum_{k \geq 2} \gamma^k B_k, \text{ what is } B_2?$$

For me (10.95) was a conjecture:

$$\text{if } (b_n)^2 \rightharpoonup \sigma \text{ in } L^\infty(\Omega) \text{ weak } \star, \text{ and } B_2 = b_* I,$$
$$\text{then } b_* = -\frac{\sigma}{N a_*}.$$

For a mixture of two isotropic conductors, it follows from the fact that, if $\beta = \alpha + \gamma$ and γ is small, the two Hashin–Shtrikman bounds have the same Taylor expansion at order γ^2.

From previous hints, I quickly saw which new mathematical tool I should develop, and I called it H-measures: they are a natural way to summarize the compensated compactness method which François Murat and I had developed.

Appendix M *Compensated Compactness*

In 1976, Jacques-Louis Lions asked François Murat to generalize the div–curl lemma, which we had proved earlier,

if $E^n \rightharpoonup E^\infty$, $D^n \rightharpoonup D^\infty$ in $L^2_{\text{loc}}(\Omega)^N$ weak,

if $\text{curl}(E^n)$ stays in a compact of $H^{-1}_{\text{loc}}(\Omega)^{N(N-1)/2}$ strong,

if $\text{div}(D^n)$ stays in a compact of $H^{-1}_{\text{loc}}(\Omega)$ strong, $\hspace{2em}$ (10.96)

then $\displaystyle\sum_{i=1}^{N} E^n_i D^n_i \rightharpoonup \sum_{i=1}^{N} E^\infty_i D^\infty_i$ in the sense of distributions,

and he coined the (not so good) term "compensated compactness" for it, since $E^n_i D^n_i$ does not converge in general to $E^\infty_i D^\infty_i$ but there is a *compensation effect* for the sum to converge to the sum, and it "looks like a compactness argument", since one has only weak convergence but $\psi(E^n, D^n)$ converges weakly to $\psi(E^\infty, D^\infty)$ for a non-affine function ψ (although the only ψ for which this holds are $c(E,D)$ + an affine function).

I spent 1974–1975 in Madison (WI), and Joel Robbin explained to me the geometrical content of (10.96) (in terms of differential forms) and gave me another proof, using Hodge decomposition. It was not until the end of the 1980s, when I developed the theory of *H-measures*, that I understood how to put variable coefficients in the theory of compensated compactness, except for the cases with a geometrical framework.

A consequence of the div–curl lemma is that *in electrostatics there is no need for an internal energy*, since the density of energy is $\frac{1}{2}(E,D)$, and all this energy is stored at macroscopic level: E^n and D^n contain the details at all mesoscopic scales, but the weak limit E^∞ and D^∞ are macroscopic values.

Joel Robbin suggested that weak convergence is only adapted to quantities which are coefficients of differential forms,[109] and other quantities require different types of topology, adapted to their nature: for example,

---

[109] He taught me later how to consider the Maxwell–Heaviside equation in terms of differential forms.

H-convergence for matrices A corresponds to transforming exact 1-forms into $(N-1)$-forms.

It is important to notice that differential forms, probably introduced by Pfaff (1765–1825) for mathematical reasons, were later developed, probably by Poincaré and by (Élie) Cartan (1869–1951) for their suitability to changes of variables, but in my work they appear for questions of *robustness in the presence of oscillations*. Suppose that

Ω open $\subset \mathbb{R}^N$, U^n from Ω to $\mathbb{R}^p$, F from $\mathbb{R}^p$ to $\mathbb{R}$,

$U^n \rightharpoonup U^\infty$ in $L^\infty(\Omega; \mathbb{R}^p)$ weak $\star$, and $F(U^n) \rightharpoonup V^\infty$ in $L^\infty(\Omega)$ weak $\star$,

$$\sum_{j=1}^{p} \sum_{k=1}^{N} A_{ij,k} \frac{\partial U_j^n}{\partial x_k} = 0, \text{ for } i = 1, \ldots, q. \tag{10.97}$$

One wants to find which F have the property that (10.96) implies

$$V^\infty(x) \geq F(U^\infty(x)) \text{ a.e. in } \Omega. \tag{10.98}$$

A necessary condition for (10.97) to imply (10.98) is

$t \mapsto F(a + t\lambda)$ is convex for every $a \in \mathbb{R}^p$ and every $\lambda \in \Lambda$,

where Λ is defined from the characteristic set $\mathcal{V}$ by

$$\Lambda = \{\lambda \in \mathbb{R}^p \mid \exists \xi \in \mathbb{R}^N \setminus 0, (\lambda, \xi) \in \mathcal{V}\},$$

$$\mathcal{V} = \left\{(\lambda, \xi) \in \mathbb{R}^p \times \mathbb{R}^N \setminus 0 \mid \sum_{j=1}^{p} \sum_{k=1}^{N} A_{ij,k} \lambda_j \xi_k = 0, i = 1, \ldots, q\right\}.$$

For $a \in \mathbb{R}^N$, $(\lambda, \xi) \in \mathcal{V}$, this follows from using functions like

$$U^n(x) = a + \varphi_n((\xi, x)) \lambda.$$

As a corollary, if F is such that (10.97) implies

$$V^\infty(x) = F(U^\infty(x)) \text{ a.e. in } \Omega, \tag{10.99}$$

then

$t \mapsto F(a + t\lambda)$ is affine for every $a \in \mathbb{R}^p$ and every $\lambda \in \Lambda$. $\quad$ (10.100)

Another necessary condition for (10.97) to imply (10.99) is

$\nabla^r F(a) \cdot (\lambda^1, \ldots, \lambda^r) = 0$ for every $a \in \mathbb{R}^p$, whenever

$(\lambda^m, \xi^m) \in \mathcal{V}$, $m = 1, \ldots, r$, and rank$(\xi^1, \ldots, \xi^r) < r$. $\qquad$ (10.101)

If Λ spans $\mathbb{R}^p$, (10.100) implies that F is a polynomial of degree $\leq p$, and (10.101) then implies that its degree is $\leq N$.

Let Q be a real quadratic form on $\mathbb{R}^p$ satisfying

$$Q(\lambda) \geq 0 \text{ for all } \lambda \in \Lambda, \tag{10.102}$$

and let U^n be a sequence (from Ω to $\mathbb{R}^p$) satisfying

$$U^n \rightharpoonup U^\infty \text{ in } L^2_{loc}(\Omega; \mathbb{R}^p) \text{ weak,}$$

and for $i = 1, \ldots, q$

$$\sum_{j=1}^{p} \sum_{k=1}^{N} A_{i,j,k} \frac{\partial U_j^n}{\partial x_k} \text{ belongs to a compact of } H^{-1}_{loc}(\Omega) \text{ strong.}$$

Then, if a subsequence satisfies

$$Q(U^m) \rightharpoonup \mu \text{ in } \mathcal{M}(\Omega) \text{weak} \star,$$

for a Radon measure μ, then one has

$$\mu \geq Q(U^\infty) \text{ in } \Omega \text{ (in the sense of measures).}$$

Our proof uses localization, a Fourier transform and the Plancherel formula.[110] Applying the result to $\pm Q$ gives

if $Q(\lambda) = 0$ for all $\lambda \in \Lambda$, then $Q(U^n) \rightharpoonup Q(U^\infty)$ in $\mathcal{M}(\Omega)$ weak $\star$.

For the div–curl lemma, $p = 2N$, $U_i = E_i$ and $U_{N+i} = D_i$ for $i = 1, \ldots, N$, and information on $curl(E^n)$ and $div(D^n)$ gives

$$\mathcal{V} = \big\{ (E, D, \xi) \mid \xi \neq 0, \ E \text{ parallel to } \xi, \ D \text{ orthogonal to } \xi \big\},$$
$$\Lambda = \{ U = (E, D) \in \mathbb{R}^N \times \mathbb{R}^N \mid (E, D) = 0 \}.$$

Differential forms: If a^n is a sequence of p-forms, b^n is a sequence of q-forms, with good *exterior derivatives*, then the *exterior product* behaves well: with m_r the dimension of r-forms,

$a^n \rightharpoonup a^\infty$ in $L^2(\Omega)^{m_p}$ weakly, and da^n has bounded coefficients in $L^2(\Omega)^{m_{p+1}}$,
$b^n \rightharpoonup b^\infty$ in $L^2(\Omega)^{m_q}$ weakly, and db^n has bounded coefficients in $L^2(\Omega)^{m_{q+1}}$,
then $a^n \wedge b^n \rightharpoonup a^\infty \wedge b^\infty$ in $L^1(\Omega)^{m_{p+q}}$ weak $\star$.

Joel Robbin's proof using Hodge decomposition applies, but also in the case where a^n and da^n have bounded coefficients in $L^\alpha(\Omega)$, b^n and db^n have bounded

---

[110] François Murat had first used another method for the case where $\pm Q$ satisfies (10.102), which imposed a constant rank condition, i.e. for each $\xi \neq 0$, the associated set of λ is a subspace, whose dimension is assumed to be independent of ξ.

coefficients in $L^\beta(\Omega)$, with $1 < \alpha, \beta < \infty$ and $\frac{1}{\alpha} + \frac{1}{\beta} \leq 1$: in this case the Hodge decomposition part relies on the Calderón–Zygmund theorem.

François Murat had also proved the div–curl lemma in the L^α–L^β setting, using the Hörmander–(Mikhlin) theorem for Fourier multipliers, and it is usually for applying this theorem that one adds a hypothesis of constant rank.

Equipartition of hidden energy: Applying the div–curl lemma to the wave equation gives what one usually calls "equipartition of energy", but I prefer to mention that it only applies to the energy hidden at mesoscopic levels: if ρ and A are independent of t, and A is a symmetric tensor, the solutions of

$$\rho \frac{\partial^2 u}{\partial t^2} - \operatorname{div}(A\operatorname{grad}(u)) = 0 \text{ in } \Omega \times (0, T) \qquad (10.103)$$

satisfy the conservation law

$$\frac{\partial}{\partial t}\left(\frac{1}{2}\rho\left|\frac{\partial u}{\partial t}\right|^2 + \frac{1}{2}(A\operatorname{grad}(u), \operatorname{grad}(u))\right) - \operatorname{div}\left(A\operatorname{grad}(u)\frac{\partial u}{\partial t}\right) = 0,$$

expressing the conservation of total energy (assuming ρ positive and A positive definite, for (10.103) to be a wave equation): the density of kinetic energy is $\frac{1}{2}\rho\left|\frac{\partial u}{\partial t}\right|^2$, and the density of potential energy is $\frac{1}{2}(A\operatorname{grad}(u), \operatorname{grad}(u))$. If a sequence of solutions of (10.103) converges to 0 in $H^1(\Omega \times (0, T))$ weak, then

$$\frac{1}{2}\rho\left|\frac{\partial u_n}{\partial t}\right|^2 - \frac{1}{2}(A\operatorname{grad}(u_n), \operatorname{grad}(u_n)) \rightharpoonup 0 \text{ in } L^1 \text{ weak } \star. \qquad (10.104)$$

(10.104) says that there is an equipartition: the density of kinetic energy and the density of potential energy are the same.[111]

For the Maxwell–Heaviside equation, equipartition of hidden energy requires more of compensated compactness than just the div–curl lemma: the quadratic quantities in B, D, E, H which are sequentially weakly continuous for sequences of solutions of the Maxwell–Heaviside equation are linear combinations of

$$(D, H); \ (B, E); \ (B, H) - (D, E). \qquad (10.105)$$

The third quantity in (10.105) is about equipartition of hidden energy between the magnetic part $\frac{1}{2}(B, H)$ and the electric part $\frac{1}{2}(D, E)$. Joel Robbin explained to me the list (10.105).

B and E are coefficients of a 2-form ω_E defined as

$$\omega_E = B_1 dx_2 \wedge dx_3 + B_2 dx_3 \wedge dx_1 + B_3 dx_1 \wedge dx_2 - (E_1 dx_1 + E_2 dx_2 + E_3 dx_3) \wedge dt,$$

---

[111] H-measures can express this common limit, but one needs more than weak limits for the sequence of initial data.

which satisfies

$$d(\omega_E) = 0, \tag{10.106}$$

corresponding to

$$\text{div}(B) = 0, \quad \frac{\partial B}{\partial t} + \text{curl}(E) = 0.$$

D and H are coefficients of a 2-form ω_H defined as

$$\omega_H = D_1\,dx_2 \wedge dx_3 + D_2\,dx_3 \wedge dx_1 + D_3\,dx_1 \wedge dx_2 + (H_1\,dx_1 + H_2\,dx_2 + H_3\,dx_3) \wedge dt,$$

which satisfies

$$d(\omega_H) = \omega_3,$$

corresponding to

$$\text{div}(D) = \rho, \quad -\frac{\partial D}{\partial t} + \text{curl}(H) = j,$$

with

$$\omega_3 = \rho\,dx_1 \wedge dx_2 \wedge dx_3 - (j_1\,dx_2 \wedge dx_3 + j_2\,dx_3 \wedge dx_1 + j_3\,dx_1 \wedge dx_2) \wedge dt,$$

which satisfies

$$d(\omega_3) = 0,$$

corresponding to the *conservation of electric charge*

$$\frac{\partial \rho}{\partial t} + \text{div}(j) = 0.$$

With this way of writing the Maxwell–Heaviside equations in terms of differential forms, (10.105) is natural since

$$\omega_H \wedge \omega_H = (D,H)\,dx_1 \wedge dx_2 \wedge dx_3 \wedge dt,$$
$$\omega_E \wedge \omega_E = -(B,E)\,dx_1 \wedge dx_2 \wedge dx_3 \wedge dt,$$
$$\omega_H \wedge \omega_E = \big((B,H) - (D,E)\big)\,dx_1 \wedge dx_2 \wedge dx_3 \wedge dt.$$

A consequence of (10.106) is

$$d(\omega_1) = \omega_E,$$

with

$$\omega_1 = -A_1\,dx_1 - A_2\,dx_2 - A_3\,dx_3 + U\,dt,$$

corresponding to

$$B = \text{curl}(A), \quad E = -\frac{\partial A}{\partial t} - \text{grad}(U).$$

One may obtain information on the weak limits of polynomials, for example for (10.103), by using conservation laws containing quadratic quantities: if

$\rho, a_{ij}, i, j = 1, \ldots, N$, depend only on x (with $a_{ji} = a_{ij}$ a.e.), conservation of energy holds,

$$\frac{\partial}{\partial t}\left(\frac{\rho}{2}\left|\frac{\partial u^n}{\partial t}\right|^2 + \frac{1}{2}\sum_{i,j=1}^{N} a_{ij}\frac{\partial u^n}{\partial x_i}\frac{\partial u^n}{\partial x_j}\right) - \sum_{i,j=1}^{N}\frac{\partial}{\partial x_i}\left(a_{ij}\frac{\partial u^n}{\partial x_j}\frac{\partial u^n}{\partial t}\right) = 0,$$

and if $u^n \rightharpoonup 0$ in $W^{1,3}_{loc}(\Omega \times (0, T))$ weak, then the div–curl lemma in the L^3–$L^{3/2}$ setting implies

$$\frac{\partial u^n}{\partial t}\left(\frac{\rho}{2}\left|\frac{\partial u^n}{\partial t}\right|^2 - \frac{1}{2}\sum_{i,j=1}^{N} a_{ij}\frac{\partial u^n}{\partial x_i}\frac{\partial u^n}{\partial x_j}\right) \rightharpoonup 0 \text{ in } L^1(\Omega \times (0, T)) \text{weak } \star. \quad (10.107)$$

If the limit of u^n is not 0, one needs the weak limits of $\frac{\rho}{2}\left|\frac{\partial u^n}{\partial t}\right|^2 + \frac{1}{2}\sum_{i,j=1}^{N} a_{ij}\frac{\partial u^n}{\partial x_i}\frac{\partial u^n}{\partial x_j}$ and of $\sum_{j=1}^{N} a_{ij}\frac{\partial u^n}{\partial x_j}\frac{\partial u^n}{\partial t}$ for each i for computing what will appear on the right side of (10.107), and these weak limits can be obtained with H-measures.

If $\rho, a_{ij}, i, j = 1, \ldots, N$ are independent of x_k, and conservation of linear momentum in x_k holds,

$$\frac{\partial}{\partial t}\left(\rho\frac{\partial u^n}{\partial t}\frac{\partial u^n}{\partial x_k}\right) - \sum_{i,j=1}^{N}\frac{\partial}{\partial x_i}\left(a_{ij}\frac{\partial u^n}{\partial x_j}\frac{\partial u^n}{\partial x_k}\right) - \frac{\partial}{\partial x_k}\left(\frac{\rho}{2}\left|\frac{\partial u^n}{\partial t}\right|^2 - \sum_{i,j=1}^{N}\frac{a_{ij}}{2}\frac{\partial u^n}{\partial x_i}\frac{\partial u^n}{\partial x_j}\right) = 0,$$

and if $u^n \rightharpoonup 0$ in $W^{1,3}_{loc}(\Omega \times (0, T))$ weak, then the div–curl lemma in the L^3–$L^{3/2}$ setting implies

$$\frac{\partial u^n}{\partial x_k}\left(\frac{\rho}{2}\left|\frac{\partial u^n}{\partial t}\right|^2 - \frac{1}{2}\sum_{i,j=1}^{N} a_{ij}\frac{\partial u^n}{\partial x_i}\frac{\partial u^n}{\partial x_j}\right) \rightharpoonup 0 \text{ in } L^1(\Omega \times (0, T)) \text{ weak } \star.$$

$$(10.108)$$

Again, if the limit of u^n is not 0, one needs the weak limits of $\rho\frac{\partial u^n}{\partial t}\frac{\partial u^n}{\partial x_k}$ and of $\sum_{j=1}^{N} a_{ij}\frac{\partial u^n}{\partial x_j}\frac{\partial u^n}{\partial x_k}$ for each i for computing what will appear on the right side of (10.108), and these weak limits can be obtained with H-measures.

For the equation $u_{tt} - c^2\Delta u = 0$, both (10.107) and (10.108) apply, but these are not new sequentially weakly continuous functionals, since the preceding results need $u^\infty = 0$, and without this assumption, one needs *H-measures* for identifying the limits, but although these results are true, the hypothesis that the sequence stays bounded in $W^{1,3}_{loc}(\Omega \times (0, T))$ is questionable since the wave equation is not well posed in an L^p setting for $p \neq 2$.

Is there a way to give a meaning to the quantities involved?

What I called the *compensated compactness method* was a first step towards understanding the interaction of differential constraints (like the balance laws in continuum mechanics) and pointwise constraints (like the constitutive relations in continuum mechanics), and I used *Young measures* for taking into account the pointwise constraints. I had heard about Young measures in seminars on "control theory" as *parametrized measures*, which was the name used by Ghouila-Houri (1939–1966), and I was the first to use them in a PDE context in continuum mechanics, but they only served in expressing the (new) information given by the quadratic compensated compactness theorem.

Since I consider good evolution models in continuum mechanics to be hyperbolic,[112] one has to address the challenge that solutions of quasi-linear hyperbolic systems of conservation laws develop shocks, and that one does not know how to get good bounds, and my method seemed to be able to answer questions without enough bounds, but I had to check that it was applicable in interesting cases, and I tried it on the Burgers equation

$$\frac{\partial u_n}{\partial t} + \frac{1}{2}\frac{\partial (u_n^2)}{\partial x} = 0, \text{ in } \mathbb{R} \times (0,T). \tag{10.109}$$

Assuming just an L^∞ bound, I considered that

$$(u_n)^k \rightharpoonup U_k \text{ in } L^\infty\big(\mathbb{R} \times (0,T)\big) \text{ weak } \star.$$

Assuming u_n smooth, I multiplied (10.109) by $2u_n$:

$$\frac{\partial u_n^2}{\partial t} + \frac{2}{3}\frac{\partial (u_n^3)}{\partial x} = 0, \text{ in } \mathbb{R} \times (0,T), \tag{10.110}$$

and this was a technical point to improve, since because of shocks the right-hand side is not 0 but a Radon measure $\mu_n \leq 0$, and one must check that it stays in a compact of H_{loc}^{-1} (and the Radon measures are not all elements of H_{loc}^{-1}). Using the div–curl lemma for (10.109), (10.110) gives

$$u_n\frac{2u_n^3}{3} - \frac{u_n^2}{2}u_n^2 \rightharpoonup U_1\frac{2U_3}{3} - \frac{U_2}{2}U_2 \text{ in } L^\infty\big(\mathbb{R} \times (0,T)\big) \text{ weak } \star,$$

so that it proves the relation

$$U_4 = 4U_1U_3 - 3U_2^2 \text{ a.e. in } \mathbb{R} \times (0,T).$$

Then one takes the weak $\star$ limit of $(u_n - U_1)^4$,

$$(u_n - U_1)^4 = u_n^4 - 4u_n^3U_1 + 6u_n^2U_1^2 - 4u_nU_1^3 + U_1^4$$
$$\rightharpoonup U_4 - 4U_1U_3 + 6U_1^2U_2 - 3U_1^4 \geq 0 \text{ in } L^\infty\big(\mathbb{R} \times (0,T)\big) \text{ weak } \star,$$

giving $U_2 = U_1^2$, and $u_n \to U_1$ in $L_{\text{loc}}^q\big(\mathbb{R} \times (0,T)\big)$ strong for every $q < \infty$.

---

[112] The evolution equation in elasticity is a conservative system, and it is unphysical to think that there is no time and that elastic materials minimize their potential energy!

The system (10.109), (10.110) has an elliptic character, and this is the essence of the *compensated compactness method*, that by using supplementary conservation laws one makes the system behave in a much nicer way!

Appendix N *H-measures*

If

$$A^* \in \mathcal{M}(\alpha, \beta) \text{ and } B^n \rightharpoonup 0 \in L^\infty\big(\Omega; \mathcal{L}(\mathbb{R}^N; \mathbb{R}^N)\big) \text{ weak } \star, \qquad (10.111)$$

then after extraction of a (Cantor diagonal) subsequence

$$A^m = A^* + \gamma\, B^n \text{ H-converges to } A^* + \gamma^2 C^* + \cdots \text{ for } |\gamma| \text{ small}, \qquad (10.112)$$

and I introduced H-measures for expressing C^*. I do not know of a general tool for expressing the next terms in the power expansion (which has a positive radius of convergence).[113]

H-measures are essentially a more precise formulation than the quadratic compensated compactness theorem: using all the (real) quadratic functions Q which are ≥ 0 on the characteristic set Λ (projection of the characteristic set $\mathcal{V}$) is equivalent to the observation that

$$\text{if } U^n \rightharpoonup U^\infty \text{ and } U^n \otimes U^n \rightharpoonup U^\infty \otimes U^\infty + R,$$

$$\text{then } R(x) \in \text{closed convex hull of } \Lambda \otimes \Lambda \text{ a.e.},$$

with an obvious change of words if the components of R are Radon measures and not locally integrable functions. An explicit representation of R is given by H-measures.

The proof of the quadratic theorem consists of using the Fourier transform $\mathcal{F}$ and looking at how $\mathcal{F}\, U^n$ behaves at infinity.[114]

In the scalar case, one starts from a sequence

$$u_n \rightharpoonup 0 \text{ in } L^2_{\text{loc}}(\Omega) \text{ weak},$$

and the basic result about H-measures is that

there exists a subsequence u_m,

there exists a non-negative Radon measure μ on $\Omega \times S^{N-1}$,

such that for all $\varphi \in C_c(\Omega), \psi \in C(S^{N-1})$,

$$\int_{\mathbb{R}^N} |\mathcal{F}(\varphi\, u_m)|^2 \psi\left(\frac{\xi}{|\xi|}\right) d\xi \to \langle \mu, |\varphi|^2 \otimes \psi \rangle,$$

---

[113] For $A \in \mathcal{M}(\alpha, \beta)$ and $||B||_{\mathcal{L}(\mathbb{R}^N;\mathbb{R}^N)} \leq \delta < \alpha$, one has $A + B \in M\big(\alpha - \delta, \frac{\alpha\beta - \delta^2}{\alpha - \delta}\big)$.

[114] I could have replaced the unit sphere S^{N-1} by the quotient of $\mathbb{R}^N \setminus \{0\}$ by the equivalence relation x equivalent to sx for all $s > 0$, or by a sphere at infinity for compactifying $\mathbb{R}^N$.

which one usually writes $\int_{\Omega \times S^{N-1}} |\varphi(x)|^2 \psi(\xi) \, d\mu(x,\xi)$; I call μ the H-measure associated with the subsequence.

In the vectorial case, the H-measure is a $p \otimes p$ non-negative Hermitian symmetric matrix of Radon measures:

$$\text{if } U^n \rightharpoonup 0 \text{ in } L^2_{\text{loc}}(\Omega; \mathbb{C}^p) \text{ weak,}$$

there is a subsequence U^m, a $p \otimes p$ non-negative, Hermitian symmetric matrix of Radon measures,

μ on $\Omega \times S^{N-1}$: for all $\varphi_1, \varphi_2 \in C_c(\Omega), \psi \in C(S^{N-1})$, and $j,k \in \{1, \ldots, p\}$,

$$\int_{\mathbb{R}^N} \mathcal{F}(\varphi_1 U_j^m) \overline{\mathcal{F}(\varphi_2 U_k^m)} \psi\left(\frac{\xi}{|\xi|}\right) d\xi \rightarrow \langle \mu_{jk}, \varphi_1 \overline{\varphi_2} \otimes \psi \rangle, \tag{10.113}$$

which one usually writes $\int_{\Omega \times S^{N-1}} \varphi_1(x) \overline{\varphi_2(x)} \psi(\xi) \, d\mu_{jk}(x,\xi)$; I call μ the H-measure associated with the subsequence.

For proving this, I developed a kind of "pseudo-differential" calculus *modulo compact operators*:

for $a \in L^\infty(\mathbb{R}^N), P_a$ defined by $\mathcal{F}P_a v = a \mathcal{F} v$ for $v \in L^2(\mathbb{R}^N)$

satisfies $P_a \in \mathcal{L}(L^2(\mathbb{R}^N); L^2(\mathbb{R}^N))$, with $||P_a|| = ||a||_{L^\infty(\mathbb{R}^N)}$,

but one only uses $a \in L^\infty(S^{N-1})$, extended as $a\left(\frac{\xi}{|\xi|}\right)$.

For $b \in L^\infty(\mathbb{R}^N)$, M_b defined by $M_b v = b v$ for $v \in L^2(\mathbb{R}^N)$

satisfies $M_b \in \mathcal{L}(L^2(\mathbb{R}^N); L^2(\mathbb{R}^N))$, with $||M_b|| = ||b||_{L^\infty(\mathbb{R}^N)}$,

However, I cannot take symbols in L^∞ because I use a *first commutation lemma*

for $a \in C(S^{N-1})$, $b \in C_0(\mathbb{R}^N)$, the commutator $P_a M_b - M_b P_a$

is a compact operator from $L^2(\mathbb{R}^N)$ into itself.

$C = P_a M_b - M_b P_a$ has norm $\leq 2||a||_{L^\infty}||b||_{L^\infty}$. One approaches b uniformly by b_n with $\mathcal{F}b_n$ having compact support, and it suffices to show that the C_n are compact, since $||C_n - C|| \to 0$

$$\mathcal{F}(C_n u)(\xi) = \int_{\mathbb{R}^N} K_n(\xi, \eta) \mathcal{F}u(\eta) \, d\eta,$$

with $K_n(\xi, \eta) = \mathcal{F}b_n(\xi - \eta)\left(a\left(\frac{\xi}{|\xi|}\right) - a\left(\frac{\eta}{|\eta|}\right)\right)$. $\tag{10.114}$

Since $\mathcal{F}b_n$ has compact support, one uses only $|\xi - \eta| \leq \rho$ in (10.114). For $|\xi|$ and $|\eta|$ large, $\frac{\xi}{|\xi|}$ and $\frac{\eta}{|\eta|}$ are uniformly near and a is uniformly continuous on S^{N-1}, giving a small contribution, and what remains has ξ and η bounded and gives the kernel of a Hilbert–Schmidt operator, which is compact.

The left side of (10.113) can be written as

$$\int_{\mathbb{R}^N} \mathcal{F}(\varphi_1 U_j^m) \overline{\mathcal{F}(\varphi_2 U_k^m)} \psi\left(\frac{\xi}{|\xi|}\right) d\xi$$
$$= \int_{\mathbb{R}^N} \mathcal{F}(P_\psi M_{\varphi_1} U_j^m) \overline{\mathcal{F}(M_{\varphi_2} U_k^m)} \, d\xi \qquad (10.115)$$
$$= \int_{\mathbb{R}^N} P_\psi M_{\varphi_1} U_j^m \overline{M_{\varphi_2} U_k^m} \, dx,$$

the first equality by the definition of the operators P_ψ and M_φ, the second by Plancherel's theorem, and then

$$P_\psi M_{\varphi_1} U_j^m - M_{\varphi_1} P_\psi U_j^m \to 0 \text{ in } L^2 \text{ strong},$$

since a compact operator transforms weak convergence into strong convergence.

It then shows that the limit depends (linearly) only on $\varphi_1 \overline{\varphi_2}$,[115] and since it is linear in ψ, the limit defines a linear operator from $C(S^{N-1})$ into the dual of $C_0(\mathbb{R}^N)$, which is $\mathcal{M}_b(\mathbb{R}^N)$.[116]

By the kernel theorem of Laurent Schwartz,[117] this operator has a kernel which is a distribution. If $j = k$, using $\varphi_1 = \varphi_2$,[118] and $\psi \geq 0$, one deduces that this distribution is ≥ 0, which by a theorem of Laurent Schwartz implies that μ_{jj} is a Radon measure; applying the argument to $U_j^m \pm U_k^m$ and to $U_j^m \pm i U_k^m$ shows that μ_{jk} is a Radon measure, and that the matrix μ of components μ_{jk} is Hermitian non-negative.

The class of symbols I use are of the form

$$s(x, \xi) = \sum_n a_n(\xi) b_n(x) \text{ with } a_n \in C(S^{N-1}), \ b_n \in C_0(\mathbb{R}^N)$$

$$\text{and } \sum_n \|a_n\|_{C(S^{N-1})} \|b_n\|_{C_0(\mathbb{R}^N)} < +\infty,$$

and I call the *standard operator* of symbol s the operator

$$S = \sum_n P_{a_n} M_{b_n} :$$

---

[115] One first uses the separability of $C_0(\mathbb{R}^N)$ and $C(S^{N-1})$ for extracting a sequence for which the limit exists for all $\varphi_1, \varphi_2 \in C_0(\mathbb{R}^N)$ and all $\psi \in C(S^{N-1})$.

[116] One must be careful with $C_b(\mathbb{R}^N)$, the space of bounded continuous functions, or $\mathrm{BUC}(\mathbb{R}^N)$, the space of bounded uniformly continuous functions, because these spaces are not separable, and their duals are not spaces of distributions in $\mathbb{R}^N$, since some elements "live at infinity".

[117] In my article (in the Proceedings of the Royal Society of Edinburgh), I used only elementary results from functional analysis. I should have quoted the kernel theorem of Laurent Schwartz and mentioned that Jacques-Louis Lions had told me that he had written a simpler proof with Lars Garding (1928–2005).

[118] For example equal to $\sqrt{\varphi}$ for $\varphi \in C_0(\mathbb{R}^N)$ with $\varphi \geq 0$.

since

$$(\mathcal{F}P_{a_n}M_{b_n}u)(\xi) = a_n\left(\frac{\xi}{|\xi|}\right)(\mathcal{F}(b_nu))(\xi) = a_n\left(\frac{\xi}{|\xi|}\right)\int_{\mathbb{R}^N} b_n(x)\,u(x)\,e^{-2i\pi\,(x,\xi)}\,dx,$$

one deduces that

$$\mathcal{F}Su(\xi) = \int_{\mathbb{R}^N} s\left(x,\frac{\xi}{|\xi|}\right)u(x)\,e^{-2i\pi\,(x,\xi)}\,dx,$$

so that it does not depend upon which decomposition of S one chooses in (10.115). In the theory of pseudo-differential operators, which started with the work of Joseph Kohn and Louis Nirenberg, one uses C^∞ symbols and I did not want to be limited to equations with C^∞ coefficients, but one also uses

$$L = \sum_n M_{b_n}P_{a_n},$$

which gives

$$Lu(x) = \int_{\mathbb{R}^N} s\left(x,\frac{\xi}{|\xi|}\right)\mathcal{F}u(\xi)\,e^{+2i\pi\,(x,\xi)}\,d\xi;$$

I preferred the definition of S, since in a domain Ω one must first multiply by a function of x before one can use the Fourier transform, which needs functions defined on $\mathbb{R}^N$. Anyway, by the first commutation lemma, the difference $S-L$ is a compact operator, so that I chose to define an operator of symbol s as any operator differing from S by a compact operator.

If L_j and L_k are two operators of symbols s_j and s_k, then

$$L_jU_j^m\overline{L_kU_k^m} \rightharpoonup v \text{ weakly } \star \text{ in the sense of measures,}$$
$$\text{with } \langle v,\varphi\rangle = \langle\mu_{jk},\varphi\,s_j\overline{s_k}\rangle \text{ for all } \varphi\in C_c(\Omega).$$

One may write formally that v is the integral of $s_j\overline{s_k}\mu_{jk}$ on the sphere S^{N-1}, but technically it is a projection.

This shows what H-measures are: besides the weak $\star$ limits of all $U_j^m\overline{U_k^m}$, they can give the weak $\star$ limits when one applies a natural class of operators of order 0 (mapping L^2 into L^2).

One may want to consider the real parts and the imaginary parts of U^m, and one should notice that if U^m are real-valued, then the H-measure has a symmetry: changing ξ into $-\xi$ results in μ being changed into its complex conjugate, but for $N\geq 2$ the off-diagonal components of μ may be complex.

In the definition, I used sequences converging weakly to 0, but saying that a sequence U^n converging weakly to U^∞ corresponds to an H-measure μ will mean that μ is the H-measure associated with $U^n - U^\infty$.

An important property is what I called the *localization principle*:[119] if U^m converges weakly to U^∞ and corresponds to an H-measure μ, then

if $\sum_{j,k} \frac{\partial (A_{jk} U_j^m)}{\partial x_k}$ stays in a compact of H_{loc}^{-1} strong, with A_{jk} continuous for

$j = 1, \ldots, p$, $k = 1, \ldots, N$, then $\sum_{j,k} A_{jk} \xi_k \mu_{j\ell} = 0$ for all $\ell = 1, \ldots, p$.
$$(10.116)$$

Localizing in x and using the Riesz operators $R_j = P_{i\xi_j/|\xi|}$, the hypothesis in (10.116) is equivalent to

$$\sum_{j,k} R_k A_{jk} \varphi \, (U_j^m - U_j^\infty) \to 0 \text{ in } L^2(\mathbb{R}^N) \text{ strong,}$$

for all $\varphi \in C_c^1(\Omega)$, and the proof follows from the construction of the calculus of operators, the continuity of the A_{jk} being related to the hypothesis for the first commutation lemma.

The continuity of b in the first commutation lemma may be relaxed a little, and the theory of H-measures can be developed with $b \in L^\infty(\Omega) \cap$ VMO(Ω),[120] thanks to a result of Raphaël Coifman, Rochberg, and Weiss, that the commutators $[M_b, R_j]$ map $L^p(\mathbb{R}^N)$ into itself for $1 < p < \infty$ when $b \in \text{BMO}(\mathbb{R}^N)$, with a norm depending only upon the BMO($\mathbb{R}^N$) semi-norm of b;[121] it seems that Uchiyama (1948–1997) proved that $b \in \text{VMO}(\mathbb{R}^N)$ is necessary for the commutators to be compact.

If u_n is a scalar sequence converging weakly to u_∞ in H_{loc}^1, then $U^n = \text{grad}(u_n)$ satisfies $\text{curl}(U^n) = 0$ and the localization principle tells us that the H-measure of U^m (which is an $N \times N$ matrix) must have the form $(\xi \otimes \xi) \nu$, where ν is a scalar non-negative Radon measure. If u_n satisfies the wave equation

$$\frac{\partial}{\partial t} \left(\rho \frac{\partial u_n}{\partial t} \right) - \sum_{i,j} \frac{\partial}{\partial x_i} \left(A_{ij} \frac{\partial u_n}{\partial x_j} \right) = 0,$$

with ρ and all A_{ij} continuous, and we also have a sequence of solutions u_n converging in H_{loc}^1,[122] then, using $t = x_0$ as a simplifying notation, the H-measure of $\left(\frac{\partial u_m}{\partial t}, \text{grad}(u_m) \right)$ (which is a $(N+1) \times (N+1)$ matrix) must have

---

[119] I chose the term localization because in the scalar case, $\sum_j b_j \frac{\partial u_m}{\partial x_j}$ bounded in L^2 implies that the support of μ is included in the zero set of $\sum_j b_j(x) \xi_j$.

[120] The space VMO (vanishing mean oscillation) is the closure of smooth functions with compact support in BMO (bounded mean oscillation), a space which was introduced by Fritz John (1910–1994).

[121] They also proved the converse, that if the commutators map $L^p(\mathbb{R}^N)$ into itself, then b must belong to BMO($\mathbb{R}^N$).

[122] For applying the classical existence results for solutions of a wave equation, ρ will be uniformly positive and independent of t (or smooth in t) and A will be symmetric uniformly positive definite and independent of t (or smooth in t).

the form $(\xi \otimes \xi)\,\nu$, where the scalar non-negative Radon measure ν satisfies

$$\left(\rho\,\xi_0^2 - \sum_{i,j} A_{ij}\xi_i\xi_j\right)\nu = 0.$$

Proving a result of propagation of energy (and momentum) along the intuitive light rays, which are the bicharacteristic rays of the function $\rho\,\xi_0^2 - \sum_{i,j}A_{ij}\xi_i\xi_j$,[123] requires more regularity.

The reason is that I used a second commutation lemma for writing a first-order PDE satisfied by ν. My approach differs from the classical one (with an amplitude and a phase): H-measures can tell us where energy goes, which is the role of amplitude, but since they use no characteristic length, they cannot see a phase; since the PDE for ν is written in weak form, it is not bothered by caustics, but caustics appear if one is interested in the regularity of the density of the measure ν.[124]

Another important difference is that the classical approach constructs one solution of the wave equation by an asymptotic expansion for an amplitude and a phase (in terms of a frequency parameter) and shows that outside caustics the amplitude satisfies a transport property along bicharacteristic rays,[125] while I show that for all sequences of solutions of the wave equation the high frequencies modes have energy propagating along bicharacteristic rays, not being bothered by caustics.

Although I had introduced H-measures for the question of *small amplitude homogenization*, for which no characteristic length is needed, I presented my results in a different order (in my article), because I found that the more important result for H-measures was to define (at last) what a beam of light is, and that they tell how energy propagates.

My approach with H-measures is adapted to a large class of hyperbolic systems,[126] with smooth enough coefficients, so that it does not extend to questions with discontinuous coefficients along a smooth interface, like refraction effects.

---

[123] Bicharacteristic rays of a function Q are usually defined as $\frac{dx_j}{ds} = \frac{\partial Q}{\partial \xi_j}$, $\frac{d\xi_j}{ds} = -\frac{\partial Q}{\partial x_j}$, with Q homogeneous in ξ for inducing an ODE for half-lines in ξ, but since I chose to take here S^N as a way to select one point in each half-line, one must correct the equation of bicharacteristic rays.

[124] Propagation of energy is crucial for physics, but not regularity, which is a question interesting mathematicians. I know no connection with physics of the notion of wave front sets of distributions introduced by Lars Hörmander: it is useful for propagation of microlocal regularity, and is not even about propagation of singularities, as his followers usually say.

[125] It is a logical error, sometimes made by physicists, to conclude a general rule because one has checked it on a few examples.

[126] They must have a quadratic conservation law extending to complex solutions (as a sesqui-linear conserved quantity).

I first thought that the reason was that refraction effects are frequency dependent, but this dependence results from light interacting with matter, which is not taken into account in the Maxwell–Heaviside equation (nor in a scalar wave equation, which is a naive model for light). I now think that it is due to a difficulty with grazing rays for which the conjectures come from GTD (the geometrical theory of diffraction developed by Joe Keller in the 1950s), which provides a better intuition about why light rays near the sun are deflected than the naive theory by Einstein,[127] but such conjectures must be proved.

Since the mapping which associates the solution with initial data of a wave equation is not a pseudo-differential operator, and it is a reason why Lars Hörmander developed the theory of Fourier integral operators, it is then surprising that, with ideas similar to pseudo-differential operators, I was able to show the propagation of "hidden energy" along "light rays".

Actually, due to the formal theory of geometrical optics, no one would have thought of introducing a microlocal object without any characteristic length for proving results about the wave equation, but I did not think in this way.

I introduced H-measures for questions of small amplitude homogenization, and once I had proved interesting results with them, I wondered if they could be used for proving that for a first order scalar PDE the H-measures (which describe questions of oscillations and concentration effects) propagate along the same bicharacteristic rays as for the propagation of microlocal regularity in the work of Lars Hörmander, although his question does not seem of any use for physics!

It then is important to develop one's own ideas, and sometimes without reading some previous results, in order not to be influenced too much in the discovery of new lines of attack for some well-known problems (already solved or not).

For computing C^* in (10.112) when (10.111) holds, one solves

$$-\mathrm{div}\big((A^* + \gamma B^n)\,\mathrm{grad}(u_n)\big) = -\mathrm{div}\big(A^*\mathrm{grad}(u_*)\big), u_n \in H_0^1(\Omega),$$

with $u_* \in H_0^1(\Omega)$ given. The solution is analytic in γ, so that

$$\mathrm{grad}(u_n) = \mathrm{grad}(u_*) + \gamma\,\mathrm{grad}(v_n) + o(|\gamma|), v_n \in H_0^1(\Omega),$$

with v_n a solution of

$$-\mathrm{div}\big(A^*\mathrm{grad}(v_n) + B^n\mathrm{grad}(u_*)\big) = 0. \tag{10.117}$$

---

[127] Probably because he had heard about Hilbert working on gravitation in a setting of Riemannian manifolds, and that a weight on an elastic membrane deforms the geodesics, Einstein thought that this effect would be responsible for bending light rays, but the Maxwell–Heaviside equations (which he should have used instead of a scalar wave equation) are quite different from the equations for linearized elasticity, out of which one deduces the equations for thin membranes.

Since B^n grad $(u_*) \rightharpoonup 0$ in $L^2(\Omega)^N$ weak, one deduces that

$$v_n \rightharpoonup 0 \text{ in } H_0^1(\Omega) \text{ weak,}$$
$$B^n \text{grad}(v_n) \rightharpoonup C^* \text{grad}(u_*) \text{ in } L^2(\Omega)^N \text{ weak.} \tag{10.118}$$

There are technical details,[128] but the idea is that (10.117) defines a mapping $B^n \mapsto \text{grad}(v_n)$ whose components are in the class of symbols used, so that the H-measure of B^n permits one to express $C^* \text{grad}(u_*)$ in (10.118), hence gives a formula for C^*.

For example, if $B^n = b_n I$, hence the H-measure μ is a scalar non-negative measure, the formula is

$$\int_\Omega C_{ij}^* \varphi \, dx = -\left\langle \mu, \frac{\varphi(x)\xi_i \xi_j}{(A^*(x)\xi, \xi)} \right\rangle \text{ for } \varphi \in C_c(\Omega), \ i,j = 1,\ldots,N, \tag{10.119}$$

which one may write as $C_{ij}^*(x) = -\int_{S^{N-1}} \frac{\xi_i \xi_j}{(A^*(x)\xi, \xi)} \, d\mu(x, \xi)$.

If $A^* = a_* I$, then the trace of C^* is $-\frac{1}{a_*} \int_{S^{N-1}} d\mu(x, \xi)$, but since $\int_{S^{N-1}} d\mu(x, \xi)$ is the weak limit of b_n^2, one finds that

if $A^* = a_* I$, $B^n = b_n I$, with $b_n \rightharpoonup 0$, $b_n^2 \rightharpoonup \sigma$ in $L^\infty(\Omega)$ weak $\star$,

then Trace$(C^*) = -\frac{\sigma}{a_*}$, $\qquad\qquad\qquad\qquad\qquad\qquad$ (10.120)

which I had guessed in 1986 from a formula in a book by L. Landau & Lifshitz; in 1974, I had stopped reading a few pages before a similar formula, concluding that physicists had not understood homogenization: I still think so, but they formulate interesting conjectures, and it was for proving formulas like (10.120) that I developed the theory of H-measures.

For a different mathematical reason,[129] Patrick Gérard introduced similar objects which he called microlocal defect measures, but since he was following Lars Hörmander he thought that only the support is important.[130] I dislike the term defect he used: the energy of light hidden in high frequencies is not a defect, it is a crucial aspect of how the Universe functions!

---

[128] Since the class of symbols uses continuous functions, one may begin by taking $u^* \in C_c^1(\Omega)$ and A^* continuous. Then, since one can approach L^∞ coefficients strongly in norm L^r for any $r < \infty$, one uses a regularity theorem of Meyers, that there is for some $p > 2$ a uniform bound for grad(v_n) in $L^p(\Omega)^N$.

[129] His purpose was to give a different approach to the question of compactness by averaging: I had tried to prove such results with my H-measures, but I had not found the way.

[130] Hence he was not thinking of proving formulas like my results on small amplitude homogenization, nor my results of propagation of energy by writing a PDE for μ.

Many wrongly call propagation of singularities the results of Lars
Hörmander, which concern the propagation of microlocal regularity: since
I consider that the worst sin of a teacher is to induce error in students, I teach
that I do not know any connection to physics for mathematical questions of
regularity!

Since I knew of problems with small length scales, I proposed to analyze
a sequence U^n for a sequence of characteristic lengths ε_n tending to 0 by
considering the H-measure of the sequence

$$V^n(x, x_{N+1}) = U^n(x) \cos \frac{2\pi \, x_{N+1}}{\varepsilon_n},$$

and Patrick Gérard introduced a slightly different variant.

He called semi-classical measure (associated with a subsequence) a Radon
measure in $\Omega \times \mathbb{R}^N$ giving the limits

$$\lim_{m \to \infty} \int_{\mathbb{R}^N} |\mathcal{F}(\varphi \, u_m)|^2 \psi(\varepsilon_m \xi) \, \mathrm{d}\xi,$$

for the case of a scalar sequence, taking $\psi \in \mathcal{S}(\mathbb{R}^N)$.[131] He noticed two
problems with his definition, one at infinity because ψ tends to 0 (which my
variant does not have),[132] and one at 0 because ψ is continuous (which my
variant also has).[133]

Later, Pierre-Louis Lions and Thierry Paul introduced some objects which
they called Wigner measures, because they used the Wigner transform in their
definition. Since they made (in the preprint I received) the mistaken claim that
it permits one to recover H-measures. I did not bother to read their lengthy
computations; I would be more indulgent now, since I heard the Nobel laureate
in Physics's remarks about problems in physics being simpler than those of
biology as physics has only one scale.

I later asked Patrick Gérard what they were doing with[134]

$$W_n(x, \xi) = \int_{\mathbb{R}^N} u_n\left(x + \frac{\varepsilon_n y}{2}\right) \overline{u_n\left(x - \frac{\varepsilon_n y}{2}\right)} \, \mathrm{e}^{-2\mathrm{i}\pi \, (y, \xi)} \, \mathrm{d}y,$$

and I immediately saw a way to avoid their computations by using two-point
correlations and the Bochner theorem, which we checked; later, Patrick Gérard

---

[131] ψ smooth permits a more general localization principle.

[132] I could take ψ in (a separable subspace of) $\mathrm{BUC}(\mathbb{R}^N)$.

[133] I proposed later to have ψ like $\psi_0\left(\frac{\xi}{|\xi|}\right)$ near 0.

[134] Although the Wigner transform changes sign, they had to show that a rescaled limit is
non-negative. They used an argument they attributed to a Japanese author, but Marc Feix
(1928–2005) mentioned to me that he had told them that Wigner (1902–1995, 1963 Nobel
Prize in Physics) had done it before.

adopted this way of using two-point correlations for introducing semi-classical measures.

I knew that one needs at least one characteristic length for defining correlations for a general sequence u_n, but before that discussion I did not find anything interesting to say about that.

If $u_n \rightharpoonup 0$ in $L^2(\mathbb{R}^N)$ (and $\varepsilon_n \to 0$), then for $y, z \in \mathbb{R}^N$ the sequence $u_n(\cdot + \varepsilon_n y)\,\overline{u_n(\cdot + \varepsilon_n z)}$ is bounded in $L^1(\mathbb{R}^N)$, so that

for fixed $y, z \in \mathbb{R}^N$, there is a subsequence such that
$$u_m(\cdot + \varepsilon_m y)\,\overline{u_m(\cdot + \varepsilon_m z)} \rightharpoonup C_{y,z} \text{ in } \mathcal{M}_b(\mathbb{R}^N)\text{weak } \star,$$

and the basic observation is that

$$C_{y+h,z+h} = C_{y,z} \text{ for all } h \in \mathbb{R}^N.$$

Using a Cantor diagonal subsequence u_m for all $y, z \in \mathbb{Q}^N$,

$$\left|\sum_{j=1}^{q} \lambda_j u_m(\cdot + \varepsilon_m y^j)\right|^2 \rightharpoonup \sum_{j,k=1}^{q} \lambda_j \overline{\lambda_k} C_{y^j, y^k} \text{ in } \mathcal{M}(\mathbb{R}^N) \text{ weak } \star,$$

for all $y^1, \ldots, y^q \in \mathbb{Q}^N$ and $\lambda_1, \ldots, \lambda_q \in \mathbb{C}$, one has

$$\sum_{j,k=1}^{q} \lambda_j \overline{\lambda_k} \Gamma(y^j - y^k) \geq 0, \qquad (10.121)$$

where $\Gamma \in \mathcal{M}(\mathbb{R}^N)$ is defined by $\Gamma(h) = C_{\frac{h}{2}, \frac{-h}{2}}$.

If Γ is continuous, (10.121) is true for all $q \in \mathbb{N}$, $y^1, \ldots, y^q \in \mathbb{R}^N$, $\lambda_1, \ldots, \lambda_q \in \mathbb{C}$, so that Γ is (by definition) a function of positive type (in y), and by the Bochner theorem it is the Fourier transform of a non-negative Radon measure.

Without continuity of $C_{y,z}$ in (y, z), one must use Laurent Schwartz's extension of the Fourier transform to $\mathcal{S}'(\mathbb{R}^N)$.

I checked with Patrick Gérard a situation where 3-point correlations can be used, for a diffusion equation with a small coefficient of diffusion: in this case, there is no analog of Bochner's theorem, and since one must start with a bounded sequence in L^3, the application to questions like the wave equation, the Maxwell–Heaviside equation, or other hyperbolic systems is ruled out. Is there a replacement for 3-point correlations?

Graeme Milton had told me that my H-measures express the scale-invariant part of the 2-point correlations, which I first could not understand, since there were no correlations used in my definition of H-measures, but I knew another reason before checking the use of Bochner's theorem with Patrick Gérard.

If one considers the small amplitude homogenization formula (10.119) in the case when A^* is independent of x, $B^n = b_n I$ with b_n periodic (i.e. $b_n(x) = b\left(\frac{x}{\varepsilon_n}\right)$) with average 0 on the unit cube, one has

$$C_{ij}^* = \sum_{m \in \mathbb{Z}^N \setminus \{0\}} |\beta_m|^2 \frac{m_i m_j}{(A^* m, m)}, \tag{10.122}$$

where the β_m are the Fourier coefficients of b, i.e.

$$\beta_m = \int_{[0,1]^N} b(y) e^{-2i\pi (m,y)} \, dy \text{ for } m \in \mathbb{Z}^N.$$

Since $b \in L^\infty(cube)$ implies $b \in L^2(cube)$, which is equivalent to $\sum_m |b_m|^2 < +\infty$, the series in (10.122) is absolutely convergent.

If for b real, one wants to make the 2-point correlation function of b appear, i.e. the (even) function Corr_2 defined by

$$\text{Corr}_2(h) = \int_{[0,1]^N} b(y) b(y+h) \, dy \text{ for } h \in \mathbb{R}^N,$$

one observes that (using $b(z) = \overline{b(z)}$, and then $z = y + h$)

$$|\beta_m|^2 = \int_{[0,1]^N \times [0,1]^N} b(y) b(z) e^{-2i\pi (m, y-z)} \, dy \, dz$$

$$= \int_{[0,1]^N} \text{Corr}_2(h) e^{2i\pi (m,h)} \, dh \text{ for } m \in \mathbb{Z}^N,$$

so that (10.122) transforms into a formal integral

$$C_{ij}^* = \int_{[0,1]^N} \text{Corr}_2(h) K_{ij}(h) \, dh, \text{ with the singular kernel}$$

$$K_{ij}(h) = \sum_{m \in \mathbb{Z}^N \setminus \{0\}} \frac{m_i m_j}{(A^* m, m)} e^{2i\pi (m,h)} \, dh,$$

since K_{ij} is a Fourier series with coefficients in ℓ^∞, which might be summed in the sense of distributions of Laurent Schwartz, or with the Calderón–Zygmund theory of singular integrals.

In the periodic case, one can write formulas giving each coefficient of γ^k, and if a term $|b_m|^2$ appears in the coefficients of γ^2, one could write it as $b_m b_{-m}$, and for $k = 3$ there are terms like $b_m b_n b_p$ with $m + n + p = 0$, but I was not able to guess from such formulas what general mathematical tool to define.

Actually, maybe a Taylor expansion is not the right approach, and one could think of Padé approximants, for example, or some geometrical methods, or use adapted diagrams.

If H-measures are natural objects after Young measures, the question of bounds for effective coefficients involve both, but I have obtained some bounds using the H-measure for A^n and some bounds using the H-measure for $(A^n)^{-1}$, so that it seems that a better (non-linear) microlocal tool should be developed!

I have introduced some multi-scale H-measures [7], which I shall not describe here, since nothing fundamental was achieved with them yet, and I find it better to leave young researchers to have new ideas, and be bold enough to imagine completely new ways of attacking some of the problems which I mentioned.

References

[1] Amirat, Youcef, Hamdache, Kamel & Ziani, Abdelhamid, Homogénéisation d'équations hyperboliques du premier ordre et application aux écoulements miscibles en milieu poreux, *Ann. Inst. H. Poincaré Anal. Non Linéaire* **6** (1989), no. 5, 397–417. Étude d'une équation de transport à mémoire, *C. R. Acad. Sci. Paris Sér. I Math.* **311** (1990), no. 11, 685–688.

[2] Enoch, Jay M., History of Mirrors Dating Back 8000 Years, *Optometry & Vision Science*, vol. 83, **10**, October 2006, 775–781.

[3] Feynman, Richard P., Leighton, Robert B. & Sands, Matthew, *The Feynman lectures on physics: the definitive and extended edition*, 3 vols, Addison-Wesley, 2005.

[4] Gondran, Michel & Gondran, Alexandre, *Mécanique Quantique, Et si Einstein et de Broglie avaient aussi raison?*, Éditions Matériologiques, Paris 2014.

[5] Rashed, Roshdi, A Pioneer in Anaclastics: Ibn Sahl on Burning Mirrors and Lenses, *Isis*, **81** (1990), 464–491.

[6] Tartar, Luc, *The General Theory of Homogenization, A Personalized Introduction*, Lecture Notes of the Unione Matematica Italiana 7, Springer-Verlag, Berlin, 2010.

[7] Tartar, Luc, Multi-scales H-measures, *Discrete and Continuous Dynamical Systems - Serie S*. Special Issue on Numerical Methods based on Homogenization and Two-Scale Convergence, Vol. 8, **1**, February 2015, 77–90.

* Carnegie Mellon University, Pittsburgh, PA.
luctartar@gmail.com